D0022775

Introduction to Mathematical Structures

Introduction to Mathematical Structures

Steven Galovich
Carleton College

Harcourt Brace Jovanovich, Publishers
and its subsidiary, Academic Press

San Diego New York Chicago Austin Washington, D.C.
London Sydney Tokyo Toronto

For Alex, Anna,
and Liz

ISBN: 0-15-543468-3
Library of Congress Catalog Card Number: 88–82149
Printed in the United States of America

Preface

The mission of this textbook is twofold: to provide an introduction to mathematical structures and to provide an introduction to mathematical thinking. To accomplish our goal, an attempt is made to portray as accurately as possible the nature of mathematics, illustrating explicitly certain features of the mathematical enterprise: (1) structures that are basic to the study of mathematics; (2) the nature of the formal arguments that establish the validity of theorems; (3) the kinds of thinking that lead to the formulation of conjectures that, when proved, become theorems; (4) strategies that are used by mathematicians to solve problems to conduct research, and (5) analogies that exist among various mathematical concepts.

This textbook presents many of the basic mathematical topics and techniques that every student should know before moving on to upper-level mathematics and computer science courses. The core material is contained in Chapters 1–3 and covers proof techniques, sets, relations, and functions. In these chapters, special care has been taken to motivate definitions and proofs and to make explicit the reasoning processes by which conjectures are reached. In Chapters 4–7, material from several upper-level mathematics courses is presented, including the foundations of mathematics (Chapters 4 and 7), combinatorics (Chapter 5), and modern algebra (Chapter 6).

The book begins with a discussion of some of the tools needed to study mathematical structures. The topics presented include logic, the general notion of an axiomatic system, and an axiomatic development of the real numbers. Various techniques for proving mathematical statements are then discussed in detail. Finally, the chapter closes with a description (based on the work of George Polya) of some strategies that are useful for attacking difficult problems.

Set theory provides the setting in which modern mathematics takes place, and Chapter 2 introduces basic set theory. The text presents an informal introduction to sets and set operations and supplements the presentation with a discussion of the Zermelo–Fraenkel axioms of set theory. Chapter 3 continues the development of set theory with a detailed study of relations that emphasizes order relations, equivalence relations, and functions. The treatment of sets concludes, in Chapter 4, with a discussion of the concept of size of a set. Also discussed are the properties of finite sets and countable sets, and Cantor's beautiful proof that the set of real numbers is uncountable. The chapter closes with an informal introduction to cardinal numbers.

Chapter 5 deals with discrete or combinatorial structures. Combinatorics deals with questions concerning the counting and the arranging of finite sets of objects. The chapter begins with a description of some basic counting principles, then considers some more advanced techniques based on recurrence relations, and closes with a quick introduction to graph theory.

Chapter 6 presents some of the basic structures of abstract algebra. Our goal is to motivate concepts such as group, ring, and field through such familiar examples as the integers, the rational numbers, and the real numbers.

Finally, in Chapter 7 the set of real numbers is constructed almost from scratch. The existence of the set of natural numbers is assumed, but on the basis of this set, the integers, the rationals, and the reals are defined precisely. The chapter closes with a discussion of the many properties of the set of real numbers, complementing the axiomatic treatment of the reals presented in Chapter 1.

The problems appearing at the end of each section constitute an integral part of the textbook. They come in a variety of shapes and sizes including (1) exercises that are concrete illustrations of definitions, theorems, and techniques; (2) problems that require the combination of two or more ideas for solution; (3) questions that demand experimentation, conjecture, and justification; (4) exercises of the "prove, or disprove and salvage" type, which require either a proof or a counterexample with a modification of the original statement and a proof of the modified statement. Each chapter contains a selection of summary exercises which serve to tie together many of the ideas presented in the chapter. Solutions and hints to solutions of selected exercises are presented at the end of the text.

At the conclusion of each chapter, various comments, historical remarks, and bibliographical references are presented. These notes should provide the reader with additional perspective on the material and some directions for further study.

Finally, a word about prerequisites. The minimal prerequisite for a course based on this book is a solid background in one-variable calculus. It is advisable, but certainly not mandatory, that anyone taking this course also be conversant with multivariable calculus and linear algebra. At several points in the text, examples from linear algebra and multivariable calculus are used. These examples are not, however, central to the text and students who are, for example, not acquainted with linear algebra should be able to read and understand all of the material presented here.

Throughout the text the symbol ■ marks the end of a formal proof.

Note to Students

When reading this or any other mathematics book, you should keep in mind several thoughts. First, read *carefully* and *critically*. When studying a proof, be sure you read slowly enough to understand a given sentence or step before moving on. A mathematical proof is like a house of cards: One missing piece and the whole enterprise will fall apart. The same remarks apply to the reading of any example or definition. Second, read *actively*. After reading the statement of a theorem, for example, try to sense the reasonableness of the theorem. You can do this by drawing appropriate pictures, by finding examples that illustrate the theorem, and/or by finding analogous results that were previously proved. Then, before reading the proof of a theorem, try to come up with a proof, or the outline of a proof, of your own.

In general, read this book with a sharp pencil in hand, plenty of paper nearby, and several questions in mind. A list of questions might include the following:

1. What is the essence of this theorem, example, or definition?
2. Can I come up with an outline of a proof of the theorem?
3. Do I understand all the steps in the proof?
4. What is the main idea of the argument?
5. Can I draw a picture illustrating the situation?
6. Am I stuck? If so, where am I stuck? Can I review some material to help me get unstuck? Are there examples that I can consider to help me understand the given situation?

There is no guarantee that by following this advice, you will attain a complete understanding of the material presented in this or any textbook. Nonetheless, by doing so, you will most certainly learn the material better than you would through a casual reading. By engaging yourself with the book, you are more likely to be able to use the material successfully to solve problems, and you will increase both your proficiency and your enjoyment of the subject.

Note to Instructor

This text can be used in a variety of ways. For example, a basic one-quarter course could cover Sections 1–3 of Chapter 1, Chapters 2, 3, and 4, and, if time permits, selected sections from the remaining chapters. A course emphasizing discrete structures might replace Chapter 4 (except for Section 2) with Chapter 5. A one-semester course could include the basic one-quarter material and either Chapter 5 or Chapter 6 or Chapter 7, depending on the interest of the instructor. Alternatively, a one-semester course could cover the basic one-quarter course and Sections 5.1–5.4, Sections 6.1–6.2, and Sections 7.1–7.4. Most of the text can be covered in a two-quarter course or in a year-long course.

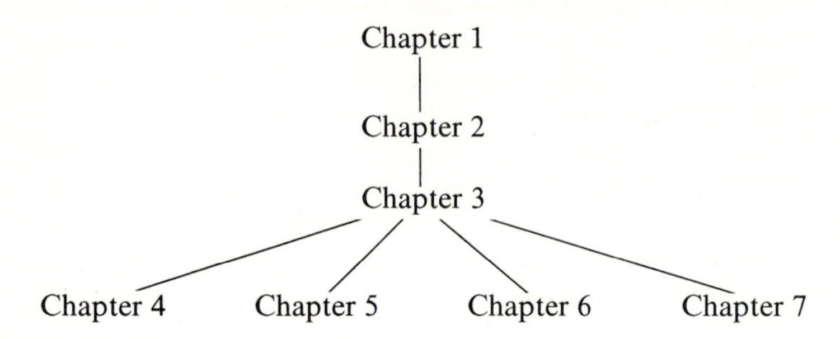

A word should be said about Section 1.4. I have found it useful to assign this material as reading and to cover it only briefly in class. Throughout the remainder of the course, this section can be referred to when appropriate. By the end of the course, students should be sufficiently conversant with the Polya–Schoenfeld scheme to be able to use it in subsequent courses.

Acknowledgments

I would like to thank the many people and institutions who contributed to the writing and production of this text. Early work on the manuscript was generously supported by the Vaughn Foundation Fund. During the 1984–85 academic year, I was able to spend a sabbatical leave in northern California with the support of a generous grant from the Alfred P. Sloan Foundation administered through Carleton College. While in California I benefited from the gracious hospitality of the Institute of Cognitive Studies at the University of California at Berkeley and the Mathematics Department of the University of California at Davis.

Several individuals provided assistance in various ways. Dean Peter Stanley and especially Dean Roy Elveton of Carleton College were extremely generous in providing both encouragement and technical support. The list of colleagues who read the manuscript at various stages and offered insightful suggestions includes Don Chakerian and Sherman Stein of the University of California at Davis; Joe Gallian of the University of Minnesota at Duluth; and Christopher Ennis, Mark Krusemeyer, and Sy Schuster of Carleton College. Professors Ennis and Krusemeyer taught from earlier versions of the manuscript and contributed many valuable suggestions.

In addition, I received the benefit of many valuable comments from reviewers of the manuscript, and I wish to thank the following people for their suggestions: Richard Chandler (North Carolina State University); Sherralyn Craven (Central Missouri State University); Robert Etter (California State University, Sacramento); Jim Fey (University of Maryland); Michael Hoffman (California State University, Los Angeles); Barbara K. Kaiser (Gustavus Adolphus College); Laura Mayer (Beloit College); Robert C. Penner (University of Southern California); Barbara J. Pence (San Jose State University); Thomas Robertson (Occidental College); and Richard B. Thompson (University of Arizona).

I also want to express my gratitude to the many students at Carleton College who made numerous suggestions that led to improvements in the manuscript. Without trying to list all the names, let me mention five who were particularly helpful: Stephen Lindblad, Robert MacIntyre, Darwin Serkland, Jim Shope, and Gloria Stier.

For technical and typing help, I owe many thanks: first to James Flaten and Professor Richard Brown, who exhibited their computer wizardry in implementing and maintaining the T^3 mathematical word processing system; to Borden Armstrong and Tom McCabe, who devised the means of coaxing mathematical text out of an impact printer; and to Gloria Stier, who, at a critical moment, helped to sort out difficulties with the printing of Chapter 4. Most of all, I want to thank Barbara Jenkins, who, with boundless energy and enthusiasm, mastered two word processing systems and spent countless hours typing the text and the many revisions thereof. Thanks again, Barbara.

In addition, I wish to thank my editors Richard Wallis and Nancy Evans at Harcourt Brace Jovanovich for their valuable assistance.

Last but certainly not least, I want to express my appreciation to Elizabeth Culver for her constant support and encouragement during the writing of this book.

Contents

Chapter One
Tools 1

1.1. Logic 3
1.2. The Real Numbers and Axiomatic Systems 20
1.3. Formal Thinking: Methods of Proof 37
1.4. Informal Thinking: Methods of Inquiry 69

Chapter Two
Set Theory 93

2.1. Sets 95
2.2. Operations on Sets 105
2.3. Ordered Pairs and Cartesian Products 127

Chapter Three
Relations and Functions 139

3.1. Relations 139
3.2. Order Relations 156
3.3. Equivalence Relations 166
3.4. Functions 183

Chapter Four
Cardinality 207

4.1. Equinumerous Sets 209
4.2. Finite Sets 219
4.3. Denumerable Sets 227
4.4. Uncountable Sets 237
4.5. Cardinal Numbers 244

Chapter Five
Discrete Structures 255

5.1. An Overview of Discrete Structures 255
5.2. Basic Combinatorial Principles 260
5.3. Permutations and Combinations 270
5.4. Binomial Coefficients and the Binomial Theorem 278
5.5. Recurrence Relations 289
5.6. Graph Theory 300

Chapter Six
Algebraic Structures 321

6.1. An Overview of Algebraic Structures 321
6.2. Algebraic Properties of the Integers 329
6.3. Rings 337
6.4. Fields 349
6.5. Groups 361
6.6. Homomorphisms and Isomorphisms 378

Chapter Seven
The Real Number System 389

7.1. The Natural Numbers 391
7.2. Arithmetic and Order Properties of the Natural Numbers 395
7.3. The Integers 403
7.4. The Rational Numbers 411
7.5. The Real Numbers 418
7.6. Algebraic Properties of the Real Numbers 427
7.7. Order Properties of the Real Numbers 435
7.8. The Completeness Property of the Real Numbers 441

Hints and Solutions to Selected Exercises 449

Index 481

CHAPTER ONE

Tools

In this chapter we present some of the basic "tools" that mathematicians use when constructing and analyzing mathematical structures. Unlike the devices employed in the physical and life sciences, the tools of mathematicians are not physical machines that provide extensions of our senses. Rather, they are processes of the mind that enable mathematicians to perceive, articulate, and verify the existence of orderly behavior in the phenomena that underlie our physical, social, and intellectual world.

Given the (perhaps arguable) fact that mathematical items do not exist in the physical world, we may be forgiven for referring to the concepts discussed in this chapter as "tools." But, indeed, when exploring mathematical entities, mathematicians repeatedly use certain kinds of investigatory reasoning; when justifying their findings, they employ certain specific forms of argument. These types of argument and reasoning (to be described in detail in Sections 1.3 and 1.4) are among the tools of mathematics.

To pursue this vein a bit further, we might divide the tools of mathematics into two categories:

1. Modes of thinking.
2. Modes of expression.

As philosophers have noted over the ages, thought and language are intimately related. Whether they are inseparable is an issue that need not concern us here. Our goal is simply to draw attention to mathematical thinking and language as crucial aspects of the mathematical enterprise. Proficiency in a mathematical field comes from a blending of several characteristics including knowledge of the subject matter, skill at attacking problems, and the ability to communicate clear and correct arguments. In this text we provide a quick introduction to several areas of mathematics while simultaneously encouraging the efficient and wise use of mathematical tools.

Our first tool is logic. Logic gives us the rules of inference by which valid arguments that lead from assumptions to conclusions can be constructed. Next we discuss the basic properties of the real number system and the general nature of axiomatic systems. We then turn to the structure of specific types of arguments that are used to prove statements in mathematics. We refer to these methods of proof as the *formal* tools of the mathematical trade. Next we consider the *informal* tools—the ways of thinking that tend to help us to solve problems, to create proofs, and to devise procedures. In contrast to the formal tools, which are part of the process of verification, the informal tools (also called *heuristics*) are part of the discovery process. As we shall attempt to show, however, the two processes (i.e., proof and discovery) are complementary and should be regarded neither as unrelated nor as adversary.

We hasten to note that our discussions of all the topics we raise in this chapter will be brief. Volumes can be written on many of these subjects, and in some cases at least (e.g., logic and problem solving) volumes have been written. In most instances our treatment will barely scratch the surface. Nonetheless, a frank discussion of the tools and methodology of mathematics is valuable. By making explicit aspects of mathematical activity such as problem solving and proof writing (above and beyond the "content" of the subjects to be covered), we aim to present a relatively faithful picture of what is involved in doing mathematics.

§1.1 Logic

Logic or symbolic logic is the study of valid reasoning processes. As such, logic provides the basic rules by which reasoning in mathematics is carried out. As Davis and Hersh point out in *The Mathematical Experience*, logic is the "nuts and bolts" of mathematics.

In this section we present a brief, utilitarian summary of logic. We concentrate primarily on the so-called propositional calculus, that part of logic which focuses on the analysis of statements that are either true or false. We also venture slightly into the domain of predicate calculus, which is concerned with "indefinite" statements (i.e., those that are not necessarily true or false) that describe properties of objects and relationships among objects. We emphasize that our purpose is not to study logic for its own sake but to use it as a tool that enables us to carry out mathematical investigations.

Propositional Calculus

A *proposition* is a sentence that is either true or false but is not both true and false. For example,

Mary Jones has black hair,

John Jones is at least six feet tall,

are examples of propositions. To each proposition, we assign a *truth value*, true or false, depending on whether the proposition is true or false. As an example of a sentence to which a truth value cannot be assigned, consider the statement:

This statement is false.

A moment's thought reveals that if the statement is true, then it is false, and if the statement is false, then it is true. Thus, no truth value can be assigned to the statement; hence the statement is not a proposition.

We now consider how new propositions can be formed from existing propositions. We list five ways of creating new propositions out of previously given propositions: conjunction, disjunction, negation, conditional, and biconditional. These five methods allow for the creation of new propositions through the use of the words "and," "or," "not," "if-then," and "if and only if," respectively. These items are called *logical connectives*.

Definition 1.1.1 *Logical Connectives*

Let P and Q be propositions.

(i) The *conjunction* of P and Q is the statement "P and Q."
(ii) The *disjunction* of P and Q is the statement "P or Q."
(iii) The *negation* of P is the statement "not-P," which is often phrased as "It is not the case that P."
(iv) The *conditional* formed from the pair P and Q is the statement "If P then Q."
(v) The *biconditional* formed from the pair P and Q is the statement "P if and only if Q."

1. The *conjunction* of P and Q is the sentence "P and Q." The conjunction of P and Q is defined to be true if P and Q are both true and is defined to be false in all other cases.

A convenient way of conveying this information is that via a device called a *truth table*. The truth table for "P and Q" expresses the truth or falsity of "P and Q" in terms of the truth values of the individual sentences P and Q (Table 1.1.1). Reading horizontally we see that if P is true (T) and Q is true, then "P and Q" is true. If P is true and Q is false (F), then "P and Q" is false, etc.

P	Q	P and Q
T	T	T
T	F	F
F	T	F
F	F	F

Table 1.1.1

In many texts on mathematical logic, the symbol $\wedge$ is used in place of the word "and." However, we shall usually write "P and Q" in place of $P \wedge Q$.

As a quick exercise, let us make a truth table for the sentence "Q and P" (Table 1.1.2). Observe that the propositions "P and Q" and "Q and P" have the same truth values for each pair of possible truth values for Q and P.

P	Q	Q and P
T	T	T
T	F	F
F	T	F
F	F	F

Table 1.1.2

2. The *disjunction* of P and Q is the sentence "P or Q"; "P or Q" is sometimes written $P \vee Q$. The disjunction of P and Q is defined to be true if at least one of P and Q is true and to be false if both P and Q are false. The truth table for "P or Q" is shown in Table 1.1.3.

P	Q	P or Q
T	T	T
T	F	T
F	T	T
F	F	F

Table 1.1.3

Observe that "Q or P" has the same truth values as "P or Q" for each pair of truth values of the propositions P and Q.

The disjunction "P or Q" is also known as the "inclusive-or." This is in contrast to the "exclusive-or" (xor) that in everyday discourse is often used in place of "or." The exclusive-or is defined as follows: "P xor Q" is true when exactly one of P and Q is true. In mathematics, the inclusive-or is used almost exclusively.

3. For a given proposition P, the *negation* of P, written "not-P," is defined to be true when P is false and to be false when P is true. The truth table for "not-P" is shown in Table 1.1.4. The proposition "not-P" is expressed symbolically in various ways including $\neg P$, $\sim P$, or $-P$.

P	not-P
T	F
F	T

Table 1.1.4

For each pair of propositions P and Q, no matter how complex, the propositions "P or Q" and "P and Q" are easy to phrase grammatically.

However, an accurate and literate phrasing of "not-*P*" can be difficult to render. A cautious but legitimate way to state "not-*P*" is: "It is not the case that *P*." Thus, if *P* is the proposition "John is a boy," then "not-*P*" can be read as:

It is not the case that John is a boy,

which can in turn be translated into:

John is not a boy.

As another example suppose *P* is the sentence:

John is a boy or Jane is a girl.

Then not-*P* reads:

It is not the case that John is a boy or Jane is a girl,

which evidently can be accurately restated as:

John is not a boy and Jane is not a girl.

Generalizing from this observation, we might conjecture (i.e., guess) that for any pair of propositions *P* and *Q*, the proposition "not-(*P* or *Q*)" has the same "meaning" as the proposition "(not-*P* and not-*Q*)." More precisely, we conjecture that "not-(*P* or *Q*)" and "not-*P* and not-*Q*" have the same truth tables (Table 1.1.5).

P	*Q*	*P* or *Q*	not-(*P* or *Q*)	not-*P*	not-*Q*	not-*P* and not-*Q*
T	T	T	F	F	F	F
T	F	T	F	F	T	F
F	T	T	F	T	F	F
F	F	F	T	T	T	T

Table 1.1.5

Thus, no matter what the truth values of *P* and *Q*, the truth values of "not-(*P* or *Q*)" and "not-*P* and not-*Q*" coincide. This observation supports the interpretation of "not-(*P* or *Q*)" that was made in the previous paragraph.

We have thus observed two examples in which two different statements (in this example "not-(*P* or *Q*)" and "(not-*P*) and (not-*Q*)") have identical truth tables. This phenomenon can be formalized into a definition.

Definition 1.1.2 *Logical Equivalence*

Suppose A and B are propositions formed from a given collection of propositions $P, Q, R, S, \ldots$ using "and," "or," "not," and the conditional and biconditional to be defined below. The propositions A and B are *logically equivalent* if A and B have the same truth values for all possible truth values of the propositions $P, Q, R, S, \ldots$.

Table 1.1.5 demonstrates that "not-(P or Q)" and "(not-P) and (not-Q)" are logically equivalent.

Table 1.1.6 shows the logical equivalence of "not-(P and Q)" and "(not-P) or (not-Q)." This result tells us how to translate the negation of a conjunction. For example, the negation of

John is a boy and Jane is a girl

is

John is not a boy or Jane is not a girl.

P	Q	P and Q	not-(P and Q)	not-P	not-Q	not-P or not-Q
T	T	T	F	F	F	F
T	F	F	T	F	T	T
F	T	F	T	T	F	T
F	F	F	T	T	T	T

Table 1.1.6

We saw earlier that the propositions "P and Q" and "Q and P" are logically equivalent and that "P or Q" and "Q or P" are logically equivalent. Here are some other examples:

(i) P is logically equivalent to "P and P."
(ii) P is logically equivalent to "P or P."
(iii) "(P and Q) and R" is logically equivalent to "P and (Q and R)."

4. The next method of combining a pair of propositions to form a new proposition is that via the conditional statement. Let P and Q be propositions. The *conditional statement* "If P then Q" is true when P and Q are both true or when P is false; in the only remaining case, which occurs when P is true and Q is false, "If P then Q" is false. The truth table for "If P then Q" is shown in Table 1.1.7.

P	Q	If P then Q
T	T	T
T	F	F
F	T	T
F	F	T

Table 1.1.7

The conditional "If P then Q" is also called an *implication* and is often written "P implies Q." Symbolically the implication "If P then Q" is written $P \Rightarrow Q$. In an implication or conditional $P \Rightarrow Q$, we refer to the proposition P as the *hypothesis* of the implication and the statement Q as the *conclusion* of the implication. In some texts P is called the *antecedent* and Q the *consequent* of the conditional $P \Rightarrow Q$.

Notice that the implication $P \Rightarrow Q$ is defined to be true whenever P is false. This practice might seem puzzling, but there are good reasons for it. First, why should we consider $P \Rightarrow Q$ to be true when P is false and Q is true? Perhaps the most reasonable response is that any implication with a true conclusion should be regarded as true no matter what the hypothesis. Thus,

If I have 1,000,000 dollars, then I have 5 cents

is true since I do indeed have 5 cents even though I do not have 1,000,000 dollars. What about when P and Q are both false? One justification for defining $P \Rightarrow Q$ to be true in this case is that we want the implication $P \Rightarrow P$ to be true for any proposition P. Thus,

If I have 1,000,000 dollars, then I have 1,000,000 dollars

is true even if I do not have 1,000,000 dollars. Thus, to ensure that $P \Rightarrow P$ is true for all propositions P, we define $P \Rightarrow Q$ to be true when P and Q are both false.

Let "If P then Q" be a given implication. The *contrapositive* of "If P then Q" is defined to be the conditional "If not-Q then not-P." For example, the contrapositive of the implication

"If $x = 2$, then $x^2 = 4$"

is the implication

"If $x^2 \neq 4$, then $x \neq 2$."

How does the contrapositive of a conditional compare logically with the conditional itself? To answer this question we form a truth table (Table 1.1.8). As is evident in the table, for each pair of possible truth values of P and Q, the truth values of $P \Rightarrow Q$ and (not-Q) $\Rightarrow$ (not-P) coincide. Thus any implication is logically equivalent to its contrapositive.

P	Q	$P \Rightarrow Q$	not-Q	not-P	(not-Q) $\Rightarrow$ (not-P)
T	T	T	F	F	T
T	F	F	T	F	F
F	T	T	F	T	T
F	F	T	T	T	T

Table 1.1.8

This remark is more than merely an idle curiosity. Suppose that we are trying to prove that $P \Rightarrow Q$ is true. This implication holds exactly when the contrapositive (not-Q) $\Rightarrow$ (not-P) holds. Moreover, in a given case, the latter implication might be easier to establish than the former. Thus the logical equivalence of $P \Rightarrow Q$ and (not-Q) $\Rightarrow$ (not-P) actually provides us with a method of proving that $P \Rightarrow Q$ is true. As we shall see in Section 1.3, this proof technique will become one of the tools we use to analyze mathematical statements.

Our next observation concerning implications leads to a question. Notice that "If P then Q" is true for three of the possible four pairs of truth values of P and Q. Similarly, the disjunction of two propositions is true for three of the four pairs of truth values of the component propositions. Thus a connection between implications and disjunctions is suggested. Specifically, we ask: Is there a disjunction formed somehow from P and Q that is logically equivalent to the implication "If P then Q"? It is certainly not "P or Q." (Why not?) What then is it? Since "If P then Q" is false when P is true and Q is false, it is reasonable to consider a disjunction that is false when P is true and Q is false; "not-P or Q" is such a disjunction. We leave as an exercise the verification of the claim that "If P then Q" and "not-P or Q" are logically equivalent.

Let $P \Rightarrow Q$ be a given implication. The *converse* of $P \Rightarrow Q$ is the conditional $Q \Rightarrow P$. For instance, the converse of the implication

$$\text{If } x = 2, \text{ then } x^2 = 4,$$

is the implication

$$\text{If } x^2 = 4, \text{ then } x = 2.$$

Are the implications "If P then Q" and its converse "If Q then P" logically equivalent? Let us form a truth table (Table 1.1.9). Because $P \Rightarrow Q$ and $Q \Rightarrow P$ have different truth values when P is true and Q is false, *these propositions are not logically equivalent*. This remark cannot be emphasized too heavily!

P	Q	$P \Rightarrow Q$	$Q \Rightarrow P$
T	T	T	T
T	F	F	T
F	T	T	F
F	F	T	T

Table 1.1.9

This observation actually opens the way for a wealth of mathematical questions. Whenever we prove a statement of the form "If P then Q," we can ask: Is the converse "If Q then P" also true? For certain pairs of statements, P and Q, the converse will indeed hold; in other cases it will not. For example, the statement "For a real number x, if $x + 4 = 8$ then $x = 4$" and its converse are both true. On the other hand, the following are true statements whose converses are false:

(i) "For a real number x, if $x = 2$, then $x^2 = 4$," and
(ii) "For a function f on the real numbers, if f is a function that is differentiable at a, then f is a function that is continuous at a."

Thus the moral is that whenever we prove or even attempt to prove an implication $P \Rightarrow Q$, we should also consider its converse $Q \Rightarrow P$. Both $P \Rightarrow Q$ and $Q \Rightarrow P$ might be true, in which case P and Q provide two ways to express the same concept. For example, to say about a real number x that $x = 4$ is equivalent to saying that $x + 4 = 8$. If only one of the implications $P \Rightarrow Q$ and $Q \Rightarrow P$ is true, then we obtain a clearer understanding of the relationship between P and Q.

5. The last logical connective to be considered is the *biconditional*. By definition the biconditional of P and Q is the statement "(If P then Q) and (if Q then P)." In logical symbols the biconditional is $(P \Rightarrow Q) \wedge (Q \Rightarrow P)$. Symbolically, the biconditional of P and Q is written $P \Leftrightarrow Q$ and is read "P if and only if Q." The statement "P if and only if Q" is sometimes abbreviated "P iff Q."

Under what conditions is $P \Leftrightarrow Q$ true? By definition $P \Leftrightarrow Q$ is true precisely when the implication $P \Rightarrow Q$ and its converse are true. We consider two cases: (i) P is true, (ii) P is false.

(i) If P is true, then since $P \Rightarrow Q$ is true, Q must also be true. (ii) If P is false, then since $Q \Rightarrow P$ is true, Q must also be false. Thus, if the implications "If P then Q" and "If Q then P" are both true, then P and Q must have the same truth values, i.e., P and Q must be logically equivalent. On the other hand, if P and Q are either both true or both false, then $P \Rightarrow Q$ and $Q \Rightarrow P$ are both true. Thus, to summarize our findings, the implications $P \Rightarrow Q$ and $Q \Rightarrow P$ are both true precisely when P and Q have the same truth values. In other words, $P \Leftrightarrow Q$ is true when P and Q are logically equivalent and false when P and Q are not logically equivalent. The truth table for the biconditional is shown in Table 1.1.10.

P	Q	$P \Leftrightarrow Q$
T	T	T
T	F	F
F	T	F
F	F	T

Table 1.1.10

Both conditionals and biconditionals are often phrased in other ways. Each of these different forms occurs frequently enough to warrant mentioning at this time.

(i) "P only if Q" means "if P then Q."
(ii) "P if Q" means "if Q then P."
(iii) "P is a necessary condition for Q" means "if Q then P."
(iv) "P is a sufficient condition for Q" means "if P then Q."
(v) "P is a necessary and sufficient condition for Q" means "P iff Q."

Let us state some mathematical results from calculus and linear algebra in this language. Let f be a function defined on the set of real numbers. In elementary calculus it is proved that if f is differentiable at a, then f is continuous at a. However, the converse of this implication is false. Thus, for a function f, differentiability at a is a sufficient but not necessary condition for continuity at a. As another example, consider a function that is differentiable at each real number. Then a necessary and sufficient condition for f to be an increasing function is that $f'(x) \geq 0$ for each real number x. From linear algebra we have the following result: For any $n \times n$ matrix A, A is invertible iff $\det(A) \neq 0$. Thus a necessary and sufficient condition for the invertibility of A is that A have a nonzero determinant.

Examples

Let us now apply the rules concerning truth values of propositions to some specific cases.

Example 1.1.1 Write out a truth table for $(P \Rightarrow Q) \Rightarrow (Q \Rightarrow P)$ (Table 1.1.11). Notice that $(P \Rightarrow Q) \Rightarrow (Q \Rightarrow P)$ and $Q \Rightarrow P$ are logically equivalent.

P	Q	$P \Rightarrow Q$	$Q \Rightarrow P$	$(P \Rightarrow Q) \Rightarrow (Q \Rightarrow P)$
T	T	T	T	T
T	F	F	T	T
F	T	T	F	F
F	F	T	T	T

Table 1.1.11

Example 1.1.2 Show that $P \Leftrightarrow Q$ and "$(P$ and $Q)$ or (not-P and not-Q)" are logically equivalent.

To show the logical equivalence of the given proposition, we construct a truth table for each (Table 1.1.12).

P	Q	$P \Leftrightarrow Q$	P and Q	(not-P) and (not-Q)	(P and Q) or (not-P and not-Q)
T	T	T	T	F	T
T	F	F	F	F	F
F	T	F	F	F	F
F	F	T	F	T	T

Table 1.1.12

Example 1.1.3 Construct a truth table for $[P \vee ((\text{not-}P) \wedge Q)] \vee [\text{not-}P \wedge \text{not-}Q]$ (Table 1.1.13) Thus, the given proposition is true no matter what the truth values of P and Q.

P	Q	$P \vee ((\text{not-}P) \wedge Q)$	$(\text{not-}P) \wedge (\text{not-}Q)$	$[P \vee ((\text{not-}P) \wedge Q)] \vee [\text{not-}P \wedge \text{not-}Q]$
T	T	T	F	T
T	F	T	F	T
F	T	T	F	T
F	F	F	T	T

Table 1.1.13

Definition 1.1.3 *Tautology and Contradiction*

Let A be a proposition A formed from propositions $P, Q, R, \ldots,$ using the logical connectives.

(i) A is called a *tautology* if A is true for every assignment of truth values to $P, Q, R \ldots$.

(ii) A is called a *contradiction* if A is false for every assignment of truth values to $P, Q, R, \ldots$.

For example, as Table 1.1.13 shows, $[P \vee ((\text{not-}P) \wedge Q)] \vee [\text{not-}P \wedge \text{not-}Q]$ is a tautology. On the other hand, from Example 1.1.1 we see that

$(P \Rightarrow Q) \Rightarrow (Q \Rightarrow P)$ is not a tautology. Notice that two propositions A and B (formed from propositions $P, Q, R, \ldots$) are logically equivalent if and only if the biconditional $A \Leftrightarrow B$ is a tautology. Finally, observe that A is a tautology if and only if not-A is a contradiction.

The examples given above provide some practice with truth tables. Our goal, however, is not to make us all truth-table wizards. Instead, we should acquire solid knowledge of Tables 1.1.1, 1.1.3, 1.1.4, 1.1.7, and 1.1.10 (to work problems such as the ones given above helps us to develop this skill) and we should develop the ability to form the negation of a statement and the converse and contrapositive of an implication.

Example 1.1.4 To illustrate these remarks, consider a previously cited statement from calculus:

> If f is a function that is differentiable at a, then f is a function that is continuous at a.

First, one might attempt to prove or simply to attain a better understanding of this statement by considering its (logically equivalent) contrapositive:

> If f is a function that is not continuous at a, then f is a function that is not differentiable at a.

In forming the contrapositive of an implication, we are forced, of course, to take the negation of both the hypothesis and the conclusion of the implication. Once we have given a proof of the given implication, we can probe further into the relationship between its hypothesis and conclusion by considering its converse:

> If f is a function that is continuous at a, then f is a function that is differentiable at a.

In other words, we know that differentiability at a forces continuity at a. But does the continuity of f at a imply the differentiability of f at a? (As mentioned previously, the converse is false; hence continuity is a necessary but not a sufficient condition for differentiability.) For now, we wish to underscore the process of analyzing a mathematical implication by considering its contrapositive and its converse.

Predicate Calculus

While the propositional calculus is an essential tool in mathematical work, not all mathematical statements are propositions. For example, the sentence "$x^2 + 2x - 3 = 0$" is neither true nor false as it stands. If $x = 1$, then the sentence becomes a true proposition; if $x = 2$, then the resulting proposition is false. In the sentence "$x^2 + 2x - 3 = 0$" the symbol "x" represents a number, perhaps an arbitrary real number or perhaps a number chosen from a restricted collection of real numbers. In order to deal with sentences involving a symbol or symbols that represent objects chosen from a collection of objects, we need a new language. This language is provided by the predicate calculus.

The predicate calculus is concerned with arguments and inferences involving sentences. In addition to the logical connectives used in propositional calculus, the predicate calculus involves *variables*, *predicates*, and *quantifiers*. We now define and illustrate each of these concepts.

Definition 1.1.4 *Variable*

A *variable* is a symbol representing an unspecified object that can be chosen from some collection of objects.

Consider the statements:

1. If x is a nonzero real number, then $x^2 > 0$.
2. If f is a function that is differentiable at 2, then f is a function that is continuous at 2.

In the first example, x represents a number chosen from the collection of all nonzero real numbers; x can be *any* nonzero real number. In statement 2, f is taken from the collection of all functions that are differentiable at 2, and the symbol f represents any function in this collection.

Definition 1.1.5 *Predicate*

A *predicate* is a sentence $P(x_1, \ldots, x_n)$ involving variables $x_1, \ldots, x_n$ with the property that when specific values from a domain are assigned to $x_1, \ldots, x_n$, the resulting statement is either true or false.

For example, the sentence "x is less than y" is a predicate that is true when $x = 2$ and $y = 3$ but is false when $x = 3$ and $y = 2$.

One property of a variable that must be specified in order to avoid ambiguity is its *scope*; in other words one must state exactly the collection from which the value of the variable is chosen. For example, to emphasize the fact that the symbol x in statement 1 can be an arbitrary nonzero real number, that statement can be rewritten as:

3. For every real number x, if $x \neq 0$, then $x^2 > 0$, or
4. For every x, if x is a nonzero real number, then $x^2 > 0$.

The phrase "for every" prescribes the scope of the variable x. Because of its frequent appearance in logic and mathematics, this phrase has a name.

Definition 1.1.6 *Quantifiers*

The phrase "for every" is called a *universal quantifier*. The phrase "there exists" is called an *existential quantifier*.

Statements 1, 3, and 4 are interpreted to mean that for each choice of a nonzero real number x, it is the case that $x^2 > 0$. Statement 2 can be written with a universal quantifier as:

5. For every f, if f is a function that is differentiable at 2, then f is continuous at 2.

In general, a statement involving a universal quantifier has the form "For every x, $P(x)$," where $P(x)$ is a sentence asserting that x possesses some property. In logic texts (and in some mathematics books) the symbol "$\forall$" is used to represent the phrase "for every." With this notation the sentence "For every x, $P(x)$" becomes "$\forall x\, P(x)$." We shall rarely use the symbol "$\forall$" after this section. Finally, the phrase "for every" is often written as "for each" and "for all." We shall use these three terms interchangeably.

The basic form of a statement involving an existential quantifier (with a single variable) is "There exists x such that $P(x)$." The symbol "$\exists$" is often used as an abbreviation for "There exists." The phrases "some x," "for some x," and "for at least one x" also denote existential quantifications. For example, the statement

6. There exists x such that x is a nonzero real number, and $x^2 > 0$

means that for *at least one* choice of a nonzero real number x, it is the case that $x^2 > 0$. Assertion 6 can also be phrased as:

7. $x^2 > 0$ for some nonzero real number x.

It is almost always the case that a clearer understanding of a statement can be obtained by rewriting it using variables and quantifiers. Here are some examples:

8. Every real number is either positive, negative, or zero.
8′. For every x, if x is a real number, then either x is positive, x is negative, or x is zero.
9. Some continuous function is differentiable.
9′. There exists f such that f is continuous and f is differentiable.
10. The sum of two rational numbers is a rational number.
10′. For every x and every y, if x and y are rational, then $x + y$ is rational.

More examples are provided in the exercises. In many cases, ambiguity arises when proper quantification is omitted from a statement: Consider the equation $x + 1 = 0$. The quantification on x is missing. By supplying quantification we obtain either

(i) For every x, $x + 1 = 0$, or
(ii) There exists x such that $x + 1 = 0$.

But even these forms are not complete since we have not specified the collection from which x is to be chosen. For example, if x is allowed to be an arbitrary real number, then (i) is false while (ii) is true. But if x is taken only from the collection of positive numbers, then both (i) and (ii) are false. On the other hand, if x is taken from the collection of numbers whose only member is the number -1, then both (i) and (ii) are true. Thus the moral is: In any statement involving variables, be sure that each variable is properly quantified and remember the scope of each variable.

Frequently, it is necessary to form the negation of a statement involving variables and quantifiers. Such negations can be tricky. Let us look at a few examples.

Example 1.1.5 Form the negation of the statement: "If x is a nonzero real number, then $x^2 > 0$."

The safest way to begin is to add the appropriate quantification: "For every x, if x is a nonzero real number, then $x^2 > 0$." Perhaps the most prudent next step is to state the negation by adding the phrase "it is not the case that" at the beginning of the given statement:

It is not the case that for every x, if x is a nonzero real number, then $x^2 > 0$.

Thus we must negate the statement "If x is a nonzero real number, then $x^2 > 0$." But the implication $P \Rightarrow Q$ is logically equivalent to "(not-P) or Q," and hence "not-($P \Rightarrow Q$)" is logically equivalent to "not-

[(not-P) or Q]," which is logically equivalent to "P and not-Q." Therefore, we can state the negation of the original statement as:

There exists x such that x is a nonzero real number and $x^2 \ngtr 0$.

We can generalize this example as follows: The negation of a statement of the form "For every x, $P(x)$" is the statement "There exists x such that not-$P(x)$." On the other hand, the negation of "There exists x such that $Q(x)$" is the statement "For each x, not-$Q(x)$" (i.e., for each x, it is not the case that $Q(x)$ holds). We summarize these remarks:

Statement	**Negation**
For every x, $P(x)$. $(\forall x\, P(x))$	There exists x such that not-$P(x)$. $(\exists x(\text{not-}P(x)))$
There exists x such that $Q(x)$. $(\exists x\, Q(x))$	For every x, not-$Q(x)$. $(\forall x(\text{not-}Q(x))$

The statements $P(x)$ and $Q(x)$ might themselves be quite involved and difficult to negate. The best way to proceed is to write $P(x)$ and $Q(x)$ in terms of logical connectives and use the rules for negation discussed earlier in this section. This procedure was followed in Example 1.1.5. Let us illustrate by negating the quantified statement: $(\forall x)(P(x) \wedge Q(x))$. If this statement does not hold, then there exists x such that it is not the case that $P(x) \wedge Q(x)$ holds. Therefore, the negation of the given statement is $(\exists x)(\text{not-}(P(x) \wedge Q(x)))$, which becomes $(\exists x)((\text{not-}P(x)) \vee (\text{not-}Q(x)))$.

A final word about quantification. We often encounter two or more quantifiers in a given statement. Here are some examples (x and y are real numbers):

For all x and for all y, $x + y = y + x$;

There exists (a number) 0 such that for all y, $0 + y = y$; and

For all x there exists y such that $x + y = 0$.

It is important for us to observe that the order of the quantifiers is crucial. For instance, the statements "For all x there exists y such that $x + y = 0$," and "There exists y such that for all x, $x + y = 0$" have entirely different meanings. The first asserts that for any given real number x, a real number y can be found for which the equation $x + y = 0$ holds. The number y depends on the given number x. The second statement claims that there exists a single number y for which the equation $x + y = 0$ holds for all numbers x. Thus we must take care when both an existential and a universal quantifier appear in the same sentence.

Exercises §1.1

1.1.1. Let P, Q, and R denote propositions. Construct truth tables for each of the following propositions.
(a) $P \Rightarrow (P \Rightarrow Q)$.
(b) $(P \Rightarrow Q) \Rightarrow (P \wedge \text{not-}Q)$.
(c) $(P \Rightarrow \text{not-}(Q \text{ and } R))$.
(d) $(P \Rightarrow (Q \text{ and } R))$ or (not-P and Q).
(e) $(P \text{ and } Q) \Rightarrow [((Q \text{ and } (\text{not-}Q)) \Rightarrow (R \text{ and } Q))]$.
(f) $(P \Rightarrow Q) \Rightarrow (\text{not-}P \Rightarrow \text{not-}Q)$.

1.1.2. Show that for any statement P, P is logically equivalent to each of the following: (a) P and P. (b) P or P. (c) not-(not-P).

1.1.3. (a) Suppose P and $P \Rightarrow Q$ are true. Show Q is true.
(b) Suppose P and Q are propositions for which "P and Q" is false and "P or Q" is true. What can be said about the truth values of P and Q?
(c) If $P \Rightarrow Q$ is true for all propositions Q, then what can be said about the truth value of P?
(d) If $P \Leftrightarrow Q$ is true, what is the truth value of $P \Leftrightarrow (\text{not-}Q)$?

1.1.4. Are $P \Rightarrow Q$ and $(\text{not-}P) \Rightarrow (\text{not-}Q)$ logically equivalent?

1.1.5. Prove that "not-$P \vee Q$" is logically equivalent to $P \Rightarrow Q$.

1.1.6. (a) Show that not-$(P \Rightarrow Q)$ is logically equivalent to "P and (not-Q)."
(b) Show that $P \Rightarrow (Q \vee R)$ is logically equivalent to $(P \Rightarrow Q) \vee (P \Rightarrow R)$.

1.1.7. Let P, Q, and R be propositions. Show that the following pairs of statements are logically equivalent. (In every case your truth table will have eight rows.)
(a) "$(P \wedge Q) \wedge R$" and "$P \wedge (Q \wedge R)$."
(b) "$(P \vee Q) \vee R$" and "$P \vee (Q \vee R)$."

1.1.8. Show that each of the following is a tautology:
(a) P or not-P.
(b) If P then P.

1.1.9. Show that $(P \Leftrightarrow Q) \Leftrightarrow [(P \wedge Q) \vee ((\text{not-}P) \wedge (\text{not-}Q))]$ is a tautology.

In Exercises 1.1.10–1.1.21 each of the statements is a theorem of calculus or precalculus. In every case (a) write the contrapositive of the statement, and (b) write the converse and state if the converse is true.

1.1.10. For all real numbers a and b, if $a > 0$ and $b > 0$, then $a \cdot b > 0$.

1.1.11. For all triangles T_1 and T_2 in the plane, if T_1 and T_2 are congruent, then corresponding angles of T_1 and T_2 are equal.

1.1.12. If Σa_n is a convergent infinite series, then $\lim_{n \to \infty} a_n = 0$.

1.1.13. If f is continuous on $[a, b]$, then $\int_a^b f(x)\,dx$ exists.

1.1.14. If f is constant on $[a, b]$, then $f'(x) = 0$ for all x in (a, b).

1.1.15. Let T be a triangle having sides a, b, c with c being the longest side. If T is a right triangle, then $c^2 = a^2 + b^2$.

1.1.16. If $\Sigma|a_n|$ converges, then Σa_n converges.

1.1.17. If f is continuous on $[a, b]$, then f attains a maximum value on $[a, b]$.

1.1.18. If f has a local (relative) maximum at the real number a or a local (relative) minimum at a, then $f'(a) = 0$ or $f'(a)$ does not exist.

1.1.19. Suppose for every positive integer n, a_n and b_n are real numbers such that $0 \leq a_n \leq b_n$. If Σa_n diverges, then Σb_n diverges.

1.1.20. With the same hypotheses on a_n and b_n as in Exercise 1.1.19, if Σb_n converges, then Σa_n converges.

1.1.21. If T is an equilateral triangle, then T is an equiangular triangle.

In Exercises 1.1.22–1.1.30 (a) write each of the statements using variables and quantifiers, and (b) negate each statement.

1.1.22. Some integers are perfect squares.

1.1.23. Every rational number is a real number.

1.1.24. The product of two rational numbers is rational.

1.1.25. The cube root of any integer is irrational.

1.1.26. The square root of any irrational number is irrational.

1.1.27. No solution of $ax = b$ where a and b are arbitrary integers is irrational.

1.1.28. The derivative of a constant function is 0.

1.1.29. If x and y are irrational, then $x + y$ is irrational.

1.1.30. $|x + 7| = 3$ if and only if $x = -10$ or $x = -4$.

1.1.31. Form a negation of each of the following quantified statements:
(a) $(\forall x)(P(x) \vee Q(x))$.
(b) $(\forall x)(P(x) \Rightarrow Q(x))$.
(c) $(\forall x)(P(x) \Leftrightarrow Q(x))$.
(d) $(\exists x)(P(x) \wedge Q(x))$.
(e) $(\exists x)(P(x) \vee Q(x))$.
(f) $(\exists x)(P(x) \Leftrightarrow Q(x))$.

§1.2 The Real Numbers and Axiomatic Systems

This section has two principal goals. The first is to describe the real number system in both an intuitive and a formal way. Because the real number system is such a fundamental part of mathematics, a working knowledge of the real numbers is necessary for anyone wishing to use or to understand most mathematical topics. In this sense the real number system can be regarded as a basic mathematical tool. Our formal description of the real numbers leads into the second theme of this section, the notion of an axiomatic system. Although axiomatic systems have been part of mathematics since the time of Euclid (300 B.C.), only in the past 100 years have they assumed a significant role in mathematical research and teaching. Axiomatic thinking has provided mathematicians with a new way of looking not only at mathematical theories but also at mathematics itself. Our purpose is to discuss the nature of axiomatic systems and to illustrate our remarks by using the real numbers as an example.

The Real Numbers

Our discussion of the real numbers begins with an intuitive description familiar to any student of calculus. Our informal summary of the real number system will serve as a background for the first six chapters of this text. In Chapter 7 we shall outline a construction of the real numbers based on the theory of sets.

The second part of this section presents an axiomatic description of the real numbers. Specifically, we give a list of statements that describe properties satisfied by the real numbers. We show how many other "facts" about the real numbers can be derived from the basic list of properties or axioms. When we construct the real numbers from set theory in Chapter 7, we shall show that the real numbers do indeed satisfy the list of axioms presented in this section.

Perhaps the simplest way of envisioning the real number system is geometrically. Imagine a straight line stretching without bound in both directions. The idea is to match a real number to each point on this line. Henceforth, this line is referred to as a *number line*.

We begin by marking a point on the line and labeling it 0. Next choose a unit distance, mark off one unit to the right of 0, and label this point 1 (Figure 1.2.1). The point one unit to the right of 1 is labeled 2. The set of numbers that label the points obtained by starting with 0 and repeatedly taking points one unit to the right of the previously labeled point as in

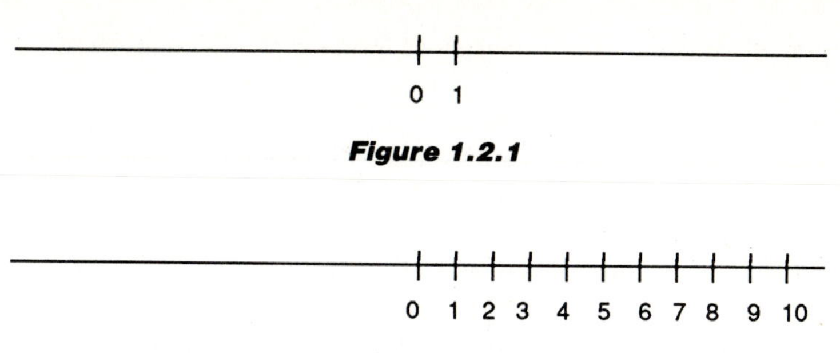

Figure 1.2.1

Figure 1.2.2

Figure 1.2.2 is called the set of *natural numbers* and is denoted by **N**. The members of the set **N** are the numbers $0, 1, 2, 3, \ldots$.

If we mimic this geometric construction on the left side of 0, then we obtain numbers that are labeled $-1, -2, -3, \ldots,$ as shown in Figure 1.2.3. These numbers taken together with the the set of natural numbers form a new collection, called the set of *integers* and denoted by **Z** (**Z** is an abbreviation for the German word for numbers, *Zahlen*). The members of the set **Z** are the numbers $0, \pm 1, \pm 2, \pm 3, \ldots$. The integers $1, 2, 3, \ldots$ are called the *positive integers* and the numbers $-1, -2, -3, \ldots$ are called the *negative integers*. From the definition of **Z**, it follows that any natural number is an integer.

Next we define the set of *rational numbers*, denoted by **Q**. A rational number can be represented as a ratio or quotient of integers, m/n, where $n \neq 0$. Geometrically the numbers m/n are obtained from the following process: We consider two cases, the first being when n is a positive integer. Divide the segment from 0 to 1 into n equal pieces. Take the first of these segments, the one whose left end is the point 0. If m is positive, then reproduce this segment m times to the right of 0. The right endpoint of the segment so obtained corresponds to m/n. If m is negative, then reproduce the segment $-m$ times to the left of 0 to obtain m/n. (See Figure 1.2.4.) The second case occurs when n is negative. In this case we correspond m/n

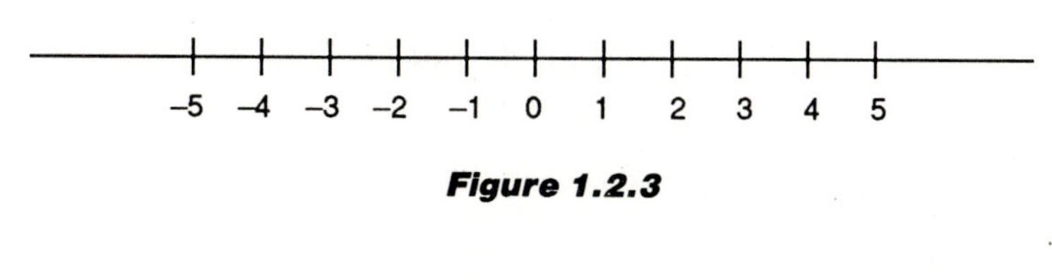

Figure 1.2.3

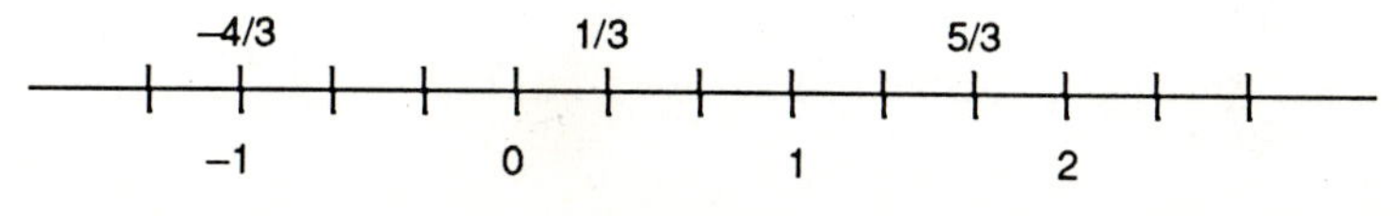

Figure 1.2.4

to the point $(-m/(-n))$ as constructed in the previous case. Note that any integer is also a rational number.

Clearly this process allows us to attach numerical labels to many points on the line. In fact, if a and b are points with rational labels, then between a and b there lies a point corresponding to a rational number. Thus the rational numbers are not separated geometrically in the way the integers are; the rationals are distributed quite densely along the number line. One might even ask: Does every point on the line correspond to a rational number? Is every point on the line obtained by the geometric construction described in the previous paragraph? Many early Greek mathematicians, especially the followers of the mathematician Pythagoras, believed, primarily on philosophical and religious grounds, that indeed every point on the line is matched to a rational number. The idea that all distances are rational (commensurable was the word the Greeks used) was central to the Pythagorean conception of the order of the universe. Thus they were amazed and dismayed by the discovery of Hippasus, himself a Pythagorean, who lived in the fifth century B.C., that there are distances, i.e., points on the line, that are not rational. Specifically, Hippasus showed that $\sqrt{2}$ is irrational, i.e., not rational. (We give Hippasus's proof that $\sqrt{2}$ is irrational in Section 1.3.) Legend has it that Hippasus announced that $\sqrt{2}$ is not rational while on a ship with other Pythagoreans and was rewarded for his blasphemous discovery by being tossed overboard.

The fact that not all points on the line correspond to rational numbers opens the way for the introduction of even more numbers. We show how to represent each point on the line, rational or irrational, by the so-called decimal expansion. Given a point on the line, first find the integer point immediately to its left; for x, as in Figure 1.2.5, this integer is 3. Next, on the segment from 3 to 4, find the one-tenth mark immediately to the left of x. In Figure 1.2.6, this is the number $3\,1/10 = 31/10$, which we write in decimal representation as 3.1. Repeat this procedure for the segment between $3\,1/10$ and $3\,2/10$. In Figure 1.2.7 x lies between the points $3\,14/100$ and $3\,15/100$, which are written in decimal form as 3.14 and 3.15,

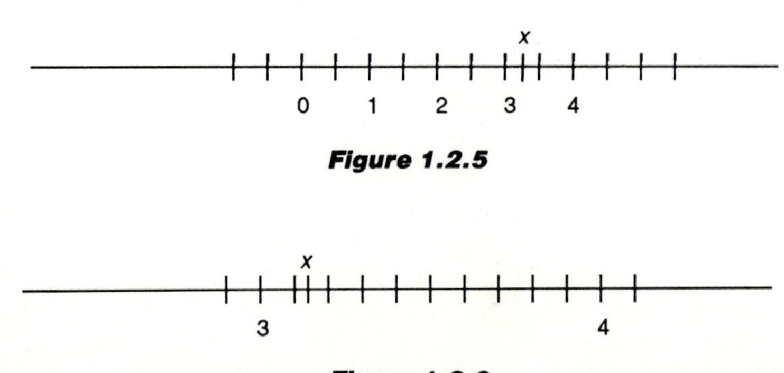

Figure 1.2.5

Figure 1.2.6

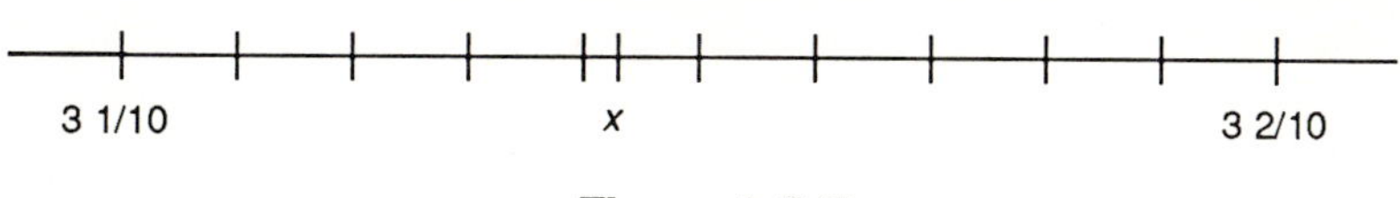

Figure 1.2.7

respectively. Continuing this process, we obtain the decimal representation of x:

$$x = 3.1415926535\ldots.$$

Each of the rational numbers $3, 3.1, 3.14, 3.141, 3.1415, \ldots$ lies to the left of x on the number line. This procedure can be repeated for any arbitrary point x on the number line. Thus each point on the line is assigned a decimal representation. Most numbers have unique decimal expansions: If $x = .a_1a_2a_3\ldots = .b_1b_2b_3\ldots$ are two decimal expansions of x, then $a_1 = b_1$, $a_2 = b_2$, $a_3 = b_3, \ldots$ with the exception of numbers terminating in 9's or 0's. For instance, $.12999\ldots = .13000\ldots = .13$.

The set of such decimal expansions is called the *set of real numbers* and is denoted by **R**. Observe that any rational number is a real number. In fact, we can show that a real number x is rational if and only if the decimal expansion of x terminates (such as $1/8 = .125000\ldots$) or repeats (such as $12/99 = .121212\ldots$).

We have then two informal perspectives from which we can regard the real numbers. First, from the geometric perspective we view the set of real numbers as an indefinitely extended line on which each point corresponds to a unique real number. This geometric viewpoint has several benefits. For example, it provides us with a way of grasping the real numbers as a whole. Often certain properties of the real numbers (for instance, the transitivity property of $<$ proved in Theorem 1.2.2 below) are obvious when considered geometrically. Second, we have the decimal representation of a real number. Each real number has a decimal expansion. This expansion is unique except for those reals whose decimal expansion can be represented from some point on only the digit 9. The decimal representation is useful whenever precise calculations with real numbers are necessary.

Axiomatic Systems

As noted earlier in this section, the first axiom system to appear in the history of mathematics was Euclid's development of plane geometry. Euclid's treatment begins with *primitive* or *undefined* terms. Next, a list of statements that are assumed to be true are added. Some of these statements are called *axioms*; some are called *postulates*. The collection of undefined terms, axioms, and postulates forms the *axiom system* for Euclidean geometry. Let us look at Euclid's system more closely.

The undefined terms in Euclidean geometry are *points* and *lines*. The *axioms*, or self-evident truths as they were sometimes called, are general statements whose truth cannot presumably be disputed. Examples are:

The whole is greater than the part.

and

Equals added to equals yield equals.

The *postulates* are statements that describe relationships among the undefined terms of the system. Hence the scope of a postulate is more limited than that of an axiom. Examples of postulates are:

Two points determine exactly one line.

and

Through a point not on a given line, one and only one line can be drawn that does not meet the given line.

From this collection of undefined terms, axioms, and postulates, one can deduce through logical reasoning a large number of statements (called *theorems*, *propositions*, or *lemmas*) about plane geometry, a notable example being: The sum of the angles of a triangle is π radians where π is the ratio of the circumference of any circle to the diameter of that circle. In summary, the Euclidean axiom system consists of a collection of undefined terms, axioms, and postulates. The body of statements that can be deduced from the axioms and postulates constitutes the Euclidean theory of plane geometry.

Until the middle of the nineteenth century, Euclidean geometry held a unique place in mathematics. First, Euclidean geometry was regarded by most learned people as the true geometry of the universe. Its axioms and postulates were indeed basic truths of the physical world. Thus all the theorems of Euclidean geometry were also true descriptions of physical reality. Second, until this time no other portion of mathematics was treated axiomatically. Whenever possible, mathematicians attempted to deduce results in other areas of mathematics (such as calculus) from Euclidean geometry. Otherwise, they used some seemingly obvious results (such as $x^2 - y^2 = (x - y)(x + y)$ where x and y are real numbers) without a careful regard for the status of such statements: Is $x^2 - y^2 = (x - y)(x + y)$ a postulate of the real number system, does it follow from some other postulates and axioms for the real numbers, or is it an empirical fact about real numbers?

During the nineteenth century two important developments occurred. First, for reasons both internal and external to mathematics, mathematicians began to question the status of Euclidean geometry. Are the postulates and theorems actually truths of nature, or are the theorems just statements deducible from the axioms and postulates which are in turn merely assumptions made by mathematicians? These questions led a few mathematicians, notably Lobatchevsky, Bolyai, and Gauss, to modify the postulates of Euclidean geometry to obtain other geometric theories. Although these non-Euclidean geometries produced results that seemed counterintuitive, they did appear to be logically consistent. Thus the existence of axiom systems in mathematics other than Euclidean geometry became a mathematical reality.

The second major development in the nineteenth century was the rapid expansion of several other areas of mathematics. Because of pressures both inside and outside mathematics, disciplines such as number theory, algebra, and analysis grew significantly in extent and sophistication. Mathematicians were compelled to clarify and to organize the results of each of these complex subjects. To do so, mathematicians such as H. Grassman, R. Dedekind, K. Weierstrass, and G. Peano turned to the axiomatic method. Their goal was to develop each mathematical theory from a collection of primitive terms, axioms, and postulates. Initially, these mathematicians were interested primarily in clarifying the development of previously existing bodies of mathematical knowledge. As mathematicians became more adept at using axiomatic systems, they came to use them as more than merely organizing devices. In fact, during the twentieth century, many mathematical theories have from their inception been developed on an axiomatic basis. Thus the axiomatic method has become a mathematical research tool.

The modern view of axiomatic systems is remarkably close to the Euclidean conception of axiomatic geometry. One important change is that the distinction between axioms and postulates has disappeared in the past 100 years. This conceptual shift has been yet another result of the advances made in mathematics during this century. Many of Euclid's "self-evident" truths have been seen to be not at all evident. For example, as we shall see in Chapter 4, there are situations in which a whole is not necessarily greater than one of its parts. According to the modern view, then, an *axiom system* consists of primitive or undefined terms together with axioms, which are statements describing relationships among and properties of the primitive concepts. (To emphasize the arbitrariness of the undefined term, the influential mathematician David Hilbert once asserted: "One must be able to say at all times—instead of points, straight lines, and planes—tables, chairs, and beer mugs.") The axiom system and all the theorems deducible from the axioms together form the *axiomatic theory*.

An axiomatic theory can be likened to the everyday concept of a game. Consider any familiar game such as chess or checkers. The pieces, the gameboard, and the players of the game constitute the primitive terms. The rules of each game describe how the pieces interact; thus the rules play the role that the axioms play in an axiomatic theory. All the possible legal maneuvers and configurations in each game are derivable from the rules; hence these maneuvers and configurations are analogous to the theorems of an axiomatic theory.

A note about the names of statements that are deducible from the axioms: The most important of these statements is called a *theorem*. A statement that is derivable from axioms and that is considered less significant or perhaps more obvious than theorems is often called a *proposition* (not to be confused with the propositions of propositional calculus); a statement that is technical in nature and of a limited scope is called a *lemma*. Finally, a statement that follows immediately from a theorem, in some cases being merely a special case of a theorem, is called a *corollary*. From a logical point of view, theorems, lemmas, propositions, and corollaries are on an equal footing. The labels, however, enable us to clarify their position in the development of a particular subject.

We now give an axiomatic description of the real number system. The axioms for **R** fall into three classes according to the types of properties which they describe: field properties, order properties, and the completeness property.

The *real numbers* consist of a collection or set, **R**, of objects called *numbers* satisfying the following fourteen conditions.

Field Axioms

1. Closure for Addition For each pair x, y in **R**, there exists a unique object in **R**, written $x + y$ and called the *sum* of x and y.
2. Associative Law for Addition For all x, y, z in **R**, $(x + y) + z = x + (y + z)$.
3. Additive Identity There is an object 0 in **R** such that for all x in **R**, $x + 0 = 0 + x = x$.
4. Additive Inverse For all x in **R**, there exists y in **R** such that $x + y = y + x = 0$.
5. Commutative Law for Addition For all x and y in **R**, $x + y = y + x$.
6. Closure for Multiplication For each pair x, y in **R**, there exists a unique object in **R**, written $x \cdot y$ and called the *product* of x and y.
7. Associative Law for Multiplication For all x, y, z in **R**, $(x \cdot y) \cdot z = x \cdot (y \cdot z)$.
8. Multiplicative Identity There exists an object 1 in **R** such that $1 \neq 0$ and for all x in **R**, $x \cdot 1 = 1 \cdot x = x$.
9. Multiplicative Inverse For each x in **R** such that $x \neq 0$, there exists y in **R** such that $x \cdot y = y \cdot x = 1$.
10. Commutative Law for Multiplication For all x and y in **R**, $x \cdot y = y \cdot x$.

11. **Distributive Law for Multiplication over Addition** For all x, y, z in $\mathbf{R}$, $x \cdot (y + z) = x \cdot y + x \cdot z$.

Remark on inverses One can show that for any x in $\mathbf{R}$ there exists a unique y such that $x + y = y + x = 0$ (see Theorem 1.2.1). This real number y is called *the additive inverse of* x, denoted by $-x$. In addition we define $x - y$ to be $x + (-y)$. Similarly one can show that each x in $\mathbf{R}$ such that $x \neq 0$ has a unique *multiplicative inverse*. This element is denoted by x^{-1}. For $y \neq 0$ we define x/y to be $x \cdot y^{-1}$.

Order Axioms

There exists a subcollection $\mathbf{R}^+$ of $\mathbf{R}$ called the *positive numbers* satisfying the following conditions:

12. If x and y are in $\mathbf{R}^+$, then $x + y$ and $x \cdot y$ are also in $\mathbf{R}^+$.
13. For each x in $\mathbf{R}$, exactly one of the following three conditions holds: (i) x is in $\mathbf{R}^+$, (ii) $x = 0$, or (iii) $-x$ is in $\mathbf{R}^+$.

If x is in $\mathbf{R}$ and $-x$ is in $\mathbf{R}^+$, then x is called a *negative number*.

Axioms 12 and 13 allow us to propose the following definition, which describes how the members of $\mathbf{R}$ can be compared or ordered.

Definition 1.2.1 *Ordering of* $\mathbf{R}$

Let x and y be in $\mathbf{R}$. We say that x *is greater than* y, written $x > y$, if $x - y = x + (-y)$ is in $\mathbf{R}^+$. We say that x *is greater than or equal to* y, written $x \geq y$, if $x > y$ or $x = y$. We say that x *is less than* (resp. *less than or equal to*) y, written $x < y$ (respectively $x \leq y$), if $y > x$ (resp. $y \geq x$).

This concept of ordering in turn allows us to introduce the notion of boundedness, which is needed for the statement of the completeness property.

Definition 1.2.2 *Boundedness*

A collection A of real numbers is called *bounded from above* if there exists a real number x (not necessarily in A) such that for all a in A, $x \geq a$. Such an element x is called an *upper bound* for A.

Note the order of the quantification. A collection A is bounded if there exists at least one number x that is greater than or equal to all the members of A.

Definition 1.2.3 *Least Upper Bound*

Let A be a set of real numbers that has an upper bound. An element x_0 of $\mathbf{R}$ is called a *least upper bound* of A if (i) x_0 is an upper bound for A, and (ii) for any upper bound x of A, $x_0 \leq x$.

As an example consider the set A of all real numbers that are less than 2. A is bounded from above by 2, 3, or 4, for instance. Observe that 2 is a least upper bound for A.

With these concepts in hand, we can state the final axiom for $\mathbf{R}$.

Completeness Axiom

14. Any *nonempty* collection of real numbers having an upper bound has a least upper bound.

At this point let us summarize formally our definition of the real numbers.

Definition 1.2.4 *The Real Numbers*

The *real numbers* are a collection of objects, denoted by $\mathbf{R}$, which satisfies Axioms 1–14.

This definition suggests an important question: Is there actually a collection of objects that satisfies Axioms 1–14? It is at least conceivable that the axioms are somehow self-contradictory and hence can be satisfied by no mathematical entity. If we picture the real numbers as points along a line or if we represent real numbers in decimal form, then perhaps we can be convinced that the real numbers do satisfy all the axioms given above. Such an argument can be suggestive but is hardly convincing. To demonstrate mathematically that the real numbers, $\mathbf{R}$, do exist, we must

(i) define the collection of objects that constitute $\mathbf{R}$,
(ii) define, for all x and y in $\mathbf{R}$, the sum $x + y$,

(iii) define, for all x and y in **R**, the product $x \cdot y$,
(iv) define the objects that constitute the collection $\mathbf{R}^+$ of positive numbers, and
(v) show that Axioms 1–14 hold for **R** with addition, multiplication, and positivity as defined in (i)–(iv).

In Chapter 7 we show how a collection that satisfies Axioms 1–14 can actually be defined. In that language of mathematical logic, this result shows that the axiom system consisting of Axioms 1–14 has a *model*. Moreover, it can be proved that in a sense there is only one collection that satisfies these axioms: If two collections of objects $\mathbf{R}_1$ and $\mathbf{R}_2$ are defined and if $\mathbf{R}_1$ and $\mathbf{R}_2$ both satisfy Axioms 1–14, then the objects in $\mathbf{R}_1$ and $\mathbf{R}_2$ can be matched exactly in such a way that addition, multiplication, and positivity are preserved. In this case we say that $\mathbf{R}_1$ and $\mathbf{R}_2$ are *isomorphic*. Thus $\mathbf{R}_1$ and $\mathbf{R}_2$ are identical in terms of their properties although the labelings of their elements may differ. In technical terms this result shows that Axioms 1–14 have a unique model. Therefore, we can legitimately call any collection satisfying the fourteen postulates *the* real numbers.

Until Chapter 7 we assume that the real numbers do exist. In Chapters 2–6 we work with the elements of **R** as we always have: We represent real numbers either geometrically as points on a line or numerically via their decimal representations and we assume that Axioms 1–14 hold for **R**. For now let us derive some elementary results from the axioms. Our first theorem asserts that each element has at most one additive inverse. This result and Axiom 4 imply that each element of **R** has a unique additive inverse.

Theorem 1.2.1 *Let x be any real number. Suppose y_1 and y_2 are elements of* **R** *that satisfy Axiom* 4, *i.e.*, $x + y_1 = y_1 + x = x + y_2 = y_2 + x = 0$. *Then* $y_1 = y_2$.

Proof By assumption $x + y_1 = 0$. Therefore $y_2 + (x + y_1) = y_2 + 0 = y_2$ by Axiom 3. By Axiom 2, $y_2 + (x + y_1) = (y_2 + x) + y_1$. Thus $y_2 = (y_2 + x) + y_1 = 0 + y_1 = y_1$ by our assumption on y_2 and by Axiom 3. ■

As noted earlier, because for each x in **R** there exists exactly one y in **R** such that $x + y = y + x = 0$, we can denote y by $-x$ without fear of ambiguity. We also observe that this proof uses only Axioms 1–3; hence our use of $-x$ in later axioms is legitimate. The analogous fact for multiplication can be derived in a similar way: If $x \neq 0$, then there exists exactly one y in **R** such that $x \cdot y = y \cdot x = 1$. This element y is denoted by x^{-1}.

Next we derive some order properties of **R**.

Theorem 1.2.2 (i) *Let x and y be in* **R**. *Then exactly one of the following conditions holds*: (a) $x > y$, (b) $x = y$, *or* (c) $x < y$.
(ii) *Let x, y, and z be in* **R**. *If $x < y$ and $y < z$, then $x < z$.*

Proof (i). Consider $x - y$. By Axiom 13, exactly one of the following conditions holds: (i) $x - y$ is in $\mathbf{R}^+$, (ii) $x - y = 0$, or (iii) $-(x - y)$ is in $\mathbf{R}^+$.

If $x - y$ is in $\mathbf{R}^+$, then by definition $x > y$. If $x - y = 0$, then $x = y$. Finally, note that $-(x - y) = y - x$ (see Exercise 1.2.4); hence, if $-(x - y) = y - x$ is in $\mathbf{R}^+$, then $x < y$.

(ii). If $x < y$ and $y < z$, then $y - x$ and $z - y$ are in $\mathbf{R}^+$. By Axioms 2–4, $(z - y) + (y - x) = ((z - y) + y) + (-x) = (z + (-y + y)) + (-x) = (z + 0) + (-x) = z + (-x) = z - x$. By Axiom 12, $z - x = (z - y) + (y - x)$ is in $\mathbf{R}^+$, and therefore $x < z$. ■

The real number system is, of course, a very familiar object. After having worked with this system for so many years, we might find it strange to be proving "obvious" facts about it. At the same time it might be unclear exactly what can be assumed about **R** and what must be proved about **R**. For example, how do we know that if a, b, and c are real numbers and $a = b$, then $a + c = b + c$? Is this a property that one merely assumes? Or does it follow from the axioms? The answer is that it follows from the axioms, for the pair of numbers a and c and the pair b and c are identical. Therefore by Axiom 1, the number $a + c$ must coincide with $b + c$.

In general, any of the familiar properties of the real numbers can be deduced from Axioms 1–14.

We now consider the completeness property. This property is clearly the most complex of all given in the axioms. For starters, its statement is technical in nature, requiring the concepts of upper bound and least upper bound. Nonetheless, the completeness property can be visualized rather easily as we shall see at the end of this section. For now let us deduce from the completeness axiom an important property of the real numbers, namely the fact that the square root of 2 exists in **R**. (For the record, the element 2 in **R** is defined to be $1 + 1$ where 1 is the multiplicative identity in **R**.)

Theorem 1.2.3 *There exists a real number u such that $u^2 = 2$.*

The idea behind the following proof is to define a collection of real numbers A such that A is bounded and the least upper bound of A (which exists by Axiom 14) is a number whose square is 2. The proof is rather subtle and must be read carefully. At least twice in the course of the argument, a proof by contradiction is used. This method of proof will be discussed thoroughly in the next section. Briefly, here is the logic behind a proof by contradiction: To show that a possible conclusion P does not

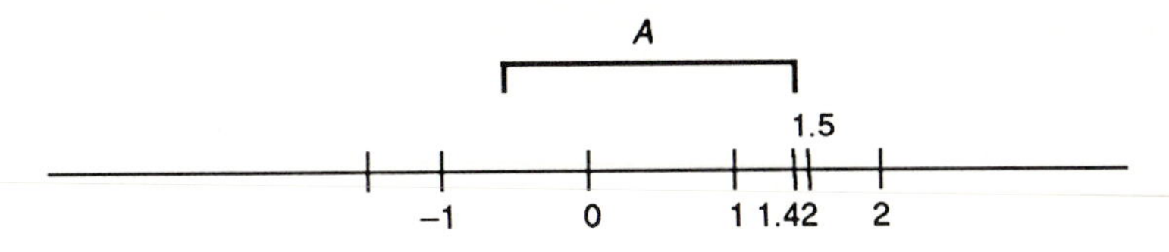

Figure 1.2.8

hold, we show that a false statement Q can be derived from P. Thus the implication $P \Rightarrow Q$ is valid where Q is false. Therefore, P must be false.

Proof Let A be the collection of all real numbers whose square is less than 2. Thus x is in A if and only if $x^2 < 2$. First note that A is nonempty since $0 \in A$. Next observe that the collection A has an upper bound: For example, if x is in A, then $x < 2$. (Using the axioms for $\mathbf{R}$, we can show that if $x \geq 2$, then $x^2 \geq 4$; hence x is not in A.) Hence 2 is an upper bound for A. A similar argument shows that 1.5 and 1.42 are also upper bounds for A (Figure 1.2.8).

By Axiom 14, A has a least upper bound, call it u. We claim that $u^2 = 2$.

By Theorem 1.2.2 exactly one of the following possibilities holds: (i) $u^2 = 2$, (ii) $u^2 < 2$, (iii) $u^2 > 2$. We show that each of the last two possibilities is impossible since each leads to a contradiction. For future reference we note that $1.41 < u < 1.42$.

First we suppose $u^2 < 2$. By finding a number that is in A and that is slightly larger than u, we show that, contrary to our assumption, u is not an upper bound for A.

Let $a = 2 - u^2$. Since $u > 1.41$, $0 < a = 2 - u^2 < 2 - (1.41)^2 < .02$. Now consider the real number $u + a/6$. We show that $u + a/6$ is in A and $u < u + a/6$. This conclusion will imply that u is not an upper bound for A, which is contrary to the definition of u. Since this invalid conclusion is implied by the assumption that $u^2 < 2$, this assumption cannot be true.

We show $u + a/6$ is in A by proving that $(u + a/6)^2 < 2$. By Exercises 1.2.5 and 1.2.6,

$$(u + a/6)^2 = u^2 + ua/3 + a^2/36.$$

Our strategy is to estimate the numbers $ua/3$ and $a^2/36$ in order to show that $(u + a/6)^2 < 2$. Since $u < 1.42$, $u/3 < 1/2$, and hence $ua/3 < a/2$. Also $a/36 < .02/36 < 1/2$ and $a^2/36 < a/2$. Thus

$$(u + a/6)^2 = u^2 + ua/3 + a^2/36 < u^2 + a/2 + a/2 = u^2 + a = 2.$$

Therefore, $u + a/6$ is in A. Finally, since $a > 0$, $a/6 > 0$ and $u + a/6 > u$. We now know that $u^2 \not< 2$.

Suppose next that $u^2 > 2$. Let $b = u^2 - 2$. Since $u < 1.42$, $0 < b < (1.42)^2 - 2 < .02$. Consider the real number $u - b/6$. We show that $(u - b/6)^2 > 2$, which implies that $u - b/6$ is an upper bound for A; but since $u - b/6 < u$, it follows that u is not a least upper bound for A.

We have

$$(u - b/6)^2 = u^2 - ub/3 + b^2/36$$

$$> u^2 - ub/3 > u^2 - 2b/3 > u^2 - b = 2.$$

Thus $u - b/6$ is an upper bound for A that is less than u. Therefore $u^2 \ngtr 2$.

As a result we conclude $u^2 = 2$. ■

Theorems 1.2.1–1.2.3 illustrate the nature of axiomatic systems. Each theorem is deduced from certain of the axioms of the real number system. For instance, in proving Theorem 1.2.3, we did not explicitly describe a member of **R** whose square is 2. Instead we showed how the existence of that element follows from the axioms of **R** and the rules of logic. Thus, once we know that a collection of objects satisfying Axioms 1–14 exists, then we know that in that collection an element whose square is 2 exists.

We remarked earlier that **R** is in a sense the only collection that satisfies Axioms 1–14. In the technical language of the foundations of mathematics, this claim asserts that the axiom system given by Axioms 1–14 has a unique *model*. If we consider only the field axioms, namely Axioms 1–11, then we can find many models that are essentially different. (The technical word is *nonisomorphic*.) More specifically there are collections of objects, F_1 and F_2, which satisfy Axioms 1–11 with the property that the elements of F_1 and F_2 cannot be matched in such a way that addition and multiplication are preserved by the matching. For now let us give one other example of a field, namely the collection of rational numbers **Q**.

By definition a real number x is rational if $x = m/n$ where m and n are integers and $n \neq 0$. Rational numbers are added and multiplied according to the following well-known formulas:

$$m/n + m'/n' = (mn' + m'n)/nn'.$$

$$(m/n) \cdot (m'/n') = mm'/nn'.$$

From these definitions it follows that the sum of two rational numbers is rational and the product of two rational numbers is rational. One says that Q is *closed* under addition and multiplication. (For example, the collection of all rational numbers with denominators at most two, when put in lowest terms, is closed under addition but is not closed under multiplication.) Also the sum (product) of m/n and m'/n' as defined above coincides with the sum (product) of m/n and m'/n' within the real number system. Thus addition and multiplication of rational numbers satisfy the associative, commutative, and distributive properties. Moreover, the real numbers $0 = 0/1$ and $1 = 1/1$ are also rational. Finally, for any x in **Q**, $x = m/n$, the additive inverse of x, $-x = -m/n$, is also in **Q** and, for $x = m/n \neq 0$, the multiplicative inverse of x, $x^{-1} = n/m$, is also in **Q**. Therefore, the collection **Q** of rational numbers satisfies Axioms 1–11. In other words, Axioms 1–11 are valid if **R** is replaced by **Q** throughout and addition and multiplication are defined as above. The collection of rational numbers then provides another example of a mathematical object that satisfies the field axioms. In other words, **Q** is a model for the axiom system consisting of Axioms 1–11.

Also notice that Axioms 12 and 13 hold when **R** is replaced by **Q**. As a result **Q** is called an *ordered field*. At this point our luck runs out. **Q** does not satisfy Axiom 14. Our argument is based on the fact that $\sqrt{2}$ is not in **Q**, which we shall prove in Section 1.3. Therefore, the set A of all rationals such that $x^2 < 2$ has an upper bound in **Q** (for example, 2 is an upper bound for A) but does not have a least upper bound in **Q**. (Remember that $\sqrt{2}$ is a least upper bound for A in **R**.) To sum up, while **Q** is an ordered field, **Q** is not a complete ordered field.

We close this section by connecting the informal and formal developments for the real numbers given above. We give a special emphasis to geometric interpretations of the axioms for **R**.

Logically speaking, it is unnecessary for us to bother with geometric interpretations of the axioms: We need only our deductive faculties and not our intuition in order to prove theorems about the real numbers. What then is the role of the concrete, informal representations of **R** presented earlier in this section? First, the geometric and decimal representations of real numbers provide a working grasp of **R** that is adequate for most of the remainder of this text. More important, the intuitive ways of looking at **R** are helpful (if not necessary) when we seek *to discover new properties* of **R**. At times, more insight is gained from a crude but general picture of a mathematical entity than from a precise but narrow description of it. In any case, by combining the informal representations with the formal axiomatic development, we gain a deeper understanding of the real number system.

First we interpret positivity geometrically. The positive real numbers correspond to those points on the number line lying to the right of the point 0. In general, $x < y$ means that the point x lies to the left of y on the number line (Figure 1.2.9).

Next, if x and y are positive real numbers, then we can interpret $x + y$ as follows:

Take a copy of the line segment from 0 to x and move the segment so that its left end is at y. The right end point of the segment is at the point $x + y$ (Figure 1.2.10).

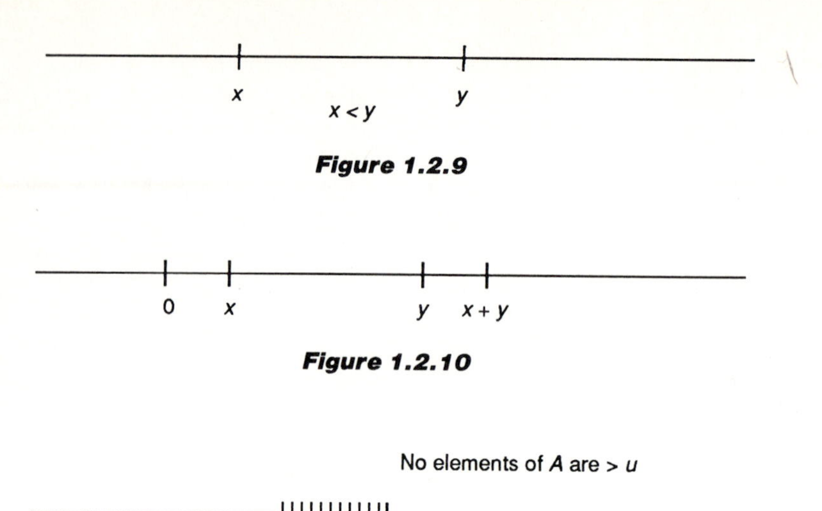

Figure 1.2.9

Figure 1.2.10

Figure 1.2.11

What about the concepts of upper bound and least upper bound? Recall that an upper bound z for a set of numbers A has the property that $z \geq x$ for all x in A. Thus z lies to the right of A. The least upper bound u of A is essentially the *first* real number to the right of A. This statement, however, requires some clarification. Since u is the least upper bound of A, no real number less than u is an upper bound for A: If $y < u$, then y is not an upper bound for A; thus there exists an x in A such that $y < x \leq u$. If u is in A, then for any $y < u$, we can always take $x = u$. Thus, if u is in A, then u is the right end of A. If u is not in A, then u lies to the right of A and there are points of A arbitrarily close to u just to the left of u (Figure 1.2.11).

Our axiomatic presentation of the real number system is now complete. As noted earlier, in the next five chapters we assume that a collection of objects satisfying Axioms 1–14 exists. In Chapter 7 we explicitly define the collection of real numbers and show that this collection satisfies Axioms 1–14.

Exercises §1.2

1.2.1. Prove that if x is in **R** and $x \neq 0$, then there exists a unique element y in **R** such that $x \cdot y = y \cdot x = 1$.

1.2.2. (a) Prove: For all x in **R**, $x \cdot 0 = 0$.
(b) Prove: For all x in **R**, $-(-x) = x$.
(c) Prove: For all x, y in **R**, $x \cdot (-y) = (-x) \cdot y = -(x \cdot y)$.
(d) Prove: For all x in **R**, $(-1) \cdot x = -x$.

1.2.3. (a) Prove: For all x in **R**, $-1(-x) = x$.
(b) Prove: $(-1) \cdot (-1) = 1$.
(c) Prove: For all x in **R**, $(-x) \cdot (-x) = x \cdot x$.

1.2.4. Prove: For all x, y in **R**, $-(x - y) = y - x$.

1.2.5 Prove: For all x, y in **R**, $(x + y)^2 = x^2 + 2 \cdot x \cdot y + y^2$. (Just for the record, $a^2 = a \cdot a$ and $2 \cdot a = (1 + 1) \cdot a = a + a$.)

1.2.6. (a) Prove: For x, y in **R**, if $x < y$, then $-x > -y$.
(b) Prove: For x, y, z, w in **R**, if $x < y$ and $z < w$, then $x + z < y + w$.

1.2.7. Let $\mathbf{R}^+$ be the collection of positive reals.
(a) Using Axioms 12 and 13, show that 1 is in $\mathbf{R}^+$ and -1 is not in $\mathbf{R}^+$.
(b) Show that, if x is in $\mathbf{R}^+$, then x^{-1} is in $\mathbf{R}^+$.

1.2.8. In parts (a)–(c) propose a definition for each of the following concepts:
(a) A collection A of numbers *bounded from below*.
(b) A *lower bound* for a set of numbers A.
(c) A *greatest lower bound* for a collection of numbers A.
(d) Give an example of a collection A of real numbers which has a lower bound. Does this collection have a greatest lower bound?

1.2.9. (a) Let A be any collection of real numbers having a lower bound. Let B consist of all those real x such that $-x$ is in A. Prove that B has an upper bound.
(b) Using the axioms of **R**, prove that any nonempty set A of real numbers having a lower bound has a greatest lower bound.

1.2.10. Let $\mathbf{Q}(\sqrt{2})$ denote all real numbers of the form $a + b\sqrt{2}$ where a and b are rational. Prove: $\mathbf{Q}(\sqrt{2})$ is an ordered field. (Note: You may assume that if $x = a + b\sqrt{2} \neq 0$ for a and b rational, then $a^2 - 2 \cdot b^2 \neq 0$.)

1.2.11. Let F be any set satisfying Axioms 1–13 (i.e., let F be an ordered field). Prove that if x is in F with $x \neq 0$, then x^2 is in F^+, the set of positive elements of F.

1.2.12. Which of the Axioms 1–14 hold when **R** is replaced by **Z**?

1.2.13. Which of the Axioms 1–14 hold when **R** is replaced by **N**?

1.2.14. Prove: There exists a real number x such that $x^2 = 3$.

1.2.15. Show that if x_0 and x_1 are both least upper bounds for a collection A of real numbers, then $x_0 = x_1$.

1.2.16. (a) Show that 5 is the least upper bound of the collection of all real numbers that are less than 5.
(b) Let A be the collection of real numbers x such that $0 < x < 2$ and $x \neq 1$. Show that A does not satisfy the completeness axiom. In other words, show that there is a collection of elements of A which is bounded above in A but which does not have a least upper bound in A.

1.2.17. Consider the collection F having only two members, 0 and 1. Addition and multiplication are defined in F by the following tables:

+	0	1
0	0	1
1	1	0

·	0	1
0	0	0
1	0	1

Show that F satisfies the field axioms for $+$ and $\cdot$ as defined in these tables.

1.2.18. Let F be the collection of objects consisting of only 0, 1, and 2. Addition and multiplication are defined in F by the following tables:

+	0	1	2
0	0	1	2
1	1	2	0
2	2	0	1

·	0	1	2
0	0	0	0
1	0	1	2
2	0	2	1

Show that F satisfies the field axioms for $+$ and $\cdot$ as defined in these tables.

1.2.19. Consider the following axiomatic system. The primitive or undefined terms are *person* and *collection*. On the basis of these terms, the following concepts are defined:

(a) A *committee* is a collection of one or more persons.
(b) A person in a committee is called a *member* of that committee.
(c) Two committees are equal if and only if every member of the first committee is a member of the second and vice versa.
(d) Two committees having no members in common are called *disjoint* committees.

The axioms of the system are:

(a) Every person is a member of at least one committee.
(b) For every pair of distinct persons there is one and only one committee of which both are members.
(c) For every committee there is one and only one committee that is disjoint from it.
    (i) Prove: Every person is a member of at least two committees.
    (ii) Prove: There are at least four persons in the entire collection, provided there is at least one person in the collection.

§1.3 Formal Thinking: Methods of Proof

If we look upon the decisive moment in the development of mathematics, the moment when we took its first step and when the ground on which it is based came into being—I have in mind logical proof— ...

I. R. Shafarevich

The object of mathematical rigor is to sanction and legitimate the conquests of intuition...

Jacques Hadamard

A good proof is one which makes us wiser.

Yu. Manin

The opinions presented by Shafarevich and Hadamard represent two extreme positions on the role of proof and rigor within mathematics and indeed on the nature of mathematics itself. Both Hadamard (1865–1963) and Shafarevich (1923–) have made many significant contributions to mathematics; thus we are compelled to look carefully at the comments of each of them. To gain a better understanding of the place that proof occupies within mathematics, we explore the positions of Hadamard and Shafarevich in greater detail.

According to Shafarevich, the existence of proofs of mathematical statements distinguishes mathematics among all intellectual endeavors. In other scientific fields, investigators can at best propose conjectures about the essence of things and attempt to confirm these conjectures with experimental data or observation of natural phenomena. In fact, most scientific theories "explain" the events of nature in terms of other scientific or mathematical theories. For example, geologists account for earthquakes via the theory of plate tectonics, and physicists describe the motion of planets in our solar system in terms of a system of differential equations. However, because of the complex and haphazard way of the world, the fallibility of human observational methods, or many other factors, the possibility always exists that a scientific theory will have to be abandoned or modified because of empirical evidence.

In mathematics the situation is different. Mathematicians working in a particular field

1. agree on certain fundamental rules, i.e., the axioms or postulates,
2. introduce some auxiliary concepts, called definitions, and

3. deduce statements, called (depending on their importance) theorems, propositions, lemmas, or corollaries relating the items that are presented in the axioms or definitions.

A particular statement becomes a theorem (proposition, lemma) when it is justified or proved. The proof of a statement consists of a sequence of statements, beginning with an axiom or previously established statement, and proceeding through a sequence of statements where each is deducible by means of rules of logic from axioms, previous theorems, or previous statements in the sequence. Without the concept of proof, mathematics would be another empirical science. But, with this notion, mathematics becomes *the* deductive science; hence its uniqueness and its justification.

Hadamard, on the other hand, believes that the essence of mathematics lies in its discovery of interesting and striking relationships among the objects of the mathematician's fancy—numbers, functions, curves, surfaces, groups, Boolean algebras, etc. Once the theorems are articulated, proofs must be given, but only to legalize the theorems. Their truth is visible to the experienced mathematician before the proofs are given. Some philosophers of mathematics extend this position to entire mathematical theories: They argue that in practice the axioms of a particular mathematical discipline are actually created last, as a postscript to the discovery of the theorems of the subject, even though in any formal exposition of the discipline (in a textbook or research paper) the axioms head the parade, and the theorems and their attendant proofs march faithfully behind. From this viewpoint, mathematics is closely akin to the sciences. Results are discovered, and then justified and organized into a comprehensible framework.

The debates ranging from the nature of mathematics to the role of proof within mathematics rage to this day. Of course, issues of this sort may never be resolved to the satisfaction of all parties. However, discussions such as the one outlined in the previous paragraph help us understand the purpose, practice, and essence of mathematics. From these discussions we obtain a more detailed portrait of mathematics and a clearer fix on its position relative to the other intellectual activities of humans.

On the basis of the preceding discussion, what can we conclude about the role of proof in mathematics?

On the one hand, no one familiar with mathematics will deny the importance of rigor and proof. Experience has convinced every mathematician that any assertion must be supported by a rigorous argument. Countless novice, apprentice, and master mathematicians have attempted to prove an intuitively obvious "theorem" only to watch it disintegrate into a false statement. Thus every mathematician must become proficient with proofs. He or she must learn to read mathematical proofs critically as they are presented in texts or research papers and must develop the ability to write valid proofs. Proofs are indeed one of the focal points of mathematical activity; hence they need to be mastered by any student of formal mathematics.

On the other hand, activities such as experimentation, discovery, and reasoning by analogy are important aspects of a mathematician's experience. Mathematicians investigate abstract phenomena in search of order and pattern. Employing a combination of informal reasoning and experimentation, they conjure up conjectures that describe the patterns they observe. At this point they either prove their conjectures formally, or disprove them by constructing counterexamples, in which case the process of investigation and conjecture repeats itself. From this point of view, mathematics encompasses a broad range of activities, requiring mathematicians to guess and to prove, to hypothesize and to criticize.

At the same time we can perhaps see the line between formal and informal mathematics begin to blur. Proof itself often acts as an aid to discovery. Using methods of proof to be discussed in this section, a mathematician can investigate several possible avenues toward a solution and eliminate those outcomes that cannot actually arise. At times, a mathematician will set out to prove a particular statement and will choose what appears to be a natural method of proof only to find that the proof forces a modification of the statement being proved. What arises is a kind of dialogue between two faces of mathematics, one being that of rigor, logic, and criticism, the other that of intuition, creativity, and discovery. In this dialogue, proof serves as a tool for both investigation and verification.

Perhaps now the wisdom of Manin's words quoted at the start of this section becomes apparent. (Manin, like Shafarevich a contemporary Soviet mathematician, has conducted important research in areas of pure mathematics and mathematical physics.) A good proof makes us wiser in several ways: The proof convinces us of the validity of the statement that it demonstrates. Once we have secured a proof, we can be certain that the result is valid and can be used whenever appropriate. The proof helps us understand the statement by providing insight into its nature. For instance, sometimes the proof of a theorem will "require" results and techniques that are not suggested by the statement of the theorem. Finally, the proof often directs us toward the result itself. Thus a good proof confirms, clarifies, and creates mathematical results.

In the remainder of this section, we discuss the structure and mechanics of mathematical proofs. We describe and illustrate several kinds of mathematical proofs. Our overall goal is to become familiar with various proof techniques and to acquire a feeling for the appropriateness of a given proof technique in a particular situation. Admittedly, the latter skill is not readily acquired. Most mathematicians probably agree that it develops from experience with reading and writing proofs. Nevertheless, the presentation of proof techniques in this section and the providing of reasons for choosing a specific proof method throughout the book can nurture the student's ability to select a reasonable proof strategy and to develop this strategy into a complete proof.

Techniques

We now outline five general and commonly used proof techniques:

1. Direct proof.
2. Proof by contrapositive.
3. Proof by contradiction.
4. Mathematical induction.
5. Case analysis.

Our goal is to describe the essence of each method and to illustrate each one in action. In addition, we indicate situations in which a given method might be especially appropriate.

Several of the applications that we give to illustrate the various techniques will be used in the remainder of the text. The most important of these results will be labeled as "Theorems" rather than merely as "Examples." Before diving into a discussion of proof techniques, we should ask a simple question: What is the structure of the assertions that one encounters in mathematics?

Most statements of mathematical theorems are of the "If-then" form:

1. *If f is differentiable at a, then f is continuous at a.*
2. *If a, b are real numbers and $ab = 0$, then either $a = 0$ or $b = 0$.*
3. *If G is a finite simple group, then G has even order.*

Of course, there are some noteworthy exceptions to this claim about the form of statements in mathematics. For example, consider this statement:

4. *There exists a function that is continuous at every real number and differentiable at no real number.*

Statement 4 can probably be rewritten accurately in an "If-then" form. But any such rephrasing is likely to be awkward; hence most mathematicians prefer a statement similar to the one given above. Nonetheless, implicit in statement 4 are some assumptions—to be specific, the definitions of function, continuity of a function, and differentiability of a function. Statement 4 asserts that given the definitions of these concepts, a function f can be defined which is continuous everywhere and differentiable nowhere. Another example is the statement:

5. *There are an infinite number of prime numbers.*

Implicit in this sentence are the definitions of "prime number" and "infinite number of." Perhaps the simplest way to write this statement as an implication is as follows:

> If a prime number is a positive integer greater than 1, divisible by only 1 and itself, then, for any positive integer n, there are at least n prime numbers.

Again, because of the awkwardness of this statement, we prefer the more concise and elegant version given in statement 5. Nonetheless, it is important to write or at least to understand statement 5 as an implication in order to construct a proof of it.

Now to the basic topic of this section: Given a mathematical statement in the form of an implication, "If P, then Q," how do we go about proving that the statement is true or showing that it is false?

Direct Proof

A direct proof proceeds from the hypothesis P and deduces that Q must hold. In concocting a direct proof, we are free to use all the hypotheses that are given and any statement that has been previously established. Here is an elementary example.

Example 1.3.1 Prove: The product of two odd integers is an odd integer.

First, let us translate this statement into an implication. Our initial attempts will not contain any symbols.

(i) If the product of two odd integers is taken, then that product is an odd integer.
(ii) If two odd integers are given, then their product is an odd integer.

Now let us state a version using some variables:

(iii) If a and b are odd integers, then $a \cdot b$ is an odd integer.

To make the quantification on a and b explicit, we have the following version:

(iv) For all integers a and b, if a and b are odd, then $a \cdot b$ is an odd integer.

All these statements describe a property of the set of integers; hence none is any more "mathematical" than any of the others. Although the third and fourth sentences might appear to be the most mathematical of the four, they differ from the others mainly in their conciseness and sophistication. For example, the use of "$a \cdot b$" in the conclusion must be interpreted by the reader to mean "the product of a and b." Now to the proof. We take as our definition of an odd integer the following: An integer a is odd if there exists an integer m such that $a = 2 \cdot m + 1$.

Proof By definition, since a and b are odd, there exist integers m and n such that $a = 2 \cdot m + 1$ and $b = 2 \cdot n + 1$. We must show that there exists an integer k such that $a \cdot b = 2 \cdot k + 1$.

By the properties of addition and multiplication of real numbers described in Section 1.2,

$$\begin{aligned} a \cdot b &= (2m + 1)(2n + 1) \\ &= (2m + 1) \cdot 2n + (2m + 1) \cdot 1 \\ &= 2m \cdot 2n + 2n + 2m + 1 \\ &= 4mn + 2m + 2n + 1 \\ &= 2(2mn + m + n) + 1. \end{aligned}$$

Thus $a \cdot b = 2k + 1$ where k is the integer $2mn + m + n$. This proves that $a \cdot b$ is odd if a and b are odd. ■

Remark Here is an incorrect proof of the previous result: Let a and b be odd integers. For example, let $a = 5$ and $b = 7$. Then $a \cdot b = 35$, which is odd. Therefore, $a \cdot b$ is an odd integer.

This "proof" is incorrect because it ignores the quantifications on a and b. One must show that *for all* odd integers a and b, $a \cdot b$ is odd. All this argument shows is that there exist odd integers a and b such that $a \cdot b$ is odd. Thus, to repeat what was said earlier, the moral is: Remember the scope of the variables.

Next we give a direct proof of an important property of the real number system.

Theorem 1.3.1 Cancellation Law for Multiplication. *If a, b, and c are real numbers such that $a \neq 0$ and $ab = ac$, then $b = c$.*

Proof Suppose a, b, and c are arbitrary real numbers such that $a \neq 0$ and $ab = ac$. Since $a \neq 0$, the multiplicative inverse of a, a^{-1}, exists. Thus

$$a^{-1} \cdot (a \cdot b) = a^{-1} \cdot (a \cdot c).$$

Hence, by the associative law for multiplication,

$$(a^{-1} \cdot a) \cdot b = (a^{-1} \cdot a) \cdot c.$$

Thus

$$1 \cdot b = 1 \cdot c$$

or

$$b = c. \qquad \blacksquare$$

The following result can be either derived from Theorem 1.3.1 or proved via a similar argument. We leave the proof as an exercise. (See Exercise 1.3.2.)

Corollary 1.3.2 *If a and b are real numbers such that $ab = 0$, then either $a = 0$ or $b = 0$.*

Example 1.3.2 Show: If x is a real number and $x^2 - 4x + 6 = x$, then $x = 2$ or $x = 3$.

Remark Throughout this text we use the words "prove" and "show" interchangeably.

Before reading on, try to prove the given statement.

Proof If $x^2 - 4x + 6 = x$, then subtracting x from both sides (i.e., adding $-x$ to both sides) we find that $x^2 - 5x + 6 = 0$. Upon factoring the left side, we see that $(x - 2) \cdot (x - 3) = 0$. By Corollary 1.3.2, either $x - 2 = 0$ or $x - 3 = 0$. Therefore, either $x = 2$ or $x = 3$. ■

Conversely, we can easily check that 2 and 3 are solutions to the given equation. Thus, for a real number x, $x^2 - 4x + 6 = x$ if and only if $x = 2$ or $x = 3$.

Our next example comes from differential calculus. We assume familiarity with some basic properties of limits and functions.

Example 1.3.3 Let f and g be functions defined on the set of real numbers. If f and g are continuous at a number a, then $f + g$ is continuous at a.

Proof To show that $f + g$ is continuous at a, we must verify that

(1) $f + g$ is defined at a, and

(2) $\lim_{x \to a}(f + g)(x)$ exists and equals $(f + g)(a)$.

By assumption, f and g are continuous at a; thus properties (1) and (2) hold for the individual functions f and g in place of $f+g$.

Property (1): By definition of addition of functions, $f+g$ is defined at a if both f and g are defined at a. Furthermore, $(f+g)(a)=f(a)+g(a)$. Thus (1) holds.

Property (2): By a theorem on limits, $\lim_{x\to a}(f+g)(x)$ exists if $\lim_{x\to a} f(x)$ and $\lim_{x\to a} g(x)$ both exist, and in this case $\lim_{x\to a}(f+g)(x)$ equals $\lim_{x\to a} f(x)+\lim_{x\to a} g(x)$. Since f and g are continuous at a, $\lim_{x\to a} f(x)$ and $\lim_{x\to a} g(x)$ both exist and equal $f(a)$ and $g(a)$, respectively. Thus $\lim_{x\to a}(f+g)(x)$ exists and

$$\begin{aligned}\lim_{x\to a}(f+g)(x) &= \lim_{x\to a} f(x)+\lim_{x\to a} g(x)\\ &= f(a)+g(a)=(f+g)(a).\end{aligned}$$

Since (1) and (2) hold, $f+g$ is continuous at a. ■

Our final example describes a basic property of integers. Let a and b be integers; then a is a *factor* of b if there is an integer c such that $b=a\cdot c$. Thus, since $6=2\cdot 3$ and 3 is an integer, 2 is a factor of 6. Notice that 2 is not a factor of 7 in spite of the equality $7=2\cdot(7/2)$, since $7/2$ is not an integer and for each integer c, $2\cdot c\neq 7$. Also observe that an integer n is even if and only if 2 is a factor of n. Several other ways of expressing factorability are commonly used. Instead of saying a is a *factor* of b, one can say *a divides b*, a is a *divisor* of b, b is *divisible* by a, or b is a *multiple* of a.

Lemma 1.3.3 *Let a, b, c be integers. If a is a factor of b and of c, then a is a factor of $b+c$.*

Proof Since a is a factor of b, there exists an integer n such that $b=a\cdot n$. Similarly, $c=a\cdot m$ for some integer m. Thus $b+c=a\cdot n+a\cdot m=a\cdot(n+m)$. Since $n+m$ is an integer, a is a factor of $b+c$. ■

Notice the similarity among these five examples. In each case we reasoned directly from the hypothesis to the conclusion, using relevant definitions, facts, and rules at our disposal.

Proof by Contrapositive

By contrast the next two techniques of proof to be discussed are methods of indirect proof, for in each case, rather than starting with the hypothesis of the given implication, one begins with the negation of its conclusion.

We first discuss the *method of contraposition* or *proof by contrapositive*. This technique establishes the validity of an implication "If P then Q" by showing that its logically equivalent contrapositive "If not-Q then not-P" holds. Thus, to prove that "If P then Q" is valid, one assumes that not-Q holds and derives the conclusion that not-P is valid.

Example 1.3.4 Let n be an integer. Prove: If n^2 is an odd integer, then n is an odd integer.

Proof To prove the statement, we prove the contrapositive: If n is an integer that is not odd, then n^2 is an integer that is not odd.

Suppose n is an integer that is not odd. Then n is even; hence $n = 2k$ for some integer k. Thus $n^2 = (2k)^2 = 4k^2 = 2(2k^2)$ is also an even integer. Therefore, n^2 is not odd if n is not odd. ∎

Observe that in this example the proof of the contrapositive of the original statement is a direct proof. We take the meaning of the assumption that n is not odd, mix in some simple arithmetic, and conclude that n^2 is not odd. With a more complex statement, it might be necessary to use other pieces of knowledge; such complications, however, need not cloud over the basic structure of the proof by contrapositive.

In spite of the logical equivalence of the statements "If P then Q" and "If not-Q then not-P," the application of a proof by contrapositive must be handled with care. The danger lies in the formation of negations. If P is a complex sentence, then not-P might be difficult to formulate. For example, as we have seen, the negation of a sentence involving universal and existential quantification can be especially challenging. With these words of warning in mind, we move to a related proof technique—proof by contradiction.

Proof by Contradiction

To establish an implication "If P then Q" via the *method of contradiction*, we assume that the statement "If P then Q" is false and attempt to derive a contradiction, namely a statement that is always false no matter what the truth values of its components. Thus we begin with the statement "P and not-Q," and from this statement we deduce a contradiction, usually a statement of the form "R and not-R." Thus we show that the implication $(P \wedge \text{not-}Q) \Rightarrow (R \wedge \text{not-}R)$ is true. Since $R \wedge \text{not-}R$ is false, it follows that $P \wedge \text{not-}Q$ is also false or equivalently "not-(If P then Q)" is false and "If P then Q" is true. In most cases the statement R is either a result established in the course of the proof, a previously derived result, or an assumption made at some point in the proof.

Example 1.3.5 Show that the circle whose equation is $x^2 + y^2 = 2$ and the line with equation $y = x + 4$ do not intersect.

Proof We rewrite the statement as an implication: If C is the circle with equation $x^2 + y^2 = 2$ and L is the line with equation $y = x + 4$, then C and L do not intersect. We give a proof by contradiction. Assume C and L do intersect. Then there exists a point (x_0, y_0) on both C and L. Then $y_0 = x_0 + 4$ and $x_0^2 + y_0^2 = 2$. Therefore, $x_0^2 + (x_0 + 4)^2 = 2$ from which it follows that $x_0^2 + 4x_0 + 7 = 0$ or that $(x_0 + 2)^2 + 3 = 0$. Thus, if (x_0, y_0) lies on the circle $x^2 + y^2 = 2$ and the line $y = x + 4$, then $(x_0 + 2)^2 = -3$. However, for any real number z, $z^2 \geq 0$; hence $z^2 \neq -3$. We have achieved a contradiction, thus confirming that the given circle and line do not intersect. ∎

We now prove the theorem of Hippasus mentioned in Section 1.2.

Example 1.3.6 Prove that $\sqrt{2}$ is irrational.

Proof We recall that a number x is rational if there exist integers m and n with $n \neq 0$ such that $x = m/n$. Therefore, to say that $\sqrt{2}$ is irrational is to say that for all integers m and n, $\sqrt{2} \neq m/n$. A direct proof that $\sqrt{2}$ is irrational would evidently require us to begin with an arbitrary pair of integers m and n and to show that $\sqrt{2} \neq m/n$. This task appears to be difficult. Thus a proof by contradiction is worth trying.

Our goal is to show that for all integers m, n with $n \neq 0$, $\sqrt{2} \neq m/n$. We assume that the negation of this statement holds; hence we assume that there exists a pair of integers m and n for which $\sqrt{2} = m/n$. Our aim is to derive a contradiction. If m and n are both even integers, then $m = 2m'$ and $n = 2n'$ where m' and n' are integers and $\sqrt{2} = m/n = 2m'/2n' = m'/n'$. If m' and n' are even, then we repeat this procedure. In fact, we continue to repeat it until we find integers a and b (which are factors of m and n, respectively), *at least one of which is odd*, such that $\sqrt{2} = a/b$.

Squaring both sides and multiplying both sides of the result by b^2 gives $2b^2 = a^2$. Thus a^2 is even; hence a is even. (See Example 1.3.1.) Therefore, $a = 2k$ and $2b^2 = a^2 = (2k)^2 = 4k^2$. By cancellation $b^2 = 2k^2$ from which it follows that b is even.

We conclude that both a and b are even, a contradiction of the fact that at least one of a and b is odd. This completes the proof that $\sqrt{2}$ is irrational. ∎

As was mentioned at the start of the argument, trying a direct proof in this case would seem to be futile. On the other hand, by assuming the negation of the desired conclusion (i.e., that $\sqrt{2}$ is rational), we obtain something quite tangible to manipulate: We are given integers m and n such that $\sqrt{2} = m/n$. In general, we should consider a proof by contradiction if negating the conclusion yields a statement that can be manipulated and exploited. For instance, in Example 1.3.5, by assuming that the given curves did intersect, we obtain a point (x_0, y_0) on both curves, and by substituting the numbers x_0 and y_0 in the equations for the curves, we derive a contradiction.

Our next example comes from the theory of numbers, the discipline of mathematics in which the behavior of whole numbers, with respect to the operations of addition and multiplication, is studied. An example of a statement of number theory is the two-hundred-year-old, as-yet-unproved Goldbach conjecture:

> Every even number larger than two is
> expressible as a sum of two primes.

We illustrate the method of proof by contradiction by proving a truly ancient result, one that goes back to Book 7 of Euclid's *Elements*. Recall that a positive integer $n > 1$ is *prime* if the only positive integer factors of n are 1 and n. For example, the primes less than 25 are 2, 3, 5, 7, 11, 13, 17, 19, and 23. The only even positive integer that is prime is 2. Notice that if n is not prime, then n can be factored as $n = ab$ where a and b are both greater than 1. In other words, nonprimes can be written as a product of two smaller positive integers.

Once prime numbers are defined, several questions concerning primes come to mind. For instance, one can ask: Is every integer greater than 1 divisible by a prime? The next result that is used to prove Theorem 1.3.5 assures us that indeed this is the case.

Lemma 1.3.4 *Let n be any integer greater than* 1. *Let a be the smallest factor of n that is greater than* 1. *Then a is prime.*

Proof We take n and a as in the statement of Lemma 1.3.4. We must show that a is prime. We give a proof by contradiction. Suppose a is not prime. Then $a = b \cdot c$ where $b > 1$ and $c > 1$. Since b is a factor of a and a is a factor of n, b is a factor of n. (See Exercise 1.3.3(b).) But b is also greater than 1 and less than a. To summarize, b is a factor of n such that $1 < b < a$. Therefore, a is not the smallest factor of $a > 1$. This contradiction (of the definition of a) means that a is indeed prime. ∎

Another question one might ask about primes is the following: Suppose we begin to list the prime numbers: $2, 3, 5, 7, 11, \ldots$. Does this list ever end? Or must it continue without stopping? The next theorem answers these questions by assuring us that the list of primes does not stop after a finite number of entries.

Theorem 1.3.5 *There exist infinitely many prime numbers.*

Proof Suppose that the statement is false. Then only a finite number of primes exist. List them in increasing order: $2, 3, 5, \ldots, p$ where p is the largest prime number. Our goal is to produce a prime number that is not one of the primes on this list. Since this list allegedly contains all prime numbers, a contradiction is achieved.

To create a new prime, we exhibit a number that has no prime factor from the assumed complete list of primes, $2, 3, 5, \ldots, p$. But, by Lemma 1.3.4, this number must have at least one prime factor, call it q. The prime number q will then be a prime number that is not on the list.

We seek a number that has no factors among the primes $2, 3, \ldots, p$. We might be tempted to consider $p + 1$, but as examples suggest (and as is easily proved), every prime factor of $p + 1$ is on the list $2, 3, \ldots, p$. Instead, let us consider $M = 2(2 \cdot 3 \cdot \cdots \cdot p) + 1$.

The number M might or might not be prime, but in any case M has a prime factor by Lemma 1.3.4. Let q be a prime factor of M. For instance, if q is the smallest factor of M greater than 1, then q is prime by Lemma 1.3.4. We claim that q is not one of the primes $2, 3, \ldots, p$, for, if so, then q is a factor of $2 \cdot 3 \cdot \cdots \cdot p$ and of $M = (2 \cdot 3 \cdot \cdots \cdot p) + 1$. It follows from Lemma 1.3.3 that q is a factor of $M - (2 \cdot 3 \cdot \cdots \cdot p) = 1$. But no integer greater than one can be a factor of one. Thus q is not one of the primes $2, 3, \ldots, p$. (Note that we have just given a proof by contradiction within a proof by contradiction.)

Therefore, we have produced a prime number that is different from each of the primes $2, 3, \ldots, p$. But this conclusion contradicts the assumption that the list $2, 3, \ldots, p$ contains every prime number. ■

Remarks 1. Instead of taking $M = 2 \cdot 3 \cdot \cdots \cdot p + 1$, we could have chosen $M = p! + 1 = (p \cdot (p - 1) \cdot (p - 2) \cdot \cdots \cdot 3 \cdot 2 \cdot 1) + 1$. The important point is that no integer larger than 1 can divide two consecutive integers. Thus none of the primes $2, 3, \ldots, p$ divides $p! + 1$; hence any prime factor of $p! + 1$ is different from $2, 3, \ldots, p$.

2. One might ask if a more precise version of Theorem 1.3.5 exists. For instance, for each positive integer n, can we give a formula for the number of primes $\leq n$? Following custom, we let $\pi(n)$ denote the number of primes $\leq n$. Thus we are asking if there is a nice formula for the function $\pi(n)$. This question was first considered by the great German mathematician C. F. Gauss in 1791 when he was 14 years old. Gauss did not give a precise formula for $\pi(x)$, but he did discover the nature of the long-term growth of $\pi(n)$. By inspecting tables of primes, Gauss conjectured that $\pi(n)$ is approximately $n/\log(n)$ where $\log(n)$ is the natural logarithm of n. Specifically, Gauss conjectured that

$$\lim_{x \to \infty} \pi(n)/[n/\log(n)] = 1.$$

This intriguing result, known as the Prime Number Theorem, was first proved in 1896 by two French mathematicians, Hadamard (whom we met at the start of this section) and de la Vallée Poussin. Note that one corollary of the Prime Number Theorem is that

$$\lim_{n \to \infty} \pi(n) = \infty; \textit{ hence the number of primes is infinite.}$$

Our final illustration comes from differential calculus. In the argument we make use of Rolle's Theorem, which is a special case of the Mean Value Theorem.

Example 1.3.7 Let f be a function defined on the set of real numbers. Prove that if $f'(x) > 0$ for any real number x, then there is at most one real number x_0 such that $f(x_0) = 0$.

Proof Suppose to the contrary that there are at least two real numbers, x_1 and x_2, such that $f(x_1) = f(x_2) = 0$. Then, by Rolle's Theorem, there is a number a between x_1 and x_2 such that $f'(a) = 0$. This conclusion, however, contradicts the assumption that $f'(x) > 0$ for all x. Thus the assumption that at least two numbers, x_1 and x_2, exist for which $f(x_1) = f(x_2) = 0$ is false, and the original statement is proved. ■

The reader might have noticed that the argument in Example 1.3.7 is actually a proof by contrapositive: We began by negating the conclusion and ended by deducing the negation of the hypothesis. Of course, we can argue that this or any proof by contrapositive is actually a proof by contradiction: Assume P and not-Q, then derive not-P. Thus both P and not-P hold and we have a proof by contradiction.

Although this observation is technically correct, we find it pedagogically useful to separate the two methods. In fact, contradiction and contraposition possess notable similarities and differences. In both methods the negation of the conclusion of the implication "If P then Q," namely not-Q, is assumed to hold. In a proof by contrapositive, the goal is to derive the statement not-P from the statement not-Q. In a proof by contradiction, we attempt to extract some contradiction from the statements P and not-Q. In a proof by contradiction, we have more to work with—both P and not-Q—than we have in a proof by contrapositive—only not-Q. But we pay a price for the extra weapon. In a proof by contrapositive, we have the advantage of aiming for a specific goal, not-P. By contrast, with the contradiction method, we shoot only for some contradiction. Very often we do not know what shape this contradiction will take when we start the proof. Because of this fact, a proof by contradiction is sometimes difficult to understand (upon reading) and to create.

As is always the case in mathematics and science, a technique, once mastered, must be applied judiciously. A common question is: When should a particular proof technique be applied? For better or worse, there seems to be no definitive answer to this question, but, as far as the contradiction and contrapositive methods are concerned, there are some useful guidelines to observe.

1. Consider a contradiction or contrapositive argument if the conclusion is difficult to grasp or manipulate as is, and more information and material to play with are obtained by negating the conclusion.

This advice was followed with success in all the examples presented. For example, in the proof of Theorem 1.3.4, the negation of the existence of infinitely many primes enables us to manipulate the finite set of all primes. We can then use addition and multiplication along with properties of these operations to derive a contradiction.

2. Consider a contradiction or contrapositive argument if the conclusion is the negation of some other statement.

Thus, if the conclusion has the form not-$P(x)$, then we would assume that $P(x)$ holds and would derive a contradiction. For example, to show that $\sqrt{2}$ is not rational, we assume that $\sqrt{2}$ is rational and derive a contraction. In Example 1.3.5, we assume that the given circle and line intersect and derive the contradiction that there exists a real number z such that $z^2 = -3$.

3. Consider a contradiction or contrapositive argument if the conclusion demands either a unique object or at most one object having a given property.

For a problem asking for a unique object that has a given property, we usually break the argument down into two steps:

(1) Show that there is at least one object having the property;
(2) Show that there is at most one object having the property.

In establishing these steps, especially step (2), contradiction or contraposition can be useful. The advice given in guideline 3 is perhaps implicit in guideline 1. Nevertheless, since statements asserting the uniqueness of an object are common in mathematics, it is worth noticing that such statements tend to be susceptible to either contraposition or contradiction.

Mathematical Induction

Our next method of proof, the principle of mathematical induction, arises in a variety of situations throughout mathematics and computer science. Often, in the course of analyzing a mathematical phenomenon, we encounter statements of the following form:

1. For every positive integer n, $1 + 2 + \cdots + n = n(n + 1)/2$.
2. For any positive integer n, $1 + x + \cdots + x^n = (1 - x^{n+1})/(1 - x)$ for any real number x not equal to 1.

Each of these statements entails an assertion about every positive integer. For example, when $n = 5$, statement 1 says that $1 + 2 + \cdots + 5 = 5 \cdot 6/2$, which happens to be true. When $n = 2$, statement 2 reads: $1 + x + x^2 = (1 - x^3)/(1 - x)$ for each real number $x \neq 1$. This equality does indeed hold for all $x \neq 1$.

Returning to statement 1, we ask: Is it true that for each positive integer n, $1 + 2 + \cdots + n = n(n + 1)/2$? To investigate the matter further, we can calculate $1 + 2 + \cdots + n$ for $n = 1, 2, \ldots, 10$. Doing so, we obtain the sequence 1, 3, 6, 10, 15, 21, 28, 36, 45, 55. If we calculate $n(n + 1)/2$ for $n = 1, 2, \ldots, 10$, then we obtain the identical list of ten integers. By now we probably believe that statement 1 is true for every positive integer n. To confirm our belief a bit more, we can check more cases such as $n = 11$ and $n = 12$ and we find that statement 1 is also true in these instances. Thus, for any positive integer $n \leq 12$, $1 + 2 + \cdots + n = n(n + 1)/2$. But how can we show that this equality holds *for every positive integer n*?

To prove statements 1 and 2, we appeal to a property of the set of positive integers known as the *Principle of Mathematical Induction* (PMI). We state PMI in a general form and show how it applies to statements 1 and 2. Later we consider a slightly different (but logically equivalent) version of mathematical induction called the Principle of Strong Induction.

Principle of Mathematical Induction. Let $S(1), S(2), \ldots, S(n), \ldots$ be a list of statements, one for each positive integer. Then every statement on the list is true if the following two conditions hold:

(i) $S(1)$ is a true statement,
(ii) For each positive integer k, if $S(k)$ is true, then $S(k + 1)$ is true.

PMI can be regarded as a proof technique that is applicable whenever one is attempting to prove that each statement in an infinite list of statements is true. Why are conditions (i) and (ii) sufficient to imply that $S(n)$ is true for all positive integers n? The first statement merely says that statement $S(1)$ is true. The second asserts that if any particular statement is true, then the next statement on the list is also true. Therefore, knowing by (i) that $S(1)$ is true and using (ii), we can conclude that $S(2)$ is true. From (ii) it follows that since $S(2)$ is true, $S(3)$ is true. By another application of (ii), one sees that $S(4)$ is true and so on.

Analogously, we might imagine an infinite staircase. Evidently, we can climb the staircase indefinitely if we know that (i) we can step on the first step and (ii) whenever we are on a given step, we can climb to the next step.

Henceforth, we refer to the first step of PMI, the assertion that $S(1)$ is true, as the *basis step* of the proof by mathematical induction, and the second step, the proof that for each positive integer k, $S(k)$ implies $S(k+1)$, as the *inductive step* of the proof by mathematical induction. In most, but not all, proofs by mathematical induction, the basis step is the easier of the two steps to establish. In proving that the inductive step holds, two general strategies are usually used. The first is to produce a direct proof that $S(k)$ implies $S(k+1)$. The second strategy is to reduce $S(k+1)$ back to $S(k)$. Some examples will illustrate these two strategies.

Example 1.3.8 Show that for each positive integer n, $1 + \cdots + n = n(n+1)/2$.

Proof For each positive integer n, let $S(n)$ be the statement $1 + \cdots + n = n(n+1)/2$.

Basis step. $S(1)$ is the statement: $1 = 1(1+1)/2$. Thus $S(1)$ is true.

Inductive step. We suppose that $S(k)$ is true and prove that $S(k+1)$ is true. Thus we assume that

$$1 + \cdots + k = k(k+1)/2$$

and prove that $1 + \cdots + k + (k+1) = (k+1)(k+1+1)/2$. If we add $k+1$ to both sides of the equality in $S(k)$, then on the left side of the sum we obtain the left side of the equality in $S(k+1)$. Our hope is that the right of the sum equals the right side of $S(k+1)$. Let us check: Adding $k+1$ to both sides of $1 + \cdots + k = k(k+1)/2$, we see that

$$\begin{aligned} 1 + \cdots + k + (k+1) &= k(k+1)/2 + (k+1) \\ &= (k+1)(k/2+1) \\ &= (k+1)(k/2+2/2) \\ &= (k+1)(k+2)/2 \\ &= (k+1)(k+1+1)/2. \end{aligned}$$

Hence, if $S(k)$ is true, then $S(k+1)$ is true.

We can prove this statement by induction. Let $S(n)$ be the statement: For each positive integer n, $1 + \cdots + (2n-1) = n^2$.

Basis step. $S(1)$ holds since $1 = 1^2$.

Inductive step. Suppose $S(n)$ is true. We must show that $S(n+1)$ is true. Thus we must show that

$$1 + \cdots + (2(n+1)-1) = (n+1)^2.$$

Since $S(n)$ is true,

$$1 + \cdots + (2n - 1) = n^2.$$

Hence

$$\begin{aligned} 1 + \cdots + (2n - 1) + (2(n + 1) - 1) &= n^2 + (2(n + 1) - 1) \\ &= n^2 + 2n + 2 - 1 \\ &= n^2 + 2n + 1 = (n + 1)^2. \end{aligned}$$

Therefore, $S(n + 1)$ is true.

The inductive proof is now complete. ∎

In the inductive steps of the three examples presented thus far, we have established $S(k + 1)$ by arguing directly from $S(k)$. Now we present an example in which it is not at all obvious how to manipulate $S(k)$ to produce $S(k + 1)$, but in which it is natural to reduce $S(k + 1)$ back to $S(k)$.

Example 1.3.11 Prove that for each positive integer n, $|\sin(nx)| \le n(\sin(x))$ for $0 \le x \le \pi$. (Here, $|u|$ is the absolute value of the real number u: $|u| = u$ if $u \ge 0$ and $|u| = -u$ if $u < 0$.)

Proof For each positive integer n, define $S(n)$: $|\sin(nx)| \le n \sin(x)$ for $0 \le x \le \pi$.

Basis step. We must show that $|\sin(1 \cdot x)| = |\sin(x)| \le 1 \cdot \sin(x) = \sin(x)$ for $0 \le x \le \pi$. But for $0 \le x \le \pi$, $\sin(x) \ge 0$; hence $|\sin(x)| = \sin(x)$.

Inductive step. We show that for each positive integer k, $S(k)$ implies $S(k + 1)$. Therefore, we assume that for $0 \le x \le \pi$, $|\sin(kx)| \le k \sin(x)$ and deduce that $|\sin(k + 1)x| \le (k + 1)\sin(x)$.

Although it is easy to obtain $(k + 1)\sin(x)$ from $k \cdot \sin(x)$ and vice versa, it is not clear how to build $|\sin(k + 1)x|$ from $|\sin(kx)|$. But the addition formula for the sine function, $\sin(u + v) = (\sin u)(\cos v) + (\cos u)(\sin v)$, can help us express $\sin(k + 1)x$ in terms of $\sin(kx)$. Thus we shall attempt to manipulate the left side of $S(k + 1)$ to the point where we can use statement $S(k)$. From the addition formula for the sine function,

$$\begin{aligned} |\sin(k + 1)x| &= |\sin(kx + x)| \\ &= |\sin(kx) \cdot \cos(x) + \cos(kx) \cdot \sin(x)| \\ &\le |\sin(kx) \cdot \cos(x)| + |\cos(kx) \cdot \sin(x)|, \end{aligned}$$

since for all real numbers u and v, $|u + v| \leq |u| + |v|$. (This is the so-called *triangle inequality*.) Moreover, since $|u \cdot v| = |u| \cdot |v|$ for all real u and v and since $|\cos(u)| \leq 1$ for all real u, we have

$$\begin{aligned} |\sin(k+1)x| &\leq |\sin(kx) \cdot \cos(x)| + |\cos(kx) \cdot \sin(x)| \\ &\leq |\sin(kx)| \cdot |\cos(x)| + |\cos(kx)| \cdot |\sin(x)| \\ &\leq |\sin(kx)| + |\sin(x)|. \end{aligned}$$

Recall that we are assuming that $S(k)$ holds. Thus $|\sin(kx)| \leq k \cdot \sin(x)$. Also, since $0 \leq x \leq \pi$, $|\sin(x)| = \sin(x)$. It follows that

$$\begin{aligned} |\sin(k+1)x| &\leq |\sin(kx)| + |\sin x| \\ &\leq k \cdot \sin(x) + \sin(x) \leq (k+1) \cdot \sin(x). \end{aligned}$$

Hence the inductive step is established.

By PMI, $|\sin(nx)| \leq n \cdot \sin(x)$ for each positive integer n and each real number x such that $0 \leq x \leq \pi$. ■

Let us recap the basic strategy in the last argument. We took an important portion of the inequality that constituted statement $S(k + 1)$, namely $|\sin(k + 1)x|$, and manipulated it (using properties of the sine, cosine, and absolute value functions) to a point where statement $S(k)$ could be used. In this sense we "reduced" $S(k + 1)$ to $S(k)$. This kind of reduction argument stands in contrast to the proofs of inductive steps in Examples 1.3.8–1.3.10. In those examples, statement $S(k + 1)$ was "constructed" from statement $S(k)$: Statement $S(k)$ is manipulated in an appropriate way to obtain $S(k + 1)$. Both of these techniques are used frequently in proofs of the inductive step.

We emphasize that, in the inductive step, statement $S(k + 1)$ is always deduced from statement $S(k)$. Such was the case in each of Examples 1.3.8–1.3.11. In some cases it is appropriate to begin with statement $S(k)$ and to conclude that statement $S(k + 1)$ holds after a sequence of intermediate steps. In other cases it is more appropriate to use statement $S(k)$ at some intermediate point in the argument that establishes statement $S(k + 1)$. Whether or not to begin with statement $S(k)$ is a matter of tactics. But the *logic* of the inductive step is the same in every case.

Our final illustration of PMI yields one of the most important properties of the system of integers. This result, called the Division Algorithm, will be used repeatedly throughout this text. At the same time, the proof is rather sophisticated since the statement of the Division Algorithm involves two integer variables instead of only one integer variable as all the previous examples have.

Theorem 1.3.6 Division Algorithm. *Let a and b be positive integers. There exist integers q and r such that $a = b \cdot q + r$, where $0 \leq r \leq b - 1$.*

Remark The statement asserts that b can be "taken out" of a several times (q times to be exact) in such a way that the remainder is less than b but is at least 0. Here is a quick example: Let $a = 61$ and $b = 13$. Then $61 = 13 \cdot 4 + 9$. Here, $q = 4$ and $r = 9$.

Proof For each positive integer a let $S(a)$ be the statement: For each positive integer b there exist integers q and r such that $a = b \cdot q + r$ where $0 \leq r \leq b - 1$.

Basis step. To prove $S(1)$ we let b be an arbitrary positive integer, and we must find integers q and r such that $1 = b \cdot q + r$ where $0 \leq r \leq b - 1$. We observe that if $b = 1$, then we can satisfy the desired condition by taking $q = 1$ and $r = 0$. If $b > 1$, then we satisfy the conditions by setting $q = 0$ and $r = 1$.

Inductive step. Suppose $S(a)$ is true. We show $S(a + 1)$ is true. Because $S(a)$ is true, there exist integers q and r such that $a = b \cdot q + r$ where $0 \leq r \leq b - 1$. To show that $S(a + 1)$ holds, we must find integers q_1 and r_1 such that $a + 1 = b \cdot q_1 + r_1$ where $0 \leq r_1 \leq b - 1$.

We know that $a = b \cdot q + r$ where $0 \leq r \leq b - 1$. Thus $a + 1 = b \cdot q + r + 1$. If $0 \leq r \leq b - 2$, then $1 \leq r + 1 \leq b - 1$, and $a + 1 = b \cdot q_1 + r_1$ where $q_1 = q$, $r_1 = r + 1$, and $0 \leq 1 \leq r_1 \leq b - 1$. On the other hand, if $r = b - 1$, then $r + 1 = b$ and $a + 1 = bq + r + 1 = bq + b = b \cdot (q + 1) = b \cdot q_1 + r_1$ where $q_1 = q + 1$ and $r_1 = 0$. Therefore, in every possible case, $S(a + 1)$ holds.

The inductive step is thus established. ∎

One minor technical point arises on occasion. In some cases it is natural to label the statements on the list so that the first statement is $S(2)$, $S(5)$, or in general $S(k_0)$ for some integer $k_0 \geq 0$. For such cases we have a slightly more general version of PMI.

Principle of Mathematical Induction (Generalized). Let k_0 be a nonnegative integer and let $S(k_0), S(k_0 + 1), \ldots, S(n), \ldots$ be a list of statements, one for each integer $n \geq k_0$. Then every statement on the list is true if the following two conditions hold:

(i) $S(k_0)$ is true.
(ii) For each integer $k \geq k_0$, if $S(k)$ is true then $S(k + 1)$ is true.

There is a different form of mathematical induction that possesses two significant virtues: It has the same effect as PMI, and it is applicable in many situations in which PMI cannot be so readily used. We refer to this form of mathematical induction as the Principle of Strong Induction (PSI), since the hypothesis in the inductive step of PSI is stronger than that of PMI.

Principle of Strong Induction. Let k_0 be a nonnegative integer and let $S(k_0)$, $S(k_0 + 1), \ldots, S(n), \ldots$ be a list of statements, one for each positive integer greater than or equal to k_0. Then every statement on the list is true if the following two conditions hold:

(i) $S(k_0)$ is a true statement.
(ii) For each positive integer $k \geq k_0$, if $S(k_0)$, $S(k_0 + 1), \ldots, S(k)$ are all true, then $S(k + 1)$ is true.

First note that if PSI (i) and (ii) are established, then $S(k)$ is true for each positive integer $k \geq k_0$, for, if (i) is proved, then $S(k_0)$ is true. By (ii) applied in the case $k = k_0$, $S(k_0 + 1)$ is true. Since $S(k_0)$ and $S(k_0 + 1)$ are true, it follows from PSI (ii) that $S(k_0 + 2)$ is true. Since $S(k_0)$, $S(k_0 + 1)$, and $S(k_0 + 2)$ are true, $S(k_0 + 3)$ is true by PSI (ii). Again by PSI (ii), we deduce that $S(k_0 + 4)$ holds and so on. Thus PSI enables us to prove that $S(k)$ is true for each $k \geq k_0$.

We emphasize that PMI and PSI are similar in that both are proof techniques used to prove that every statement in an infinite list of statements is true. However, PMI and PSI differ in a significant way. In the inductive step of PSI, we assume that all of the first k statements on the list are true and deduce that the $(k + 1)$st statement is true. In the inductive step of PMI, $S(k + 1)$ is deduced from $S(k)$ only. Because more is assumed in the inductive step of PSI than in the inductive step of PMI, the hypothesis of PSI (ii) is stronger than the hypothesis of PMI (ii). We can summarize the two forms of induction in the following chart:

	Basis Step		**Inductive Step**
PMI	Show: $S(k_0)$ is true	Show:	$S(k) \Rightarrow S(k + 1)$
PSI	Show: $S(k_0)$ is true	Show:	$\begin{array}{l} S(k_0) \\ S(k_0 + 1) \\ \vdots \\ S(k) \end{array} \Rightarrow S(k + 1)$

The important question remains: When should we use PSI instead of PMI? The basic idea in the inductive step of PMI is that $S(k + 1)$ can be deduced from or reduced back to $S(k)$. This is precisely what occurred in Examples 1.3.8–1.3.11. In some cases, though, $S(k + 1)$ cannot be readily deduced from $S(k)$, but can be derived from all of the previous statements $S(k_0), \ldots, S(k)$. In these cases we appeal to PSI in order to conclude that all the statements $S(k_0), \ldots, S(k), \ldots$ are true. Our next example illustrates this point.

Theorem 1.3.7 *Each integer greater than* 1 *is either prime or is a product of primes.*

Proof For an integer $n \geq 2$, define the statement $S(n)$: n is either prime or is a product of primes.

Basis step. $S(2)$ is true since 2 is a prime number.

Inductive step. Suppose statements $S(2), \ldots, S(n)$ are all true and show that $S(n + 1)$ is true. Without doubt the integer $n + 1$ either is prime or is not prime. If $n + 1$ is a prime, then $S(n + 1)$ is true. If $n + 1$ is not prime, then $n + 1 = a \cdot b$ where $a, b < n + 1$; hence, $a, b > 1$. But by hypothesis $S(a)$ and $S(b)$ are both true. Thus, each of a and b either is prime or is a product of primes. Write $a = p_1 \cdot \cdots \cdot p_r$ and $b = q_1 \cdot \cdots \cdot q_s$ where $p_1, \ldots, p_r, q_1, \ldots, q_s$ are all prime. (For example, if $r = 1$, then a itself is prime.) Therefore,

$$n + 1 = a \cdot b = (p_1 \cdot \cdots \cdot p_r) \cdot (q_1 \cdot \cdots \cdot q_s)$$

and $a \cdot b$ is a product of primes.

By PSI, the proof is complete. ■

In the inductive step in the proof of Theorem 1.3.7, we used the fact that a and b are expressible as a product of primes to conclude that $n + 1 = a \cdot b$ is a product of primes. It is not at all obvious that this conclusion can be derived from the assumption that n is expressible as a product of primes.

To summarize our work, we have introduced two forms of mathematical induction, PMI and PSI. We described each principle, illustrated each with examples, and discussed when it is more appropriate to use PSI instead of PMI. We have not, however, addressed the question of when a proof by mathematical induction should be used.

The obvious answer appears to be: Use mathematical induction whenever an infinite list of statements is given like those in Examples 1.3.8 or 1.3.9. However, problems susceptible to inductive proof do not always come so neatly and clearly packaged. (Theorems 1.3.5 and 1.3.6 bear witness to this observation.) Thus when should a proof by mathematical induction be used?

As with most matters involving the use of a proof technique, no definitive answer to this question exists. Nonetheless, *it is appropriate to look for a proof by mathematical induction of any statement that involves an integer variable*. For example, in Theorem 1.3.7 the variable is the integer to be factored. The integer variable appearing in the problem statement can then be used to number the statements on the infinite list of statements. Of course, we are not claiming that every statement in which an integer variable appears must be proved by mathematical induction. But, as a general policy, induction should be considered as a proof technique for such a statement.

Case Analysis

The final method of proof to be discussed is proof by *case analysis*. In a case analysis proof of a given statement, the entire argument is divided into a collection of cases. Each of the cases is then resolved, thereby establishing the original statement.

For example, in proving some statement involving a single integer variable, call it x, one might consider two cases: (1) $x \geq 0$ and (2) $x < 0$. Or in proving a statement involving a real number x, one can analyze the following cases: (1) $x =$ integer, (2) $x =$ rational, (3) $x =$ arbitrary real.

We distinguish two types of case analysis proofs: *divide-and-conquer* and *bootstrap*. In a divide-and-conquer proof, the original problem is divided into a number of separate cases that are established (or conquered) independently of each other. In a bootstrap proof, the original problem is divided into a sequence of cases in which a given case after the first one is established with the use of some or all of the previous cases. We now illustrate each of these types of case analysis proofs. We begin with an example of a divide-and-conquer argument.

Example 1.3.12 (i) Find all real solutions to the inequality $|x - 1| < |x - 3|$.

Recall that

$$|x| = \begin{cases} x & \text{if } x \geq 0 \\ -x & \text{if } x < 0. \end{cases}$$

Thus,

$$|x - 1| = \begin{cases} x - 1 & \text{if } x - 1 \geq 0, \text{ i.e., if } x \geq 1 \\ -(x - 1) = 1 - x & \text{if } x - 1 < 0, \text{ i.e., if } x < 1 \end{cases}$$

and

$$|x - 3| = \begin{cases} x - 3 & \text{if } x \geq 3 \\ 3 - x & \text{if } x < 3. \end{cases}$$

Thus we consider three cases: (1) $x \geq 3$, (2) $1 \leq x < 3$, (3) $x < 1$.

Case 1. $x \geq 3$. Then $|x - 1| = x - 1$ and $|x - 3| = x - 3$ and the original inequality reduces to $x - 1 < x - 3$, which holds if and only if $-1 < -3$. Thus $|x - 1| < |x - 3|$ holds for no $x \geq 3$.

Case 2. $1 \leq x < 3$. In this case $|x - 1| = x - 1$ and $|x - 3| = 3 - x$ and the inequality $|x - 1| < |x - 3|$ becomes $x - 1 < 3 - x$, which holds if and only if $2x < 4$. Therefore, for $1 \leq x < 3$, $|x - 1| < |x - 3|$ is valid if $1 \leq x < 2$.

Case 3. $x < 1$. Then the given inequality becomes $1 - x < 3 - x$, which is equivalent to $1 < 3$. Thus $|x - 1| < |x - 3|$ holds for all $x < 1$.

From Cases 1–3 we conclude that $|x - 1| < |x - 3|$ holds if and only if $x < 2$.

Notice that the original problem is divided into three cases or subproblems. These cases are independent in that the solution to any particular case does not depend on the solutions to the other cases. Thus the order in which the cases are considered is irrelevant. Also it is perhaps obvious but important to note that every possible value of x is handled by one of the cases.

(ii) Let us prove the triangle inequality:

$$\text{For all } x, y \text{ in } \mathbf{R}, |x + y| \leq |x| + |y|.$$

Again we consider several separate cases:

Case 1. $x \geq 0$ and $y \geq 0$. Then $x + y \geq 0$ and $|x + y| = x + y = |x| + |y|$.

Case 2. $x \geq 0$ and $y < 0$. Then $|x| = x$ and $|y| = -y$. The value of $|x + y|$ depends on whether $x + y \geq 0$ or $x + y < 0$. Let us therefore consider subcases:

Subcase 1. $x + y \geq 0$. Then

$$\begin{aligned} |x + y| &= x + y \leq x + (-y) \qquad \text{since } y < 0 < -y \\ &= |x| + |y|. \end{aligned}$$

Subcase 2. $x + y < 0$. Then

$$|x + y| = -(x + y) = (-x) + (-y) \leq x + (-y) = |x| + |y|.$$

Case 3. $x < 0$ and $y \geq 0$. This case is handled by an argument similar to that given in Case 2.

Case 4. $x < 0$ and $y < 0$. Then $x + y < 0$ and $|x + y| = -(x + y) = (-x) + (-y) = |x| + |y|$.

Next we illustrate a bootstrap-type case analysis proof.

Example 1.3.13 Suppose f is a function on the real numbers with the property that $f(x + y) = f(x) + f(y)$ for all real numbers x and y. (We call this the additive property of f.) Prove that for all real numbers x and all rational numbers r

$$f(r \cdot x) = r \cdot f(x).$$

Proof The bootstrap proof is divided into five cases: (1) $r = 0$, (2) $r =$ positive integer, (3) $r =$ negative integer, (4) $r = 1/n$ where n is a nonzero integer, (5) $r = m/n$ where m is an integer and n is a nonzero integer.

Case 1. We show $f(0) = 0$: $f(0) = f(0 + 0) = f(0) + f(0)$; hence $f(0) = 0$. Thus, for any real x, $f(0 \cdot x) = f(0) = 0 \cdot f(x)$.

Case 2. We show $f(n \cdot x) = n \cdot f(x)$ for any real x and any positive integer n. To demonstrate this claim we use PMI. Let $S(n)$ be the statement: $f(n \cdot x) = n \cdot f(x)$ for any real x and any integer n.

Basis step. $f(1 \cdot x) = f(x) = 1 \cdot f(x)$ for any real x. Thus $S(1)$ holds.

Inductive step. We show $S(n)$ implies $S(n + 1)$. Suppose $f(n \cdot x) = n \cdot f(x)$ for all real x. We show $f((n + 1) \cdot x) = (n + 1) \cdot f(x)$ for all real x.

By the additive property of f,

$$\begin{aligned} f((n + 1) \cdot x) &= f(n \cdot x + x) = f(nx) + f(x) \\ &= n \cdot f(x) + f(x) \qquad \text{by inductive hypothesis} \\ &= (n + 1) \cdot f(x). \end{aligned}$$

Therefore, by PMI $f(n \cdot x) = n \cdot f(x)$ for all real x and all positive integers n.

Case 3. Let n be a negative integer. We show $f(n \cdot x) = n \cdot f(x)$. By Case 1 and the additive property of f,

$$0 = f(0) = f(n \cdot x + (-n \cdot x)) = f(nx) + f(-n \cdot x).$$

Since $-n$ is a positive integer, $f(-n \cdot x) = -n \cdot f(x)$ by Case 2. Hence

$$f(n \cdot x) + f(-n \cdot x) = f(n \cdot x) - n \cdot f(x) = 0$$

and

$$f(n \cdot x) = n \cdot f(x).$$

We now know that $f(n \cdot x) = n \cdot f(x)$ for all real x and any integer n.

Case 4. We show that if $r = 1/n$ where n is a nonzero integer, then $f(r \cdot x) = r \cdot f(x)$ for any real x. We have

$$f(x) = f(n \cdot (1/n) \cdot x) = n \cdot f((1/n) \cdot x)$$

by Case 2 or Case 3.

Thus

$$f((1/n) \cdot x) = 1/n \cdot f(x).$$

Case 5. Let r be any rational and x any real. Then $r = m/n$ where m and n are integers and $n \neq 0$. Thus, by Cases 2, 3, and 4,

$$\begin{aligned} f(r \cdot x) &= f(m/n \cdot x) = f(m \cdot x/n) = m \cdot f(x/n) \\ &= m/n \cdot f(x) \\ &= r \cdot f(x). \end{aligned}$$

■

Observe that every case in this proof after the first two refers back to earlier cases. In general a bootstrap proof is built up in a sequence of layers with a given layer relying on previous layers.

Quantification and Counterexamples

We close this section with some general comments on the logical structure of mathematical proofs of quantified statements. Our remarks have been implicit throughout this section, but it is useful to state them explicitly at this point.

To prove a statement of the form "For all x $P(x)$," one must show that for each x in the given domain of the problem (e.g., for each real number x or for each integer x), the statement $P(x)$ is valid. This task can be carried out in several ways: via either a direct proof, an indirect proof (contraposition or contradiction), mathematical induction, or a case analysis. Several statements of the type "For all x $P(x)$" were proved in this section.

To prove a statement of the form "There exists x such that $P(x)$," we can choose one of two methods. The first is to construct directly an object x_0 such that the statement $P(x_0)$ is valid. For instance, to show that there exists a pair (x, y) of real numbers satisfying the system of equations

$$2x + y = 2$$
$$4x - 2y = 0,$$

we can use standard techniques to show that the pair $(1/2, 1)$ does satisfy the stated property.

A second method of demonstrating "There exist x such that $P(x)$" is an argument by contradiction: Assume that the negation of this statement holds and derive a contradiction. Thus we suppose that it is not the case the "There exists x such that $P(x)$." In other words, we suppose that "For all x, not-$P(x)$." Then, from this statement, we derive a contradiction that implies that the original statement is valid, and therefore an object x satisfying $P(x)$ does indeed exist. Here is an example.

Example 1.3.14 Let a be a positive real number. Prove: There exists a real number x_0 such that $x_0^2 = a$.

Proof We use properties of continuous functions to derive a contradiction. Suppose that for all real x, $x^2 \neq a$. Consider the function $f(x) = x^2 - a$. The function f is continuous on $\mathbf{R}$. Note that $f(a + 1) = (a + 1)^2 - a = a^2 + a + 1 > 0$ while $f(0) = -a < 0$. Thus f is a continuous function taking on both positive and negative values. But, since $x^2 \neq a$ for all real x, $f(x) = x^2 - a = 0$ has no real solution. This contradiction completes the argument. ∎

Finally, let us discuss how the statement "For all x $P(x)$" can be *disproved*. As noted above, to prove such a statement, one must show that for each x, the statement $P(x)$ is valid. Thus, to disprove "For every x $P(x)$," one must know that "There exists x_0 such that not-$P(x_0)$" is valid; i.e., one must find at least one x_0 for which $P(x_0)$ does not hold. An object x_0 with this property is called a *counterexample* to $P(x)$. Thus disproving "For every x $P(x)$" amounts to finding a counterexample to $P(x)$.

Example 1.3.15 Consider the statement: For every 2×2 matrix A with real number entries, if $A \neq \begin{bmatrix} 0 & 0 \\ 0 & 0 \end{bmatrix}$, then there exists a 2×2 matrix B such that $AB = I = \begin{bmatrix} 1 & 0 \\ 0 & 1 \end{bmatrix}$.

The given statement is true if one assumes that A and B are real numbers (i.e., A and B are 1×1 matrices). But, for 2×2 matrices, matters are more complicated. In fact, the matrix $A = \begin{bmatrix} 1 & 0 \\ 0 & 0 \end{bmatrix}$ is not the zero matrix, yet for any 2×2 matrix $B = \begin{bmatrix} a & b \\ c & d \end{bmatrix}$, $AB = \begin{bmatrix} a & b \\ 0 & 0 \end{bmatrix}$. Thus, for each 2×2 matrix B, $AB \neq \begin{bmatrix} 1 & 0 \\ 0 & 1 \end{bmatrix} = I$, and $A = \begin{bmatrix} 1 & 0 \\ 0 & 0 \end{bmatrix}$ provides a counterexample to the given statement.

As this example illustrates, to find a counterexample to an implication involving a variable, "If $P(x)$ then $Q(x)$," one must find a specific value of x, call it x_0, for which $P(x_0)$ is true and $Q(x_0)$ is false. In Example 1.3.15

the matrix $A = \begin{bmatrix} 1 & 0 \\ 0 & 0 \end{bmatrix}$ is a 2×2 matrix with real entries such that $A \neq \begin{bmatrix} 0 & 0 \\ 0 & 0 \end{bmatrix}$ and for each 2×2 matrix B, $AB \neq I$.

Once a counterexample to a statement has been obtained, our work is not finished. We should seek ways of modifying the original statement so as to obtain a true statement. Specifically, we can try to determine some additional conditions that, when added to the hypotheses of the statement, allow us to deduce the conclusion. For instance, in Example 1.3.15, if we modify the hypothesis by insisting that the determinant of A is not 0, then the resulting statement is true:

> If A is a 2×2 matrix with real entries such that the determinant of A is not 0, then there is a matrix B such that $AB = \begin{bmatrix} 1 & 0 \\ 0 & 1 \end{bmatrix}$.

Let us consider an example that illustrates a type of exercise that will appear periodically throughout this text.

Example 1.3.16 Prove, or disprove and salvage: For any integer $n > 1$, $\sqrt{n}$ is irrational.

There are two available options: (i) Prove the given statement, or (ii) disprove the statement by giving a counterexample, and then modify the statement and prove the new statement.

Since the given statement involves an integer variable, an approach through mathematical induction is suggested. The first case to consider is $n = 2$ and by Example 1.3.6, $\sqrt{2}$ is indeed irrational. This result constitutes the basis step in an inductive proof. Rather than trying to prove the inductive step, we can search for more evidence in support of the statement. For instance, $\sqrt{3}$, $\sqrt{6}$, and $\sqrt{8}$ can all be proved to be irrational. But by this point a counterexample has perhaps surfaced: $\sqrt{4} = 2$, which is a rational number; moreover, for any positive integer n, $\sqrt{n^2} = n$, which is rational. Thus we must modify the original statement.

There are at least two ways of approaching this task. The first is to consider other specific cases ($\sqrt{5}$, $\sqrt{7}$, $\sqrt{11}$, or $\sqrt{12}$, for example), and then formulate a conjecture based on the finding in these cases. The second approach is simply to revise the original statement so as to exclude the counterexamples observed above ($\sqrt{n^2} = n$). In the latter case we come up with the statement:

(*) If n is a positive integer that is not the square of another integer, then $\sqrt{n}$ is irrational.

Even though (*) contains an integer variable, that variable ranges only over the nonsquare positive integers. Thus a proof by mathematical induction does not seem feasible. What kind of proof should be attempted? Is the result true? Does (*) require further modification? We shall be able to settle these questions using the results of Chapter 6. For the present we offer some exercises that provide more evidence concerning the validity of (*).

Exercises §1.3

In Exercises 1.3.1–1.3.9 use direct proof and Axioms 1–14 of **R** to establish the given statement.

1.3.1. (a) Prove: The sum of two even integers is an even integer.
(b) Prove: The product of two even integers is an even integer.

1.3.2. Prove Corollary 1.3.2.

1.3.3. (a) Prove: If a and b are integers such that a is a factor of b, then, for any integer c, a is a factor of $b \cdot c$.
(b) Prove: If a, b, and c are integers such that a is a factor of b and b is a factor of c, then a is a factor of c.
(c) Prove: If a and b are integers and a is a factor of b, then a^2 is a factor of b^2.

1.3.4. (a) Prove: For all real x and y, $x \cdot y = (-x) \cdot (-y)$. (You may use Exercises 1.2.2(d) and 1.2.3(b).)
(b) Prove: The product of two negative real numbers is positive.

1.3.5. Prove: The product of a positive real and a negative real is a negative real. (You may use Exercise 1.2.2(c).)

1.3.6. Let a and b be integers. Prove: If a is odd and b is even, then $a + b$ is odd.

1.3.7. Let x and y be real numbers such that $x > 0$ and $y > 0$. Prove: If $x^2 = y^2$, then $x = y$.

1.3.8. Let a and b be positive real numbers. Prove: $\sqrt{a \cdot b} = \sqrt{a} \cdot \sqrt{b}$.

1.3.9. Suppose a, b, and c are real numbers.
(a) Prove: If $a > b$ and $c > 0$, then $a \cdot c > b \cdot c$.
(b) Prove: If $a > b > 0$ and $c > 1$, then $a \cdot c > b$.

In Exercises 1.3.10–1.3.15 use indirect proof (either proof by contrapositive or proof by contradiction) and Axioms 1–14 of **R** to establish the given result.

1.3.10. Prove: If a is a positive real number, then $a \cdot (-1) < 0$.

1.3.11. (a) Let a be an integer. Prove: If a^2 is even, then a is even.
(b) Let a and b be integers. Show: If $a \cdot b$ is odd, then a is odd and b is odd.

1.3.12. Prove: $\sqrt{8}$ is irrational.

1.3.13. (a) Prove: $\sqrt{6}$ is irrational.
(b) Prove: $\sqrt{10}$ is irrational.

1.3.14. (a) Prove: $\log_2(3)$ is irrational.
(b) Prove: $\log_2(5)$ is irrational.

1.3.15. Prove: If a, b, and c are real numbers such that $a > 1$, $b > 1$, and $a \cdot b = c$, then $a < c$ and $b < c$.

In each of the following ten exercises, use mathematical induction to establish the given statement.

1.3.16. Prove: For each positive integer n, $1 + 4 + 7 + \cdots + (3n - 2) = (3n^2 - n)/2$.

1.3.17. Prove: For each positive integer n, $1 + 5 + 9 + \cdots + (4n - 3) = 2n^2 - n$.

1.3.18. Prove: For each positive integer n and each real $x \geq -1$, $(1 + x)^n \geq 1 + nx$.

1.3.19. Let α be a real number such that α/π is not an integer. Prove that for each positive integer n,

$$1/2 + \sum_{1}^{n} \cos(k\alpha) = \frac{\sin((n + 1/2) \cdot \alpha)}{2 \sin(\alpha/2)}.$$

1.3.20. Prove: $1^2 + 3^2 + \cdots + (2n - 1)^2 = (4n^3 - n)/3$ for each positive integer n.

1.3.21. Show: $n^2 \leq 2^n$ for each integer $n \geq 4$.

1.3.22. Prove: $(1/5)x^5 + (1/3)x^3 + (7/15)x$ is an integer for each positive integer x.

1.3.23. Prove: For each positive integer n,

$$\frac{1}{n+1} + \cdots + \frac{1}{2n} = (1 - 1/2) + (1/3 - 1/4) + \cdots + \left(\frac{1}{2n-1} - \frac{1}{2n}\right).$$

1.3.24. Generalized Distributive Law. Let $a, b_1, \ldots, b_n$ be real numbers. Show: $a \cdot (b_1 + \cdots + b_n) = a \cdot b_1 + \cdots + a \cdot b_n$.

1.3.25. Prove: For any integer $n \geq 2$, the product of n odd integers is an odd integer.

1.3.26. For $n \geq 1$ let $s_n = 1 + (8 \cdot 1 + \cdots + 8 \cdot n)$.
(a) Compute s_1, s_2, s_3, s_4, s_5.
(b) Conjecture a closed formula for s_n for an arbitrary value of n.
(c) Prove your conjecture by mathematical induction.

1.3.27. Let $a_n = 1/(1 \cdot 2) + 1/(2 \cdot 3) + \cdots + 1/(n(n + 1))$ for each positive integer n.
(a) Compute a_1, a_2, a_3, a_4, a_5.
(b) Conjecture a closed formula for a_n for an arbitrary value of n.

(c) Prove your conjecture by mathematical induction.
(d) Show that the infinite series $\Sigma_{k=1}^{\infty} 1/k(k+1)$ converges and determine its sum.

1.3.28. Let $b_n = 1/3 + 1/15 + \cdots + 1/(4n^2 - 1)$.
(a) Compute b_1, b_2, b_3, b_4, b_5.
(b) Conjecture a closed formula for b_n.
(c) Prove your conjecture by mathematical induction.

1.3.29. Let $c_n = 1^3 + \cdots + n^3$.
(a) Compute c_1, c_2, c_3, c_4, c_5.
(b) Conjecture a closed formula for c_n.
(c) Prove your conjecture.

1.3.30. For each positive integer k, define the integer $k!$ (called "k factorial") by $k! = k \cdot (k-1) \cdot \cdots \cdot 2 \cdot 1$. Let $e_n = 1 \cdot 1! + 2 \cdot 2! + 3 \cdot 3! + \cdots + n \cdot n!$ Conjecture and prove a closed formula for e_n.

1.3.31. Let $d_n = \Sigma_{k=1}^{n} k/(k+1)!$.
(a) Compute d_1, d_2, d_3, d_4, d_5.
(b) Conjecture a closed formula for d_n.
(c) Prove your conjecture.

1.3.32. Conjecture and prove a closed formula for the product $(1 - 1/4) \cdot (1 - 1/9) \cdot \cdots \cdot (1 - 1/n^2)$.

1.3.33. Let f be a function that has derivatives of every order at each real number x; i.e., $f'(x), f''(x), \ldots, f^{(n)}(x) \ldots$ exist for each positive integer n and each real number x. Let $g(x) = x \cdot f(x)$. Conjecture and prove a formula for $g^{(n)}(x)$ that expresses $g^{(n)}(x)$ in terms of $f^{(n)}(x)$ and x.

1.3.34. (a) Prove: For every positive integer n, 3 divides $n^3 - n$.
(b) Prove: For every positive integer n, 5 divides $n^5 - n$.
(c) Formulate a statement that contains the statements in (a) and (b) as special cases.

1.3.35. Define a sequence of positive integers m_n, $n \geq 1$, by $m_1 = 2$, and for $n \geq 2$, $m_n = m_{n-1} \cdot (m_{n-1} - 1) + 1$.
(a) Compute m_2, m_3, and m_4.
(b) Show that for $n \geq 1$, $m_{n+1} = (m_1 \cdot \cdots \cdot m_n) + 1$.

1.3.36. Male bees, which have a mother but no father, hatch from unfertilized eggs. Female bees hatch from fertilized eggs. How many ancestors does a male bee have in the tenth generation back? How many of these ancestors are male?

1.3.37. Prove that every positive integer has a base 2 or *binary* expansion. Specifically prove that if n is a positive integer, then there exist integers $a_0, a_1, \ldots, a_k$ such that each $a_i = 0$ or 1 and $n = a_0 + a_1 \cdot 2 + \cdots + a_k \cdot 2^k$.

In each of the following exercises use proof by case analysis to establish the result.

1.3.38. Solve the inequality $|x + 1| < |x - 1|$.

1.3.39. Solve the inequality $|x + 1| < |x^2 - 1|$.

1.3.40. (a) Show that if a is an integer, then there exists an integer q such that $a = 3 \cdot q + r$ where $r = 0$, 1, or 2. (You will not need a case analysis; just use the Division Theorem.) Now use this result together with a proof by contradiction and case analysis in part (b).
(b) Show that if 3 divides a product $a \cdot b$ where a and b are integers, then either 3 divides a or 3 divides b.

1.3.41. Can one form a ten-digit integer by putting a digit between 0 and 9 in the empty boxes in the given table as follows: The digit in the box labeled 0 indicates the number of times 0 appears in the number, the digit in box 1 indicates the number of times 1 appears in the number, etc.?

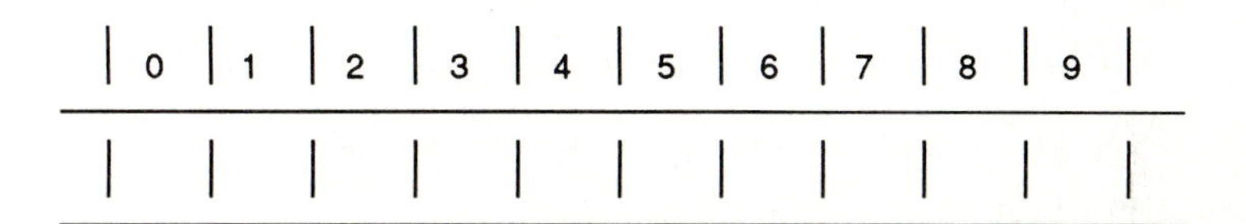

0	1	2	3	4	5	6	7	8	9

(For instance, if 9 is placed in box 0, then the remaining boxes must be filled with 0 in order that the nine zeros actually appear. In this case, however, we reach a contradiction, since box 9 cannot contain 0 as at least one 9 appears in the number. Thus the desired ten-digit number cannot have 9 as its leftmost digit.) (Hint: Consider the possible ways of filling the box labeled 0.) How many such numbers exist?

1.3.42. The natural logarithm function, log, has the property that $\log(xy) = \log(x) + \log(y)$ for all positive reals x and y. Show that $\log(x^r) = r \log(x)$ for all positive real numbers x and all rational numbers r.

1.3.43. Let f be a function on the real numbers such that $f(0) \neq 0$ and $f(x + y) = f(x) \cdot f(y)$ for all real x and y. Conjecture and prove a formula for $f(r \cdot x)$ valid for all rational r and real x.

Give a counterexample to each of the following statements.

1.3.44. The sum of two irrational numbers is irrational.

1.3.45. The product of two irrational numbers is irrational.

1.3.46. If a and b are positive integers and $a \cdot b$ is a perfect square, then a and b are perfect squares.

1.3.47. The square root function has the additive property.

In each of the following exercises, write the given statement as an implication (a) without using variables, and (b) using variables and using quantifiers.

1.3.48. The product of an even integer and an odd integer is an even integer.

1.3.49. The square root of an irrational number is irrational.

1.3.50. The cube root of 2 is irrational.

1.3.51. The derivative of a constant function is 0.

1.3.52. The determinant of an invertible matrix is nonzero.

1.3.53. The product of two irrational numbers is irrational.

1.3.54. The rank of a singular $n \times n$ matrix is less than n.

1.3.55. The sum of two irrational numbers is irrational.

1.3.56. For a function to be constant, it is sufficient that its derivative be identically zero.

For each of the following statements indicate the technique or techniques that you would use when trying to devise a proof. Do not attempt a proof.

1.3.57. $dx^n/dx = nx^{n-1}$ where n is a positive integer.

1.3.58. Let c be a nonzero real number. If $\Sigma(ca_n)$ converges, then $\Sigma\, a_n$ converges.

1.3.59. If f is differentiable at a, then f is continuous at a.

1.3.60. If x is irrational, then $\sqrt{x}$ is irrational.

1.3.61. The sum of two rational numbers is rational.

1.3.62. $\sqrt[3]{2}$ is irrational.

1.3.63. Let n be a positive integer. If $2^n - 1$ is prime, then n is prime.

1.3.64. The sum of the cubes of the first n integers is always a perfect square.

1.3.65. If a function has an inverse, then it has only one inverse.

1.3.66. The sum of the interior angles of a convex n-gon is $(n - 2)180°$ or $(n - 2)\pi$ radians.

1.3.67. The product of an odd integer and an even integer is even.

1.3.68. If the product of two integers is odd, then each of the integers is odd.

1.3.69. $1^2 + 2^2 + \cdots + b^2 = b(b + 1)(2b + 1)/6$ for each positive integer b.

1.3.70. The equation $x^4 + x^2 + 1 = 0$ has no real number solution.

1.3.71. The product of a nonzero rational number and an irrational number is irrational.

1.3.72. The product of four consecutive positive integers is always one less than a perfect square.

1.3.73. Extend Theorem 1.3.6 by proving the following statement: Let a and b be integers with $b > 0$. Then there exist integers q and r such that $a = b \cdot q + r$ where $0 \leq r \leq b - 1$.

§1.4 Informal Thinking: Methods of Inquiry

The heart of mathematics is its problems.

P. R. Halmos

If Paul Halmos, a well-known twentieth-century mathematician, is indeed correct, then the ability to solve mathematical problems, especially those that differ appreciably from any previously encountered exercise, is the foremost skill that a mathematician must possess. The typical person on the street probably believes that this ability is granted at birth to a chosen few and denied to the rest of the population. According to the popular folklore, either you are a mathematical genius or you can't do math at all.

On closer inspection, however, the simplicity of this viewpoint becomes apparent. Mathematical aptitudes and abilities vary over a seemingly continuous range. Some people prefer algebra to geometry in high school; others enjoy geometry far more than algebra. Some find high school mathematics easy and college mathematics difficult. Some are attracted more by "pure" mathematics than by "applied" mathematics. Others choose computational mathematics over "theorem-proving" mathematics. Moreover, as people study and practice mathematics, they generally do become more adept at learning new mathematics and more skillful at solving problems. In addition, many of us have encountered an inspiring mathematics teacher who aroused our interest and increased our performance in mathematics beyond what we thought possible. While no one can deny the existence of innate differences in mathematical ability among individuals, it is clear that environmental influences and personal characteristics are important in determining the mathematical performance of each individual.

Our question is, then: What can we as individuals do to improve our own mathematical performance? Certainly we can learn as much mathematics as possible. For example, the more we know, the more likely we are to be able to answer a particular question. But beyond that, how can we become better at solving problems, proving theorems, writing computer programs, building mathematical models of phenomena? In short, how can we become better at doing mathematics?

The usual answer to this question is: Practice. If you sit down and try to solve many mathematical problems or to prove many theorems, eventually you will become good at problem solving or theorem proving. The experience that you gain from failure, partial success, and success teaches you, perhaps slowly and painfully, how to do mathematics. But are "practice and experience" the only answers to these questions?

Our response to this question is a qualified No. Practice and the experience born of practice are necessary for the improvement of one's mathematical performance. But they are not sufficient to guarantee acceptable results. In fact, we assert that proficiency in mathematics requires

1. knowledge of subject matter,
2. awareness of strategies and tactics for solving problems and learning new material, and
3. ability to monitor and to control one's activities when doing mathematics.

Most mathematical textbooks have been concerned only with item 1. Indeed most of this text deals with the subject matter of a variety of topics. But in this section we focus on items 2 and 3. Our goal is to introduce and to discuss briefly some of the considerations involved in 2 and 3. Throughout the text frequent mention will be made of the ideas raised in this section.

What we present are some fundamental rules of thumb that might be of assistance in the problem-solving process. No one of the principles is guaranteed to work on any given problem. Nevertheless, they can serve to help a person discover a solution to a problem or a proof of a theorem. Principles of this sort—those that aim to help in the discovery of a solution but that are not certain to lead to a solution—are called *heuristic* principles. We now turn to a discussion of the heuristics, i.e., the rules of discovery, of problem solving.

Our exposition is based on the work of George Polya and Alan Schoenfeld. Both Polya and Schoenfeld are professional mathematicians who have devoted considerable effort to the study of the activity of mathematical problem solving by humans. Polya's principal method was introspection: He carefully observed his own actions as he was solving mathematical problems, noting particularly the steps that enabled him to solve a difficult problem. In a well-known book, *How To Solve It*, Polya organized his findings into a scheme that divides the problem-solving processing into four parts:

1. Understand the Problem.
2. Devise a Plan.
3. Carry Out the Plan.
4. Look Back.

In contrast to Polya, Schoenfeld has carried out a number of controlled experiments with college students and professional mathematicians. His work has led him to modify Polya's scheme somewhat. In Schoenfeld's scheme there are three major steps:

1. Analysis.
2. Exploration.
3. Verification.

We present a hybrid version of the Polya and Schoenfeld schemes in Table 1.4.1.

Table 1.4.1

I. Understand the Problem (Analysis).
   1. Identify the unknown.
   2. Isolate the hypotheses and the data.
   3. Develop a representation of the problem.
      a. Draw a figure.
      b. Introduce suitable notation.
   4. Examine special cases.
      a. Select special values to acquire a "feel" for the problem.
      b. Consider extreme cases.
      c. Evaluate integer parameters at $n = 1, 2, 3, \ldots$ and look for a pattern.
   5. Try to simplify the problem.
      a. Exploit symmetry.
      b. Choose appropriate units.

II. Devise a Plan (Exploration).
   1. Consider essentially equivalent problems.
      a. Replace conditions by equivalent conditions.
      b. Recombine the elements of the problem in various ways.
      c. Introduce auxiliary elements.
      d. Reformulate the problem.
         i. Change perspective or notation.
         ii. Argue by contradiction or contrapositive.
         iii. Take the problem as solved; i.e., assume you have a solution and determine its properties.
   2. Consider slightly modified problems.
      a. Aim for subgoals.
      b. Relax a condition, then restore it.
      c. Consider case analysis.
   3. Consider broadly modified problems.
      a. Construct an analogous problem with fewer variables.
      b. Generalize the problem.

***Table 1.4.1** Continued.*

c. Hold all but one variable fixed to determine that variable's impact.
d. Try to exploit any problem that has a similar form, hypothesis, or conclusion.

III. Carry Out the Plan (Verification).
  1. Check each step.
  2. Prove that each step is correct.

IV. Look Back (Verification).
  1. Apply these specific tests to your solution:
    a. Does it use all the pertinent data?
    b. Does it conform to reasonable estimates or predictions?
    c. Does it withstand tests of symmetry, dimension analysis, or scaling?
  2. Apply these general tests:
    a. Can the result be obtained differently?
    b. Can the result be verified in special cases?
    c. Can the result be reduced to known results?
    d. Can the result be used to derive other known results?

The remainder of this section will be devoted to comments on certain aspects of this outline. Our treatment will be very sketchy. (More details can be found in Polya's books *How to Solve It*, *Mathematical Discovery*, and *Mathematics and Plausible Reasoning* and in Schoenfeld's *Mathematical Problem Solving*.) We highlight certain features of this scheme: focusing on the unknown, problem representation, related problems, and looking back.

Focusing on the Unknown

When presented with a problem, the problem solver's first task is to understand the problem. Understanding a problem (or concept) involves several activities and stages; however, Polya's clear and forceful advice is to begin with the unknown. In Polya's chart (*How to Solve It*, pp. xvi–xvii) the first line begins with the question: What is the unknown? Focusing on the unknown provides coherence to the problem and directs the efforts of the problem solver toward a specific aim. In other words, if you know what you are looking for, then you are free to delve deeper into other aspects of the problem in order to obtain a more thorough understanding of it.

The unknown itself can take various forms. For example, it can be a specific mathematical object—a real number, a point in the plane, the point of intersection of two lines, a certain function, or a circle. In the latter case, for instance, to find the unknown circle, one must find its center (a point in the plane) and its radius (a positive real number). In other instances, the unknown is not so much a specific object as a task to perform or a goal to achieve: Prove that $\sqrt{2}$ is irrational, or find an efficient algorithm that will yield the solution to a given system of equations.

Whether looking for a specific object or a general task, by focusing on the unknown, the problem solver is forced to begin with the goal and to understand the meaning of all the terms in the unknown. What must be done to find a circle? What is an irrational number? As obvious as it might seem, the advice to focus on the unknown and to know the meaning of all the terms involved in the unknown (and the known, for that matter) is often overlooked.

As an exercise, take any mathematics book that you used in a previous course. Open the book to any section that you studied in that course and read over the exercises given at the end of the section. In each case identify the unknown and the meaning of the unknown. Repeat this procedure for several sections of the text and for the exercises in Section 1.3 of this chapter. Concentrate only on the identification and the meaning of the unknown. Ignore the techniques used in solving the problems. Does concentrating only on the unknown make the problems seem clearer and less forbidding now than when you first met them?

George Polya has much more to say on the issue of focusing on the unknown. See the section "Look at the Unknown" in the "Dictionary" portion of *How to Solve It*. We need not repeat Polya's remarks here. However, the moral of the story does bear repeating: In any problem, identify the unknown and understand its meaning.

Problem Representation

Suppose that in a given problem you have isolated the unknown and the hypotheses or conditions. Suppose also that the meanings of all the terms involved in the problem are clear. What next? At this point the formulation of an adequate representation of the problem and its components will probably be most useful. By a representation of the problem, we mean a method of describing the problem that (i) is internally coherent or consistent, (ii) corresponds closely to the problem, and (iii) is closely connected with the knowledge of the problem solver about the items of the problem and related concepts. The representation might take any one of several forms. It might involve a picture, a graph, an algebraic equation or a system of algebraic equations, a differential equation or a system of differential equations, a vector diagram, or a mathematical system such as a vector space, a group, or a Boolean algebra. In summary, a representation of a problem amounts to a view of the problem that is sensible and relevant

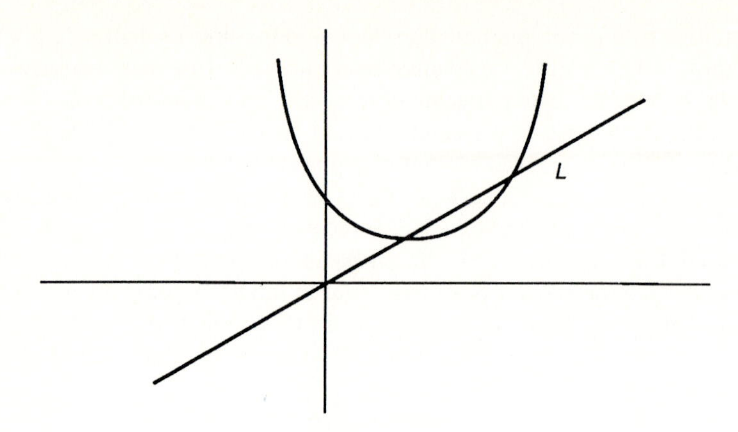

Figure 1.4.1

and can be manipulated in order to achieve a solution. Let us consider some examples of representations.

Perhaps the most familiar kind of representation is the pictorial or geometric representation of algebraic equations involving real numbers. We reconsider some previous examples to illustrate how a picture can be helpful.

Example 1.3.2 (revisited) If x is a real number and $x^2 - 4x + 6 = x$, then $x = 2$ or $x = 3$.

To represent this problem pictorially, recall that the equation $y = x^2 - 4x + 6$ has as its graph in the plane the parabola P with vertex at the point $(2, 2)$ and focus at $(2, 3)$. (The line $x = 2$ is the axis of symmetry of the parabola.) The graph of $y = x$ is the straight line L with slope 1 passing through the origin. The values of x for which $x^2 - 4x = x$ are precisely the first coordinates of the points at which the curves P and L intersect. The picture (Figure 1.4.1) makes the result appear extremely plausible. In fact, on the basis of the picture, we might

(i) guess that $x = 2$ and $x = 3$ are solutions to the original question,
(ii) check that $x = 2$ and $x = 3$ actually satisfy the equation,
(iii) show that no other solutions exist.

On the other hand, even if we do not use the picture (in this or any similar question) as a means of discovering solutions to the equation, we can use it as a check on a solution obtained by another method.

Example 1.3.8 (revisited) Show that for each positive integer n, $1 + 2 + \cdots + n = n(n + 1)/2$.

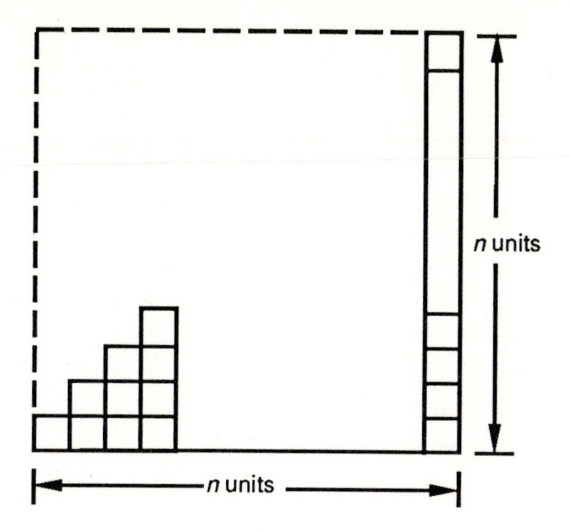

Figure 1.4.2

How can an equality of this sort be pictured? In this case we view the numbers on the left and right of the equality as different representations of a given geometric quantity. Of the common geometric quantities—length, area, volume—we consider area.

Can we represent $1 + 2 + \cdots + n$ as the area of a geometric figure? The sum suggests that we find figures of area $1, 2, \ldots, n$, respectively; then the area of the totality of these figures is $1 + 2 + \cdots + n$. Since a simple figure of area 1 is a square, let us consider squares arranged in the step-like fashion shown in Figure 1.4.2. The area of this figure is indeed $1 + 2 + \cdots + n$. Note that the figure sits inside an $n \times n$ square and apparently covers slightly more than half the square. Therefore,

$$1 + 2 + \cdots + n > n^2/2.$$

Since $n(n + 1)/2$ is also slightly greater than $n^2/2$, the equality $1 + 2 + \cdots + n = n(n + 1)/2$ is at least plausible.

A minor modification of this geometric representation makes the equality even more plausible. The rectangle in Figure 1.4.3 has area $n(n + 1)$ and is divided into two "step-like" regions of equal area, namely the gridded region and the nongridded region; the area of each figure is therefore $n(n + 1)/2$. On the other hand, the area of the lower region is $1 + \cdots + n$. Hence the equality $1 + \cdots + n = n(n + 1)/2$.

Drawing a figure provides one method for representing a mathematical problem or situation. However, there are other kinds of representations. For example, introducing symbols to stand for unknown quantities constitutes a form of representation. In max-min or related rate problems in

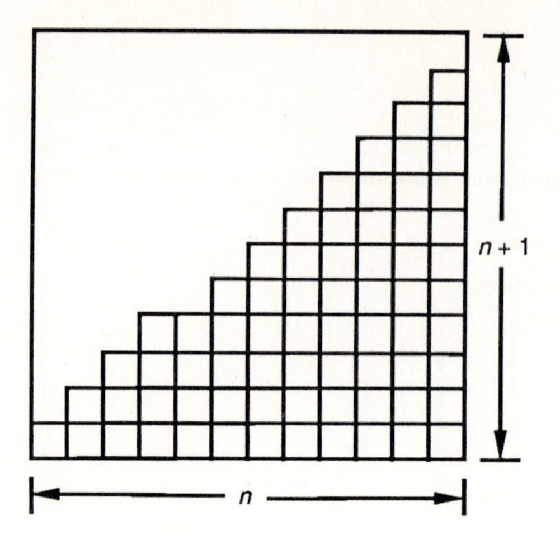

Figure 1.4.3

calculus, one obtains a functional relationship among two or more relevant variables which constitutes in effect a representation of the problem.

In some cases a representation entails the introduction of a new idea. For instance, the system of linear equations

$$ax + by = e$$

$$cx + dy = f$$

(a, b, c, d, e, and f are given real numbers; x and y are unknown real numbers that, if they exist, satisfy both equations) can be *represented* in matrix form $AX = B$ where

$$A = \begin{bmatrix} a & b \\ c & d \end{bmatrix}, \quad X = \begin{bmatrix} x \\ y \end{bmatrix}, \quad \text{and } B = \begin{bmatrix} e \\ f \end{bmatrix}.$$

To describe this representation we need the notions of matrix and matrix multiplication. The advantage of the matrix representation is that it suggests an analogy with real numbers: The matrix equation $AX = B$ is like the equation $ax = b$ where a and b are given real numbers and x is an unknown real number. From this analogy it is natural to consider the idea of an invertible matrix and to solve the matrix equation in case A is invertible. The solution to the matrix equation can then be interpreted to yield a solution to the original linear system.

Examples of other kinds of representations will appear throughout the remainder of the text.

Finding Related Problems

The heart of the problem-solving process by Polya and Schoenfeld lies in the second stage, Devising a Plan or Exploration. At this point the would-be problem solver possesses a firm understanding of the elements of the problem, can construct an adequate representation of the problem, and perhaps can verify the result in special cases or can argue for the plausibility of a certain solution. In other words, the problem solver has a solid grip on the problem.

Even so, a solution might not be readily apparent. In fact, with most substantive problems, solutions rarely appear at the understanding stage of the problem-solving process. What then should one do in order to find a solution?

Polya's advice is clear: Find a related problem. The basic strategy is the following: Given a problem P,

1. find a related problem P',
2. solve problem P',
3. use the solution to P' or the method of solution of P' to solve P.

This plan faces several apparent obstacles. First, how does one actually find a related problem? Second, the related problem itself might be difficult to solve; it might be necessary to devise and solve a problem P'', related to P', and use the solution to P'' to solve P'. Finally, the transition from a solution to P' to a solution to P may be difficult to perform.

For the present we concentrate primarily on the first step of the plan outlined in the previous paragraph: How does one find a related problem? A general answer to this question is: You find a related problem by playing with the hypotheses and conclusion of the given problem. Not surprisingly, this general answer does not cover all cases. We give several examples illustrating the process of finding related problems. Our examples are organized under the three parts of phase II of Table 1.4.1.

1. *Essentially equivalent problems*. We have already encountered one method of finding an essentially equivalent problem: To prove that an implication "If P then Q" is true, consider the contrapositive, "If not-Q then not-P." The original implication holds if and only if the contrapositive holds. When we consider the contrapositive, we are replacing the original problem by a logically, hence essentially, equivalent problem.

It is often very natural to replace a condition in a problem by an equivalent condition. For example, to show that the linear system

$$\begin{aligned} ax + by &= e \\ cx + dy &= f \end{aligned}$$

has a unique solution for any given e and f, it is necessary and sufficient to show that the matrix $A = \begin{bmatrix} a & b \\ c & d \end{bmatrix}$ is invertible. Furthermore, showing that A is invertible amounts to showing that $\det(A) = ad - bc \neq 0$. Thus in two cases a condition has been replaced by an equivalent condition.

2. *Slightly modified problems*. The next example reveals how effective use can be made of subgoals. A subgoal is a statement that lies "between" the hypothesis and conclusion of the given problem. If the original problem is an implication "If P then Q," then a subgoal is a statement R that the solver (1) hopes to derive from P and (2) intends to use to derive Q. We take an example from vector geometry.

Example 1.4.1 Find two unit vectors perpendicular to the vectors $\mathbf{v} = (1, 2, 3)$ and $\mathbf{w} = (1, -1, 4)$.

Solution Our goal is to find two vectors $\mathbf{u}_1$ and $\mathbf{u}_2$ such that $|\mathbf{u}_1| = |\mathbf{u}_2| = 1$ and $\mathbf{u}_1 \cdot \mathbf{v} = \mathbf{u}_2 \cdot \mathbf{v} = \mathbf{u}_1 \cdot \mathbf{w} = \mathbf{u}_2 \cdot \mathbf{w} = 0$. ($|\mathbf{v}|$ denotes the length of the vector $\mathbf{v}$. A unit vector is a vector $\mathbf{v}$ such that $|\mathbf{v}| = 1$.) A sketch of the situation should make it clear that the vectors $\mathbf{u}_1$ and $\mathbf{u}_2$ actually exist.

We aim for a somewhat less stringent goal. Subgoal (1): Find one unit vector perpendicular to $\mathbf{v}$ and $\mathbf{w}$. Thus we have created two problems from one:

(i) Find one unit vector perpendicular to $\mathbf{v}$ and $\mathbf{w}$.
(ii) Using this unit vector, find two unit vectors perpendicular to $\mathbf{v}$ and $\mathbf{w}$.

To solve (i) we again relax a requirement in order to obtain a subgoal. Subgoal (2): Find a nonzero vector perpendicular to $\mathbf{v}$ and $\mathbf{w}$.

From the original problem we have constructed a chain of three problems: Given vectors $\mathbf{v} = (1, 2, 3)$ and $\mathbf{w} = (2, -1, 4)$,

(a) find a nonzero vector perpendicular to $\mathbf{v}$ and $\mathbf{w}$,
(b) use the solution to (a) to find a unit vector perpendicular to $\mathbf{v}$ and $\mathbf{w}$,
(c) use the solution to (b) to find two unit vectors perpendicular to $\mathbf{v}$ and $\mathbf{w}$.

We leave as an exercise the task of solving problems (a)–(c). Notice that in order to create the subgoals, we relaxed one of the conditions required of the solution to the problem: We ask for one unit vector, not two; we look for a nonzero vector perpendicular to $\mathbf{v}$ and $\mathbf{w}$ rather than a unit vector perpendicular to $\mathbf{v}$ and $\mathbf{w}$. Subgoals are usually created in this way. The restoration of the conditions amounts to moving from the subgoal to the goal. Also observe that the listing of subgoals provides a manageable and organized way of attacking the problem. In effect an apparently long leap has been reduced to a sequence of short leaps.

3. *Broadly modified problems*. Thus far our examples of related problems have in each case been closely connected with the original problems.

We now look at examples in which the related problem is obtained by modifying the original problem significantly. These examples reveal two basic methods for creating broadly modified problems: analogy and generalization.

Analogy

Given a problem A, an analogous problem B has the same "structure" as A but takes place in a different setting. For example, A might be a problem in three-dimensional geometry involving lines and planes while B is a problem in two-dimensional geometry having to do with points and lines, or A might be a problem involving a function of three real variables while B is a problem involving a function of one real variable. We consider in detail an example from algebra.

Example 1.4.2 Show that if a, b, c, d are real numbers strictly between 0 and 1, then $(1 - a)(1 - b)(1 - c)(1 - d) > 1 - a - b - c - d$.

Solution Our first impulse might be to multiply out $(1 - a)(1 - b)(1 - c)(1 - d)$ and show that the resulting mess is greater than $1 - a - b - c - d$. However, the extent of the mess involved encourages us to seek another approach.

Let us look for a similar or analogous problem involving fewer variables. Consider the statement: If a and b are real numbers strictly between 0 and 1, then $(1 - a)(1 - b) > 1 - a - b$. This statement, while having the same form as the original, involves only two variables. Moreover it is easy to check: Since $a \cdot b > 0$,

$$(1 - a)(1 - b) = 1 - a - b + a \cdot b > 1 - a - b.$$

We now add back a variable to obtain yet another analogous statement: If a, b, c are real numbers strictly between 0 and 1, then

$$(1 - a)(1 - b)(1 - c) > 1 - a - b - c.$$

To prove this statement, perhaps we can use the result in the two-variable case considered above:

$$(1 - a)(1 - b) > 1 - a - b.$$

Multiplying both sides by $(1 - c)$, we have

$$\begin{aligned}(1 - a)(1 - b)(1 - c) &> (1 - a - b)(1 - c)\\ &= 1 - a - b - c + (a + b)c > 1 - a - b - c\end{aligned}$$

since $(a + b)c > 0$.

Now what about the original problem? As can be checked, this argument extends to provide a proof that $(1-a)(1-b)(1-c)(1-d) > 1-a-b-c-d$ if a, b, c, d are strictly between 0 and 1. In fact, a pattern is emerging from these cases. As an exercise, try to describe that pattern and prove your description of it correct.

We mention in passing that several features of the Polya–Schoenfeld scheme are illustrated in the previous example. In studying the given problem, we found two analogous problems involving fewer variables. (See Table 1.4.1, II.3(a).) In answering the original problem, we were able to use the results of the analogous problems and the method used to solve these problems. Also each of the analogous problems can be regarded as extreme cases of the original problem: $c = d = 0$ for the first and $d = 0$ for the second (Table 1.4.1, I.4(b)).

In many ways analogies are central to mathematics and mathematical thinking. Very often mathematicians find themselves working in a situation A that is reminiscent of a previously encountered situation B. In this case B can be used as a model for the investigation of A. For instance, the study of functions of two or three variables is usually carried out in analogy with functions of one variable. Many concepts and theorems about functions of several variables are motivated by analogous concepts and theorems about functions of one variable. We shall encounter more analogies later in this text. For now let us mention some important analogies in mathematics.

1. Functions of one variable:	Functions of several variables:
$y = f(x)$	$y = f(x_1, \ldots, x_n)$
2. Vectors in the plane:	Vectors in 3-space:
$\mathbf{v} = (x, y)$	$\mathbf{v} = (x, y, z)$
3. Sequences of real numbers:	Functions of one variable:
$a_1, a_2, \ldots, a_n, \ldots$	$y = f(x)$
4. Sums of real numbers:	Integrals of functions:
$a_1 + \cdots + a_n$	$y = \int_a^b f(x)dx$
5. Infinite series:	Improper integrals:
$\Sigma_1^\infty a_n$	$\int_1^\infty f(x)dx$

Generalization

A *generalization* of a given statement A is a statement B that includes A as a special case. In other words, the instances in which B applies include the instances in which A applies. Mathematicians are especially notorious for their tendency to generalize. One might reasonably question this practice. What are the advantages to generalization?

First, the process of generalizing often clarifies a problem. The essential features of a question are often obscured in a special case and are illuminated in the general situation. Details that seem important in a particular case are revealed to be less relevant when placed in a broader context. As a result a generalization of a statement is sometimes easier to handle than the original statement itself.

Second, through generalization we learn the limits of a given concept or method. By seeking to know the full extent to which a given statement is true, we develop a keener understanding of the concepts involved in it. If we can determine the instances in which a certain procedure applies, we can make maximum use of it while avoiding the pitfalls that arise from inappropriate use.

Finally, generalization can serve as a tool in research. Often a given idea can be generalized from a specific situation to a more encompassing setting in several ways. Each of these generalizations provides a starting point for further research and each such research effort creates a new body of mathematical knowledge. The process of generalization has the effect of extending and clarifying the concept that was initially considered.

To summarize in a few words, we generalize in order to know, to understand, to predict, and to create. The act of generalizing has one other notable benefit: It's fun.

Example 1.4.2 (revisited) In working out this example, we observed that a certain type of inequality is valid in three distinct cases. Are these three cases (involving two, three, or four variables) part of a more general pattern? Suppose we allow any fixed finite number of real numbers in the inequality: If n is a positive integer greater than 1 and $a_1, \ldots, a_n$ are real numbers strictly between 0 and 1, then $(1 - a_1)(1 - a_2) \cdots (1 - a_n) > 1 - a_1 - a_2 - \cdots - a_n$. Note that when $n = 4$, the statement in Example 1.4.2 is recaptured.

Is this generalization true for all $n \geq 2$? Indeed, it is for $n = 2, 3$, or 4. This finding suggests a proof by mathematical induction. For each integer $n \geq 2$, let $S(n)$ be the statement: If $a_1, \ldots, a_n$ are real numbers strictly between 0 and 1, then $(1 - a_1)(1 - a_2) \cdots (1 - a_n) > 1 - a_1 - a_2 - \cdots a_n$.

To prove that $S(n)$ holds for $n \geq 2$, we must show that

(i) $S(2)$ is true,
(ii) if $S(m)$ is true, then $S(m + 1)$ is true.

We have already checked that $S(2)$ is true, so we consider the inductive step. We assume that $S(m)$ holds for some m and prove that $S(m + 1)$ is valid.

We must prove that if $a_1, \ldots, a_{m+1}$ are real numbers between 0 and 1, then

$$(1 - a_1) \cdots (1 - a_{m+1}) > 1 - a_1 - \cdots - a_{m+1}.$$

From the inductive hypothesis we know that

$$(1 - a_1) \cdot \cdots \cdot (1 - a_m) > 1 - a_1 - \cdots - a_m.$$

Since $1 - a_{m+1} > 0$, we obtain from the latter inequality

$$\begin{aligned}(1 - a_1) \cdot \cdots \cdot (1 - a_m)(1 - a_{m+1}) &> (1 - a_1 - \cdots - a_m)(1 - a_{m+1}) \\ &= 1 - a_1 - \cdots - a_m - a_{m+1} \\ &\quad + (a_1 + \cdots + a_m) \cdot a_{m+1} \\ &> 1 - a_1 - \cdots - a_m - a_{m+1}.\end{aligned}$$

Thus the inductive step is established and $S(n)$ is true for all $n \geq 2$.

The following example comes from calculus.

Example 1.4.3 Recall the following theorem from one-variable calculus: If f is a differentiable function on an interval $[a, b] = \{x | a \leq x \leq b\}$ where a and b are fixed real numbers and if f achieves a relative maximum or a relative minimum at a point c where $a < c < b$, then $f'(c) = 0$. Problem: Generalize this statement to real-valued functions of two or more variables.

Let $f(x_1, \ldots, x_n)$ be a function of n variables. We seek a statement that relates the fact that f achieves a relative extremum (maximum or minimum) at a point to the vanishing of the "derivative" of f at that point. How can the notion of derivative be extended to functions of several variables? One method is that through partial derivatives. Specifically we take the differentiability of f to mean that the partial derivatives $\partial f/\partial x_1, \ldots, \partial f/\partial x_n$ all exist. But where do they exist? To be precise, we need to generalize the notion of interval to $\mathbf{R}^n$ and we have to define the concept of relative extremum of a function on $\mathbf{R}$. With these remarks in mind, we present a generalization of the one-variable theorem presented above.

Let R be a generalized rectangle in $\mathbf{R}^n$: $R = \{(x_1, \ldots, x_n) | a_1 \leq x_1 \leq b_1, \ldots, a_n \leq x_n \leq b_n\}$ where $a_1, \ldots, a_n, b_1, \ldots, b_n$ are fixed real numbers. Suppose f is a real-valued function on R whose partial derivatives $\partial f/\partial x_1, \ldots, \partial f/\partial x_n$ exist on R. If f has a relative maximum or relative minimum at a point c inside R, then $\partial f/\partial x_1(c) = 0, \ldots, \partial f/\partial x_n(c) = 0$.

Other generalizations are possible. For example, other kinds of regions can be considered. Also at least one other definition for the notion of differentiability is possible. Details can be found in textbooks on advanced calculus. In any case, we have achieved our goal of generalizing the original statement about a function of one variable to a statement about functions of several variables. Question: Is the generalization true?

Looking Back

Polya's admonition that we look back over our work has two thrusts. First, he encourages us to check our work by applying various tests to the solution—use of data, symmetry, and alternative derivations. Second, he points us toward future problems by having us try to derive new results from the result just established. In other words, looking back has both a backward and a forward orientation. In this section we concentrate on the first feature. However, we shall comment briefly upon the second and shall illustrate both the backward and forward nature of checks throughout the text.

Example 1.4.2 (revisited) In the inductive proof of the inequality

$$(1 - a_1)(1 - a_2) \cdots (1 - a_n) > 1 - a_1 - \cdots - a_n,$$

the fact that $0 < a_i < 1$ was used at a key point. A cruder check of the inequality follows from the observation that the left side, being a product of numbers between 0 and 1, is itself between 0 and 1, while the right side is clearly less than 1 and perhaps less than 0. Thus the inequality does conform to reasonable estimates. Finally, here is a challenge for readers who have studied probability theory: Check the result by giving a probabilistic interpretation of the inequality.

Example 1.4.4 The equality $1 + 3 + \cdots + (2n - 1) = n^2$ can be checked geometrically. Consider the squares

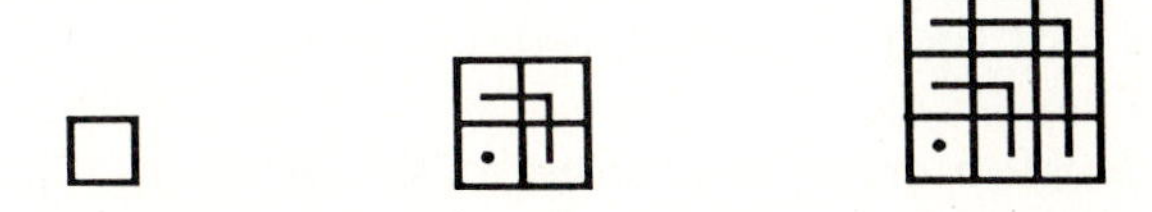

To obtain a 4×4 square from the 3×3 square, we give three squares to

the top of the 3×3 square, three squares to the right side, and one square to the upper right.

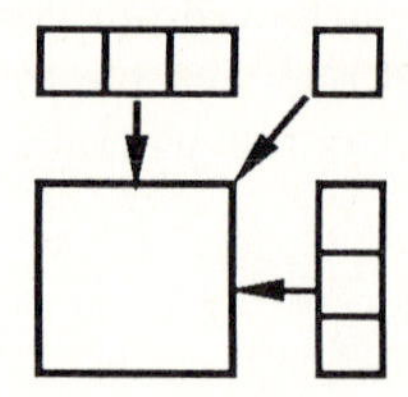

Thus $4^2 - 3^2 = 3 + 3 + 1 = 7$. In general the $(n + 1) \times (n + 1)$ square is obtained by adding $n + n + 1 = 2n + 1$ squares to the $n \times n$ square.

Notice that this check grows out of a different representation of the problem, namely a geometric representation. Thus the looking back step can also be considered a step in the understanding phase of the problem-solving process. Indeed, the distinction between the two phases can be blurry. The realization that a given maneuver can be both an attempt to understand a problem and a device to check a solution underlies Polya's suggestion that looking back entails looking ahead.

Overcoming Blocks

In this section and the preceding section, we have discussed the structure of mathematical proofs and some techniques for solving problems. In the process we have become aware of the general form of the arguments used to prove theorems and the approaches that can be used to devise a proof of a theorem or a solution to a problem. Nonetheless, all of us, no matter how proficient we are at employing problem-solving strategies or knowing proof methods, will occasionally become *stuck* on a problem. We meet what appears to be an insurmountable roadblock and we have no idea how to proceed. What do we do in these instances?

Perhaps the most useful general advice that can be offered is: STAY ACTIVE. Rather than waiting quietly for inspiration to strike, engage in steps that can help bring about a solution. Specifically, how does this advice translate into action? We offer a condensed list of suggestions, many of which have already been mentioned in this section.

1. Read the problem or theorem several times. Rephrase the problem in alternative ways; for example, use quantifiers and connectives; use different but equivalent notation; or use different but equivalent definitions of the terms involved.

2. Play with the problem by drawing pictures (or forming other representations), by checking out examples, extreme cases, and other special cases, and by generalizing in various ways.
3. As a result of this active investigation, try to develop a problem-solving or proof strategy.
4. Execute the chosen strategy and verify the result.

The moral is simple: Keep working, keep trying new things.

While you are actively working on a problem, it is wise to monitor your own activities. For instance, suppose you find an approach that seems promising. But suppose that after trying this approach for a while, you make no further progress. Then you should be able to recognize your situation and be willing to drop your current approach in favor of some other. As another example, suppose you think of two possible ways of solving your problem. Before embarking on either, try to evaluate the relative advantages of the two methods. Perhaps one method will require significantly less work than the other and hence is preferable for this reason. As these remarks indicate, a diligent problem solver should be aware of more than the proof and problem-solving methods discussed in this section and the previous section. Successful problem solving also requires an awareness of one's actions, the flexibility to change actions, and the judgment to choose appropriate methods.

One simple way to monitor and control your actions while working on a problem is to ask yourself repeatedly the following three questions:

1. What am I trying to do?
2. What am I doing?
3. Will what I am doing help me do what I want to do?

If you can answer questions 1 and 2 convincingly and can answer number 3 with a clear Yes, then proceed. Otherwise, stop and take stock of your situation.

A convincing answer to question 1 occurs only when the problem solver has a clear understanding of the problem. A convincing answer to question 2 occurs when the problem solver has a clear understanding of his or her own actions. An affirmative answer to question 3 results only when the actions of the problem solver are relevant to the goals of the problem.

If you cannot give an appropriate answer to any one of these questions, then concentrate on that question. Go back to the Polya–Schoenfeld scheme or to the advice given previously in this subsection. By actively and carefully searching for clues and insights, you greatly increase your ability to devise a successful solution strategy.

For more advice on overcoming blocks, consult *Thinking Mathematically* by J. Mason, L. Burton, and K. Stacey and/or *Conceptual Blockbusting* by J. Adams.

Summary

In this section we have presented a quick introduction to the Polya–Schoenfeld problem-solving scheme. Our goal has been to describe some tools that might help lead you to the solution of a difficult problem or the discovery of an elusive proof. We have offered only a brief tour into the vast domain of problem-solving techniques and heuristic thinking in mathematics. Our hope is that from this introduction and through repeated reference to the Polya–Schoenfeld scheme in the text and exercises of this book, you will become conversant with the heuristic strategies and will become increasingly proficient at using them. In other words, even though our presentation of problem-solving tools has been sketchy, it should at least get you started on them.

Of course, for these tools to be of value, you must be in the habit of reaching for them. As a beginning step, you might wish to use Table 1.4.1 in a very systematic way . When considering a problem, simply move through the table linearly using those steps that apply to the problem. Naturally, for some problems certain steps will not apply; we can be certain, however, that a clear understanding of the problem, which includes knowledge of the unknown and the given and an adequate representation, is absolutely necessary. As you become more adept at applying the individual steps, you may discover that the order in which the steps are taken will vary with the problem. In other words, the scheme possesses great flexibility. For example, you can use the very last step (IV.2(d)) in the analysis phase to help get a handle on the problem: Take the result of the problem as given and try to derive something known from it. If successful, then perhaps you can reverse all your steps, thereby deriving a solution to the problem. In general, you will probably find that the various checks suggested under phase IV can be applied during phases I, II, and III. Also, in the course of solving a problem, you might use a given step several times. Thus the problem-solving scheme has the virtue of being precise enough to be applicable at specific points in the problem-solving process, yet is flexible enough to be used in various forms and contexts.

Exercises §1.4

In Exercises 1.4.1–1.4.7, identify the unknown. If appropriate, describe the unknown in two or more different ways, perhaps based on two or more distinct representations of the problem.

1.4.1. Among all pairs of positive numbers whose sum is 10, which pair has the largest product?

1.4.2. Among all pairs of positive numbers whose product is 10, which pair has the smallest sum?

1.4.3. Find all points at which the tangent line to the graph of $y = x^3 - x^2 - 2x + 1$ is horizontal.

1.4.4. Find all vectors in 3-space that are perpendicular to the vector $(1, 2, 3)$.

1.4.5. Find the point(s) of intersection to the circle $x^2 + 2x + y^2 = 1$ and the parabola $y = x^2 + x + 1$.

1.4.6. Find the area under the curve $y = \sin x$ (and above the x-axis) for $0 \leq x \leq \pi$.

1.4.7. Evaluate $\Sigma_{n=1}^{\infty} 1/n(n+1)$.

In Exercises 1.4.8–1.4.15, develop an appropriate representation of the problem. The representation might be geometric or algebraic in nature. It should involve some suitable notation.

1.4.8. Let R be the region bounded by the x-axis below and the parabola $y = 4 - x^2$ above. Among all rectangles contained in R with base on the x-axis, find the one with maximum area.

1.4.9. (a) Find all vectors in $\mathbf{R}^3$ perpendicular to both $(1, 3, -1)$ and $(0, 1, 2)$.

(b) Find all vectors in $\mathbf{R}^4$ perpendicular to both $(1, 2, -1, 2)$ and $(0, 1, 2, 3)$.

1.4.10. Describe two or more ways of representing the following problem, known as the Cannibals-and-Missionaries Problem: Three cannibals and three missionaries are together on one side of a river. They wish to cross the river, and they have a boat that can carry two people. Describe a procedure for transporting all six people across the river in such a way that there are never more cannibals than missionaries on any one side at a given time.

1.4.11. Prove that in any group of six people there are either three mutual friends or three mutual strangers.

1.4.12. Prove $\Sigma_{n=1}^{\infty} 1/2^n = 1$.

1.4.13. If f and g are functions and $f(x) \geq g(x)$ for all x between a and b, then $\int_a^b f(x)\,dx \geq \int_a^b g(x)\,dx$.

1.4.14. If $0 \leq a_n \leq b_n$ for $n = 1, 2, 3, \ldots,$ and the infinite series $\Sigma_1^{\infty} b_n$ converges, then $\Sigma_1^{\infty} a_n$ converges.

1.4.15. The diagonals of a parallelogram bisect each other.

1.4.16. List several different ways of representing functions whose domain and range are part of the set of real numbers.

1.4.17. (a) Describe two ways of representing a rational number.

(b) Describe two ways of distinguishing a rational number from an irrational number.

Related Problems

In Exercises 1.4.18–1.4.24, find a related problem. The related problem can be an analogy, a generalization, or a special case of the given problem.

1.4.18. The only solution x, y, z, w to the linear system

$$x + 2y + w = 0$$

$$x - z + 2w = 0$$

$$y + z + 3w = 0$$

$$x + y + 2w = 0$$

is $x = y = z = w = 0$.

1.4.19. Find a unit vector $\mathbf{u}$ in the plane which makes an angle of $\pi/6$ with $(2, 3)$.

1.4.20. Prove that the infinite series $\Sigma_{n=1}^{\infty} 1/(n^2 + 3)$ converges.

1.4.21. The area of the ellipse $x^2/a^2 + y^2/b^2 = 1$ is πab.

1.4.22. For all real α, $\cos(\alpha/2) \cdot \cos(\alpha/4) \cdot \cos(\alpha/8) = (1/8)(\sin(\alpha)/\sin(\alpha/8))$.

1.4.23. (a) Evaluate $\Sigma_{n=1}^{\infty} 1/n(n + 1)$.

(b) Evaluate $\Sigma_{n=1}^{\infty} n/(n + 1)!$.

1.4.24. Among all rectangular solids of surface area 20 sq. units, which has the largest volume?

In each of Exercises 1.4.25–1.4.31, give a reasonable generalization of the given statement.

1.4.25. Among all rectangles of perimeter 12 the square of side 3 has the largest area.

1.4.26. Among all pentagons (5-sided polygon) with a given perimeter, the regular pentagon (the one whose sides are of equal length) has the largest area.

1.4.27. Suppose 3000 points in the plane are given having the property that no three of the points are collinear. Prove that there exist 1000 disjoint triangles having the given points as vertices.

1.4.28. Any set of three vectors in the plane $\mathbf{R}^2$ is linearly dependent.

1.4.29. Let $A(r)$ and $C(r)$ be the area and circumference, respectively, of a circle of radius r. Then $dA/dr = C(r)$.

1.4.30. Notice that $1 \cdot 2 \cdot 3 \cdot 4 = 5^2 - 1$, $2 \cdot 3 \cdot 4 \cdot 5 = 11^2 - 1$. What is the general pattern?

Notes

As a recognized intellectual discipline, logic traces its origins back to the Greek philosopher Aristotle. Aristotle placed special emphasis on two laws, the law of contradiction (P is not both true and false) and the law of the excluded middle (P is either true or false). In the nineteenth century, contributions to logic were made by Augustus DeMorgan, known primarily for DeMorgan's Laws, George Boole, who emphasized the symbolic aspects of logic, and Charles S. Peirce, who contributed to the study of propositional and predicate calculus. At the turn of the twentieth century, Gottlob Frege, Bertrand Russell, and A. N. Whitehead devoted considerable thought and energy to the place of logic in the development of mathematics. In their monumental work *Principia Mathematica* Russell and Whitehead attempted to build all of mathematics on a foundation of axiomatic logic. Their work was heavily criticized and never completed. For an interesting account of the development of logic and the foundations of mathematics, see Chapter 51 of Kline's history [9] or Chapters VII–XI of [10]. An extensive treatment of propositional and predicate calculus is found in Hamilton [6] and Stoll [18].

The axiomatic approach to the investigation of mathematical structures began to take hold near the end of the nineteenth century when several areas of mathematics were axiomatized independently by various mathematicians. The pressure to develop axiomatic treatments came primarily from within mathematics and was due both to an increase in the variety of structures considered by mathematicians and to deeper investigations of the real number system and functions of a real variable. For example, with the creation of non-Euclidean geometry by Lobatchevsky and Bolyai, the study of various geometric structures was initiated. Due to the efforts of Pasch and Hilbert, a rigorous, axiomatic approach to geometry was accepted by mathematicians. Because of the sophistication and complexity of the work in functions of a real variable (especially with respect to Fourier series), the value of an axiomatic treatment of the real number system became apparent. This axiomatization was carried out by several people including Dedekind, Peano, and Hilbert. The short article by Kennedy [7] contains an interesting discussion of this development. The Hilbert quote is found in Reid's biography [16]. A recently discovered manuscript of Herman Weyl, one of the masters of twentieth-century mathematics and physics, contains many illuminating comments on the role of axiomatic reasoning in mathematics [19].

The role of proof and rigor in the teaching of mathematics is a complex and controversial issue. In fact, the attitudes of professional mathematicians about standards of rigor are different today than from earlier eras. Grabiner documents these changing attitudes in [5].

Polya's first book on problem solving, *How to Solve It* [13], appeared in 1945. Subsequent books by Polya include *Mathematics and Plausible Reasoning* (2 vols.) [15] and *Mathematical Discovery* (2 vols.) [14]. Because of Polya's considerable influence, problem solving has become a primary focus of mathematics educators and mathematicians interested in the teaching of mathematics. Starting in the late 1950s, research in problem solving received a push from an entirely different source, the emerging discipline of artificial intelligence (AI). The definitive treatise in this field is Newell and Simon's *Human Problem Solving* [12]. In [11] Newell, himself a student of Polya and one of the founders of AI, gives an interesting account of Polya's influence (or lack thereof) in AI. Much of Schoenfeld's research on problem solving through 1984 is described in his monograph *Mathematical Problem Solving* [17]. The book by Mason et al. [10] mentioned in the text is an excellent source of valuable advice and intriguing problems. It is highly recommended for any reader of this text.

Several general references about mathematics are accessible to readers of this text. One of the best is *The Mathematical Experience* by Davis and Hersh [4] in which many philosophical and pedagogical issues in mathematics are discussed. For an overview of the disciplines of mathematics, consult either the three-volume set edited by Alexandrov et al. [1] or the three-volume set edited by Behnke et al. [3].

Bibliography

1. A. Alexandrov et al. (eds.), *Mathematics, Its Contents, Method and Meaning* (3 vols.), MIT Press, 1969.
2. E. G. Begle (ed.), *The Role of Axiomatics and Problem Solving in Mathematics*, The Conference Board of the Mathematical Sciences, Ginn & Co., 1969.
3. H. Behnke, F. Bachmann, H. Kunle, and K. Fladt (eds.), *The Fundamentals of Mathematics* (3 vols.), MIT Press, 1972.
4. P. Davis and R. Hersh, *The Mathematical Experience*, Birkhauser, 1982.
5. J. Grabiner, "Is mathematical truth time-dependent?" *Amer. Math. Monthly* 81 (1974), 354–365.
6. A. G. Hamilton, *Logic for Mathematicians*, Cambridge University Press, 1978.
7. H. Kennedy, "The origins of modern axiomatics: Pasch to Peano," *Amer. Math. Monthly* 79 (1972), 133–136.
8. M. Kline, *Mathematics: The Loss of Certainty*, Oxford University Press, 1980.
9. M. Kline, *Mathematical Thought from Ancient to Modern Times*, Oxford University Press, 1972.
10. J. Mason, L. Burton, and K. Stacey, *Thinking Mathematically*, Addison-Wesley, 1985.
11. A. Newell, "The heuristic of George Polya and its relation to artificial intelligence," in *Methods of Heuristics*, R. Groner, M. Groner, W. Bischof, and L. Erlbaum, eds., 1983.
12. A. Newell and H. Simon, *Human Problem Solving*, Prentice-Hall, 1972.
13. G. Polya, *How to Solve It*, Princeton University Press, 1945.
14. G. Polya, *Mathematical Discovery* (2 vols.), Wiley, 1962 (Vol. I), 1965 (Vol. II).
15. G. Polya, *Mathematics and Plausible Reasoning* (2 vols.), Princeton University Press, 1954 (Vol. I), 1968 (Vol. II).
16. C. Reid, *Hilbert*, Springer-Verlag, 1970.
17. A. Schoenfeld, *Mathematical Problem Solving*, Academic Press, 1985.
18. R. Stoll, *Sets, Logic, and Axiomatic Theories*, Freeman, 1961.
19. H. Weyl, "Axiomatic versus constructive procedures in mathematics," *Mathematical Intelligencer*, 7, #4 (1985), 10–17, 38, edited by T. Tonietti.
20. R. L. Wilder, *The Foundations of Mathematics*, Wiley, 1952.

Miscellaneous Exercises

1. For which values of $n \in \mathbf{N}^+ = \{1, 2, \dots\}$ are each of the following statements true?
   (a) $S(n)$: $n \leq 100{,}000$.
   (b) $S(n)$: $n = n + 1$.
   (c) $S(n)$: $n \geq 1$.
   (d) $S(n)$: $n = 1$.
   (e) $S(n)$: For all a, b in $\mathbf{R}$, $(a + b)^n = a^n + b^n$.

2. A $k \times k$ chessboard is called *defective* if exactly one square has been removed from the board. A *triomino* is an L-shaped tile that covers exactly three squares on a chessboard.

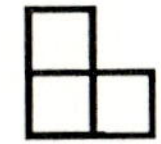

Prove that for each positive integer n, any $2^n \times 2^n$ defective chessboard can be tiled with triominoes.

3. Investigate the following question: Which positive integers can be represented as a sum of any number of consecutive positive integers? For example, $3 = 2 + 1$ can be represented as such a sum but 2 apparently cannot be.

4. Investigate the following question: How many squares are there on an $n \times n$ chessboard? For example, the 2×2 board contains 5 squares.

5. At a recent party attended by four couples, some people shook hands upon arrival. Of course, no one shook his or her own hand or the hand of his or her spouse. Mr. Edwards, being a curious person, asked each of the other seven people how many hands did he or she shake and received seven *different* answers. How many hands did Mrs. Edwards shake? Consider analogous problems. Make a general conjecture. Prove your conjecture.

6. Let a, b, and n be positive integers such that $ab = n$. Prove: $a \leq \sqrt{n}$ or $b \leq \sqrt{n}$.

7. For each positive integer n let $d(n)$ denote the number of positive divisors of n.
   (a) Calculate $d(n)$ for $1 \leq n \leq 10$.
   (b) For which natural numbers n does $d(n) = 2$?
   (c) For which natural numbers n does $d(n) = 3$?
   (d) For which natural numbers n does $d(n) = 4$?
   (e) After conducting other experiments and investigation, record any other observations that you have made about $d(n)$.

8. At Polya High there are 1000 students and 1000 lockers (numbered 1–1000). At the start of the day, all the lockers are closed; then the first student comes by and opens every locker. Following the first student, the second student goes along and closes every second locker starting with locker number 2. The third student changes the state (if the locker is open, he closes it; if the locker is closed, he opens it) of every third locker starting with locker number 3. The fourth student changes the state of every fourth locker, etc., starting with locker number 4. Finally, the thousandth student changes the state of the thousandth locker. When the last student changes the state of the last locker, which lockers are open?

9. Suppose a and b are positive integers with $a \geq b$. Suppose the rules of basketball are modified to allow a points for a field goal and b points for a free

throw. Investigate the following question: What are the possible scores attainable by a given team? (Assume no time limit on the length of a game.) *Suggestions*: Try special cases, e.g., $a = 3$, $b = 2$. Are there cases in which infinitely many scores are not attainable? Here is a somewhat related question: Which scores are attainable in American professional football? The scoring in pro football is touchdown, 6 points; extra point, 1 point; field goal, 3 points; safety, 2 points.

10. A *palindrome* is a natural number that is equal to the number formed by reversing its digits. For example, 121 is a palindrome, while 112 is not. Investigate the following assertion: Every four-digit palindrome is divisible by 11. Generalize your conclusion.

11. Do there exist subcollections A and B of $\mathbf{R}$ which are bounded above, which have the same least upper bound, and which have no numbers in common? If so, give an example of two such sets; if not, show why not.

12. Let f be a function on the collection $\mathbf{R}^+$ of positive real numbers such that $f(x) > 0$ for all x in $\mathbf{R}^+$ and

$$f(x+y) = \frac{f(x) \cdot f(y)}{f(x) + f(y)}$$

for all positive x and y.
(a) Find $f(2 \cdot x)$ and $f(3 \cdot x)$ in terms of $f(x)$.
(b) Conjecture a formula for $f(r \cdot x)$ which is valid for all positive rational numbers r and all positive real x.
(c) Prove your conjecture by a bootstrap proof.
(d) Find a function f that has the properties listed above.

13. For any real numbers x and y, the *maximum* of x and y, denoted $\max(x, y)$, is defined by

$$\max(x, y) = \begin{cases} x & \text{if } x \geq y \\ y & \text{if } x < y. \end{cases}$$

(a) Give a similar definition for the *minimum* of x and y.
(b) Show that for all real x and y, $\max(x,y) = -\min(-x,-y)$.
(c) Show that for all real x, y, and z, $\max(x,\min(y,z)) = \min(\max(x,y), \max(x,z))$.

CHAPTER TWO

Set Theory

As a discipline within mathematics, the theory of sets had its origins in the study of functions carried out during the latter half of the nineteenth century. Since then, set theory has developed into a full-blown area of specialization inside mathematics, possessing interesting and significant problems of its own, and having connections with other parts of mathematics and with related fields such as theoretical computer science and mathematical economics.

Our interest in set theory, however, stems from a different source. Over the course of the twentieth century, set theory has come to provide the basic language of mathematics. For example, many mathematicians maintain that any mathematical entity should be describable as a set. As a result, anyone wishing to study modern mathematics must become conversant with the vocabulary and concepts of set theory. Thus we study set theory not as an end in itself, but as a vehicle to carry us into the realm of modern mathematics.

In this chapter we present an informal introduction to the theory of sets. We aim only for a basic understanding of the workings of set theory, enough to allow for the subsequent use of set theory in courses such as algebra, analysis, formal language theory, geometry, and topology. Thus we emphasize a casual, intuitive development of set theory instead of a more rigorous, axiomatic treatment. By taking this tack, we do not mean to minimize the importance of an axiomatic approach to set theory; nor do we mean to suggest that the study of postulate-based development should be avoided. As indicated in Chapter 1, an axiomatic treatment of a mathematical subject establishes its logical coherence, clarifies the subject's nature, and paves the way for further research. Since we are interested in set theory primarily as a language for mathematics, we need not insist on a formal development. For the sake of completeness and for interested readers, we do discuss briefly an axiomatic approach to set theory known as the Zermelo–Fraenkel axiom system. This discussion is included at the end of each section and can be omitted by readers interested in acquiring quickly a working knowledge of set theory.

An important theme of this chapter is that of binary operations. This theme underlines the analogy between the manipulations of sets and the combining of real numbers through addition and multiplication. By presenting binary operations at this stage, we also help pave the way for the study of algebraic structures in Chapter 6.

As in Chapter 1 we use the symbols **N**, **Z**, **Q**, and **R** to represent respectively the sets of natural numbers, integers, rational numbers, and real numbers. Also we continue to use the words "prove" and "show" interchangeably.

§2.1 Sets

A *set* is a collection of objects. Examples of sets are abundant: a set (or flock) of birds, a set (or pride) of lions, a set (or herd) of cows. In everyday life we deal with all sorts of sets of all sorts of things. In their daily work mathematicians toil with sets of numbers (e.g., real or rational), sets of functions (e.g., continuous or differentiable), sets of geometric objects (e.g., curves or surfaces), or other sets of mathematical entities.

If A is a set and if x is in the set A, then we write $x \in A$ and say that x is *an element of* A or x *is a member of* A. Thus, corn is an element of the set of vegetable crops grown in the United States. If x is not in A, then we write $x \notin A$.

How can a set be specified? One method is simply to list the elements of the set. For instance, the numbers 2, 3, 4, 5 constitute the set of integers between two and five inclusive. In the usual notation reserved for sets, this set is written $\{2, 3, 4, 5\}$. The braces $\{ , \}$ are used to delineate the set.

It stands to reason that one can list all the elements of a set only when the set is rather small. In general, to describe a set, one states a property or condition that determines the elements that belong to the set.

Let P be a well-defined property of objects, mathematical or otherwise. For instance, P may be the property that a real number is positive or that a person is an American citizen. Then

$$\{x | P(x)\}$$

denotes the set of all elements having property P.

Example 2.1.1 (i) $A = \{x | x \text{ is a positive real number}\}$,
(ii) $B = \{x | x \text{ is a positive rational number}\}$, and
(iii) $C = \{x | x \text{ is a positive integer}\}$ are all examples of sets of numbers defined by properties.

One might believe that any property P defines a set. In fact, it is tempting to state:

> For any property P, there is a unique set consisting of all the elements having property P.

Much of the time, no harm arises from applying this principle. Nonetheless, this assertion cannot be taken as a basic rule or axiom of set theory, as was pointed out by the philosopher Bertrand Russell. The famous Russell paradox illustrates the danger of adopting this statement as an operating principle. We describe Russell's paradox and its importance briefly. Details are left for Exercise 2.1.11.

Define

$$S = \{A \mid A \text{ is a set and } A \notin A\}.$$

Thus S consists of all those sets that do not contain themselves as a member. Most sets do not contain themselves as a member; an example of one that does is the set of abstract ideas. If we accept the principle that every well-defined property defines a set, then we must accept S as a set. Russell's paradox arises from an attempt to answer the following question: Is S a member of itself?

One can show that answer to this question cannot be Yes and cannot be No. This rather unsettling situation leads to the following modification of the principle given above:

> For any set X and any property P of objects, there is a set whose elements are precisely those elements of X having property P.

In other words, there exists a set Y such that

$$Y = \{x \in X \mid P(x)\} = \{x \in X \mid x \text{ has property } P\}.$$

One of the most important sets in all of mathematics is the *empty set*, customarily denoted by the symbol $\varnothing$. By definition $\varnothing$ is the set having no elements. In terms of a defining property, we have the following description of $\varnothing$: Let X be any set. Then

$$\varnothing = \{x \in X \mid x \neq x\}.$$

The point is that since nothing is distinct from itself, the set $\varnothing$ contains no elements. $\varnothing$ is also referred to as the *null set*.

Very often, we consider sets with just one element. Some examples are $\{1\}$, $\{\text{Don}\}$, $\{\varnothing\}$. These sets are referred to respectively as "singleton 1," "singleton Don," "singleton $\varnothing$."

We summarize the ideas presented thus far and introduce the notion of a finite set in the following informal definition. Readers interested in a more rigorous development of set theory can consult the appendices to the sections of this chapter and the references listed at the end of the chapter.

Definition 2.1.1 *Sets*

(i) A *set* is a collection of objects.
(ii) If x is in a set A, then we say that *x is an element of A* and write $x \in A$.
(iii) The *empty set*, $\varnothing$, is the set with no elements.
(iv) *Singleton* x denotes the set $\{x\}$ whose only member is x.

(v) A set is called *finite* if it is empty or its elements can be matched up precisely with elements of the set $\{1, 2, \ldots, n\}$ where n is some positive integer. A set that is not finite is called *infinite*.

For a finite set X, the integer n appearing in the definition is called the number of elements of X. In this case we write $X = \{x_1, x_2, \ldots, x_n\}$. Thus the set of lowercase letters of the English alphabet is a finite set with 26 elements. The set of positive integers is an infinite set.

Now we consider relationships that can exist between sets. The first concerns equality of sets; the second is the notion of containment.

Definition 2.1.2 *Set Equality*

Two sets A and B are *equal*, written $A = B$, when they have precisely the same elements. We write $A \neq B$ if A and B are not equal.

In other words, two sets A and B are equal when each element of A is an element of B and each element of B is an element of A. Symbolically, $A = B$ if and only if for each x, $x \in A$ if and only if $x \in B$, or $A = B \Leftrightarrow \forall x(x \in A \Leftrightarrow x \in B)$.

Definition 2.1.3 *Set Containment*

A set A is *contained in* a set B, written $A \subseteq B$, if each element of A is an element of B. If $A \subseteq B$, then we also say that A *is a subset of* B. We call A a proper subset of B if $A \subseteq B$ and $A \neq B$; in this case we write $A \subset B$. Finally, we write $A \not\subset B$ if A is not a subset of B.

Example 2.1.2 $\{1, 2\} = \{2, 1\}$.

Example 2.1.3 For A, B, and C as defined in Example 2.1.1, $C \subset B$ and $B \subset A$.

Example 2.1.4 $\{x | x$ is a person living in the continental United States$\} \subset \{x | x$ is a person living in the Western Hemisphere$\}$.

Let A and B denote sets. How does one show that $A \subseteq B$? According to Definition 2.3, we must demonstrate that each element of A is in B: For each x, if $x \in A$, then $x \in B$. In terms of defining properties, if $A = \{x | P(x)\}$ and $B = \{x | Q(x)\}$, then to show that $A \subseteq B$ we must argue that for each x the condition $P(x)$ implies the condition $Q(x)$. This type of proof, where we take an arbitrary $x \in A$ and show that $x \in B$, is called an *element-chasing proof*. Several examples of element-chasing proofs occur in this chapter. On the other hand, to show that $A \not\subset B$, we must show that there exists an element that is in A and is not in B; this requires us to find or exhibit an element x such that $x \in A$ and $x \notin B$.

Example 2.1.5 Let $A = \{1\}$. The only subsets of A are $\varnothing$ and A. (See Theorem 2.1.1, which follows.)

Example 2.1.6 Let $A = \{\varnothing\}$ and $B = \{\varnothing, \{\varnothing\}\}$. We claim $A \subseteq B$, for if $x \in A$, then $x = \varnothing$ (note that A is not empty; it has only one element and that element is the empty set. Then $\varnothing = x \in B = \{\varnothing, \{\varnothing\}\}$. Therefore, $A \subseteq B$. Notice that the elements of the sets A and B are themselves sets. Also observe that $A = \{\varnothing\} \in B$. As it happens, many interesting sets in mathematics have other sets as their elements.

Our first theorem captures some basic facts about containment. Its proof provides our first example of element-chasing arguments.

Theorem 2.1.1 (i) *For any set A, $\varnothing \subseteq A$.*
(ii) *For any set A, $A \subseteq A$.*
(iii) *If A, B, and C are sets with $A \subseteq B$ and $B \subseteq C$, then $A \subseteq C$.*

Proof (i) We must demonstrate the truth of the statement: For each x if $x \in \varnothing$, then $x \in A$. This conditional statement has a false antecedent and hence is true. Therefore $\varnothing \subseteq A$.

(ii) We must show: For each x, if $x \in A$ then $x \in A$. But this conditional is tautology. Thus $A \subseteq A$.

(iii) We must check the following statement: For each x, if $x \in A$ then $x \in C$. By assumption, $A \subseteq B$; thus for each x, if $x \in A$ then $x \in B$. However, since $B \subseteq C$ (by assumption) and since $x \in B$, it follows that $x \in C$. Therefore, for each x, if $x \in A$ then $x \in C$. Hence $A \subseteq C$ whenever $A \subseteq B$ and $B \subseteq C$. ■

According to Definition 2.1.2, two sets are equal when they have precisely the same elements. From this definition, the next theorem, whose proof is left as an exercise, follows.

Theorem 2.1.2 *Let A and B be sets. Then $A = B$ if and only if $A \subseteq B$ and $B \subseteq A$.*

To illustrate this theorem, we dig up an exercise from high school algebra.

Example 2.1.7 Let us consider the solution set of the equality $x^2 - 2x - 4 = x - 6$. Let A denote this set:

$$A = \{x | x^2 - 2x - 4 = x - 6\}.$$

What is A?

Let us suppose $x \in A$. What can we say about x? Now if $x \in A$, then

$$x^2 - 2x - 4 = x - 6$$

or

$$x^2 - 3x + 2 = 0$$

or

$$(x - 1)(x - 2) = 0.$$

which means that $x - 1 = 0$ or $x - 2 = 0$, which in turn implies that $x = 1$ or $x = 2$. Therefore if $x \in A$, then $x = 1$ or $x = 2$, or, equivalently, $A \subseteq \{1, 2\}$.

It is now a simple matter to check that if $x \in \{1, 2\}$, then $x \in A$, i.e., that 1 and 2 are solutions of the equality $x^2 - 2x - 4 = x - 6$. Thus $\{1, 2\} \subseteq A$.

Together, these remarks show that $A = \{1, 2\}$.

Representations of Sets

We continue with a discussion of two standard methods of representing subsets of a set. The first, pictorial in nature, is applicable to an arbitrary situation. The second, being digital in nature, is primarily used to represent subsets of a finite set.

Venn Diagrams

Suppose A is a subset of a set X. Then A and X are represented as geometric figures in the plane as shown in Figure 2.1.1. The exact shapes of

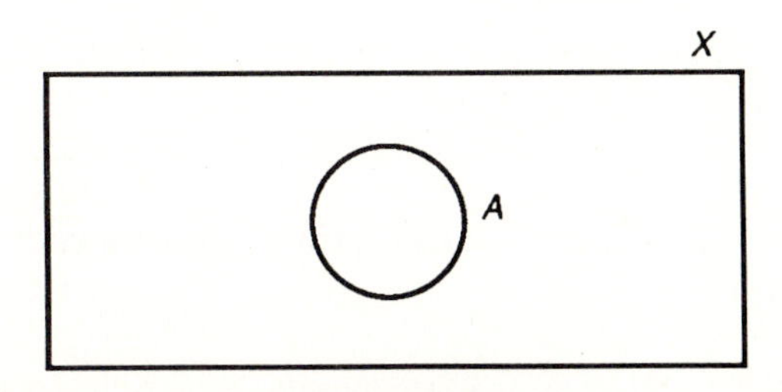

Figure 2.1.1

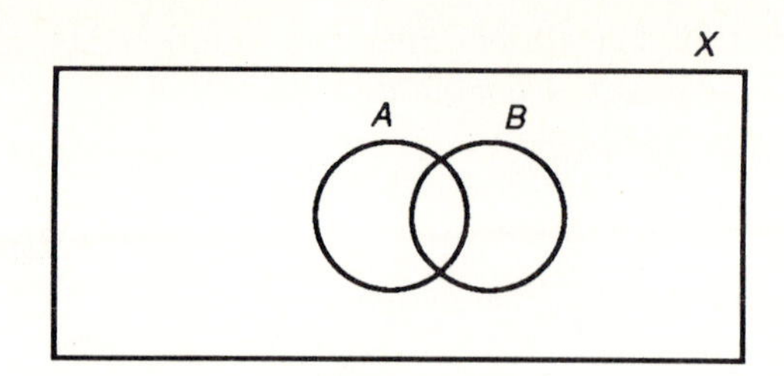

Figure 2.1.2

A and X are irrelevant; what we have is a visual representation of the general phenomenon of a subset of a set.

If B is another subset of X, then one might draw the diagram shown in Figure 2.1.2, at least under the assumption that A and B have elements in common. This geometric representation of subsets of a set provides a valuable aid to our intuition, assisting us in our investigation of properties of subsets and operations on subsets.

These visual representations of subsets are called *Venn diagrams*. We shall explore the connections between Venn diagrams and properties of sets and set operations in the next section.

Binary Sequences

Now we shall move to the digital representation of subsets of a finite set. Suppose n is a positive integer and let $X = \{x_1, \ldots, x_n\}$ be a finite set with n elements. Let A be a subset of X. The idea is to use a sequence of zeros and ones to keep track of the elements of A. Specifically form a sequence of length n, each term of which is either 0 or 1:

Step 1. If $x_1 \in A$, then the first term of the sequence is 1; if $x_1 \notin A$, then the first term is 0.

Step 2. If $x_2 \in A$, then the second term of the sequence is 1; if $x_2 \notin A$, then the second term is 0.

⋮

Continue in this fashion, until

⋮

Step *n* If $x_n \in A$, then the nth term of the sequence is 1; if $x_n \notin A$, then the nth term is 0.

Thus each subset $A \subseteq X$ corresponds to a sequence $a_1, a_2, \ldots, a_n$ where each entry a_i of the sequence is either 0 or 1. It is customary and causes no ambiguity to suppress the commas and to write the terms consecutively without spaces between them; hence, A corresponds to $a_1 a_2 \cdots a_n$. An example will illustrate:

Example 2.1.8 Let $X = \{1, 2, 3, 5\}$. If $A = \{1, 5\}$, then the sequence corresponding to A is 1001. The sequence corresponding to the empty set, $\varnothing$, is 0000; the sequence determined by X itself is 1111. Question: What subset is assigned to the sequence 1101? Note that all sequences considered in this example have four digits since the set X has four elements.

In general, a sequence of length n, $a_1 \cdots a_n$, where each a_i is either 0 or 1 is called a *binary string of length n*. Therefore, each subset of a set with n elements can be represented as a binary string of length n. (In general a *string* is a sequence of symbols chosen from a fixed finite set called the *alphabet*; the word "binary" refers to the fact that in this case the alphabet has only two elements, 0 and 1.)

The binary representation of subsets of a finite set is extremely useful. For example, it facilitates the manipulation of sets on a digital computer. Also, many set operations can be analyzed from this viewpoint. Finally, as the exercises at the end of the section will show, many counting problems having to do with finite sets can be approached successfully using the binary representation.

Axiomatic Set Theory

Set theory began to emerge as a distinct discipline within mathematics during the last quarter of the nineteenth century when important problems in the theory of functions of one real variable led to the consideration of properties of sets of real numbers. In the forefront of research on set theory was Georg Cantor, who established several important and surprising theorems about sets in general and sets of real numbers in particular. Cantor proved his stunning results (some of which will be described in Chapter 4) in spite of his taking an informal approach to set theory.

Cantor's work, however, met with severe criticism. Many mathematicians were unhappy with his methods for proving results about infinite sets, particularly with his use of proof by contradiction. At this same time some disturbing paradoxes began to appear in set theory, the culmination coming in 1901 with the appearance of Russell's paradox. It was clear to everyone involved that set theory could not continue to be developed on an informal basis.

On the other hand, set theory had scored many triumphs. Cantor's work had opened up many avenues of research that mathematicians were eagerly pursuing. In addition, some mathematicians believed that many, if not all, mathematical concepts (for example, the definition of number) could be described in terms of sets. In their view, set theory was necessary for the correct reformulation of all previous mathematics and for the proper flourishing of all future mathematics. Set theory was too important a subject to be discarded because of the appearance of a few paradoxes.

How then was the situation in set theory to be resolved? The answer to this question came in 1908 with the appearance of a paper written in German by Ernst Zermelo. (The title of Zermelo's paper in English translation is "Investigations in the Foundations of Set Theory.") Zermelo proposed an axiomatic development of set theory. He begins with certain undefined terms and with a list of axioms for a theory of sets. On the basis of his axioms, Zermelo defines other mathematical concepts (such as *function*) and proves some basic theorems about these concepts. Zermelo's axiomatic treatment did, however, have some foundational difficulties. These difficulties were cleared up beginning in 1922 through the efforts of Abraham Fraenkel. The resulting axiom system, known as the Zermelo–Fraenkel axioms for set theory (ZF for short), has been accepted by most mathematicians as a basis for set theory. We now begin our presentation of the Zermelo–Fraenkel axioms. Our discussion of ZF will continue into the next two sections. Our approach is based on that presented in *Sets: Naive, Axiomatic and Applied* by D. van Dalen, H. C. Doets, and H. de Swart.

In ZF one begins with two undefined terms: *set* and $\in$. The latter symbol can be read as "is a member of"; intuitively we think of "$x \in A$" as saying "x is a member of A." We now present the first few axioms of ZF. Whenever possible we give both an informal and a formal version of each axiom.

ZF Axioms

1. **Axiom of Extensionality** Two sets are equal if and only if they have the same members. Formally $\forall x(x \in A \Leftrightarrow x \in B) \Leftrightarrow A = B$.
2. **Axiom of the Empty Set** There exists a set that has no elements. Formally, $\exists \varnothing (\forall x(x \notin \varnothing))$.

   Notice that by the axiom of extensionality any two empty sets are equal. Thus there is exactly one empty set, $\varnothing$.
3. **Axiom of Separation** For each set A and every property P of sets, there exists a set consisting precisely of those elements of A which have property P. Formally, $\exists B\ \forall y(y \in B \Leftrightarrow y \in A \wedge P(y))$.

   For each property P and for each set A the set B, whose existence is guaranteed by the axioms of separation, is written $B = \{y \in A | P(y)\}$.
4. **Axiom of Pairing** For any sets A and B, there exists a set C whose elements are precisely A and B. Formally, $\exists C\ \forall u(u \in C \Leftrightarrow u = A \vee u = B)$.

From Axioms 1 and 4 we conclude that for any sets A and B there is a unique set whose only members are A and B; this set is denoted by $\{A, B\}$. For any set A define $\{A\} = \{A, A\}$; $\{A\}$ is called the *singleton of* A.

Our axiomatic description of set theory will continue at the end of Section 2.2.

Exercises §2.1

2.1.1. List the elements in each of the following sets:
(a) $\{x \in \mathbf{R} | x^2 - 10x + 7 = 0\}$.
(b) $\{x \in \mathbf{R} | x^2 - 8x + 5 = 0\}$.
(c) $\{x \in \mathbf{R} | 1/x + 1/x + 1 = 2\}$.
(d) $\{x \in \mathbf{R} | x^3 - 3x = 0\}$.
(e) $\{x \in \mathbf{R} | x^2 + 4x + 5 = 0\}$.

2.1.2. (a) Show: $\{1, 2, 5\} = \{5, 2, 1\}$.
(b) Show: $\{\{0\}, \{0,1\}\} \neq \{\{0\}, \{1\}\}$.
(c) Show: $\{a, a\} = \{a\}$.
(d) Show: $\{\{a\}, \{\{a\}\}\} \neq \{\{a\}\}$.
(e) Show: $\{0\} \in \{\{0\}, 1\}$ but $\{0\} \not\subset \{\{0\}, 1\}$.

2.1.3. (a) Find real numbers a and b such that $\{x \in \mathbf{R} | x^2 - x - 2 < 0\} = \{x \in \mathbf{R} | a \leq x \leq b\}$.
(b) Find real numbers a and b such that $\{x \in \mathbf{R} | |x + 1| < 5\} = \{x \in \mathbf{R} | a < x < b\}$.
(c) Show that $\{x \in \mathbf{R} | (x - 1/2)(x - 1/3) < 0\} \subseteq \{x \in \mathbf{R} | 0 < x < 1\}$.

2.1.4. (a) Prove: Theorem 2.1.2.
(b) Prove that if $A \subseteq B$ and $A \not\subset C$, then $B \not\subset C$.

2.1.5. (a) Suppose $B \subseteq A$. Draw an appropriate Venn diagram for this situation.
(b) Suppose A and B have no elements in common. Draw an appropriate Venn diagram.

2.1.6. Let $A = \{1, 2, 3, 4, 5\}$. List the binary sequences corresponding to each of the following subsets of A. (a) $\varnothing$. (b) A. (c) $\{1, 2, 3\}$. (d) $\{1, 3, 5\}$. (e) $\{2, 4, 3\}$.

2.1.7. List all subsets of (a) $\{1, 2\}$, (b) $\{1, 2, 3\}$, and (c) $\{1, 2, 3, 4\}$. Next to each subset, give its corresponding binary sequence. Note: In solving (b) (and respectively (c)), try to use your list from (a) (respectively (b)) in a systematic way.

2.1.8. A set A is called transitive if any element of A is also a subset of A. In other words, A is transitive if $x \in A$ implies $x \subseteq A$.
(a) Show that $\{\varnothing\}$ is transitive.
(b) Find a transitive set having exactly two elements.
(c) Find a transitive set having exactly three elements.
(d) Generalize your findings from (b) and (c).

2.1.9. The aim of this exercise is to arrive at a conjecture for the number of subsets of a set with n elements. Form a table as follows:

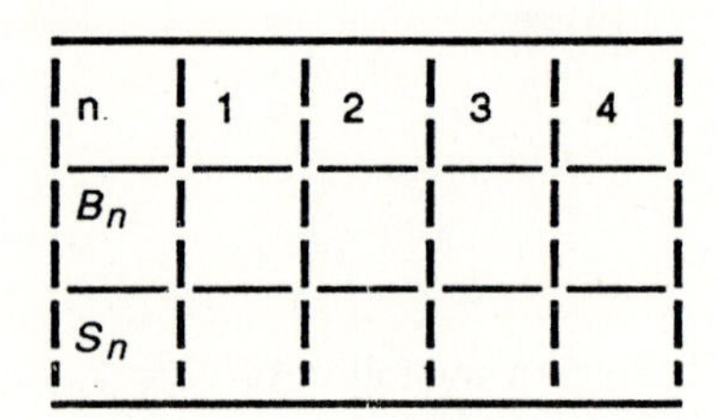

n	1	2	3	4
B_n				
S_n				

Here n denotes the number of elements of a finite set X, S_n the number of subsets of X, and B_n the number of binary strings of length n. In parts (c) and (d) you are asked to pinpoint the general pattern that is suggested in the table.

(a) How many binary strings are there of length 1? length 2? length 3? length 4?

(b) How many subsets are there in a set with one element? two elements? three elements? four elements? (See Exercise 2.1.7.)

(c) For any arbitrary positive integer n, how many binary strings of length n are there?

(d) Let n be a positive integer and let X be a set with n elements. How many subsets does X contain?

2.1.10. Let $X = \{x_1, \dots, x_n\}$. Suppose that $A \subseteq X$ and that A is represented by the binary sequence $a_1 a_2 \cdots a_n$.

(a) How can the number of elements of A be determined from $a_1 a_2 \cdots a_n$?

(b) Which sequence corresponds to $\varnothing$?

(c) Which sequence corresponds to X?

2.1.11. Russell's Paradox. Let $S = \{A \mid A \text{ is a set and } A \notin A\}$. Suppose that S itself is a set.

(a) Show that if S is a member of itself, then S cannot be a member of itself.

(b) Show that if S is not a member of itself, then S must be a member of itself.

(c) Russell's paradox is often rendered in a more folksy form: In a certain town there is a barber who shaves those and only those people who do not shave themselves. Question: Who shaves the barber?

§2.2 Operations on Sets

In this section we focus on a fundamental idea in mathematics, the notion of an operation on a set. As with many of the concepts of this chapter and the next, examples of operations have been with us since our mathematical childhood. Some illustrations from arithmetic will motivate the formal definition.

Addition of real numbers provides a way of associating to each pair of real numbers, a, b, a real number $a + b$. Similarly, subtraction and multiplication are procedures that assign a real number to each pair of real numbers. Subtraction assigns $a - b$ to the pair a, b, while multiplication associates to a and b the number ab. We now generalize this idea to an arbitrary set.

Definition 2.2.1 *Binary Operation*

Let X be a set. A *binary operation* on X is a rule, denoted $*$, which assigns to each pair of elements, a, b, of X, an element $a * b$ of X.

Example 2.2.1 (i) Each of the operations of addition, subtraction, and multiplication is a binary operation on the set of integers $\mathbf{Z}$ and on the set of rational numbers $\mathbf{Q}$.

(ii) Let $\mathbf{R}^*$ denote the set of nonzero real numbers. Then ordinary division, the rule that assigns each pair a, b in $\mathbf{R}^*$ to a/b, is an operation on $\mathbf{R}^*$. Note that, in general, the element assigned to the pair a, b, namely a/b, differs from the element assigned to b, a, namely b/a. Notice that division is *not* a binary operation on $\mathbf{Z}^* = \mathbf{Z} - \{0\}$ since the number assigned to the pair 2, 3, namely 2/3, is not in $\mathbf{Z}^*$.

Example 2.2.2 Let X be the set of real-valued functions whose domain is the set of real numbers, $\mathbf{R}$. Each of the following is an operation on the set X.

(i) Addition of Functions. If $f, g \in X$, then $f + g$ is the element of X defined by $(f + g)(x) = f(x) + g(x)$ for all $x \in \mathbf{R}$. (Note that $f(x)$ and $g(x)$ are real numbers and $f(x) + g(x)$ is the sum of the real numbers $f(x)$ and $g(x)$. Thus the operation of addition of functions depends for its definition on the operation of addition of real numbers. Strictly speaking, we should use different symbols to represent these operations. However, because of the dependence of function addition upon number addition and because of a desire to keep the number of symbols to a minimum, we use the same symbol for both operations.)

(ii) Multiplication of Functions. If $f, g \in X$, then fg is the function of $\mathbf{R}$ given by $(fg)(x) = f(x) \cdot g(x)$ for all $x \in \mathbf{R}$.

(iii) Composition of Functions. For $f, g \in X$, $f \circ g$ is the function defined by the rule: $(f \circ g)(x) = f(g(x))$ for all $x \in \mathbf{R}$.

Examples of binary operations can also be found in linear algebra. If M_n denotes the set of $n \times n$ matrices with entries in the set of real numbers, then matrix addition and matrix multiplication are binary operations on M_n.

We next turn our attention to operations in which the elements being combined are themselves sets.

Definition 2.2.2 *Union*

Let A and B be sets. The *union of A and B*, written $A \cup B$, is the set

$$A \cup B = \{x | x \in A \text{ or } x \in B\}.$$

In words, the union of A and B is the set of elements that are in at least one of the two sets A and B.

Definition 2.2.3 *Intersection*

Let A and B be sets. The *intersection of A and B*, written $A \cap B$, is the set

$$A \cap B = \{x | x \in A \text{ and } x \in B\}.$$

Thus, $A \cap B$ is the set of elements common to both A and B.

Example 2.2.3 Let $A = \{1, 2, 3\}$, $B = \{3, 4, 5\}$, and $C = \{5, 6, 7\}$. Then $A \cup B = \{1, 2, 3, 4, 5\}$, $A \cap B = \{3\}$, $A \cup C = \{1, 2, 3, 5, 6, 7\}$, and $A \cap C = \varnothing$.

The situation in which $A \cap B = \varnothing$, i.e., A and B have no elements in common, arises often enough to deserve a special name.

Definition 2.2.4 *Disjoint Sets*

Two sets A and B are *disjoint* if $A \cap B = \varnothing$.

The union of two sets is accomplished by combining or putting together the elements of the individual set into a new set. Roughly speaking, then, taking the union of two sets is analogous to adding two real numbers. We ask: Does subtraction of real numbers have a set theoretic analog? Thinking of subtraction as the taking away or removal of numbers or objects, we are led to the next definition.

Definition 2.2.5 *Set Difference*

Let A and B be sets. The *set difference of* or *difference of A and B*, written $A - B$, is defined to be:

$$A - B = \{x | x \in A \text{ and } x \notin B\}.$$

(Some authors use the symbol $A \setminus B$ in place of $A - B$.)

Example 2.2.4 (i) Let $E = \{z \in Z | z \text{ is an even integer}\}$. Then $Z - E = \{z \in Z | z \text{ is an odd integer}\}$.

(ii) Let $A = \{1, 2, 3\}$ and $B = \{2, 3, 4\}$. Then $A - B = \{1\}$ and $B - A = \{4\}$.

Let us consider a fixed set U. If A and B are subsets of U, then $A \cup B$ and $A \cap B$ are also subsets of U. Thus union and intersection are binary operations on the set of all subsets of U:

1. Union is the binary operation that assigns to each pair of subsets A, B of U (i.e., to each pair of *elements in the set of all subsets of* U) the subset $A \cup B$.
2. Intersection is the binary operation that assigns to each pair of subsets A, B of U the subset $A \cap B$.
3. Set difference is the binary operation that assigns to each pair of subsets A, B of U the subset $A - B$.

We next define another important type of operation by modifying slightly the definition of binary operation.

Definition 2.2.6 *Unary Operation*

A *unary operation* on a set X is a rule u that assigns to each $x \in X$ an element $u(x) \in X$.

Example 2.2.5 Each of the following is a unary operation on $\mathbf{R}$:

(i) $u(x) = -x$ for all $x \in \mathbf{R}$.
(ii) $u(x) = x^2$ for all $x \in \mathbf{R}$.
(iii) $u(x) = 2x$ for all $x \in \mathbf{R}$.

Example 2.2.6 Each of the following is a unary operation on the set $\mathbf{R}^* = \mathbf{R} - \{0\}$:

(i) $u(x) = 1/x$ for all $x \in \mathbf{R}^*$.
(ii) $u(x) = 1/x^2$ for all $x \in \mathbf{R}^*$.
(iii) $u(x) = 2x$ for all $x \in \mathbf{R}^*$.

Example 2.2.7 Let G_n be the set of $n \times n$ matrices having real entries and nonzero determinant:

$$G_n = \{A \in M_n | \det(A) \neq 0\}.$$

Then for any $A \in G_n$, $u(A) = A^{-1}$ (where A^{-1} denotes the inverse of A) defines a unary operation on G_n.

Definition 2.2.7 *Set Complement*

Let A be a subset of a set U. The *complement of A in U* is the set $U - A$.

If $A \subseteq U$ and $A \subseteq V$, then the sets $U - A$ and $V - A$ may be unequal, in which case the complement of A in U will differ from the complement of A in V. However, if we fix the set U throughout the discussion, then we call $U - A$ the *complement of A* and denote it by A^c. Thus the rule that assigns to A the set A^c is a unary operation on the collection of all subsets of U.

Representations of Operations

We now consider the interpretations of the set operations of $\cup$, $\cap$, $-$, and ^{c} in terms of our methods of representing sets—Venn diagrams and binary sequences.

Suppose A and B are subsets of a set U. Then Venn diagrams for $A \cup B$ and $A \cap B$ are presented as shown in Figure 2.2.1.

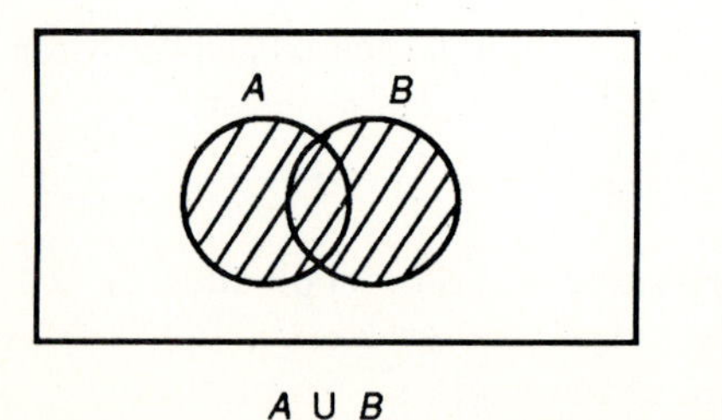

$A \cup B$

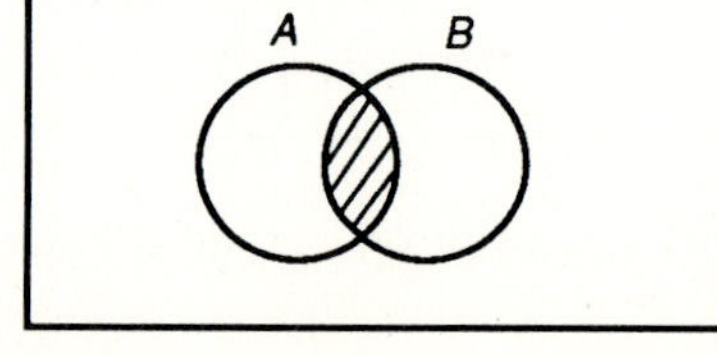

$A \cap B$

Figure 2.2.1

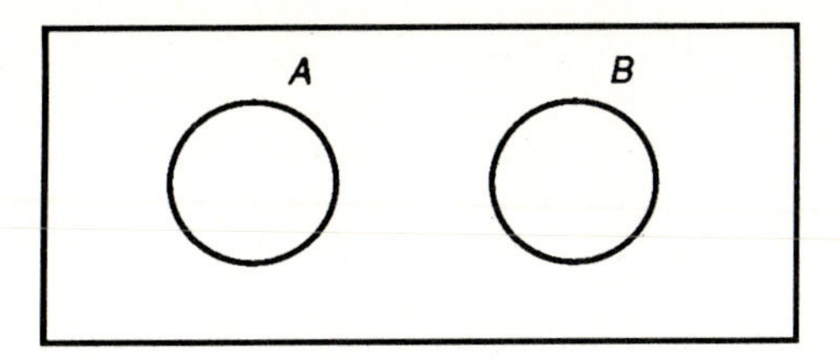

$A \cap B = \varnothing$

Figure 2.2.2

As is evident from either the definitions or the Venn diagram, $A \cap B$ is a subset of both A and B, which in turn are subsets of $A \cup B$. The proofs of these observations are left as exercises.

In drawing A and B as we have, we leave open the possibility that $A \cap B = \varnothing$. If, however, we wish to depict unambiguously the case that $A \cap B = \varnothing$, then we represent A and B as shown in Figure 2.2.2.

The Venn diagrams for the set difference, $A - B$, and the complement, A^c, are depicted in Figure 2.2.3.

We next interpret the set operations of intersection, union, and complement in terms of binary sequences. Let $U = \{x_1, \ldots, x_n\}$ be a finite set with n elements. Let A and B be subsets of U and let $a_1 \cdots a_n$ and $b_1 \cdots b_n$ be the binary sequences corresponding to A and B, respectively. How can the binary sequences of $A \cap B$, $A \cup B$, and A^c be determined from the binary sequences of A and B?

Let us consider an example. Let $U = \{1, 2, 3, 4\}$, $A = \{1, 2\}$, and $B = \{2, 3\}$. Then 1100 and 0110 are the binary sequences determined by A and B, respectively. Now $A \cap B = \{2\}$, $A \cup B = \{1, 2, 3\}$, and $A^c = \{3, 4\}$, and the binary sequences corresponding to these sets are 0100, 1110, and 0011, respectively. How are each of these sequences related to 1100 and 0110?

Perhaps the sequence belonging to A^c is the simplest to describe. The sequence 0011 (which is determined by A^c) is obtained from 1100 (which corresponds to A) by changing each 0 in 1100 to 1 and each 1 in 1100 to 0. This observation generalizes to an arbitrary subset A of an arbitrary finite set U. Let $a_1', \ldots, a_n'$ be the binary sequence of A^c. (Recall that the binary

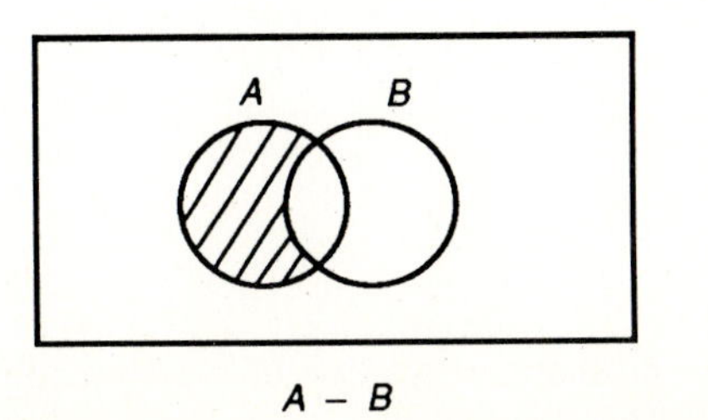

$A - B$

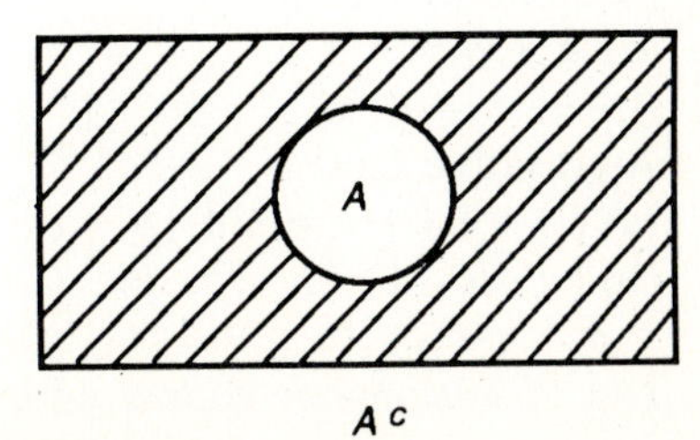

A^c

Figure 2.2.3

sequence of A is $a_1 \cdots a_n$). Then

$$a_i' = \begin{cases} 1 & \text{if } a_i = 0 \\ 0 & \text{if } a_i = 1. \end{cases}$$

Quick proof: If $a_i = 0$, then $x_i \notin A$; thus $x_i \in A^c$ and $a_i' = 1$. If $a_i = 1$, then $x_i \in A$; hence $x_i \notin A^c$ and $a_i' = 0$.

What about the binary sequence for $A \cap B$? Let $c_1 \cdots c_n$ be the binary sequence determined by $A \cap B$. Then for $1 \leq i \leq n$, c_i can be expressed in terms of a_i and b_i as follows:

$$c_i = \begin{cases} 1 & \text{if } a_i = 1 = b_i \\ 0 & \text{if } a_i = 0 \text{ or } b_i = 0 \end{cases}$$

for, if $a_i = b_i = 1$, then $x_i \in A \cap B$; hence $c_i = 1$. If $a_i = 0$ or $b_i = 0$, then either $x_i \notin A$ or $x_i \notin B$; therefore $x_i \notin A \cap B$ and $c_i = 0$. Notice that the relationship among a_i, b_i, and c_i can be expressed more concisely as $c_i = a_i \cdot b_i$. In other words the ith entry of $c_1 \cdots c_n$ is the product of the ith entries of $a_1 \cdots a_n$ and $b_1 \cdots b_n$.

Let $d_1 \cdots d_n$ denote the binary sequence of $A \cup B$. Then we leave as an exercise the proof that

$$d_i = \begin{cases} 1 & \text{if } a_i = 1 \text{ or } b_i = 1 \\ 0 & \text{if } a_i = 0 = b_i. \end{cases}$$

Properties of Operations

We now study properties of the set operations of union, intersection, difference, and complement. As a guide in this inquiry, let us recall from Section 1.2 some of the properties of the operations of addition and multiplication of real numbers. As usual, let $+$ and $\cdot$, respectively, denote these operations.

Perhaps the most familiar properties are the commutative and associative laws: for all $a, b, c \in \mathbf{R}$,

(i) Commutative Laws.
  (a) $a + b = b + a$.
  (b) $a \cdot b = b \cdot a$.
(ii) Associative Laws.
  (a) $(a + b) + c = a + (b + c)$.
  (b) $(a \cdot b) \cdot c = a \cdot (b \cdot c)$.

The two operations are connected by the distributive law of multiplication over addition:

(iii) Distributive Law. $a \cdot (b + c) = a \cdot b + a \cdot c.$

Also, the number 0 plays a special role with respect to both operations: for each $a \in \mathbf{R}$, $a + 0 = a$ and $a \cdot 0 = 0$.

The operations $\cup$ and $\cap$ possess these properties as well as a few others.

Theorem 2.2.1 *Let A, B, and C be any sets.*

(i) (a) $A \cup \varnothing = A$.
(b) $A \cap \varnothing = \varnothing$.

(ii) *Idempotent Laws.* (a) $A \cup A = A$.
(b) $A \cap A = A$.

(iii) *Associative Laws.* (a) $(A \cup B) \cup C = A \cup (B \cup C)$.
(b) $(A \cap B) \cap C = A \cap (B \cap C)$.

(iv) *Commutative Laws.* (a) $A \cup B = B \cup A$.
(b) $A \cap B = B \cap A$.

(v) *Distributive Laws.* (a) $A \cup (B \cap C) = (A \cup B) \cap (A \cup C)$.
(b) $A \cap (B \cup C) = (A \cap B) \cup (A \cap C)$.

This theorem suggests that the analogies between the operations of addition and multiplication of numbers and the operations of union and intersection of sets are strong. Commutative and associative laws hold for both the operations. If the empty set is regarded as the set analog of the number 0, then union and addition behave in similar ways, for the union of $\varnothing$ with any other set is that set while the sum of 0 and any other number is that number. Also multiplication and intersection behave in similar ways, for the product of 0 and any number is 0 while the intersection of $\varnothing$ with any set is $\varnothing$. Thus union seems to be analogous to addition, and intersection analogous to multiplication.

The analogy, however, is not exact, for neither addition nor multiplication possesses the idempotent property: It is not true that for each real number a, $a + a = a$ and $a \cdot a = a$. And while multiplication of real numbers does distribute over addition, addition does not return the favor: The equality $a + (b \cdot c) = (a + b) \cdot (a + c)$ does not hold for all $a, b, c \in \mathbf{R}$.

Where do these remarks leave us? As a heuristic principle, we can maintain and use profitably the idea that union and intersection are analogous to addition and multiplication, respectively. Hence we shall continue to use the real number system as a source in our investigation of sets. We cannot, however, thoughtlessly assert that every property of the real number system should translate directly into a property for sets or vice versa. Rather we must check each purported analogy carefully before accepting or rejecting it. In carrying out this investigation, we shall, of course, be using the standard tools of mathematics that were discussed in Chapter 1: examples to develop experience and to provide counterexamples, general and specific representations to provide various viewpoints and conjectures (in this case of sets, Venn diagrams and digital representations), and proof techniques to provide convincing demonstrations of any conjectures.

Now let us move to the verification of Theorem 2.2.1. Proofs of parts (i) and (v)(b) will be given here; the proofs of (ii)–(iv) and (v)(a) will be left as exercises. Although statement (i) is obvious, we present its proof as another example of an element-chasing proof to establish the equality of two sets.

Proof of Theorem 2.2.1(i)(a) To prove that $A \cup \emptyset = A$, we show that $A \cup \emptyset \subseteq A$ and $A \subseteq A \cup \emptyset$.

To prove that $A \cup \emptyset \subseteq A$, we take an arbitrary element x in $A \cup \emptyset$ and show that $x \in A$. If $x \in A \cup \emptyset$, then either $x \in A$ or $x \in \emptyset$. But $\emptyset$ has no elements; hence $x \notin \emptyset$. Thus, $x \in A$. Therefore, $A \cup \emptyset \subseteq A$.

Now we show $A \subseteq A \cup \emptyset$ by taking an arbitrary element $x \in A$ and proving that $x \in A \cup \emptyset$. But if $x \in A$, then either $x \in A$ or $x \in \emptyset$; in other words, if $x \in A$, then $x \in A \cup \emptyset$ and $A \subseteq A \cup \emptyset$.

We conclude that $A \cup \emptyset = A$. ■

Proof of Theorem 2.2.1(i)(b) We show that $A \cap \emptyset \subseteq \emptyset$ and $\emptyset \subseteq A \cap \emptyset$.

Because the empty set is a subset of any set (Theorem 2.1.1(i)(a)), $\emptyset \subseteq A \cap \emptyset$.

To establish the inclusion $A \cap \emptyset \subseteq \emptyset$, we argue by contradiction. Suppose $A \cap \emptyset \not\subseteq \emptyset$. Hence there exists an element $x \in A \cap \emptyset$ such that $x \notin \emptyset$. However, if $x \in A \cap \emptyset$, then $x \in A$ and $x \in \emptyset$. But since $x \in \emptyset$, $\emptyset$ is not empty, a conclusion that violates the definition of $\emptyset$. Therefore the assumption that $A \cap \emptyset \not\subseteq \emptyset$ is incorrect. Thus $A \cap \emptyset \subseteq \emptyset$. Therefore we know that $A \cap \emptyset = \emptyset$. ■

(Note that the proof by contradiction is appropriate since by negating the desired conclusion $A \cap \emptyset \subseteq \emptyset$, we immediately obtain an element $x \in A \cap \emptyset$ that we can work with.)

Proof of Theorem 2.2.1(v)(b) We show (1) $A \cap (B \cup C) \subseteq (A \cap B) \cup (A \cap C)$ and (2) $(A \cap B) \cup (A \cap C) \subseteq A \cap (B \cup C)$.

(1) We take an arbitrary $x \in A \cap (B \cup C)$ and show $x \in (A \cap B) \cup (A \cap C)$. Since $x \in A \cap (B \cup C)$, $x \in A$ and $x \in B \cup C$. Thus, if $x \in A \cap (B \cup C)$, then $x \in A$ and either $x \in B$ or $x \in C$. We have then two possible cases:

Case 1. $x \in A$ and $x \in B$.

Case 2. $x \in A$ and $x \in C$.

In Case 1, $x \in A \cap B$; in Case 2, $x \in A \cap C$. Since either Case 1 or Case 2 is true, either $x \in A \cap B$ or $x \in A \cap C$; that is, $x \in (A \cap B) \cup (A \cap C)$. Therefore $A \cap (B \cup C) \subseteq (A \cap B) \cup (A \cap C)$.

(2) We complete the proof by showing that $(A \cap B) \cup (A \cap C) \subseteq A \cap (B \cup C)$. Let $x \in (A \cap B) \cup (A \cap C)$. We must show that $x \in A \cap (B \cup C)$.

If $x \in (A \cap B) \cup (A \cap C)$, then $x \in A \cap B$ or $x \in A \cap C$. Once again we have a divide-and-conquer type case analysis. Consider two cases:

Case 1. $x \in A \cap B$.

Case 2. $x \in A \cap C$.

In Case 1 $x \in A$ and $x \in B$. Since $x \in B$, either $x \in B$ or $x \in C$. Thus, $x \in A$ and $x \in B \cup C$, which means that $x \in A \cap (B \cup C)$.

In Case 2 $x \in A \cap C$ and an argument identical to the one just presented shows that $x \in A \cap (B \cup C)$. We have proved that $x \in A \cap (B \cup C)$ whenever $x \in (A \cap B) \cup (A \cap C)$, or equivalently that $(A \cap B) \cup (A \cap C) \subseteq A \cap (B \cup C)$.

The proof of Theorem 2.2.1(v)(b) is now complete. ■

Questions concerning set difference also arise: What are the basic properties of set difference? What, if any, are the connections between set difference and the operation of union and intersection and other previously encountered concepts?

Let us investigate one reasonable question: Does set difference distribute over union? Does $A - (B \cup C) = (A - B) \cup (A - C)$ for all sets A, B, and C?

There are several ways to approach this question. One method is to search immediately for a proof of the result. A second is to analyze various examples in an effort either to find a counterexample or to develop an idea for a proof. Finally, one could seek a general "picture" of the problem by scrutinizing Venn diagrams.

Trying to prove a result before being convinced of its truth is at best unwise. So let us consider some special cases. In doing so, it usually pays to take examples that are especially simple. For example, if $A = B = C$, then $A - (B \cup C) = \varnothing = (A - B) \cup (A - C)$. A somewhat more complicated case arises when $A = B$ and $A \neq C$. For example, let us take $A = B = \{1\}$ and $C = \{2\}$. Then $A - (B \cup C) = \{1\} - \{1,2\} = \varnothing$ while $(A - B) \cup (A - C) = \varnothing \cup \{1\} = \{1\}$. Thus the conjecture that $A - (B \cup C) = (A - B) \cup (A - C)$ is not true in general.

At this point one can ask at least two questions:

1. For which sets A, B, C does $A - (B \cup C) = (A - B) \cup (A - C)$? Ideally, the answer would have the form: $A - (B \cup C) = (A - B) \cup (A - C)$ if and only if ______, where the blank is filled by a statement prescribing properties that A, B, and C satisfy.

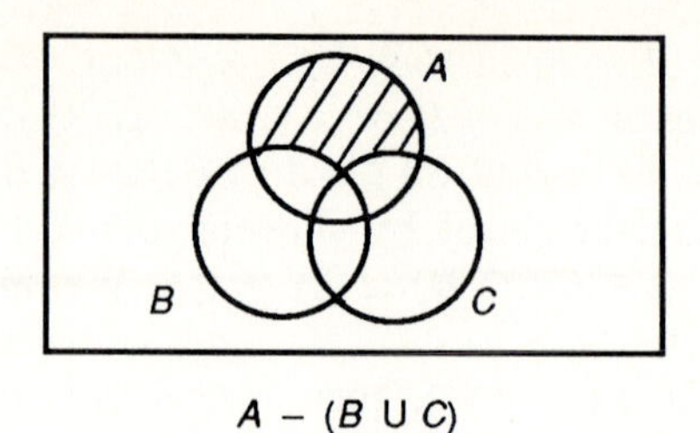

$A - (B \cup C)$

Figure 2.2.4

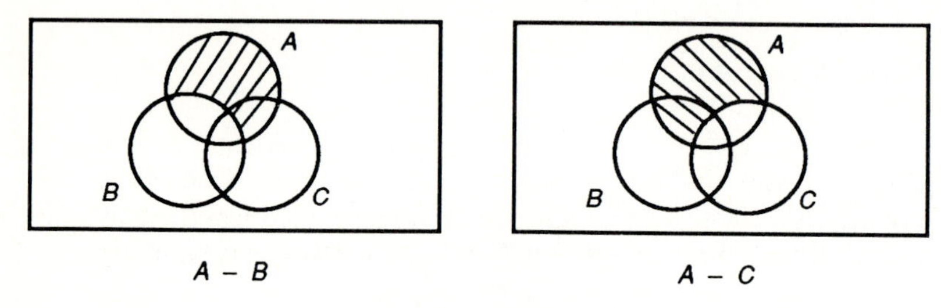

$A - B$ $\qquad$ $A - C$

Figure 2.2.5

2. Can the statement $A - (B \cup C) = (A - B) \cup (A - C)$ be modified to obtain an assertion that is true for all sets A, B, and C?

Let us investigate the second question and leave the first as an exercise. To do so, we use Venn diagrams as an investigative tool. Remember that Venn diagrams have the advantage of being general and easy to manipulate. They have the disadvantage of being imprecise. But at this stage we are interested only in making a conjecture and hence are willing to subjugate our logical inclinations to our intuition and imagination.

First, a Venn diagram for $A - (B \cup C)$ is drawn (Figure 2.2.4). Then, Venn diagrams for $A - B$ and $A - C$ are drawn as in Figure 2.2.5. Let us combine these diagrams into one, as shown in Figure 2.2.6. The region with the cross-hatching is exactly $(A - B) \cap (A - C)$. Yet this region appears to coincide with $A - (B \cup C)$. Thus we conjecture that

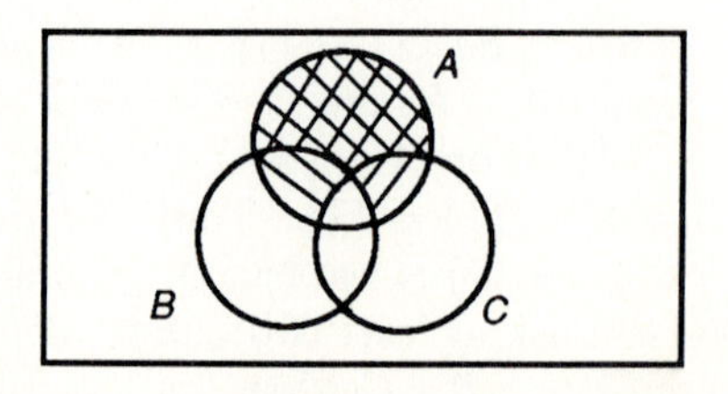

Figure 2.2.6

$A - (B \cup C) = (A - B) \cap (A - C)$. This assertion is indeed true, as we can demonstrate with an element-chasing argument.

What about other possible distributive laws such as those involving $A - (B \cap C)$, $A \cap (B - C)$, and $(A \cup B) - C$? Pause for a moment and investigate each of these sets with Venn diagrams in order to make conjectures concerning them. The answers are recorded in the next theorem.

Theorem 2.2.2 *Let A, B, and C be sets.*

(i) $A \subseteq B$ *if and only if* $A - B = \varnothing$.

(ii) $A \cap B = \varnothing$ *if and only if* $A - B = A$.

(iii) $A - (B \cup C) = (A - B) \cap (A - C)$.

(iv) $A - (B \cap C) = (A - B) \cup (A - C)$.

(v) $A \cap (B - C) = (A \cap B) - (A \cap C)$.

(vi) $(A \cup B) - C = (A - C) \cup (B - C)$.

We shall prove parts (i) and (vi). The proofs of parts (ii)–(v) are left as exercises although we shall prove special cases of parts (iii) and (iv) in the next theorem. As an exercise, before trying to prove any of the parts of the theorem, draw Venn diagrams illustrating each statement.

Proof of (i) We show: (1) If $A \subseteq B$, then $A - B = \varnothing$ and (2) if $A - B = \varnothing$, then $A \subseteq B$.

Proof of (1) To show that $A \subseteq B$ implies $A - B = \varnothing$, we argue by contraposition. Thus we assume $A - B \neq \varnothing$ and show that $A \not\subseteq B$. Since $A - B \neq \varnothing$, there exists $x \in A - B$. By the definition of $A - B$, $x \in A$ and $x \notin B$; hence there exists an element of A that is not in B. Therefore $A \not\subseteq B$.

Proof of (2) Again we argue by contraposition. We assume $A \not\subseteq B$ and show $A - B \neq \varnothing$. Since $A \not\subseteq B$, there exists $x \in A$ such that $x \notin B$. Thus $x \in A - B$ and $A - B \neq \varnothing$.

The proof of part (i) is now complete. ■

Proof of (vi) First we use element chasing to show $(A \cup B) - C \subseteq (A - C) \cup (B - C)$. Let $x \in (A \cup B) - C$. Then $x \in A \cup B$ and $x \notin C$. Therefore, ($x \in A$ or $x \in B$) and $x \notin C$; hence either ($x \in A$ and $x \notin C$) or ($x \in B$ and $x \notin C$); i.e., $x \in A - C$ or $x \in B - C$, i.e., $x \in (A - C) \cup (B - C)$.

To show $(A - C) \cup (B - C) \subseteq (A \cup B) - C$ we let $x \in (A - C) \cup (B - C)$. Then either $x \in A - C$ or $x \in B - C$. In the first instance $x \in A$ and $x \notin C$, hence $x \in A \cup B$ and $x \notin C$; in other words, $x \in (A \cup B) - C$. In the second case $x \in B$ and $x \notin C$, which implies that $x \in A \cup B$ and $x \notin C$; again we conclude that $x \in (A \cup B) - C$. Therefore $(A - C) \cup (B - C) \subseteq (A \cup B) - C$. ■

The basic properties of the complement operation are recorded in the following theorem.

Theorem 2.2.3 *Let U be a set and let A and B be subsets of U.*

(i) $\varnothing^c = U$.
(ii) $(A^c)^c = A$.
(iii) $A \subseteq B$ *if and only if* $B^c \subseteq A^c$.
(iv) (*DeMorgan's Laws*) (a) $(A \cap B)^c = A^c \cup B^c$.
(b) $(A \cup B)^c = A^c \cap B^c$.

Exercise. Before considering the proofs, draw Venn diagrams illustrating each part of the theorem.

Proof The proofs of (i) and (ii) are left as exercises.

To prove (iii) first we suppose $A \subseteq B$ and show $B^c \subseteq A^c$. This amounts to proving that if $x \in B^c$, then $x \in A^c$. Assume to the contrary that $x \notin A^c$. Then by part (ii) $x \in (A^c)^c = A$; since $A \subseteq B$, $x \in B$. But also $x \in B^c$, implying that $x \notin B$ and forcing the absurd conclusion that $x \in B$ and $x \notin B$. This contradiction means that the assumption that $x \in A$ is incorrect. Thus $x \in A^c$, which completes the proof that $B^c \subseteq A^c$.

A similar argument establishes the converse. However, a slicker argument, using part (ii) and the implication that was proved in the previous paragraph, can also be given. We have just proved: (*) If $A \subseteq B$ then $B^c \subseteq A^c$. Now we want to show that if $B^c \subseteq A^c$, then $A \subseteq B$.

By assumption $B^c \subseteq A^c$. From (*) it follows that $(A^c)^c \subseteq (B^c)^c$. But by part (ii) $(A^c)^c = A$ and $(B^c)^c = B$. Thus $A \subseteq B$.

(Notice that the previous argument makes no reference to the definition of set inclusion: We do not show that $x \in A$ implies $x \in B$. Rather, it uses previously established properties of set inclusion and set complement and thus can be thought of as taking place at a level above the definitional level. This argument has several advantages—including brevity, elegance, and clarity—over a proof proceeding directly from the definition. Now back to the proof.)

We prove (iv)(a): $(A \cap B)^c = A^c \cup B^c$. Observe that once (iv)(a) is established, (iv)(b) can be derived as follows: By parts (iv)(a) and (ii),

$$(A^c \cap B^c)^c = (A^c)^c \cup (B^c)^c = A \cup B.$$

Thus, $(A \cup B)^c = ((A^c \cap B^c)^c)^c = A^c \cap B^c$ by part (ii).

To prove $(A \cap B)^c = A^c \cup B^c$, consider first the inclusion $(A \cap B)^c \subseteq A^c \cup B^c$. Let x be an arbitrary element of U such that $x \in (A \cap B)^c$. Then $x \notin A \cap B$; that is, it is not the case that x is an element of both A and B. Thus either $x \notin A$ or $x \notin B$; equivalently, either $x \in A^c$ or $x \in B^c$. Thus, if $x \in (A \cap B)^c$, then $x \in A^c \cup B^c$.

Conversely, to show $A^c \cup B^c \subseteq (A \cap B)^c$, let x be an arbitrary element of U such that $x \in A^c \cup B^c$. We want to show $x \in (A \cap B)^c$, i.e.,

$x \notin A \cap B$. Since $x \in A^c \cup B^c$, $x \in A^c$ or $x \in B^c$; hence either ($x \in U$ and $x \notin A$) or ($x \in U$ and $x \notin B$). Thus $x \in U$ and ($x \notin A$ or $x \notin B$). (Here we use the tautology $[P \wedge (Q \vee R)] \Leftrightarrow [(P \wedge Q) \vee (P \wedge R)]$.) It follows that $x \in U$ and it is not the case that $x \in A$ and $x \in B$. Therefore, $x \in U$ and $x \notin A \cap B$, i.e., $x \in (A \cap B)^c$.

The proof of Theorem 2.2.3(iv) is now complete. ■

The Power Set of a Set

In order to represent intersection and union as operations, we had to consider the set of all subsets of a given set. This concept occurs everywhere in mathematics; the next definition supplies its official name and notation.

Definition 2.2.8

For any set A, the *power set of* A, written $P(A)$, is the set of all subsets of A. In other words, $P(A) = \{B | B \subseteq A\}$.

Example 2.2.7 Since $\mathbf{Z}, \mathbf{Q} \subseteq \mathbf{R}$, $\mathbf{Z}, \mathbf{Q} \in P(\mathbf{R})$.

Example 2.2.8 If A is any set, then $\varnothing \in P(A)$, $A \in P(A)$, and $\{a\} \in P(A)$ for each $a \in A$.

Example 2.2.9 (i) $P(\varnothing) = \{\varnothing\}$
(ii) $P(\{1\}) = \{\varnothing, \{1\}\}$.
(iii) $P(\{1,2\}) = \{\varnothing, \{1\}, \{2\}, \{1,2\}\}$.
(iv) $P(\{1,2,3\} = \{\varnothing, \{1\}, \{2\}, \{1,2\}, \{3\}, \{1,3\}, \{2,3\}, \{1,2,3\}\}$.

Let us look closely at the sets listed in Example 2.2.9. (See also Exercise 2.1.9.) First, a tight relationship seems to exist between the number of elements in a finite set and the number of subsets of the set. Let S_n denote the number of subsets in a set having n elements. (Thus, if n is the number of elements of A, then S_n is the number of elements of $P(A)$.) From Example 2.2.9 we can fill in the following table:

n	0	1	2	3
S_n	1	2	4	8

A pattern emerges, does it not? We *conjecture* that if A is a set with n elements, then $P(A)$ is a set with 2^n elements.

Instead of searching for a proof of this conjecture immediately, let us glance at one more example. Let $A = \{1, 2, 3, 4\}$. What is $P(A)$? To answer this we list the elements of $P(A)$, but let us do so in a systematic way. First list the subsets of A that are also subsets of $\{1, 2, 3\}$, in other words those subsets of A that do not contain 4. There are 8 such subsets. What about the subsets of A that contain 4? Any such subset, B, contains 4 and possibly elements from $\{1, 2, 3\}$. Thus $B = \{4\} \cup C$ where C is a subset of $\{1, 2, 3\}$. Since there are exactly 8 such sets C, there are exactly 8 subsets of A that contain 4. Since any subset of A either contains 4 or does not contain 4, all subsets of A have been counted. Therefore, $P(A)$ has 16 elements.

Armed with this added evidence for our conjecture, we embark upon a proof. Certainly we would be wise to try to base our proof on an extension of the method used in finding the power set of $\{1, 2, 3, 4\}$. Here is the theorem:

Theorem 2.2.4 *If A is a set with exactly n elements, then $P(A)$ has exactly 2^n elements.*

The proof of Theorem 2.2.4 will use the following lemma, which generalizes an observation made in our calculation of the power set of $\{1, 2, 3, 4\}$. The lemma provides a more general result since no restriction on the nature of the set A (finite or infinite) is placed.

Lemma 2.2.5 *If $A = B \cup \{x\}$ where $x \notin B$, then $P(A) = P(B) \cup \{\{x\} \cup C | C \in P(B)\}$.*

Proof Let $D = P(B) \cup \{\{x\} \cup C | C \in P(B)\}$. We show $P(A) = D$ by checking that $P(A) \subseteq D$ and $D \subseteq P(A)$.

To show that $P(A) \subseteq D$ we take $A_1 \in P(A)$. Then $A_1 \subseteq A$. Consider two cases: (1) $x \notin A_1$, and (2) $x \in A_1$.

Case 1. Suppose $x \notin A_1$. Since $A_1 \subseteq A$, $A_1 \subseteq B$, hence $A_1 \in P(B)$.

Case 2. Suppose $x \in A_1$. Let $C = \{y | y \in A_1 \text{ and } y \neq x\}$. Then $C \subseteq B$, i.e., $C \in P(B)$ and $A_1 = \{x\} \cup C$. Therefore, in this case as well, $A_1 \in D$.

Thus $P(B) \subseteq D$.

The proof of the inclusion $D \subseteq P(A)$ is left as an exercise. With that the lemma is proved. ∎

We are now ready to prove Theorem 2.2.4. Since the statement of the theorem contains a natural number parameter, a proof by mathematical induction is plausible.

Proof For $n = 1, 2, 3, \ldots$, let $S(n)$ be the statement: If A is a set with exactly n elements, then $P(A)$ has exactly 2^n elements.

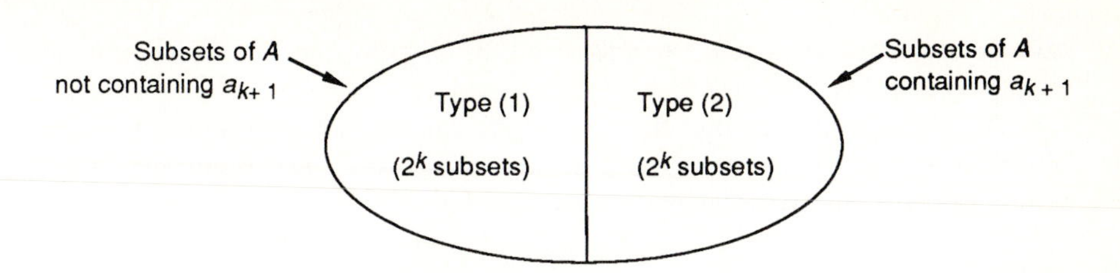

Figure 2.2.7. Set of Subsets of A.

We prove that $S(n)$ is true for all n by showing that (1) $S(1)$ is true and (2) if $S(k)$ is true, then $S(k + 1)$ is true.

(1) $S(1)$ is the statement: If A is a set with one element, then $P(A)$ has two elements. We have already verified this statement: If A has one element, then $A = \{a\}$ for some element a and $P(A) = \{\varnothing, A\}$ has two elements.

(2) Suppose $S(k)$ is true and show $S(k + 1)$ is true. Suppose A is a set with $k + 1$ elements. Then we must show that $P(A)$ has 2^{k+1} elements. Let $A = \{a_1, \ldots, a_k, a_{k+1}\}$ and let $B = \{a_1, \ldots, a_k\}$. Then $A = B \cup \{a_{k+1}\}$. By Lemma 2.2.5, every subset of A either

(a) does not contain a_{k+1} and hence is a subset of B, or

(b) contains a_{k+1} and has the form $\{a_{k+1}\} \cup C$ where C is a subset of B.

By inductive hypothesis, statement $S(k)$ is true. Thus any set with k elements has 2^k subsets. Since B has k elements, $P(B)$ has exactly 2^k subsets. Therefore, there are 2^k subsets of A of type (a). In addition, as each subset of A of type (b) has the form $\{a_{k+1}\} \cup C$ where C is a subset of B, there are precisely 2^k subsets of A of type (b). But each subset of A is either of type (a) or type (b) and no subset of A can be of both types. Thus A contains $2^k + 2^k = 2^{k+1}$ subsets. ■

Remark Notice that $P(B) \cap \{\{x\} \cup C | C \in P(B)\} = \varnothing$. In other words, each subset of A is either a subset of B or of the form $\{x\} \cup C$ where C is a subset of B, but no subset of A satisfies both these conditions.

Second Proof of Theorem 2.2.4 We outline another proof of Theorem 2.2.4, this one based on the binary sequence representation of a subset of a finite set.

Let $A = \{a_1, \ldots, a_n\}$ be a set having exactly n elements. The proof consists of two major steps:

(1) There are precisely as many subsets of A as there are binary sequences of length n.

(2) There are exactly 2^n binary sequences of length n.

The first step follows directly from the correspondence between subsets of A and binary sequences of length n defined in Section 2.1. Simply observe that under this correspondence each subset of A is matched with exactly one binary sequence and each binary sequence corresponds to exactly one subset of A.

As you might imagine, the second step is proved by induction. Each binary sequence of length n ends either with 0 or 1. But how many such sequences end with 0? How many end with 1? ■

One might wonder why we bother to give a second proof of Theorem 2.2.4 (or any other theorem for that matter). The most immediate answer is that two proofs can only increase our understanding of the result. Each of the proofs given above is based on a specific way of picturing or representing the set $P(A)$. By using each representation to construct a proof, we are demonstrating the viability and power of that representation. By viewing a given result in different ways, we gain a clearer and richer perspective on the nature of that result.

A closing remark about the name "power set" for $P(A)$. The source for the name can be seen in Theorem 2.2.4: *The number of elements in $P(A)$ is 2 to the power of the number of elements of A, when A is a finite set*. Thus a property of $P(A)$ that holds when A is finite (namely, that $P(A)$ has 2^n elements if A has n elements) is used to suggest a name for $P(A)$ for arbitrary A. This property also motivates questions concerning the size of $P(A)$ when A is infinite. For example, when n is a large integer, say $n \geq 6$, 2^n is significantly larger than n. (In fact, $2^n/n$ tends to ∞ as n tends to ∞). Thus $P(A)$ is significantly larger than A if A is a large finite set. Question: If A is infinite, is $P(A)$ larger than A? Before attempting to answer this question, we must clarify it. If A is infinite, then $P(A)$ is also infinite. (Why?) Thus what does it mean to say that one infinite set is "larger" than another infinite set? How can the concept of size of a set be extended from finite sets to infinite sets? For the time being we offer these questions as food for thought. In Chapter 4 we return to these questions and provide answers that are both surprising and beautiful.

Family of Sets

Often it is necessary to consider the union or intersection of more than two sets. To define these concepts we first consider the notion of a family of sets.

Definition 2.2.9 *Family of Sets*

Let I be any set. A *family of sets indexed by I* is a set of sets $\{A_i | i \in I\}$. The set I is called the *indexing set* for the family $\{A_i | i \in I\}$.

Note that the family $\{A_i | i \in I\}$ contains one set for each element of I; it is possible, however, that for distinct elements $i, j \in I$, the sets A_i and A_j are equal. Also the set I can be finite or infinite.

Example 2.2.10 Let H be the set of human beings alive at this moment. For each $i \in H$ let A_i denote the set of ancestors of i. Thus the members of A_i are i's parents, grandparents, great-grandparents, etc. Notice that if i and j are siblings, then $A_i = A_j$, and that if i and j are cousins, then $A_i \cap A_j \neq \emptyset$.

Example 2.2.11 For each $i \in \mathbf{Z}$, define the interval of real numbers $A_i = [i, i + 1] = \{x \in \mathbf{R} | i \leq x \leq i + 1\}$. Then $\{A_i | 1 \leq i \leq 3\} = \{[1, 2], [2, 3], [3, 4]\}$.

Definition 2.2.10 *Union and Intersection of a Family*

Let $\{A_i | i \in I\}$ be a family of sets.

(i) The *union* of the sets in this family, written $\bigcup_{i \in I} A_i$, is the set

$$\bigcup_{i \in I} A_i = \{x | \text{There exist } i \in I \text{ such that } x \in A_i\}.$$

(ii) The *intersection* of the sets in this family, written $\bigcap_{i \in I} A_i$, is the set

$$\bigcap_{i \in I} A_i = \{x | \text{For each } i \in I, \ x \in A_i\}.$$

Thus the union of a family of sets consists of all elements that are in at least one of the sets in the family and the intersection consists of all elements that are members of all the sets in the family.

Example 2.2.12 For each $i \in \mathbf{Z}$ let $A_i = [i, i + 1] = \{x \in \mathbf{R} | i \leq x \leq i + 1\}$.

(i) Consider the family $\{A_i | i \in \mathbf{Z}\}$. Then $\bigcup_{i \in \mathbf{Z}} A_i = \mathbf{R}$ while $\bigcap_{i \in \mathbf{Z}} A_i = \emptyset$.

(ii) Consider $\{A_i | i \in \{1, 2\}\} = \{[1, 2], [2, 3]\}$. Then $\bigcup_{i \in \{[1, 2]\}} A_i = [1, 3]$ while $\bigcap_{i \in \{1, 2\}} A_i = \{2\}$.

Example 2.2.13 For each $i \in \mathbf{Z}^+$, let $A_i = [0, 1/i] = \{x \in \mathbf{R} | 0 \leq x \leq 1/i\}$. Then $\bigcap_{i \in \mathbf{Z}^+} A_i = \{0\}$.

To prove this claim, we show that if $x \in \bigcap_{i \in \mathbf{Z}^+} A_i$, then $x = 0$. If not, then $x > 0$ and there exists an integer n such that $1/n < x$. Thus $x \notin A_n$, which means that $x \notin \bigcap_{i \in \mathbf{Z}^+} A_i$. Therefore $\bigcap_{i \in \mathbf{Z}^+} A_i \subseteq \{0\}$. Since the reverse inclusion clearly holds, $\bigcap_{i \in \mathbf{Z}^+} A_i = \{0\}$.

Example 2.2.14. Suppose the indexing set I is a finite set of integers, say $I = \{1, 2, \ldots, n\}$. Then we often write

$$\bigcup_{i \in I} A_i = \bigcup_{i=1}^{n} A_i$$

and

$$\bigcap_{i \in I} A_i = \bigcap_{i=1}^{n} A_i.$$

The properties of union and intersection given earlier for the union and intersection of two sets extend to the union and intersection of an arbitrary family of sets.

Theorem 2.2.6 *Let A be an arbitrary set. Suppose $\{A_i | i \in I\}$ is a family of sets such that for each i, $A_i \subseteq U$ for some set U.*
(i) $A \cap (\bigcup_{i \in I} A_i) = \bigcup_{i \in I} (A \cap A_i)$.
(ii) $A \cup (\bigcap_{i \in I} A_i) = \bigcap_{i \in I} (A \cup A_i)$.
(iii) $(\bigcap_{i \in I} A_i)^c = \bigcup_{i \in I} A_i^c$ (*where* $A_i^c = U - A_i$).
(iv) $(\bigcup_{i \in I} A_i)^c = \bigcap_{i \in I} A_i^c$.

Proof We prove parts (i) and (iii) and leave the proofs of (ii) and (iv) as exercises. Our proofs are of the element-chasing variety; however, through the use of biconditionals, we can prove both the set inclusions

$$A \cap \left(\bigcup_i A_i \right) \subseteq \bigcup_i (A \cap A_i)$$

and

$$\bigcup_i (A \cap A_i) \subseteq A \cap \left(\bigcup_i A_i \right)$$

simultaneously.

Proof of (i) For any x, $x \in A \cap (\bigcup_{i \in I} A_i)$
if and only if $x \in A$ and $x \in \bigcup_{i \in I} A_i$,
if and only if $x \in A$ and there exists $i \in I$ such that $x \in A_i$,
if and only if there exists $i \in I$ such that $x \in A \cap A_i$,
if and only if $x \in \bigcup_{i \in I} (A \cap A_i)$.

Proof of (iii) For any x, $x \in (\bigcap_{i \in I} A_i)^c$
if and only if $x \notin \bigcap_{i \in I} A_i$,
if and only if there exists $i \in I$ such that $x \notin A_i$,
if and only if there exists $i \in I$ such that $x \in A_i^c$,
if and only if $x \in \bigcup_{i \in I} A_i^c$. ■

Axiomatic Set Theory

We continue with our presentation of Zermelo–Fraenkel axiomatic set theory. Recall that the Axiom of Separation enables one to define subsets of a given set. Thus, based on that axiom, we can define the intersection and difference of two sets:

$$A \cap B = \{x \in A \mid x \in B\}$$

$$A - B = \{x \in A \mid x \notin B\}.$$

The Separation Axiom, however, does not allow us to define the union of two sets A and B unless A and B are defined as subsets of some set U. To provide for the union of an arbitrary family of sets we rely on the following axiom:

Axiom 5 Set-Union Axiom. Let $\{A_i \mid i \in I\}$ be a family of sets. Then there exists a set B whose members consist of those elements that belong to at least one of the sets A_i. Formally, $\exists B \forall x(x \in B \Leftrightarrow \exists i(x \in A_i))$.

Just as Axioms 1–4 are not sufficient to define the union of an arbitrary collection of sets, neither can they be used to define the power set of a given set. This concept is provided by the next axiom.

Axiom 6 Power Set Axiom. For every set A there exists a set B whose elements are precisely the subsets of A. Formally, $\exists B \forall x(x \in B \Leftrightarrow \forall y(y \in x \Rightarrow y \in A))$.

For each set A the set B whose existence is guaranteed by Axiom 6 is denoted by $P(A)$.

The next axiom certifies the existence of infinite sets. In this case the informal rendering of the axiom is nearly a direct translation of the formal version.

Axiom 7 Axiom of Infinity. There exists a set A such that the empty set is in A and such that if y is in A, then $y \cup \{y\}$ is in A. Formally, $\exists A(\varnothing \in A \wedge \forall y(y \in A \Rightarrow y \cup \{y\} \in A))$.

A set A that satisfies the conditions of Axiom 7 is called a *successor set*. Axiom 7 asserts that a successor set exists. In Chapter 7 we indicate how this axiom can be used to *define* the natural number system in terms of set theory.

The last axiom presented in this section is the Axiom of Regularity. This axiom helps to weed out from axiomatic set theory many (if not all) of the paradoxes (including Russell's paradox) that cropped up in informal set theory (see Exercise 2.2.20).

Axiom 8 Axiom of Regularity. Every nonempty set is disjoint from at least one of its members. Formally, $A \neq \varnothing \Rightarrow \exists x(x \in A \wedge x \cap A = \varnothing)$.

We have almost completed our list of axioms for ZF; in fact, we are down to one, the Replacement Axiom. The statement of the Replacement Axiom requires the notion of function, which in turn requires the notion of Cartesian product, which is the subject of the next section. A few exercises on axiomatic set theory are included at the end of the following list.

Exercises §2.2

2.2.1. Let $U = \{1, 2, 3, 4, 5, 6\}$, $A = \{1, 2, 3\}$, $B = \{2, 3, 4\}$, and $C = \{4, 5, 6\}$. Find each of the following sets:
(a) $A \cup B$.
(b) $A \cap B$.
(c) A^c.
(d) B^c.
(e) $B \cap C$.
(f) $A \cap (B \cup C)$.
(g) $(A \cap B) \cup (A \cap C)$.
(h) $A \cap C^c$.
(i) $A - C$.
(j) $P(A) \cap P(B)$.

2.2.2. Let A and B be arbitrary sets.
(a) Prove: $A \cup A = A$ and $A \cap A = A$.
(b) Prove: $A \cup B = B \cup A$ and $A \cap B = B \cap A$.
(c) Prove: $A \cup (B \cap C) = (A \cup B) \cap (A \cup C)$ in two ways.
(d) Prove: $\cap$ and $\cup$ are associative operations.

2.2.3. (a) Prove: $A \subseteq B$ if and only if $A \cup B = B$.
(b) Prove: $A \subseteq B$ if and only if $A \cap B = A$.
(c) Prove: If $A \subseteq B$ and $A \subseteq C$, then $A \subseteq B \cap C$.
(d) Prove: If $A \subseteq B$ and $C \subseteq D$, then $A \cap C \subseteq B \cap D$ and $A \cup C \subseteq B \cup D$.

2.2.4. Let U be a set and A, B, C be subsets of U.
(a) Prove: $A - B = A \cap B^c$ and $A - B = B^c - A^c$.
(b) Prove: $A - (A - B) = A \cap B$.
(c) Prove: $(A - B)^c = A^c \cup B$.
(d) Prove: $(A \cup B) - (A \cap B) = (A - B) \cup (B - A)$.
(e) Prove: $A \cap (B - C) = (A \cap B) - (A \cap C)$.

2.2.5. Show that $A - B = A$ if and only if $B - A = B$.

2.2.6. Give two examples of idempotent operations on $\mathbf{R}$.

2.2.7. (a) Prove or disprove: Set difference is an associative operation.
(b) Give a necessary and sufficient condition for $A - B = B - A$. (Hint: You should state and prove a result in the form: "$A - B = B - A$ if and only if ______," where the statement filling the blank gives some property satisfied by A and B. Of course, that statement should be something other than "$A - B = B - A$.")

2.2.8. List three unary operations on the set M_n of $n \times n$ matrices with real entries.

2.2.9. Prove Theorem 2.2.2, parts (ii)–(v).

2.2.10. Complete the second proof of Theorem 2.2.4.

2.2.11. Find $P(\{1\})$ and $P(P(\{1\}))$.

2.2.12. Prove: If $A \cap B = \varnothing$, then $P(A) \cap P(B) = \{\varnothing\}$. Is the converse true?

2.2.13. (a) Prove: $P(A \cap B) = P(A) \cap P(B)$. Express this result in a sentence that contains no mathematical symbols.
(b) State and prove a result concerning the power set of the union of two sets.

2.2.14. Suppose A is a set with n elements where n is a positive integer.
(a) How many subsets of A contain exactly one element?
(b) How many subsets of A contain exactly $n - 1$ elements? (Hint: Consider the binary representation of subsets.)

2.2.15. Symmetric Difference. Let A and B be subsets of a set U. The *symmetric difference of A and B*, written $A + B$, is the set $A + B = (A - B) \cup (B - A)$. Clearly symmetric difference is a binary operation on $P(U)$. (Note: Some authors write $A \triangle B$ in place of $A + B$.)
(a) Let $A = \{1,2,3\}$, $B = \{3,4,5\}$, and $C = \{4,5,6\}$. Evaluate $A + B$, $A + C$, $B + C$, $(A + B) + C$, and $A + (B + C)$.
(b) Draw a Venn diagram for $A + B$.
(c) Prove: $A + B = (A \cup B) - (A \cap B)$.
(d) Prove $+$ is commutative.
(e) Prove: For any $A \subseteq U$, $A + A = \varnothing$ and $A + \varnothing = A$.
(f) Prove: For all $A, B \in P(U)$, $A^c + B^c = A + B$.
(g) What is $A + A^c$ for an arbitrary set $A \subseteq U$?
(h) Draw a Venn diagram to convince yourself that $+$ is associative.
(i) Suppose U is finite. Describe the binary sequence of $A + B$ in terms of the binary sequences of A and B.

2.2.16. For $i \in \mathbf{Z}^+$, let $B_i = (1/2 - 1/i, 1/2 + 1/i)$. Find $\bigcap_{i \in \mathbf{Z}^+} B_i$ and $\bigcup_{i \in \mathbf{Z}^+} B_i$.

2.2.17. Prove Theorem 2.2.6, parts (ii) and (iv).

2.2.18. Use Axioms 1, 2, and 7 to deduce the existence of infinitely many sets.

2.2.19. (a) Can there exist sets x and y such that $x \in y$ and $y \in x$?
(b) Can there exist sets x, y, and z such that $x \in y$, $y \in z$, and $z \in x$?
(c) Generalize the conclusions you made in parts (a) and (b).

2.2.20. Use the Axiom of Regularity to show that the set of all sets does not exist. In other words, the Axiom of Regularity prevents Russell's paradox from arising. (Note: Also make the assumption that for any set A if $x \in A$, then x is a set.)

§2.3 Ordered Pairs and Cartesian Products

The notion of set operation provides a unifying thread for the topics discused in Section 2.2. For any set U, intersection, union, and set difference are binary operations on the set $P(U)$ and set complement is a unary operation on $P(U)$. Additionally, forming the power set *almost* yields a unary operation: Each set A is assigned to $P(A)$, the set of all subsets of A. However, the power set does not really define a unary operation on a set, for while A, B, and $A \cup B$ (or $A \cap B$ or $A - B$, if you will) are usually taken to be subsets of a set U, A and $P(A)$ are not readily pictured as subsets of a given set. Nonetheless, all these procedures have a common feature: They produce a new set from a given set or pair of sets.

This section is devoted to yet another set operation, the Cartesian product. Our purpose is to define the Cartesian product of two sets and to establish some connection between Cartesian products and previously defined set operations. We also return briefly to the definition of binary operation and close the section by completing one treatment of axiomatic set theory.

The Cartesian product is familiar to students of calculus in a special case, the Cartesian plane, $\mathbf{R}^2$. Let us recall the definition of $\mathbf{R}^2$ so as to use it as a motivating example.

By definition,

$$\mathbf{R}^2 = \{(x, y) | x \in \mathbf{R} \text{ and } y \in \mathbf{R}\}.$$

The symbol (x, y) is called the ordered pair x and y. Thus to form $\mathbf{R}^2$, one "combines" pairs of elements of $\mathbf{R}$ in a certain fashion.

Geometrically, this combination is represented as a plane stretching endlessly in every direction with two specific perpendicular lines serving to provide a method of labeling points in the plane. Symbolically, the combination (x, y) of the elements x and y of $\mathbf{R}$ is regarded as a list with two entries, x being the first and y the second. The elements x and y are often called the *first* and *second coordinates* of (x, y), respectively.

The geometric representation of $\mathbf{R}^2$ is facilitated in part by the geometric representation of $\mathbf{R}$ as a straight line whose points are labeled by real numbers (Figure 2.3.1). Suppose we wish to extend the concept of Cartesian plane by considering, in place of $\mathbf{R}$, an arbitrary set A. Conceiving of this extension in geometric terms may be difficult unless a geometric representation of A is readily available. However the symbolic description of $\mathbf{R}^2$ as a set of ordered pairs can be extended easily to an arbitrary set A: Simply take all ordered pairs (x, y) of elements of A where $x \in A$ is regarded as

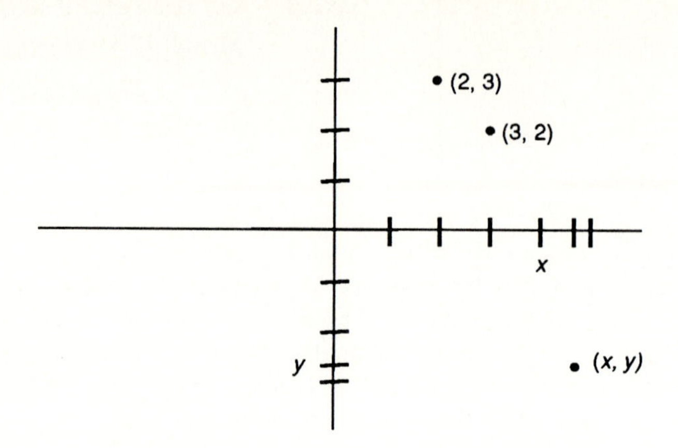

Figure 2.3.1. Geometric Representation of $\mathbf{R}^2$.

the first element and $y \in A$ is taken to be the second. Thus, for a given set A, we define

$$A^2 = \{(x, y) | x \in A \text{ and } y \in A\}.$$

Before setting this definition in concrete, let us take a closer look at the idea of ordered pair. In fact, no formal definition of the concept of ordered pair has been given; we have merely said that the symbol (x, y) is called the ordered pair x and y where x is the first member of the pair and y is the second. While the meaning of the term may be clear, the situation is unsatisfactory. Our understanding of the idea of ordered pair as "defined" above depends heavily on the written representation of it or upon concepts that are difficult to define mathematically (at least at this stage of our development) such as "list with two entries." Thus, even though we have a sound, intuitive feeling for the notion of ordered pair, we may find the idea elusive and difficult to grasp in situations where tight, rigorous reasoning is required. Thus we demand a precise, mathematical definition of the term "ordered pair."

To provide this definition, recall the dictum mentioned at the start of this chapter that any mathematical entity should be definable in terms of some set. With this in mind, we propose the following definition.

Definition 2.3.1 *Ordered Pair*

Let x and y be elements of sets A and B, respectively. The *ordered pair x and y*, written (x, y), is defined to be the set

$$(x, y) = \{\{x\}, \{x, y\}\}.$$

Observe that (x, y) is a set with two elements (well, usually; see Exercise 2.3.4(a)), and each of these elements is itself a set. This definition allows us to distinguish between (x, y) and (y, x). For example, consider the pairs (2, 3) and (3, 2):

$$(2,3) = \{\{2\},\{2,3\}\} \qquad \text{and} \qquad (3,2) = \{\{3\},\{3,2\}\}.$$

Note that $\{2\} \in (2,3)$ but $\{2\} \notin (3,2)$, hence $(2,3) \neq (3,2)$. In general, $(x, y) = (u, v)$ if and only if $x = u$ and $y = v$. (See Exercise 2.3.3.)

Now we return to the task of defining the analog of $\mathbf{R}^2$ for an arbitrary set A in place of $\mathbf{R}$. Actually, a more general concept can be defined with equal ease.

Definition 2.3.2 *Cartesian Product*

Let A and B be sets. The *Cartesian product of A and B*, written $A \times B$, is the set

$$A \times B = \{(a, b) | a \in A \text{ and } b \in B\}.$$

If $A = B$, then we write A^2 for $A \times B = A \times A$.

Example 2.3.1. Let $A = \{1,2\}$ and $B = \{3,4,5\}$. Then

$$A \times B = \{(1,3),(1,4),(1,5),(2,3),(2,4),(2,5)\}$$

$$B \times A = \{(3,1),(3,2),(4,1),(4,2),(5,1),(5,2)\}, \text{ and}$$

$$A^2 = \{(1,1),(1,2),(2,1),(2,2)\}.$$

Thus, in general, $A \times B \neq B \times A$.

The previous example does suggest some questions that we leave as exercises:

1. Under what conditions (on A and B) does $A \times B = B \times A$? Try to find some examples. Try to find all examples.
2. Let A and B be finite sets. Is there a connection between the number of elements of A and B and the number of elements of $A \times B$? Again, try some examples and make a conjecture.

What about pictures for the Cartesian product? As mentioned earlier, the Cartesian plane $\mathbf{R}^2 = \mathbf{R} \times \mathbf{R}$ has a familiar geometric realization. Be-

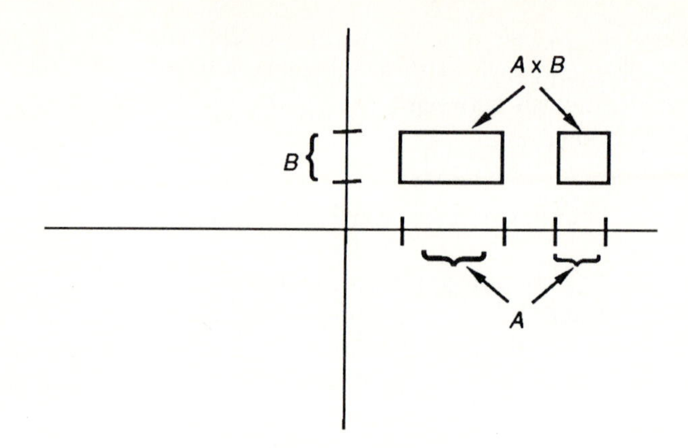

Figure 2.3.2

cause of our past experience with this set, we use $\mathbf{R}^2$ as a model example for the Cartesian product of any two sets. Thus, whenever we need a picture of the Cartesian product of two sets, we take the sets to be subsets of $\mathbf{R}$ so that their Cartesian product is a subset of $\mathbf{R} \times \mathbf{R}$ (Fig. 2.3.2).

What is the relationship between the operation of Cartesian products and each of the binary operations of union and intersection? For example, does x distribute over $\cap$ and $\cup$? Before reading the answer in the next theorem, you might wish to conjecture an answer. To help formulate a conjecture, draw a few pictures and calculate a few simple (e.g., finite) examples.

Theorem 2.3.1 *Let A, B, C, and D be sets. Then*
(i) $A \times (B \cap C) = (A \times B) \cap (A \times C)$.
(ii) $A \times (B \cup C) = (A \times B) \cup (A \times C)$.
(iii) $(A \times B) \cap (C \times D) = (A \cap C) \times (B \cap D)$.
(iv) $(A \times B) \cup (C \times D) \subseteq (A \cup C) \times (B \cup D)$.

Proof of Theorem 2.3.1(i) We show that for all x, $x \in A \times (B \cap C)$ if and only if $x \in (A \times B) \cap (A \times C)$. From the definitions of intersection and Cartesian product,

$$x \in A \times (B \cap C)$$

if and only if $x = (u, v)$ where $u \in A$ and $v \in B \cap C$,
if and only if $x = (u, v)$ where $u \in A$, $v \in B$, and $v \in C$,
if and only if $x = (u, v) \in A \times B$ and $x = (u, v) \in A \times C$,
if and only if $x \in (A \times B) \cap (A \times C)$. ∎

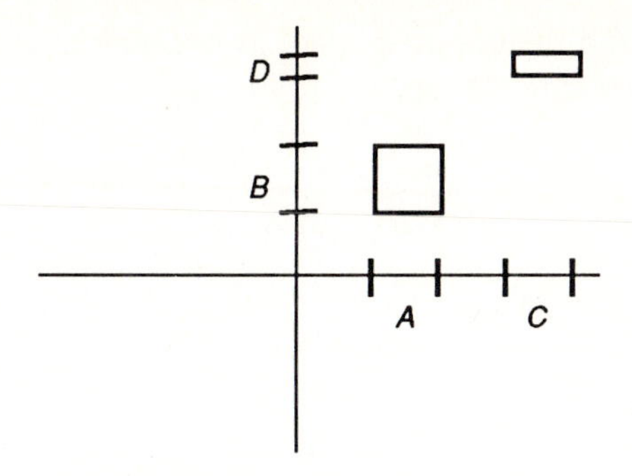

Figure 2.3.3. $(A \times B) \cup (C \times D)$.

Proof of Theorem 2.3.1(iv) To show that $(A \times B) \cup (C \times D) \subseteq (A \cup C) \times (B \cup D)$, we show that if $x \in (A \times B) \cup (C \times D)$, then $x \in (A \cup C) \times (B \cup D)$.

Since $x \in (A \times B) \cup (C \times D)$, either $x \in A \times B$ or $x \in C \times D$. If $x \in A \times B$, then $x = (a, b)$ where $a \in A$ and $b \in B$. Since $A \subseteq A \cup C$ and $B \subseteq B \cup D$, $x = (a, b) \in (A \cup C) \times (B \cup D)$.

On the other hand, if $x \in C \times D$, then $x = (c, d)$ where $c \in C \subseteq A \cup C$ and $d \in D \subseteq B \cup D$. Thus $x = (c, d) \in (A \cup C) \times (B \cup D)$. Therefore, if $x \in (A \times B) \cup (C \times D)$, then $x \in (A \cup C) \times (B \cup D)$. ■

We leave the proofs of (ii) and (iii) as exercises.

Remark Let us draw a picture to illustrate Theorem 2.3.1(iv). We take A, B, C, and D to be intervals in **R**. Then $(A \times B) \cup (C \times D)$ and $(A \cup C) \times (B \cup D)$ are pictured in Figures 2.3.3 and 2.3.4, respectively. These figures clearly illustrate the statement that $(A \times B) \cup (C \times D) \subseteq (A \cup C) \times (B \cup D)$.

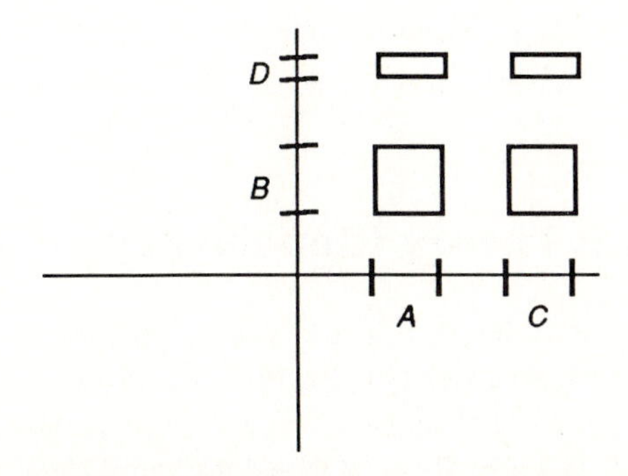

Figure 2.3.4. $(A \cup C) \times (B \cup D)$.

We now extend the notion of the Cartesian product of a set with itself. The motivating example behind the following definition is $\mathbf{R}^3$, real 3-space.

Definition 2.3.3 *n-Fold Cartesian Product*

Let A be a set. Define $A^3 = A^2 \times A$, $A^4 = A^3 \times A$, and in general for $n \in \mathbf{Z}$ with $n \geq 3$, define $A^n = A^{n-1} \times A$. The set A^n is called the *n-fold Cartesian product of A*.

A typical element of A^3 has the form $((a, b), c)$ where $a, b, c \in A$. For simplicity we write (a, b, c) in place of $((a, b), c)$. Similarly we write a typical element of A^n in the form $(a_1, \ldots, a_n)$ where $a_i \in A$ for $1 \leq i \leq n$. We call $(a_1, \ldots, a_n)$ an *n-tuple* of elements of A.

Perhaps the most important feature of the concept of Cartesian product is that it allows some important concepts to be expressed simply and clearly. In other words the Cartesian product is a significant part of the language of mathematics. For example, armed with the notion of Cartesian product, we can reformulate slightly the definition of binary operation on a set.

Definition 2.2.1 *(revised) Binary Operation*

Let A be a set. A *binary operation on A* is a rule, denoted by $*$, which assigns to each element $(a, b) \in A \times A$ an element $a * b$ of A.

The concepts of function and (more generally) relations are most easily defined in terms of Cartesian products. These topics are described in the following chapter and appear in the remainder of this book, and indeed throughout mathematics. We finish this section by wrapping up our discussion of axiomatic set theory.

Axiomatic Set Theory (Conclusion)

The concepts of ordered pair and Cartesian product can be defined in terms of the axioms presented in Sections 2.1 and 2.2. For instance, the ordered pair (x, y) is defined to be $\{\{x\}, \{x, y\}\}$. By Axiom 4 the sets $\{x, y\}$ and $\{x\}$ exist, and from another application of this axiom the set $\{\{x\}, \{x, y\}\}$ exists. Notice that $(x, y) \in P(P(\{x, y\}))$. Next let A and B be sets. From the Axioms of Pairing, Union, Power Set, and Separation we can define $A \times B = \{z \in P(P(A \cup B)) | \exists x \in A \wedge \exists y \in B$ such that $z = (x, y)\}$.

Once the Cartesian product of two sets has been defined, we can define the notion of function. Functions are, of course, among the most important of all mathematical objects. The last section of Chapter 3 is devoted entirely to properties of functions. We need the concept of function at this time in order to state the Axiom of Replacement, the last of the Zermelo–Fraenkel axioms.

Definition 2.3.4 *Function*

Let A and B be sets. A *function from A to B*, written $f\colon A \to B$, is a subset f of $A \times B$ such that for each $a \in A$ there exists at most one $b \in B$ such that $(a, b) \in f$. If $(a, b) \in f$, then we write $b = f(a)$.

For example, $f = \{(x, x^2) | x \in \mathbf{R}\}$ is a function from $\mathbf{R}$ to $\mathbf{R}$. (Observe that $f \subseteq \mathbf{R} \times \mathbf{R}$ and for each $x \in \mathbf{R}$, (x, x^2) is the only pair in f whose first coordinate is x.) Note that $f = \{(x, \pm\sqrt{x}) | x \in \mathbf{R}$ and $x \geq 0\}$ is not a function, since $(1, 1) \in f$ and $(1, -1) \in f$.

The Axiom of Replacement is the final axiom in our rendition of ZF. Essentially, this axiom asserts that the image of a set under a function is a set.

According to Definition 2.3.4, a function is a set of ordered pairs that possess the additional condition that $(x, y) \in f$ and $(x, z) \in f$ imply $y = z$. We can reasonably ask how in general can a function actually be defined? In the case of a function f from A to B, an explicit formula (such as $f(x) = x^2$) might be sufficient. But what about an arbitrary function? Zermelo's answer to this question was to define a function by stating the property by which the objects x and y are related in order for (x, y) to be in f. The idea that a function is defined by a property of the pair (x, y) is reminiscent of the Axiom of Separation in which a property of objects is used to define sets. Zermelo, however, was somewhat vague on what was meant by a property of sets. This notion was made precise by Skolem and Fraenkel in the early 1920s with their introduction of the concept of *well-formed formula*. (A *well-formed formula* is a statement constructed from the statements $x \in y$ and $x = y$, where x and y are variables representing sets, using the logical connectives and qualifiers.) Thus the function concept can be defined as follows: A *function* is a set f such that

$$f = \left\{ (x, y) \,\middle|\, \begin{array}{l} F(x, y) \text{ holds where } F(x, y) \text{ is a well-formed} \\ \text{formula with the property that if } F(x, y) \text{ and} \\ F(x, z) \text{ hold, then } y = z \end{array} \right\}$$

The Axiom of Replacement asserts that if in a function the variable x ranges over a set, then the corresponding variable y $(= f(x))$ ranges over a set. In other words, the image of a set under a function is a set.

Axiom 9 Axiom of Replacement Let f be a function defined by a property F. Let A be a set. Then there exists a set whose elements are those objects y such that there exists $x \in A$ such that $F(x, y)$ holds. Formally,

$$\forall A(\exists B(\forall y(y \in B \Leftrightarrow \exists x(x \in A \wedge F(x, y))))).$$

As above, we write $y = f(x)$ if $F(x, y)$ holds. Thus, according to Axiom 9, for any set A and any function f there exists a set B such that

$$B = \{ f(x) | x \in A \}.$$

This representation inspires the use of the word "Replacement" to describe Axiom 9. The elements $x \in A$ are replaced by $f(x)$ to form the set B.

Axiom 9 is often called an *axiom scheme* since it actually entails infinitely many axioms, one for each property $F(x, y)$. In this sense, Axiom 9 is similar to the Axiom of Separation (Axiom 3). In fact, the latter axiom can be deduced from the Axiom of the Empty Set (Axiom 2) and the Axiom of Replacement (see van Dalen et al. [1], p. 143, or Hamilton [5], pp. 125–126, for details). Thus the axioms of ZF are redundant and from a logical point of view Axiom 3 can be omitted from the list of axioms. In most treatments of the ZF axioms, however, the Axiom of Separation is included since sets are usually defined in the form prescribed in the statement of that axiom.

Exercises §2.3

2.3.1. (a) Find $A \times B$ if $A = \{1,2,3\}$ and $B = \{3,4,5\}$.
(b) Find $\{a, b\} \times \{a, b\}$.
(c) Find $\{(a, b)\} \times \{a, b\}$.

2.3.2. Draw a sketch of each of the following subsets of $\mathbf{R}^2$.
(a) $[0,1] \times [1,2]$ (Recall $[a, b] = x \in \mathbf{R} | a \leq x \leq b$).
(b) $([0,1] \cup \{2\}) \times [1,2]$.
(c) $([0,1] \cup \{2\}) \times ([1,2] \cup \{3\})$.

2.3.3. Prove: $(a, b) = (c, d)$ if and only if $a = c$ and $b = d$.

2.3.4. (a) Show: $(a, a) = \{\{a\}\}$.
(b) Show: $\{a\} \times \{a\} = \{\{\{a\}\}\}$.

2.3.5. True or false: $(a, b) \cap (b, a) = \varnothing$.

2.3.6. Fill in the blank: $A \times B = \varnothing$ if and only if ______.

2.3.7. (a) Prove: If $a, b \in A$, then $(a, b) \in P(P(A))$.
(b) Prove: If $a \in A$ and $b \in B$, then $(a, b) \in P(P(A \cup B))$.

2.3.8. Let A and B be sets. State a necessary and sufficient condition for $A \times B$ to equal $B \times A$.

2.3.9. Give a set theoretic definition of the concept "unordered pair x and y." Your definition should imply that the unordered pair x and y coincides with the unordered pair y and x.

2.3.10. (a) Draw pictures for each of the four assertions of Theorem 2.3.1.
(b) Prove Theorem 2.3.1(ii) and (iii).
(c) Prove, or disprove and salvage: If A, B, C, and D are all nonempty sets, then $(A \times B) \cup (C \times D) \subset (A \cup C) \times (B \cup D)$.

2.3.11. Let A and B be sets. Write $A^2 - B^2$ as a union of two sets each of which is a Cartesian product of two sets. (Hint: Draw a picture.)

2.3.12. Suppose A and B are finite sets.
(a) Formulate a conjecture that expresses the number of elements of $A \times B$ in terms of the number of elements of A and B. Check your conjecture in some simple cases.
(b) Prove your conjecture. (Hint: Use mathematical induction on the number of elements of A.)

2.3.13. State and prove analogs of Theorem 2.3.1(i) and (ii) for families of sets. For example: $A \times (\bigcap_{i \in I} B_i) = ?$

2.3.14. Prove, or disprove and salvage: If A, B, and C are sets with $A \times B = A \times C$, then $B = C$.

2.3.15. (a) Define the concept of "ordered triple" (a, b, c).
(b) For each positive integer n, define the concept of an ordered n-tuple $(a_1, a_2, \ldots, a_n)$.

2.3.16. Using Axioms 3 and 6 but not Axiom 4, prove that for any set A there exists a set having A as its only element.

2.3.17. Prove: For any set A there exists a set whose only members are $\{x\}$ where $x \in A$.

Notes

Set theory originated in the last quarter of the nineteenth century with the work of Georg Cantor. We shall encounter some of Cantor's major results in Chapter 4.) Cantor's approach was informal rather than axiomatic. By 1895 Cantor had encountered some paradoxes, similar to Russell's paradox, which, as we now see, arose from Cantor's nonaxiomatic development of set theory. Because of these paradoxes and because of Cantor's work on infinite sets, many mathematicians, the most notable being Leopold Kronecker, strenuously objected to many of Cantor's results. On the other hand, many of Cantor's theorems earned the appreciation of other influential mathematicians. Among the latter was Richard Dedekind, who sought to base the concept of number on the idea of set [3].

The problem of finding a rigorous development of the real number system also inspired the work of Frege, Russell, and Whitehead. Russell's discovery of his paradox in 1901 shook the believers in set theory profoundly. For example, the second volume of Frege's two-volume formalization of arithmetic was about to appear when he learned of Russell's paradox. Frege added an epilogue to his second volume and one of his particularly well-known remarks is worth quoting:

> To a scientific author hardly something worse can happen, than the destruction of the foundation of his edifice after the completion of his work. I was placed in this position by a letter of Mr. Bertrand Russell, when the printing came to a close.

Dedekind himself was moved to withhold the publication of the third edition of [3] for several years until 1911.

In the meantime Zermelo began his axiomatic development of set theory in 1908. Zermelo's project was completed independently by Fraenkel and T. Skolem in the early 1920s. Axiomatic treatments of set theory were given by (i) Russell and Whitehead and (ii) von Neumann, Bernays, and Gödel. A brief account of the early history of set theory is found in [6]. A more complete treatment can be found in the informative essay by van Dalen in [2].

Several excellent textbooks on set theory are available. An old classic is Halmos [4]. The relatively recent and somewhat advanced text by van Dalen et al. [1] contains both an informal and a formal approach to sets. The text by Hamilton [5] is carefully written and contains many insightful comments on axiomatic set theory.

Bibliography

1. D. van Dalen, H. C. Doets, and H. DeSwart, *Sets: Naive, Axiomatic, and Applied*, Pergamon, 1978.
2. D. van Dalen and A. F. Monna, *Sets and Integration: An Outline of the Development*, Walters-Noordhoff, 1972.
3. R. Dedekind, *Essays on the Theory of Numbers*, Dover, 1963.
4. P. R. Halmos, *Naive Set Theory*, Van Nostrand, 1960.
5. A. G. Hamilton, *Numbers, Sets and Axioms*, Cambridge University Press, 1982.
6. M. Kline, *Mathematical Thought from Ancient to Modern Times*, Oxford University Press, 1972.

Miscellaneous Exercises

1. Let m, n be fixed positive integers. For $a, b \in \mathbf{R}$ define $a * b$ to be $(ma + nb)/(m + n)$.
   (a) Show that the binary operation $*$ on $\mathbf{R}$ has the *self-distributive property*: $a *(b * c) = (a * b)*(a * c)$.
   (b) Show that the arithmetic mean of a and b, $a * b = (a + b)/2$ has the self-distributive property.

2. Diagonalization Principle. A standard proof technique in set theory, mathematical logic, and the theory of computation is the so-called *diagonal argument*. Prove the following general version of the diagonal argument:
   Let R be any relation on a set A. Let $D = \{a \in A | (a, a) \notin R\}$. D is called the *diagonal set* for the relation R. For each $a \in A$, define $R_a = \{b \in A | a R b\}$. Then, for all $a \in A$, $D \neq R_a$.

3. Suppose A and B are finite sets with m and n elements respectively. Prove, or disprove and salvage: $A \cup B$ is a finite set with $m + n$ elements.

4. Propose a definition of the concept of "ternary operation on a set." (A dictionary might help.) Give three examples.

5. Let A be a finite set. Show that the subsets of A (i.e., the elements of $P(A)$) can be listed in such a way that
   (a) $\varnothing$ is first on the list,
   (b) each subset of A appears exactly once,
   (c) each subset on the list after the first is obtained either by deleting an element from the previous set on the list or by adding an element to the previous set on the list.

6. Let $\{A_i | i \in I\}$ and $\{B_i | i \in I\}$ be two families of sets having the same indexing set I. Suppose that for each $i \in I$, $A_i \subseteq B_i$. Show that
   (a) $\bigcup_{i \in I} A_i \subseteq \bigcup_{i \in I} B_i$.
   (b) $\bigcap_{i \in I} A_i \subseteq \bigcap_{i \in I} B_i$.

7. For $n \in N^+$, let $A_n = (-1/n, 1/n) = \{x \in \mathbf{R} | -1/n < x < 1/n\}$. Find $\bigcap_{n=1}^{\infty} A_n$ and $\bigcup_{n=1}^{\infty} A_n$.

8. Let A be a set. Show that $\{x \in A | x \in x\} \subseteq P(A) - B$ where $B = \{\{a\} | a \in A\}$. (Hint: Use the Axiom of Regularity.)

9. Let $A \not\subset \mathbf{R}$ and $B \subseteq \mathbf{R}$. Suppose that A and B have an upper bound.
   (a) Show that $A \cup B$ has an upper bound.
   (b) Express the least upper bound of $A \cup B$ in terms of the least upper bounds of A and B.

10. Prove, or disprove and salvage: Let A and B be sets. Then $P(A - B) \subseteq P(A) - P(B)$.

CHAPTER THREE

Relations and Functions

§3.1 Relations

We continue the discussion of set theory begun in the last chapter. We concentrate on a fundamental theme, that of *relation*, and several important variations on this theme, namely, those of order relations, equivalence relations, and functions. To varying degrees these ideas are familiar to most readers: functions, of course, have been with us since high school when we analyzed linear and quadratic equations thoroughly. Ideas such as "less than" in algebra and "congruence" in geometry are examples of order relations and equivalence relations, respectively.

The idea of a relation arises out of the consideration of connections among elements of a given set or among elements of one set and elements of another set. For example, two people, A and B, living today might be connected by the fact that they are siblings. Obviously, some pairs of people are connected in this way; most are not. As for some mathematical examples, two real numbers, x and y, might be connected by the fact that $y = e^x$, or x and y might be related by the fact that x is less than or equal to y. A precise mathematical idea, called a *relation*, hides behind these illustrations. We shall propose a formal definition after considering the last example in more detail.

Recall that if x and y are real numbers, then x is *less than or equal to* y, written $x \leq y$, if $y - x$ is nonnegative. Thus, given a pair of real numbers, in fact an ordered pair of real numbers, (x, y), to determine if $x \leq y$ we simply check if $y - x$ is nonnegative; equivalently, we see whether y can be obtained by adding some nonnegative number to x. If this is not the case, then we write $x \nleq y$. For example, $e \leq \pi$, $\pi^2 \leq 10$, but $\pi \nleq e$. Hence the statement "x less than or equal to y" describes a property of the ordered pair (x, y). We can then use this property to define a set: Consider the subset, call it $R_{\leq}$, of $\mathbf{R}^2$ defined as follows:

$$\begin{aligned}\mathbf{R}_{\leq} &= \{(x, y) \in \mathbf{R}^2 | x \leq y\} \\ &= \{(x, y) | y - x \text{ is nonnegative}\}.\end{aligned}$$

The set $\mathbf{R}_{\leq}$ determines the relationship between x and y relative to the property "less than or equal"; for $x \leq y$ if and only if $(x, y) \in \mathbf{R}_{\leq}$. Generalizing from this example, we introduce the following simple yet important concept.

Definition 3.1.1 *Relation*

A *relation on a set A* is a subset, R, of $A^2 = A \times A$.

Being a subset of $A \times A$, a relation is specified in the ways that subsets of a given set are generally defined. For instance, if A is a finite set (preferably a small finite set), then a relation on A can be defined simply by listing its elements. Of course, rarely is this method practical or interesting. Usually, to define a relation we provide a statement (more precisely, a predicate) that singles out a collection of elements of $A \times A$ for membership in the relation. For any relation R, one writes, of course, $(x, y) \in R$ if the pair (x, y) satisfies the defining property of R. In practice, we usually drop this notation in favor of a more natural way of writing things. Again with the example of $\leq$ in mind, we write $x R y$ if $(x, y) \in R$ and $x \not R y$ if $(x, y) \notin R$. Let us look at several examples.

Example 3.1.1 For $x, y \in \mathbf{R}$, define $x < y$ to mean $y - x$ is positive. Thus $x < y$ if and only if $x \leq y$ and $x \neq y$.

Example 3.1.2 When first encountering a new idea, it always helps if we see "extreme" examples of it, even if these examples happen to be trivial and uninteresting. For any set A, there are two extreme examples of relations on A, namely $\varnothing$ and A^2 (and, indeed, these examples are not very interesting). In the first case, no element is related to any other, and in the second case, each element is related to all others. These examples are "extreme" in the sense that if R is any relation on A, then $\varnothing \subseteq R \subseteq A \times A$. They are "trivial" in the sense that for any set A, $\varnothing$ and $A \times A$ are relations on A.

Example 3.1.3 Let $A = \{1, 2\}$. Here are several relations on A: $R_1 = \{(1,1), (1,2)\}$, $R_2 = \{(1,1), (2,2), (1,2)\}$. $R_3 = \{(1,2)\}$, $R_4 = \{(2,1)\}$, and $R_5 = \{(1,2), (2,1)\}$. Notice that since $A \times A$ has 4 elements, $P(A \times A)$ has $2^4 = 16$ subsets; thus there exist 16 relations on $A = \{1, 2\}$.

Example 3.1.4 Here is another simple, general (and familiar) relation. For any set A let

$$I_A = \{(x, y) | x = y\}.$$

Although I_A merely provides a formal way of expressing equality, it usually goes by the name *identity relation*.

Example 3.1.5 Let $\mathbf{Z}^+$ denote the set of positive integers. (Thus $\mathbf{Z}^+ = \mathbf{N} - \{0\}$.) For $a, b \in \mathbf{Z}^+$, one says that a *divides* b, written $a|b$, if there exists $c \in \mathbf{Z}^+$ such that $ac = b$. Write $a \nmid b$ if a does not divide b. Examples: $2|8$, $7|91$, and $3 \nmid 89$. Thus the common notion of divisibility of positive integers yields an example of a relation. This relation can be extended to the set $\mathbf{Z}$ of all integers in an obvious way: For $a, b \in \mathbf{Z}$, $a|b$ means that there exists $c \in \mathbf{Z}$ such that $a \cdot c = b$. If $a|b$, then we also say b is *divisible* by a, b is a *multiple* of a, a is a *factor* of b, or a is a *divisor* of b.

Example 3.1.6 Let U be a set.

(i) Containment defines a relation, $R_{\subseteq}$, on $P(U)$: $(A, B) \in R_{\subseteq}$ if and only if $A \subseteq B$.

(ii) For $x, y \in P(U)$, define $x R y$ to mean $x \cap y \neq \varnothing$. R is a relation on $P(U)$.

Example 3.1.7 Let us recall another familiar example. Let $\mathscr{F}$ be the set of fractions of integers; $1/2$, $2/3$, $8/9$, $-13/11$, etc. Define $a/b \equiv c/d$ if $ad = bc$. Then $\equiv$ is a relation on $\mathscr{F}$. Moreover, $\equiv$ is precisely the relation that we employ when we assert that two ways of writing a number as a fraction are equivalent. For instance, when we say that $1/2 = 2/4$, we are merely saying that the pair $(1/4, 2/4)$ is a member of the relation $\equiv$.

Properties of Relations

We continue our study of relations by isolating various properties and studying those relations that have certain combinations of these properties. Several important relations happen to possess many of these properties. Actually this fact is at least partly responsible for the formulation of these concepts. At the same time we continue with our policy of representing a concept in multiple ways by interpreting the ensuing ideas geometrically whenever appropriate.

Definition 3.1.2 *Reflexive Relation*

A relation R on a set A is *reflexive* if for all $x \in A$, $x R x$.

Definition 3.1.3 *Irreflexive Relation*

A relation R on a set A is *irreflexive* if for all $x \in A$, $x \not R x$.

Definition 3.1.4 *Symmetric Relation*

A relation R on a set A is *symmetric* if, for all $x, y \in A$, $x R y$ implies $y R x$.

Definition 3.1.5 *Antisymmetric Relation*

A relation R on a set A is *antisymmetric* if, for all $x, y \in A$, $x R y$ and $y R x$ imply $x = y$.

Definition 3.1.6 *Asymmetric Relation*

A relation R on a set A is *asymmetric* if for all $x, y \in A$, $x R y$ implies $y \not R x$.

Definition 3.1.7 *Transitive Relation*

A relation R on a set A is *transitive* if, for all $x, y, z \in A$, $x R y$ and $y R z$ imply $x R z$.

Example 3.1.8 Consider the inequality relations on **R** (or if you like on **Q**, **Z**, or **N**).

(i) $\leq$ is reflexive, antisymmetric, and transitive. The relation $\geq$ has the same properties.

(ii) The relation $<$ is irreflexive, asymmetric, and transitive.

Example 3.1.9 Let U be a set and consider the inclusion relations on $P(U)$:

(i) $\subseteq$ is reflexive, antisymmetric, and transitive. The relation $\supseteq$ possesses the same properties.

(ii) The relation $\subset$ is irreflexive, asymmetric, and transitive.

Example 3.1.10 Let $A = \{a, b, c\}$.

(i) The relation $R = \{(a, a), (a, b), (b, a), (b, c)\}$ is neither reflexive, irreflexive, symmetric, asymmetric, antisymmetric, nor transitive. Nor for that matter is R very interesting; however it does illustrate the fact that simple examples of relations not having (or having) various properties can be easily constructed.

(ii) Let $R = \{(a, a), (b, b), (a, b), (c, c)\}$. Then R is a reflexive, antisymmetric, and transitive relation on A. Notice that R is not symmetric.

Example 3.1.11 The divisibility relation $|$ on $\mathbf{N}$ is reflexive, antisymmetric, and transitive. (Recall that for $a, b \in \mathbf{Z}^+$, $a|b$ if there exists $c \in \mathbf{N}$ such that $a \cdot c = b$.) For instance, to prove transitivity, suppose $a|b$ and $b|c$ and show $a|c$. If $a|b$ and $b|c$, then there exist $x, y \in \mathbf{Z}$ such that $b = a \cdot x$ and $c = b \cdot y$. Therefore, $c = b \cdot y = (a \cdot x) \cdot y = a \cdot (x \cdot y)$, which implies that $a|c$. On $\mathbf{Z}$, however, this relation is merely reflexive and transitive. Since $2|(-2)$, $(-2)|2$, and $-2 \neq 2$, the relation $|$ is not antisymmetric on $\mathbf{Z}$.

As we saw in Examples 3.1.7 and 3.1.8, the relations $\leq$ on R and $\subseteq$ on $P(U)$ are reflexive, antisymmetric, and transitive. At the same time these relations provide ways of comparing the elements of their respective sets. For instance, if $A, B \in P(U)$ and $A \subseteq B$, then we can think of A as being "smaller than" B. We now extend the notions of inequality and ordering from these examples to arbitrary sets by isolating the basic properties of these relations. Unfortunately, the terminology is not completely standard; we shall use what is probably the most common.

Definition 3.1.8 *Partial Ordering*

A relation R on a set A is a *partial ordering* if R is a reflexive, antisymmetric, and transitive relation. We write (A, R) to denote the set A and the partial ordering R.

As noted earlier, the relations $\leq$ and $\geq$ on R and $\subseteq$ and $\supseteq$ on $P(U)$ are all partial orderings. By contrast, none of the relations $<$, $>$, $\subset$, or $\supset$ is a partial ordering.

Example 3.1.12 (i) The relation $|$ on $\mathbf{N}^+$ is a partial ordering.

(a) Reflexivity: For any $a \in \mathbf{N}^+$, $a|a$ since $a \cdot 1 = a$.

(b) Antisymmetry: If $a|b$ and $b|a$, then $b = ac$ and $a = bd$ for some $c, d \in \mathbf{N}^+$; then $b = ac = b(dc)$, which means that $dc = 1$, which in turn implies that $c = d = 1$, and $b = a$.

(c) Transitivity: The transitivity of $|$ on $\mathbf{N}$ was established in Example 3.1.11. It follows easily that $|$ is transitive on $\mathbf{N}^+$.

(ii) For any $n \in \mathbf{N}^+$, let $D_n = \{d \in \mathbf{N} | d \text{ divides } n\}$. Then the relation defines a partial ordering on D_n.

We study order relations more systematically in the next section. For now, we regard partial orderings as natural and important examples of relations.

Another important type of relation is an equivalence relation. Equivalence relations will be studied in detail in Section 3.3. As we shall see, the concept of equivalence relations is significant because it provides a natural way of generalizing the notion of equality. For the time being we present the definition and note a few examples.

Definition 3.1.9 *Equivalence Relation*

A relation R on a set A is an *equivalence relation* if R is reflexive, symmetric, and transitive.

Example 3.1.13 (i) Let A be a nonempty set. Then the identity relation I_A is an equivalence relation on A. Note that $\varnothing$ is not an equivalence relation on A since $\varnothing$ is not reflexive.

(ii) The relation $\equiv$ on the set $\mathcal{F}$ of fractions of integers defined in Example 3.1.7 is an equivalence relation on $\mathcal{F}$.

In addition to order relations (including partial ordering and other relations to be defined in Section 3.2) and equivalence relations, another type of relation plays a major role in mathematics. This class of relations is actually quite familiar, having been part of our mathematical diet since high school.

Definition 3.1.10 *Function*

For any set A, a *function on A* is a relation, f, on A with the property that for all $a, b, c \in A$, if $(a, b) \in f$ and $(a, c) \in f$, then $b = c$.

What does this condition say in ordinary mathematical English? It asserts that for each $a \in A$ there is at most one $b \in A$ such that $(a, b) \in f$ or afb. Note that for a given $a \in A$ there may exist no element $b \in A$ for which $(a, b) \in f$, but if there is one such b, then there is exactly one. Since this element b is unambiguously determined by a, we can and should express this dependence notationally, and we do so by following the usual custom and writing $b = f(a)$. Note that this notation is both efficient and clear, for when we define $f(a)$ we are saying that $(a, f(a))$ is the unique element of f with first coordinate a.

In order to define a function, one must describe, for each $a \in A$ or for each a in some given subset of A, the element $f(a) \in A$ that corresponds to a. When $A = \mathbf{R}$, this correspondence can be based on concepts and constructions from arithmetic, geometry, or calculus. Here are some illustrations:

Example 3.1.14 In each of the following examples, f will be a function on $\mathbf{R}$.

(i) $f(x) = x$ for $x \in \mathbf{R}$. (Note that $f = I_{\mathbf{R}}$.)
(ii) $f(x) = x^2 + 1$ for $x \in \mathbf{R}$.
(iii) $f(x) = \sin(x)$ for $x \in \mathbf{R}$.
(iv) $f(x) = e^x$ for $x \in \mathbf{R}$.
(v) $f(x) = \int_0^x (t^2 + 1)\, dt$ for $x \in \mathbf{R}$.
(vi) $f(x) = \sum_{n=1}^{\infty} (x^n/n^n)$ for $x \in \mathbf{R}$.
(vii) $f(x) = \sqrt{x}$ for $x \geq 0$.

Perhaps one aspect of our development of relations has seemed somewhat restrictive. By our definition a relation on a set A is a subset of $A \times A$. As a general example of a relation on A, we introduced the concept of a function on A. But, as was apparent in calculus, we often work with functions, f, "from a set A to a set B" where A and B are possibly distinct sets. In this case f assigns to each $x \in A$ a unique element $f(x) \in B$; hence f can be described as the subset $\{(x, f(x)) \in A \times B \mid x \in A\}$ of $A \times B$. The next definition extends this idea to relations.

Definition 3.1.11 *Relations and Functions*

Let A and B be sets.

(i) A *relation from A to B* is a subset R of $A \times B$.

(ii) A *function* from A to B is a relation f from A to B with the property that if $(x, y) \in f$ and $(x, z) \in f$, then $y = z$. For a function f from A to B, we write $y = f(x)$ if $(x, y) \in f$.

(iii) If f is a function from A to B with the property that for each $a \in A$ there exists $b \in B$ such that $(a, b) \in f$, then we write $f\colon A \to B$.

Warning. When we write $f\colon A \to B$, we mean that for each $a \in A$, there exists $b \in B$ such that $(a, b) \in f$. When we say that f is a function from A to B, we allow for the possibility that there exists $a \in A$ such that for all $b \in B$, $(a, b) \notin f$.

Example 3.1.15 Here is an example of a relation from outside mathematics. Let F be the set of American females and M the set of American males. Define

$$R = \{(x, y) \in F \times M | x \text{ is the mother of } y\}.$$

Then R is a relation from F to M.

Representation of Relations

Is there a handy geometric way of representing relations? In posing this question, we have several goals in mind. First, we seek both concrete pictures of specific relations and general ways of viewing typical or model relations. Second, once we have pictures of relations, we can interpret various properties of relations visually, thereby obtaining a deeper understanding of these properties. For example, what does it mean pictorially for a relation to be symmetric? Presently, we shall impose other conditions upon relations and in several cases shall be able to describe these conditions in geometric terms.

By now, you might have guessed what our geometric interpretation of a relation will be. As we did when we constructed geometric realizations of Cartesian products, let us assume that A is a subset on $\mathbf{R}$. Then any relation on A, being a subset of A^2, is also a subset of $\mathbf{R}^2$. Thus we can sketch the

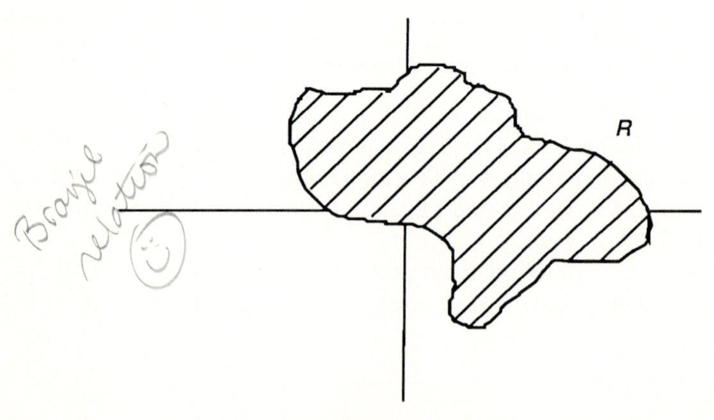

Figure 3.1.1

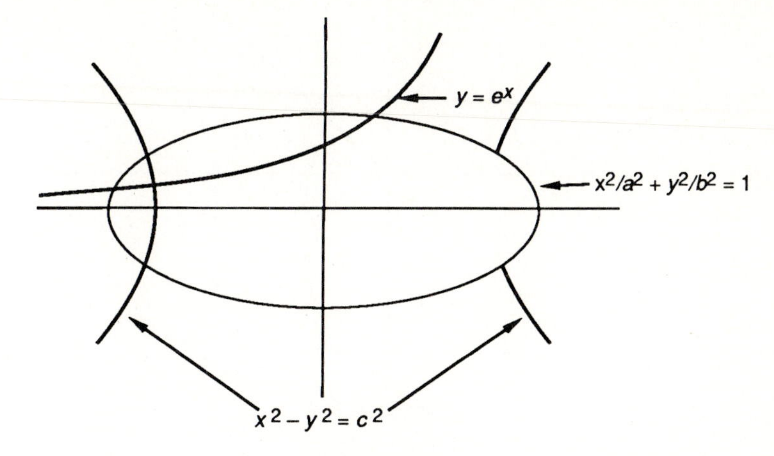

Figure 3.1.2

set of points $\{(x, y) \in \mathbf{R}^2 | (x, y) \in R\}$; this picture is called the *graph of* R. Figure 3.1.1 shows the graph of a typical, garden-variety relation R on $\mathbf{R}$.

Conversely, any geometric figure or region in the Cartesian plane represents a relation on $\mathbf{R}$. Thus the routine curves and figures of calculus and analytic geometry, which include parabolas, hyperbolas, ellipses, and graphs of functions, all yield relations on $\mathbf{R}$ (Figure 3.1.2).

Reflexivity and symmetry have nice geometric interpretations. Consider a relation R on $\mathbf{R}$. If R is reflexive, then $x R x$ for all x in $\mathbf{R}$; hence the graph of R contains the diagonal line $y = x$ in $\mathbf{R}^2$ (Figure 3.1.3).

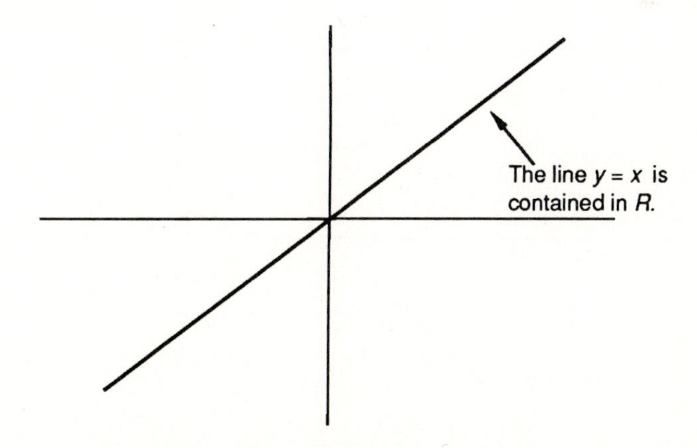

Figure 3.1.3

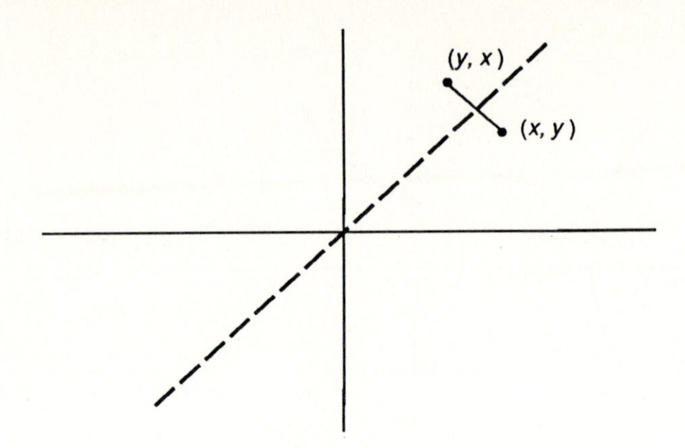

Figure 3.1.4

If R is symmetric, then for all $x, y \in \mathbf{R}$, $x R y$ implies $y R x$. Now, for any $x, y \in \mathbf{R}$, the line in $\mathbf{R}^2$ joining (x, y) to (y, x) is perpendicular to and is bisected by the line $y = x$ (Figure 3.1.4). Thus the graph of any symmetric relation, R, is symmetric in a geometric sense about the line $y = x$. Take the mirror-image of the graph of R about the line $y = x$; this image coincides with the graph of R precisely when R is symmetric (Figure 3.1.5).

Finally, a crude but effective way of representing a relation R from A to B (and in particular a function from A to B) is shown in Figure 3.1.6, which we call the *sketch* of R. Draw two closed figures to represent A and B. For $a \in A$ and $b \in B$, if $(a, b) \in R$, then label points in A and B as a

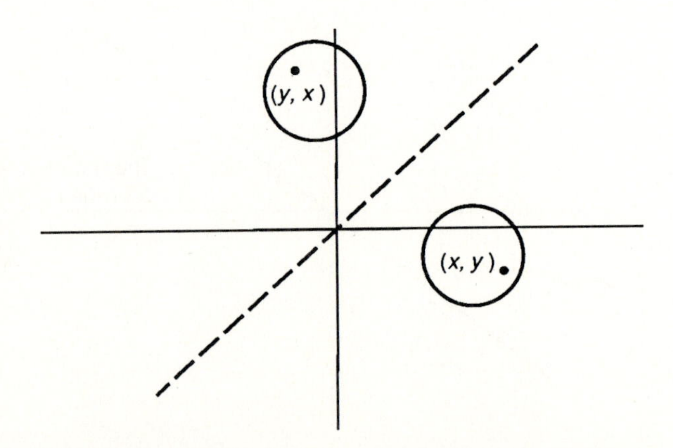

Figure 3.1.5. Graph of a symmetric relation on $\mathbf{R}$.

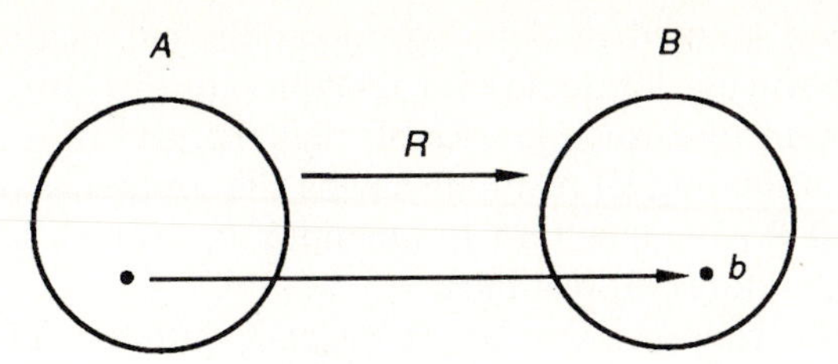

Figure 3.1.6

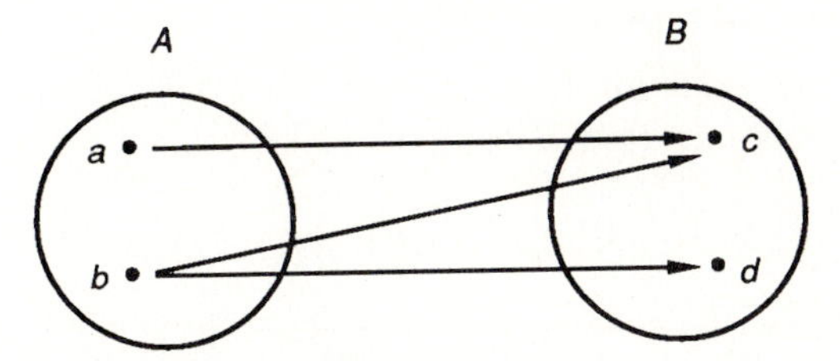

Figure 3.1.7

and b respectively and join these points by a directed segment, i.e., a segment with an arrow at b.

For example, let $A = \{a, b\}$ and $B = \{c, d\}$ and let $R = \{(a, c), (b, c), (b, d)\}$. Then R is represented as depicted in Figure 3.1.7.

Operations on Relations

We now consider the possibility of taking a given relation or a given pair of relations and producing a new relation. In other words, we look at operations on relations. There are at least two ways to approach the problem of how to define operations on relations.

First, recall that a relation from A to B is a subset of $A \times B$. Thus any concept connected with subsets can be related to relations. For instance, if R and S are relations from A to B, then $R \cap S$, $R \cup S$, and R^c are also relations from A to B called the intersection of R and S, the union of R and S, and the complement of R, respectively. Given relations R and S it is natural to investigate the extent to which properties of R and S (such as reflexivity, symmetry, etc.) are preserved under these operations. For example, we ask the following type of question: If relations R and S from A to B have property X, then does $R \cap S$ have property X? To be specific: If R and S are reflexive, is $R \cap S$ reflexive? Questions of this type appear in the exercises.

The second method of defining operations on relations comes from looking at a particular example of a relation—in this case, function—and generalizing from this example. (Recall that the idea of a partial ordering comes from considering all relations having the properties possessed by the relation $\leq$ on $\mathbf{R}$.) We now turn to functions as a source of inspiration for ideas about operations on relations.

Among the concepts that are associated with functions, two of the most important are composition and inverse. Let us first look at the inverse.

Definition 3.1.12 *Inverse Relation*

Let R be a relation from A to B. Then the *inverse of* R, written R^{-1}, is the relation from B to A defined by:

$$R^{-1} = \{(y, x) \in B \times A | (x, y) \in R\}.$$

Example 3.1.17 Let $A, B \subseteq \mathbf{R}$. Consider the following functions (relations) $f: A \to B$:

(i) $f: \mathbf{R} \to \mathbf{R}$ where $f(x) = 2x$. Then

$$f^{-1} = \{(y, x) \in \mathbf{R} \times \mathbf{R} | y = 2x\} = \{(u,(1/2)u) | u \in \mathbf{R}\}.$$

(ii) $f: \mathbf{R}^+ \to \mathbf{R}$ where $\mathbf{R}^+ = \{x \in \mathbf{R} | x > 0\}$ and $f(x) = \ln(x)$. Then $f^{-1} = \{(y, x) \in \mathbf{R} \times \mathbf{R}^+ | y = \ln(x)\} = \{(u, e^u) | u \in \mathbf{R}\}$.

Note that in both cases, the relation f^{-1} is also a function. Such is not always the case.

(iii) Let $f: \mathbf{R} \to \mathbf{R}_0 = \{x | x \geq 0\}$ be given by $f(x) = x^2$. Then $f^{-1} = \{(y, x) | y = x^2\} = \{(y, \pm\sqrt{y}) | y \geq 0\}$. Since $(4, 2) \in f^{-1}$ and $(4, -2) \in f^{-1}$, f^{-1} is not a function.

Example 3.1.18 Let $A = \{a, b, c\}$.

(i) Let $R = \{(a, a), (a, b), (b, a), (a, c)\}$. Then $R^{-1} = \{(a, a), (b, a), (a, b), (c, a)\}$. Note that R and R^{-1} are almost equal.

(ii) In fact, if we let $S = R \cup \{(c, a)\} = \{(a, a), (a, b), (b, a), (a, c), (c, a)\}$, then $S^{-1} = S$.

This example suggests a general question. Let R be a relation on a set A. When is it true that $R = R^{-1}$? (Note that R^{-1} is also a relation on A.) Our aim is to find a statement about the relation R, call it P, with the property that $R = R^{-1}$ if and only if P is true.

One way of finding such a statement is to experiment with more examples and look for a pattern. On the basis of this pattern, we can formulate a conjecture. Another method is to assume that $R = R^{-1}$ and to draw whatever conclusions are possible from this assumption. Let us try the second approach.

Suppose $R = R^{-1}$. By definition of R^{-1}, for each $(x, y) \in R$, $(y, x) \in R^{-1}$. But since $R^{-1} = R$, $(y, x) \in R$. To summarize, if $R = R^{-1}$, then for each $(x, y) \in R$, $(y, x) \in R$; i.e., if $R = R^{-1}$, then R is symmetric! It is natural to wonder if the converse is also valid. Let us try to prove it.

Suppose R is symmetric. To show $R = R^{-1}$ we show $R \subseteq R^{-1}$ and $R^{-1} \subseteq R$. To see that $R \subseteq R^{-1}$, let $(x, y) \in R$. By the definition of R^{-1}, we know that $(y, x) \in R^{-1}$; however we want to show that $(x, y) \in R^{-1}$. To do so, we can use the symmetry of R. Since $(x, y) \in R$, $(y, x) \in R$; therefore $(x, y) \in R^{-1}$, which is what we want to show. Thus $R \subseteq R^{-1}$. To show $R^{-1} \subseteq R$ we can reverse our steps. If $(x, y) \in R^{-1}$, then $(y, x) \in R$. Now, by the symmetry of R, $(x, y) \in R$, proving that $R^{-1} \subseteq R$. We have shown that if R is symmetric, then $R = R^{-1}$.

The two preceding paragraphs constitute a proof of the following theorem. Notice that in the course of investigating the condition $R = R^{-1}$, we proved one of the two implications of the theorems. Thus we used a proof technique as an exploratory tool.

Theorem 3.1.1. *A relation R on a set A is symmetric if and only if $R = R^{-1}$.*

Our next goal is to generalize the idea of composition of functions. Recall that if $f: A \to B$ and $g: B \to C$, then one can *compose* the functions f and g to obtain $g \circ f: A \to C$ where $(g \circ f)(a) = g(f(a))$. Our next definition extends this idea to relations.

Definition 3.1.13 *Composition of Relations*

Let R be a relation from A to B and S a relation from B to C. The *composition of S and R* is the relation from A to C:

$$S \circ R = \{(x, z) \mid \text{There exists } y \in B \text{ such that } x R y \text{ and } y S z\}$$

Note: Some authors write $R \circ S$ or RS to denote the composition of S and R.

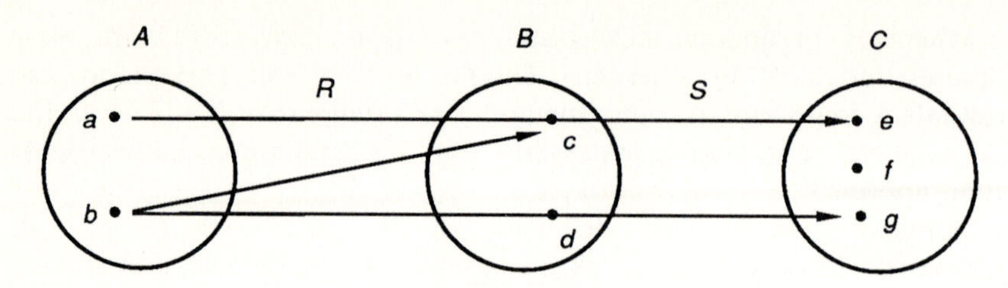

Figure 3.1.8

Example 3.1.19 Let $A = \{a, b\}$, $B = \{c, d\}$ and $C = \{e, f, g\}$.

(i) Let $R = \{(a, c), (b, c), (b, d)\}$ and $S = \{(c, e), (d, g)\}$. Then $S \circ R = \{(a, e), (b, e), (b, g)\}$.

Observe that the sketches of the relations R and S can help determine the composition $S \circ R$ (Figure 3.1.18). The composition $S \circ R$ consists of those pairs $(x, z) \in A \times C$ which are joined by segments passing through B.

(ii) Let $R = \{(a, c), (b, c)\}$ and $S = \{(d, e), (d, f), (d, g)\}$. Then $S \circ R = \varnothing$ (i.e., $S \circ R$ is the empty relation from A to C), since for no $y \in B$ does there exist $x \in A$ and $z \in C$ such that $(x, y) \in R$ and $(y, z) \in S$. Thus the composition of two relations can be quite trivial.

Example 3.1.20 (i) Once again we take $A = \{a, b, c\}$ and $R = \{(a, a), (a, b), (a, c), (b, a)\}$. Then $R \circ R = \{(a, a), (a, b), (a, c), (b, a), (b, b), (b, c)\}$.

(ii) Let $T = R \cup \{(b, b), (b, c)\} = \{(a, a), (a, b), (a, c), (b, a), (b, b)(b, c)\}$. Then $T \circ T = \{(a, a), (a, b), (a, c), (b, a), (b, b), (b, c)\}$. Notice that $T \circ T \subseteq T$. This example suggests a question: What conditions must T satisfy in order that $T \circ T \subseteq T$? Exercise: State and prove a conjecture that answers this question.

Example 3.1.21 Let us find the composition, $S \circ R$, of the relations $R = \leq$ and $S = \geq$ on $\mathbf{R}$. By definition $S \circ R = \{(x, z) | \text{There exists } y \in \mathbf{R} \text{ such that } x R y \text{ and } y S z\} = \{(x, z) | \text{There exists } y \in \mathbf{R} \text{ such that } x \leq y \text{ and } y \geq z\}$. We seek a clearer and more concise description of $S \circ R$. Consider two cases:

(i) $x \leq z$. Then, if we take $y = x$, we find that $x \leq y$ and $y \geq z$. Thus $(x, z) \in S \circ R$ if $x \leq z$.

(ii) $x > z$. Then, taking $y = x$, we see that $x \leq y$ and $y \geq z$; hence $(x, z) \in S \circ R$ if $x > z$. We conclude that $(x, z) \in S \circ R$ for all $(x, z) \in \mathbf{R} \times \mathbf{R}$; hence $S \circ R = \mathbf{R} \times \mathbf{R}$.

Example 3.1.22 On $\mathbf{R}$ let R and S be the relations given by: $x R y$ if $x \leq y$ and $x S y$ if $y = \sin(x)$. (Note that S is just the sine function.) Let us find $S \circ R$. By definition $S \circ R = \{(x, z) \in \mathbf{R} \times \mathbf{R} | \text{There exists } y \in \mathbf{R} \text{ for}$

which $x \leq y$ and $z = \sin(y)\}$. Let $(x, z) \in S \circ R$. Since $z = \sin(y)$ for some $y \in \mathbf{R}$, $-1 \leq z \leq 1$. Thus $(x, z) \in \mathbf{R} \times \{-1, 1]$, and $S \circ R \subseteq \mathbf{R} \times [-1, 1]$. Conversely, let $(x, z) \in \mathbf{R} \times [-1, 1]$. Then there exists $y \in \mathbf{R}$ such that $z = \sin(y)$. Moreover, by properties of the sine function, this number y can be chosen from the interval $[x, x + 2\pi]$. Therefore $x \leq y$ and $z = \sin(y)$. We conclude that for all $(x, z) \in \mathbf{R} \times [-1, 1]$, $(x, z) \in S \circ R$. Hence $S \circ R = \mathbf{R} \times [-1, 1]$.

As we have seen, the concept of relation has great generality, encompassing such ideas as equality, function, and order. Because of its generality, the relation concept unifies several aspects of our mathematical experience that otherwise seem unrelated. This unification enables us to understand better the various special concepts by fitting them into a larger structure where we can see comparisons and contrasts. On the other hand, we do pay a price for the breadth of the relation idea. Relations in general are too amorphous to study profitably. Additional conditions must be imposed in order to arrive at objects that can be handled with any dexterity; hence the many definitions that were presented in this section. In the remainder of this chapter, we study the most important classes of relation: order relations, equivalence relations, and functions.

Exercises §3.1

3.1.1. How are the subsets $\mathbf{R}_{\leq}$, $\mathbf{R}_{<}$, $\mathbf{R}_{\geq}$, and $\mathbf{R}_{>}$ of $\mathbf{R}^2$ related? Sketch the graphs of these relations.

3.1.2. Give several examples of relations on $\{a, b, c\}$.

3.1.3. (a) How many relations on $\{1, 2\}$ are reflexive?
(b) How many are symmetric?
(c) How many are both reflexive and symmetric?
(d) How many are neither reflexive nor symmetric?

3.1.4. (a) How many relations exist on the set $\{a, b, c\}$?
(b) Suppose A is a finite set with n elements. How many relations on A are there?

3.1.5. On a single set of your choice, give examples of relations that possess exactly one and exactly two of the following three properties—reflexivity, symmetry, and transitivity. You should give six examples.

3.1.6. Let U be a nonempty set. Which of the properties given in Definitions 3.1.2–3.1.7 do the following relations on $P(U)$ have?
(a) $A R B$ if $A \cap B = \varnothing$.
(b) $A R B$ if $A \cap B \neq \varnothing$.
(c) $A R B$ if $A + B = (A - B) \cup (B - A) = \varnothing$.
(d) $A R B$ if $A - B$ is finite.
(e) $A R B$ if $A + B$ is finite. (Note: $A + B$ is defined in part (c).)

3.1.7. (a) Can a relation be both symmetric and antisymmetric? Explain.
(b) Give an example of a relation that is neither reflexive nor irreflexive.
(c) Can a relation be both symmetric and asymmetric?
(d) Prove that any asymmetric relation is irreflexive.

3.1.8. (a) Prove: On a set A, $\varnothing$ is a symmetric and transitive relation.
(b) Prove: The identity relation on any set is a partial ordering.
(c) Prove: On a set A having at least two elements, $A \times A$ is not a partial ordering.

3.1.9. Characterize graphs of functions on $\mathbf{R}$ among graphs of relations on $\mathbf{R}$.

3.1.10. Let A be a set and let R and S be relations on A.
(a) Prove: $(R^{-1})^{-1} = R$.
(b) Prove: $(R \cap S)^{-1} = R^{-1} \cap S^{-1}$.

3.1.11. Let R be a relation from A to B. How can the sketch of R^{-1} be determined from that of R?

3.1.12. Let R and S be relations on a set A.
(a) Prove or disprove: If R and S are reflexive, then $R \cap S$ is reflexive.
(b) Prove or disprove: If R and S are reflexive, then $R \cup S$ is reflexive.
(c) Prove or disprove: If R and S are symmetric, then $R \cap S$ is symmetric.
(d) Prove or disprove: If R and S are symmetric, then $R \cup S$ is symmetric.
(e) Prove or disprove: If R and S are transitive, then $R \cap S$ is transitive.
(f) Prove or disprove: If R and S are transitive, then $R \cup S$ is transitive.

3.1.13. Let R and S be partial orderings on a set A.
(a) Prove or disprove: $R \circ S$ is a partial ordering on A.
(b) Prove or disprove: $R \cup S$ is a partial ordering on A.

3.1.14. Give the inverse of each of the following relations:
(a) $\leq$ on $\mathbf{R}$.
(b) $<$ on $\mathbf{R}$.
(c) The function on $\mathbf{R}$ given by $f(x) = x^2 + 1$.
(d) The function $f: \mathbf{R} \to \mathbf{R}^+$ given by $f(x) = e^{(e^x)}$.

3.1.15. Let $A = \{a, b, c, d\}$, $R = \{(a, a), (a, b), (a, c), (b, c), (c, a), (c, c), (c, d), (d, c)\}$, and $S = \{(a, a), (b, a), (b, c), (b, d), (d, b)\}$. Find
(a) $R \circ R$.
(b) $R \circ S$.
(c) $S \circ R$.
(d) $S \circ S$.

3.1.16. Let A be any set. Let $R = A \times A$ and $S = \varnothing$.
(a) Find R^{-1} and S^{-1}.
(b) Find $R \circ R$ and $S \circ S$.
(c) What is $R \circ S$?

3.1.17. In each case give the composition $R \circ S$.
(a) $A = \mathbf{R}$; xRy if $y = \sin(x)$; xSy if $y = x^2$.
(b) $A = \mathbf{R}$; xRy if $y = \sin(x)$; xSy if $y = x$.

3.1.18. On **R** find $S \circ R$ when:
(a) $R = \geq$ and $S = \leq$.
(b) $R = <$ and $S = >$.
(c) $R = <$ and $S = <$.

3.1.19. On **R** define $x R y$ to mean $x \leq y$ and $x S y$ to mean $y = x^2$.
(a) Find $S \circ R$ and $R \circ S$.
(b) Sketch the graphs of $S \circ R$ and $R \circ S$.

3.1.20. Give two examples of relations **R** such that $R \circ R = R$.

3.1.21. Fill in the blank and prove the resulting statement: For a relation R on a set A, $R \circ R \subseteq R$ iff ____________________.

3.1.22. In each case, fill in the blank and prove the statement:
(a) R is symmetric iff R^c ____________________.
(b) R is reflexive iff R^c ____________________.

3.1.23. Let R be a reflexive relation on A. Prove that for any relation S on A, $S \subseteq S \circ R \cap R \circ S$.

3.1.24. Let R, S, and T be relations on a set A.
(a) Does the equality $R \circ S = R \circ T$ imply that $S = T$?
(b) Does $R \circ (S \cap T) = R \circ S \cap R \circ T$?

3.1.25. Prove that composition of relations is associative. Specifically, prove that if R is a relation from A to B, S is a relation from B to C, and T is a relation from C to D, then $T \circ (S \circ R) = (T \circ S) \circ R$.

§3.2 *Order Relations*

Order relations are relations on a set that provide a way of comparing elements analogous to the way that the familiar relations $\leq$ and $<$ allow for the comparison of real numbers. Order relations are basic to mathematics, arising naturally in such diverse areas as analysis, set theory, algebra, and computer science. In this section, our purpose is to define various kinds of order relations, to present several examples of each type, and to discuss representations of order relations.

First we recall the definition of a partial ordering.

> Definition 3.2.1 *Partial Ordering*
>
> A *partial ordering* on a set A is a reflexive, antisymmetric, and transitive relation on A. If $\leq$ is a partial ordering on A, then we write $(A, \leq)$ to denote the set A and the partial ordering $\leq$. The pair $(A, \leq)$ is also called a *partially ordered set* or a *poset*.

Example 3.2.1 Let $\leq$ and $\geq$ denote the usual order relations on $\mathbf{R}$. Then $(\mathbf{R}, \leq)$ and $(\mathbf{R}, \geq)$ are posets. Moreover, if A is any nonempty subset of $\mathbf{R}$ (e.g., $A = \mathbf{Q}$ or $A = \mathbf{Z}$), then $(A, \leq)$ and $(A, \geq)$ are posets.

Example 3.2.2 Let $|$ denote the divisibility relation on $\mathbf{Z}^+ = \{z \in \mathbf{Z} | z > 0\}$: For all $a, b \in \mathbf{Z}^+$, $a|b$ if there exists $c \in \mathbf{Z}^+$ such that $a \cdot c = b$. Then $(\mathbf{Z}^+, |)$ is a poset.

Example 3.2.3 On $\mathbf{Z} \times \mathbf{Z}$ define the relation $\leq_1$ as follows: $(a, b) \leq_1 (c, d)$ if $a \leq c$ and $b \leq d$ where $\leq$ is the usual order relation on $\mathbf{Z}$. From the fact that $\leq$ is a reflexive, antisymmetric, and transitive relation on $\mathbf{Z}$, it follows that $\leq_1$ is a reflexive, antisymmetric, and transitive

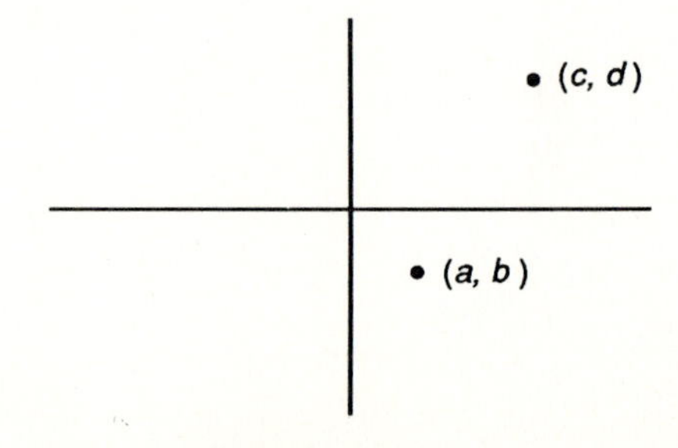

Figure 3.2.1

relation on $\mathbf{Z} \times \mathbf{Z}$. Thus $(\mathbf{Z} \times \mathbf{Z}, \leq_1)$ is a poset. Geometrically, we can describe $\leq_1$ by saying $(a, b) \leq_1 (c, d)$ if (a, b) lies both to the left of and below (c, d) (Figure 3.2.1).

Example 3.2.4 Let A be any set. Then $(P(A), \subseteq)$ and $(P(A), \supseteq)$ are posets. Let $X \subseteq P(A)$, i.e., let X be any collection of subsets of A. Then $(X, \subseteq)$ and $(X, \supseteq)$ are posets.

Another type of order relation is the so-called strict partial ordering, which takes as its motivating example the relation $<$ on $\mathbf{R}$.

Definition 3.2.2 *Strict Partial Ordering*

A *strict partial ordering* on a set A is an irreflexive, asymmetric, and transitive relation. If $<$ is a strict partial ordering on A, then we write $(A, <)$ to denote the set A and the strict partial ordering $<$.

Example 3.2.5 (i) The relations $<$ and $>$ on $\mathbf{R}$ define strict partial orderings.

(ii) For any set A the relations $\subset$ and $\supset$ define strict partial orderings on $P(A)$.

Warning We are using the symbol $\leq$ to denote a partial ordering on a set A. By using this symbol, we are not implying that A is a set of real numbers and that $\leq$ is the usual ordering on A. We are simply using the symbol $\leq$ because the idea of a partial ordering generalizes the standard ordering on $\mathbf{R}$.

As we know from our past experience, the relations $\leq$ and $<$ on $\mathbf{R}$ are closely related. In fact, for real numbers a and b, $a < b$ if and only if $a \leq b$ and $a \neq b$. In general, it is natural to ask the following questions: If $\leq$ is a partial ordering on a set A, then is the relation $<$ defined by $a < b$ if $a \leq b$ and $a \neq b$ a strict partial ordering on A? If $<$ is a strict partial ordering on A and if for $a, b \in A$, $a \leq b$ is defined to mean $a < b$ or $a = b$, then is $\leq$ a partial ordering on A? The answer is contained in the following result, which will not be used in this text and whose proof is left as an exercise. The proof is straightforward, but it does require careful application of the relevant definitions.

Proposition 3.2.1 (a) Let $\leq$ be a partial ordering on a set A. For $a, b \in A$ define $a < b$ to mean $a \leq b$ and $a \neq b$. Then $<$ is a strict partial ordering on A.

(b) Let $<$ be a strict partial ordering on A. For $a, b \in A$ define $a \leq b$ to mean $a < b$ or $a = b$. Then $\leq$ is a partial ordering on A.

Proposition 3.2.1 asserts that to each partial ordering on a given set A, there corresponds a strict partial ordering on A and vice versa. Thus to speak of a partial ordering on A is to speak of a strict partial ordering on A. For this reason we speak henceforth only of partial orderings on a given set. (We could just as easily work only with strict partial orderings, but we follow most authors in focusing on partial orderings and posets.) By the way, the relationship between the relations $\leq$ and $<$ can be expressed in set theoretic terms. Regarding $\leq$ and $<$ as subsets of $A \times A$ we can show that $\leq = < \cup \{(x, x) | x \in A\} = < \cup I_A = \{(x, y) \in A \times A | x < y$ or $x = y\}$.

Perhaps the two most important examples of partial orderings are $\leq$ on $\mathbf{R}$ and $\subseteq$ on $P(A)$ where A is a set. Notice that these relations are different in a significant way: For each $a, b \in \mathbf{R}$ either $a \leq b$ or $b \leq a$. However, if $A \neq \varnothing$ and $A \neq \{x\}$, then it is not true that for all $B, C \in P(A)$ either $B \subseteq C$ or $C \subseteq B$. Partial orderings that share with $\leq$ on $\mathbf{R}$ the property described above are given a special name.

Definition 3.2.3 *Total Ordering*

Let $\leq$ be a partial ordering on a set A. We call $\leq$ a *total ordering* if for all $a, b \in A$ either $a \leq b$ or $b \leq a$.

Also suggested by the examples described in the previous paragraph is the following idea.

Definition 3.2.4 *Comparability*

Let $(A, \leq)$ be a poset. Let $a, b \in A$; a and b are *comparable* if either $a \leq b$ or $b \leq a$; a and b are *incomparable* if $a \nleq b$ and $b \nleq a$.

Observe that a partial ordering $\leq$ on a set A is a total ordering if and only if any two elements of A are comparable.

Example 3.2.6 On $\mathbf{Z} \times \mathbf{Z}$ we define the *lexicographic ordering* $\leq'$ as follows: $(a, b) \leq' (c, d)$ if $a < c$ (in $\mathbf{Z}$) or if $a = c$ and $b \leq d$. Then $(\mathbf{Z} \times \mathbf{Z}, \leq')$ is a poset. Observe that the partial orderings $\leq_1$ of Example 3.2.3 and $\leq'$ on $\mathbf{Z} \times \mathbf{Z}$ are not identical: $(1, 2) \nleq_1 (2, 1)$ while $(1, 2) \leq' (2, 1)$. In other words, $\leq_1$ and $\leq'$ provide different ways of comparing elements of $\mathbf{Z} \times \mathbf{Z}$. The partial ordering $\leq'$ is called the lexicographic ordering since, under $\leq'$, the elements of $\mathbf{Z} \times \mathbf{Z}$ are ordered in a dictionary fashion: $x = (a, b) \leq' (c, d) = y$ if and only if the first "name" of x, i.e., a, is less than the first name of y, i.e., c, or if $a = c$ and the second name of x is less than or equal to the second name of y. Moreover, the relation $\leq'$

is a total ordering on $\mathbf{Z} \times \mathbf{Z}$: If $(a, b), (c, d) \in \mathbf{Z} \times \mathbf{Z}$, then either $a < c$ in which case $(a, b) \leq' (c, d)$; $c < a$ in which case $(c, d) \leq' (a, b)$; or $a = c$ in which case $(a, b) \leq' (c, d)$ if $b \leq d$ and $(c, d) \leq' (a, b)$ if $d < b$. The relation $\leq_1$ is not a total ordering on $\mathbf{Z} \times \mathbf{Z}$.

Exercise Describe the relation $\leq'$ geometrically. Let $(A, \leq)$ be a poset. As we have mentioned, the relation $\leq$ provides a way of comparing elements of A. Since in general not every pair of elements of A can be compared, the ordering of A provided by $\leq$ is not complete; hence the use of the word "partial" in describing $\leq$. Because the definition of a partial ordering is motivated by the example of "less than or equal" on $\mathbf{R}$, we intuitively think that for $a, b \in A$ with $a \leq b$, a is less than or equal to b or a is smaller than b. With this viewpoint in mind we introduce the following technical terms.

Definition 3.2.5 *Lower and Upper Bounds*

Let $(A, \leq)$ be a poset and let $B \subseteq A$.

(i) An element $a \in A$ is a *lower bound* for B if for all $b \in B$, $a \leq b$.

(ii) An element $a \in A$ is an *upper bound* for B if for all $b \in B$, $b \leq a$.

Definition 3.2.6 *Least Upper Bound and Greatest Lower Bound*

Let $(A, \leq)$ be a poset and let $B \subseteq A$.

(i) An element $a_0 \in A$ is a *greatest lower bound* for B if (a) a_0 is a lower bound for B, and (b) for all $a \in A$ if a is a lower bound for B, then $a \leq a_0$.

(ii) An element $a_1 \in A$ is a *least upper bound* for B if (a) a_1 is an upper bound for B, and (b) for all $a \in A$ if a is an upper bound for B, then $a_1 \leq a$.

Example 3.2.7 Consider $\mathbf{R}$ with its usual ordering $\leq$. Let $B = (0, 1) = \{x \in \mathbf{R} | 0 < x < 1\}$. Then any $a \in \mathbf{R}$ such that $a \leq 0$ is a lower bound for B while any $a \in \mathbf{R}$ such that $a \geq 1$ is an upper bound for B. Note that 0 and 1 are greatest lower bound and least upper bound, respectively, for B. Clearly $\mathbf{R}$ itself has no upper bounds and no lower bounds with respect to $\leq$.

Example 3.2.8 Let $A = \{a, b, c\}$ and define a relation $\leq$ on A as follows: $x \leq x$ for all $x \in A$, $a \leq c$, and $b \leq c$. (No other pairs of elements of A are related by $\leq$.) It follows that c is an upper bound and least upper bound for A. On the other hand, A has no lower bound and hence no greatest lower bound.

Example 3.2.9 Let $A = \{a, b, c, d\}$. Define the relation $\leq$ on A as follows: $x \leq x$ for all $x \in A$, $a \leq c$, $a \leq d$, $b \leq c$, and $b \leq d$. (No other pairs of elements of A are related by $\leq$.) Let $B = \{c, d\}$. Then a and b are both lower bounds for B; nonetheless B does not have a greatest lower bound, for if a_0 is a greatest lower bound for B, then either $a_0 = a$ or $a_0 = b$. But since $a \nleq b$, $a_0 \neq b$ and since $b \nleq a$, $a_0 \neq a$. This example shows that a greatest lower bound for a subset of a poset need not exist even though lower bounds for the subset do exist.

Example 3.2.10 Let $D_{12} = \{d \in \mathbf{N} \mid d \mid 12\} = \{1, 2, 3, 4, 6, 12\}$ be partially ordered by the relation $|$. For $a, b \in D_{12}$, we define $a \leq b$ to mean $a \mid b$. Let $A = \{2, 3\}$. Then A has greatest lower bound 1 and least upper bound 6.

As Examples 3.2.9 and 3.2.10 indicate, a subset of a poset need not have a greatest lower bound (or least upper bound). Nonetheless, as the next result shows, a subset of a poset has at most one greatest lower bound.

Proposition 3.2.2 Let $(A, \leq)$ be a poset and let $B \subseteq A$. If a and b are greatest lower bounds for B, then $a = b$.

Proof Since a is a greatest lower bound for B and since b is a lower bound for B, it follows that $b \leq a$. Reversing the roles of a and b, we find that $a \leq b$. By the antisymmetry of $\leq$, we conclude that $a = b$. ■

A corresponding result holds for least upper bounds. (See Exercise 3.2.10.) The upshot of Proposition 3.2.2 is that if a greatest lower bound (or least upper bound) of a set B exists, then it is unique. We write lub(B) and glb(B) (pronounced lub and glub, respectively) to denote the least upper bound and greatest lower bound of B when they exist.

Most of the ideas presented in this section can be captured nicely in a picture. We now describe a geometric way of representing posets which makes clear the concepts of lower and upper bound and which provides a quick way of defining posets.

The Diagram of a Poset

We introduce a pictorial method of representing a finite poset, called its *diagram*. To form the diagram of $(A, \leq)$, we do the following: Let $(A, \leq)$ be a poset.

1. We let each element of A correspond to a point in $\mathbf{R}^2$. We label the point corresponding to $x \in A$ by x itself. The point x is called a *node* of the diagram.

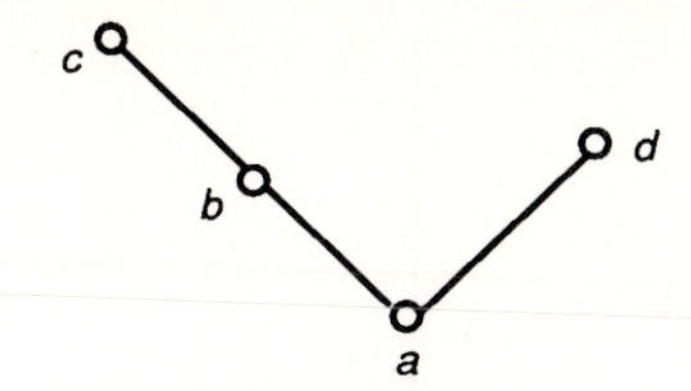

Figure 3.2.1

2. If $x, y \in A$ and $x \leq y$, then we place the node x below the node y on the page.
3. Suppose that $x \leq y$ and that there exists no other element of A "between" x and y; i.e., if $x \leq z$ and $z \leq y$, then either $z = x$ or $z = y$. Then we connect the nodes x and y by a line segment, called an *edge* of the diagram.

We denote the diagram of $(A, \leq)$ by $D(A, \leq)$. Some texts refer to the diagram of $(A, \leq)$ as the *Hasse diagram* of $(A, \leq)$.

Because $\leq$ is transitive, we know that $a \leq b$ and $b \leq c$ implies $a \leq c$. We make this fact implicit in $D(A, \leq)$ rather than having it explicitly represented in order to simplify the geometric figure. Thus $a \leq c$ if node a lies below node c and nodes a and c are joined by a sequence of edges.

Example 3.2.11 Let $A = \{a, b, c, d\}$. Define the relation $\leq$ on A as follows: $x \leq x$ for all $x \in A$, $a \leq b$, $a \leq c$, $b \leq c$, and $a \leq d$. (No other pairs of elements of A are related by $\leq$.) Then $D(A, \leq)$ is ordered as shown in Figure 3.2.1.

Example 3.2.12 Let $A = P(\{a, b, c\})$ and let us construct the diagram of $(A, \subseteq)$ (Figure 3.2.2).

Example 3.2.13 Let A be a finite set and suppose $(A, \leq)$ is totally ordered. For example, if $A = D_{16} = \{1, 2, 4, 8, 16\}$, then A is totally ordered

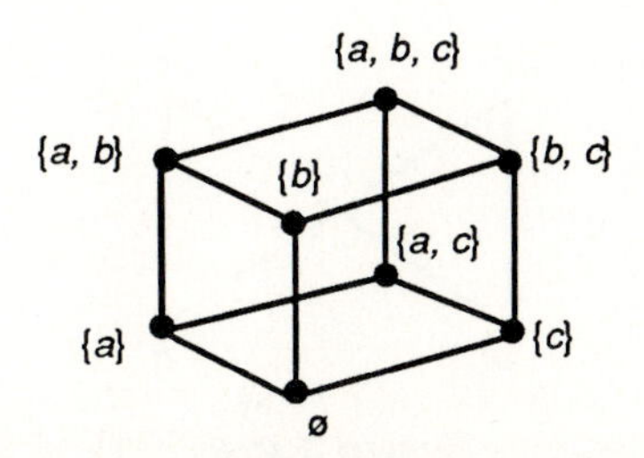

Figure 3.2.2

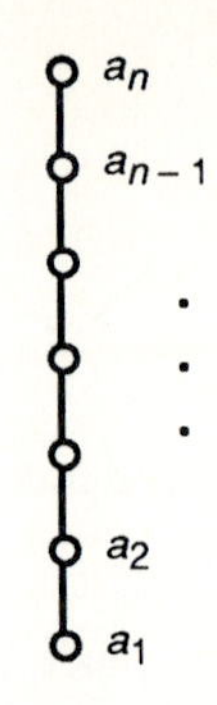

Figure 3.2.3

by division. Since any two elements of A are comparable, the members of A can be listed from smallest to largest; $A = \{a_1, a_2, \ldots, a_n\}$ where $a_1 \leq a_2 \cdots \leq a_n$; hence $D(A, \leq)$ is ordered as shown in Figure 3.2.3.

The partial ordering in Example 3.2.9 is depicted in Figure 3.2.4. (We use the elements of $A = \{a, b, c, d\}$ as labels for the nodes of $D(A, \leq)$.) It is clear that a and b are lower bounds for $\{c, d\}$, but since a and b are incomparable (there exists no line segment joining a and b), $\{c, d\}$ has no glb.

In general, if $(A, \leq)$ is a poset and $B \subseteq A$, then in $D(A, \leq)$ a lower bound for B, call it a_0, lies in or below B and each node of B is connected to a_0 by a sequence of edges. (Note that a_0 might be an element of B. Figure 3.2.5 depicts the case in which a_0 is not an element of B.)

Given a diagram, D, of nodes and edges, one can construct a poset whose diagram is D. For instance, let D be the diagram shown in Figure 3.2.6. Let $A = \{a, b, c, d, e\}$. Then $x \leq x$ for all $x \in A$, $y \leq a$ for all $y \in A$, $e \leq b$ and $e \leq c$, and all other ordered pairs of elements of A are incomparable. Thus we can pass easily from a poset $(A, \leq)$ to its diagram and back to the poset.

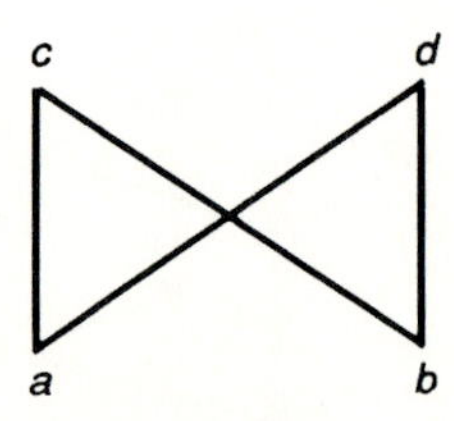

Figure 3.2.4

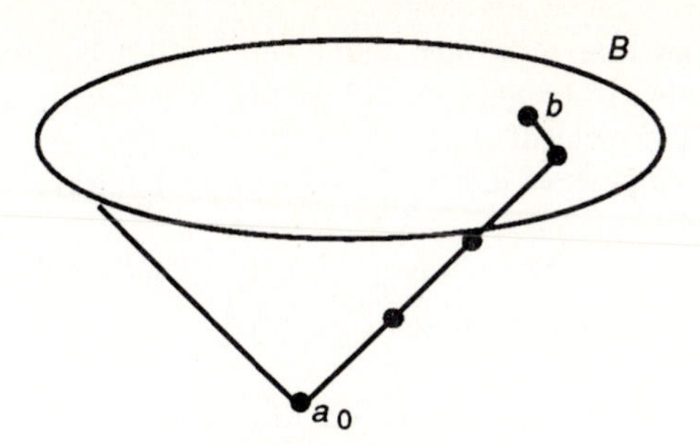

Figure 3.2.5

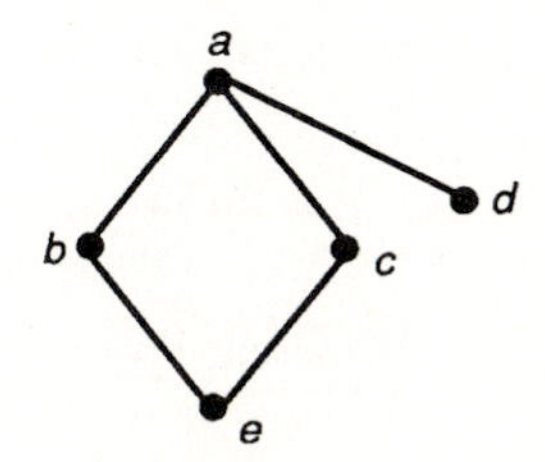

Figure 3.2.6

As with any visual representation, the diagram of a poset provides a quick way of capturing some essential features of a particular example or class of examples. For example, from the idea of the diagram of poset, we picture totally ordered sets as being "thinner" than posets that contain incomparable elements. Also, diagrams of posets can be helpful in devising proofs of statements. More important, perhaps, diagrams are a great aid in the construction of counterexamples to statements about posets.

Exercises §3.2

3.2.1. Show that for any set A, I_A is a partial ordering.

3.2.2. Which of the following relations on $\mathbf{R}$ are partial orderings?
(a) $\{(x, y) \in \mathbf{R} \times \mathbf{R} | y = x \text{ or } y = x + 1\}$.
(b) $\{(x, y) \in \mathbf{R} \times \mathbf{R} | y = x \text{ or } y \geq x + 1\}$.
(c) $\{(x, y) \in \mathbf{R} \times \mathbf{R} | y = x \text{ or } y \geq x^2\}$.

3.2.3. Is division a total ordering on $\mathbf{Z}^+$?

3.2.4. (a) Find an integer n such that $(D_n, |)$ is totally ordered and D_n has 6 elements. (In Example 3.2.13 we saw that $(D_{16}, |)$ is totally ordered and D_{16} has 5 elements.)
(b) For each positive integer k find a positive integer n such that D_n has k elements.
(c) For which positive integers n is $(D_n, |)$ totally ordered?

3.2.5. Define a relation $\leq'$ on $\mathbf{Z}^+ = \{1, 2, 3 \ldots\}$ as follows: $a \leq' b$ if $a = b$ or $a^2 \leq b$.
(a) Prove that $\leq'$ is a partial ordering on $\mathbf{Z}^+$.
(b) Find several pairs of incomparable elements for the relation $\leq'$.

3.2.6. Let $(A, \leq_1)$ and $(B, \leq_2)$ be posets. On $A \times B$ define the relation $\leq$ by $(a, b) \leq (a', b')$ if $a \leq_1 a'$ and $b \leq_2 b'$.
(a) Show that $(A \times B, \leq)$ is a poset.
(b) Suppose $\leq_1$ and $\leq_2$ are total orderings on A and B, respectively. Must $\leq$ be a total ordering on $A \times B$?

3.2.7. Let $(A, \leq)$ be a poset. Prove, or disprove and salvage.
(a) $(A, \leq^{-1})$ is a poset ($\leq^{-1}$ is the inverse of the relation $\leq$).
(b) $(A, \leq \circ \leq)$ is a poset ($\leq \circ \leq$ is the composition of $\leq$ with itself).

3.2.8. Let $(A, \leq_1)$ and $(A, \leq_2)$ be posets. Prove or disprove: $(A, \leq_1 \cap \leq_2)$ is a poset.

3.2.9. Prove Proposition 3.2.1.

3.2.10. State and prove an analog of Proposition 3.2.2 for lubs.

3.2.11. Prove, or disprove and salvage: If $(A, \leq)$ is totally ordered, then $(A, \leq^{-1})$ is totally ordered. Give some examples, as well as an argument, in support of your conclusion.

3.2.12. Let $(A, \leq)$ be a totally ordered set. One says that $(A, \leq)$ is *well ordered* if every nonempty subset $B \subseteq A$ having a lower bound has a greatest lower bound b such that $b \in B$.
(a) Show that $(\mathbf{R}, \leq)$ is not well ordered.
(b) For each $n \in \mathbf{N}^+$ let $\mathbf{N}_n = \{1, \ldots, n\} = \{k \in \mathbf{N} | 1 \leq k \leq n\}$. Show using PMI that for every $n \in \mathbf{N}^+$, $(\mathbf{N}_n, \leq)$ is well ordered. ($\leq$ is the usual total ordering on $\mathbf{N}$.)
(c) Use (b) to show that $(\mathbf{N}, \leq)$ is well ordered.

3.2.13. Prove or disprove: If $(A, \leq)$ is well ordered, then $(A, \leq^{-1})$ is well ordered.

3.2.14. For any set A, let $R(A)$ denote the set of relations on A. For $R_1, R_2 \in R(A)$, define $R_1 \leq R_2$ to mean $x R_1 y$ implies $x R_2 y$. Show that $(R(A), \leq)$ is a poset.

3.2.15. Let A be a finite set. Draw the diagram of the partial ordering I_A.

3.2.16. Draw $D(P\{1, 2, 3\}, \subseteq)$.

3.2.17. (a) Let $A = \{k \in \mathbf{N} | 1 \leq k \leq 16\}$. Draw $D(A, |)$. (Note $A \neq D_{16}$.)
(a) Draw $D(D_{30}, |)$.
(c) Draw $D(D_{24}, |)$.

3.2.18. For the posets corresponding to each of the following diagrams, write down all pairs of comparable elements.

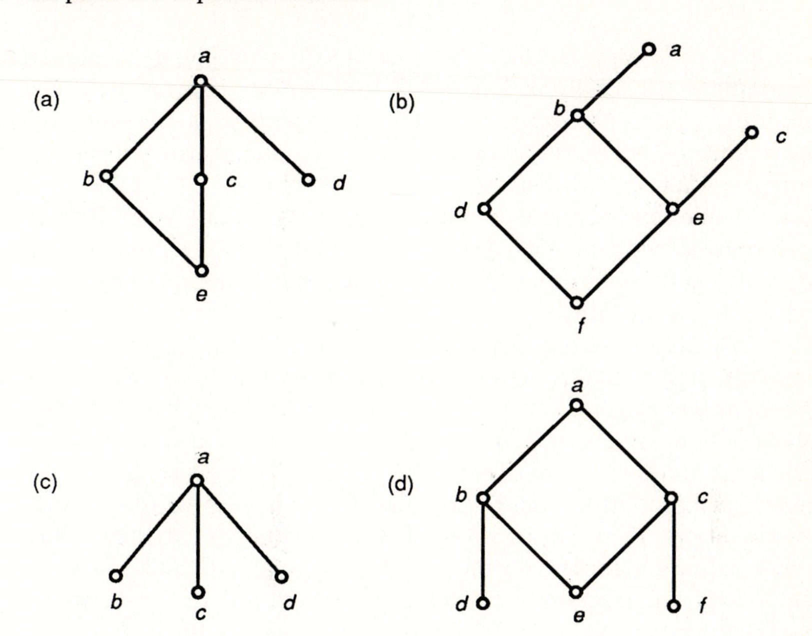

§3.3 Equivalence Relations

Let us return to the set $\mathscr{F}$ of fractions of integers mentioned in Example 3.1.7: $\mathscr{F} = \{a/b \mid a, b \in \mathbf{Z} \text{ and } b \neq 0\}$. By the way, for those desiring a more precise, set theoretic description of a fraction, we simply note that a fraction a/b can be thought of as an ordered pair (a, b) of integers where $b \neq 0$. Thus we can also say that $\mathscr{F} = \{(a, b)/a, b \in \mathbf{Z}$ and $b \neq 0\}$. We shall continue to write a typical element of $\mathscr{F}$ in the form a/b. We have all worked with the set $\mathscr{F}$ in both formal mathematics courses and informal applications of mathematics. But no matter what the setting, we do a curious thing.

Oftentimes we use elements of $\mathscr{F}$ to represent quantities—distance, amount, and weight, for example. In these situations, we *regard some fractions that appear to be distinct as being equal*. When regarded as symbols, the fractions $1/2$ and $2/4$ are not identical (the numerators of the two fractions are distinct; so are the denominators); however the fractions $1/2$ and $2/4$ represent the same quantities. If you have a half dollar and I have two quarters, then you and I have equal amounts of money. Thus, when representing quantities, we think of the fractions $1/2$ and $2/4$ as "equal." In general, the fractions $1/2$ and $a/2a$, where $a \in \mathbf{Z}$, $a \neq 0$, are regarded as equal; conversely, if $1/2$ "equals" a/b, then $a \neq 0$ and $b = 2a$.

Coincidentally, this notion of equality of fractions is captured by the relation in Example 3.1.7. If $a/b, c/d \in \mathscr{F}$, then $a/b \equiv c/d$ if and only if $ad = bc$. For example, it follows that $1/2 \equiv a/b$ iff $b = 2a$. Therefore we can say that two fractions a/b and c/d are "equal" iff $a/b \equiv c/d$. To be absolutely proper, we should probably use a word other than "equal" since "equal" is often equated with "identical." Thus we call a/b and c/d *equivalent fractions* if $a/b \equiv c/d$, i.e., if $ad = bc$.

The relation $\equiv$ is analogous to the identity relation in that $\equiv$ possesses the same properties of the identity relation:

1. $\equiv$ is a reflexive relation: For each $a/b \in \mathscr{F}$, $a/b \equiv a/b$.
2. $\equiv$ is a symmetric relation: For each $a/b, c/d \in \mathscr{F}$, if $a/b \equiv c/d$, then $c/d \equiv a/b$.
3. $\equiv$ is a transitive relation: For each $a/b, c/d, e/f \in \mathscr{F}$, if $a/b \equiv c/d$ and $c/d \equiv e/f$, then $a/b \equiv e/f$.

This example leads us to the following general definition.

Definition 3.3.1 *Equivalence Relation*

A relation $\sim$ on a set A is an *equivalence relation* if $\sim$ is a reflexive, symmetric, and transitive relation on A.

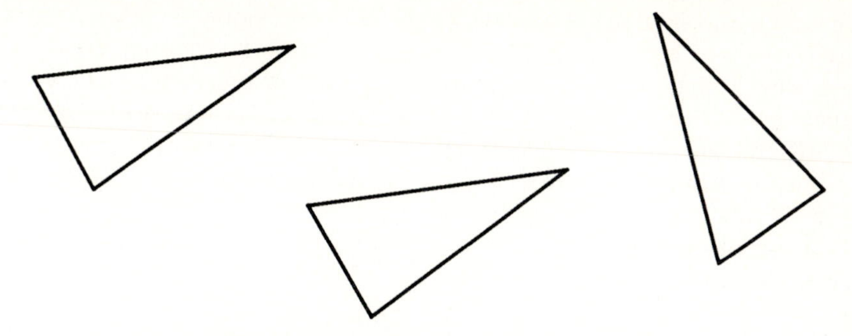

Figure 3.3.1. Three Congruent triangles.

The symbol ' $\sim$ ' is commonly used to denote an equivalence relation, probably because of its resemblance to the symbol " $=$ ". For $x, y \in A$ we read "$x \sim y$" as "x is *equivalent* to y," or in lighter moments, as "x wiggles y." We write $x \nsim y$ if x is not equivalent to y.

Example 3.3.1 As we have noted, the relation $\equiv$ on the set $\mathscr{F}$ of fractions of integers is an equivalence relation.

Example 3.3.2 Let $\mathscr{T}$ be the set of triangles in the plane. If $T_1, T_2 \in \mathscr{T}$, then we define $T_1 \text{ Cong } T_2$ to mean T_1 and T_2 are congruent triangles. (Recall that triangle T_1 is congruent to triangle T_2 if T_2 can be covered precisely by a copy of T_1. See Figure 3.3.1.) By an appeal to this definition or to theorems of Euclidean geometry, it is easy to check that Cong is an equivalence relation on $\mathscr{T}$.

Hence, equality of fractions and congruence of triangles, two of the most important relations springing from grade school and secondary school mathematics, are in fact equivalence relations. As these examples might suggest, the notion of an equivalence relation is one of the most basic concepts in mathematics. Equivalence relations will appear at several significant points later in this text—when we construct the integral, rational, and real number systems, when we measure the size of an infinite set, and when we define certain algebraic structures. Let us look at equivalence relations on some small sets.

Example 3.3.3 (i) Let us determine all equivalence relations on $A = \{a, b\}$. Let $\sim$ be an equivalence relation on A. Since $\sim$ is reflexive, $(a, a), (b, b) \in \sim$; i.e., $\sim \supseteq \{(a, a), (b, b)\} = I_A$. (Here we regard $\sim$ as a subset of $A \times A$.) We have two cases: (1) $\sim = I_A$, (2) $\sim \supset I_A$. In case 2, either $(a, b) \in \sim$ or $(b, a) \in \sim$. But if $(a, b) \in \sim$, then $(b, a) \in \sim$ by symmetry, and if $(b, a) \in \sim$, then $(a, b) \in \sim$ by symmetry. Therefore, in case 2, $\sim\, = \{(a, a), (b, b), (a, b), (b, a)\} = A \times A$. To summarize, we have shown that if $\sim$ is an equivalence relation on $A = \{a, b\}$, then either $\sim\, = I_A$ or $\sim\, = A \times A$.

(ii) As an exercise, find all equivalence relations on $A = \{a, b, c\}$. Here is one example: $\{(a, a), (b, b), (c, c), (a, b), (b, a)\}$.

Example 3.3.4 Let R be the relation on the Cartesian plane $\mathbf{R}^2$ defined by

$$(x, y)R(u, v) \quad \text{if } x^2 + y^2 = u^2 + v^2.$$

In words, two points in $\mathbf{R}^2$ are related if their respective distances to the origin are equal. A straightforward argument shows that R is an equivalence relation. For instance, to check the transitivity of R, suppose $(x, y)R(u, v)$ and $(u, v)R(z, w)$; then $x^2 + y^2 = u^2 + v^2$ and $u^2 + v^2 = z^2 + w^2$. By the transitivity of $=$, $x^2 + y^2 = z^2 + w^2$ and $(x, y)R(z, w)$. Similarly the reflexivity (resp. symmetry) of R follows from the reflexivity (resp. symmetry) of $=$. For a given point $(x, y) \neq (0, 0)$, the set of points that are equivalent to (x, y) is a circle centered at $(0, 0)$ of radius $\sqrt{x^2 + y^2}$. See Figure 3.3.2.

Thus, when looking at $\mathbf{R}^2$ from the point of view of this relation, we pay no attention to the direction from $(0, 0)$ to a point (x, y); what matters is the distance from $(0, 0)$ to (x, y).

Example 3.3.5 As in Example 3.3.2, let $\mathscr{T}$ denote the set of triangles in the Euclidean plane. Define a relation Sim on $\mathscr{T}$ by saying T_1 Sim T_2 if triangle T_1 is similar to triangle T_2. (By definition T_1 is similar to T_2 if a magnification or shrinking of T_1 transforms T_1 into a triangle that is

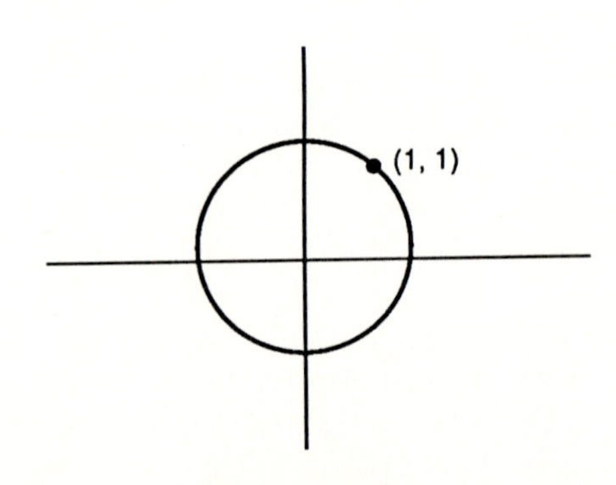

Figure 3.3.2

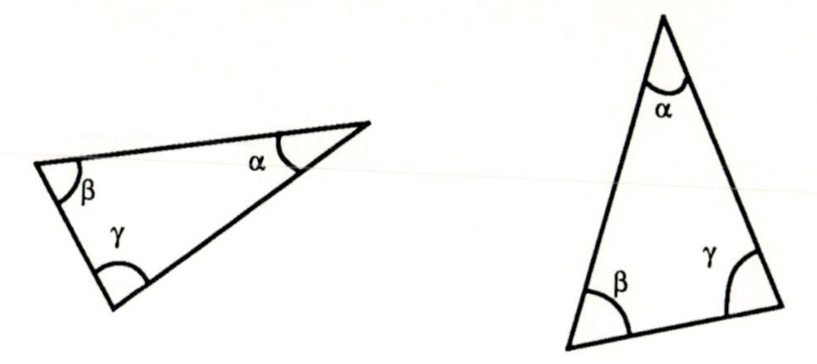

Figure 3.3.3

congruent to T_2.) With another appeal to high school geometry, one can show that Sim is an equivalence relation on $\mathscr{T}$.

Examples 3.3.2 and 3.3.5 drum home the observation that several natural equivalence relations can be defined on a given set. Each equivalence relation isolates a characteristic property relative to which the elements of the set are to be regarded as equivalent or, if we like, relative to which they are to be distinguished. For instance, informally we might say that two triangles are similar if they have the same shape, while two triangles are congruent if they have the same shape and size (Figure 3.3.3).

Example 3.3.6 Let $\mathscr{L}$ be the set of straight lines in $\mathbf{R}^2$. Define a relation $\|$ on $\mathscr{L}$ by the rule: For $\ell_1, \ell_2 \in \mathscr{L}$,

$$\ell_1 \parallel \ell_2 \quad \text{iff } \ell_1 = \ell_2 \text{ or } \ell_1 \text{ is parallel to } \ell_2.$$

You can convince yourself that $\|$ is an equivalence relation either formally by referring to theorems of Euclidean geometry or informally by expressing reflexivity, symmetry, and transitivity geometrically. See Figure 3.3.4.

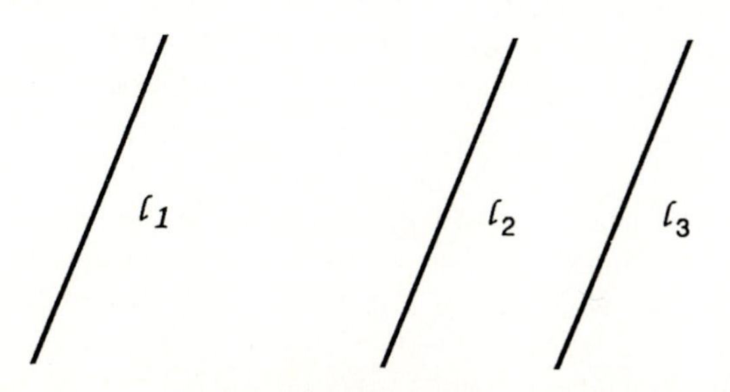

Figure 3.3.4

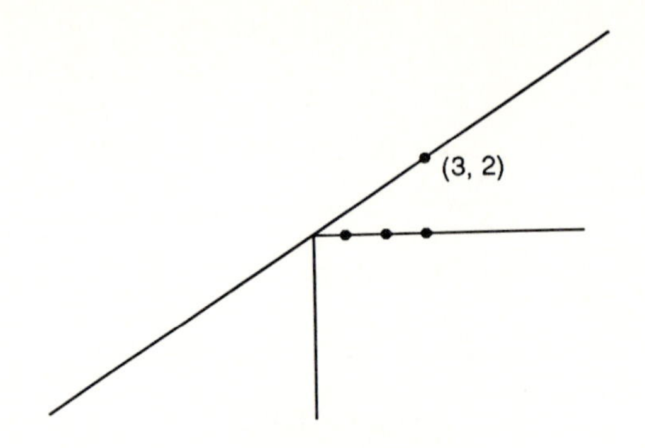

Figure 3.3.5

Example 3.3.7 Let $A = \mathbf{R}^2 - \{(0,0)\}$. For $(x, y), (u, v) \in A$, define $(x, y) \operatorname{Line}(u, v)$ if $(x, y) = (u, v)$ or if the line joining (x, y) to (u, v) passes through the origin. The relation Line can be expressed in the following equivalent fashion:

$$(x, y) \operatorname{Line}(u, v) \quad \text{if either } x = u = 0 \text{ or } y/x = v/u.$$

We leave it as an exercise to check that Line is an equivalence relation. (Notice that the relation Line is also an equivalence relation on the set $\mathbf{R}^2$. However, any point (x, y) is equivalent with respect to Line to the point $(0, 0)$; hence $(x, y) \operatorname{Line}(u, v)$ for all $(x, y), (u, v) \in \mathbf{R}^2$.) See Figure 3.3.5.

Example 3.3.8 Let E be the relation on $\mathbf{Z}$, the set of integers, defined as follows: $x \mathrm{E} y$ iff x and y are both even or both odd. We claim that E is an equivalence relation on $\mathbf{Z}$.

(i) E is reflexive: If $x \in \mathbf{Z}$, then $x \mathrm{E} x$ since x and x are either both even (if x is even) or both odd (if x is odd).

(ii) E is symmetric: If $x \mathrm{E} y$, then x and y are either both even or both odd, and thus y and x are either both even or both odd. Hence $y \mathrm{E} x$.

(iii) E is transitive: If $x \mathrm{E} y$ and $y \mathrm{E} z$, then either x and y are both even in which case (since $y \mathrm{E} z$) y and z are both even, and hence x and z are both even, or x and y are both odd, which implies that y and z are both odd (again since $y \mathrm{E} z$); thus x and z are both odd. Thus, if $x \mathrm{E} y$ and $y \mathrm{E} z$, then $x \mathrm{E} z$.

Therefore, E is an equivalence relation. As far as the relation E is concerned, all that matters about a number is its parity, i.e., whether it is even or odd. If x is even, then $x \mathrm{E} 0$; if x is odd, then $x \mathrm{E} 1$:

$$\{x \mid x \mathrm{E} 0\} = \{x \mid x \text{ is even}\} \quad \text{and} \quad \{x \mid x \mathrm{E} 1\} = \{x \mid x \text{ is odd}\}.$$

Thus, under the relation E, each element of $\mathbf{Z}$ is equivalent to either 0 or 1.

The next equivalence relation first appeared in the latter half of the eighteenth century and was formally christened in *Disquisitiones Arithmeticae* (Arithmetical Investigations), written by Carl Frederick Gauss (1777–1855) and published in 1801. Gauss is regarded by many as the greatest mathematician of all time because of both the breadth of his work in mathematics, physics, and astronomy and the depth of his insights into every subject that he addressed. For example, in his *Disquisitiones*, Gauss set the principal course for the developments in number theory (which is, to quote *Time* magazine, "the abstruse specialty concerned with properties of whole numbers") that were to occur over the following 150 years. Several sections of the *Disquisitiones* are devoted to an analysis of properties of the following concept.

Definition 3.3.2 *Congruence Modulo n*

Let n be a positive integer. For $a, b \in \mathbf{Z}$, *a is congruent to b modulo n* if n divides $a - b$. If a is congruent to b modulo n, then we write either $a \equiv_n b$ or $a \equiv b \pmod{n}$.

Referring back to Example 3.3.8, suppose $a \mathrm{E} b$ where $a, b \in \mathbf{Z}$. Then a and b are either both even or both odd, and, in either case, $a - b$ is even, i.e., 2 divides $a - b$. Therefore, if $a \mathrm{E} b$, then $a \equiv_2 b$. Conversely, if $a \equiv_2 b$, then $a \mathrm{E} b$, for otherwise exactly one of a and b is even (and exactly one is odd) and $a - b$ is not divisible by 2. Thus the equivalence relation E coincides with the relation of congruence modulo 2. The following theorem provides a fairly complete description of the relation $\equiv_n$. Note that its conclusions have already been checked for the relation $\equiv_2$.

Theorem 3.3.1 *For each positive integer n, $\equiv_n$ is an equivalence relation. Moreover, each integer is congruent modulo n to exactly one of the elements in the set $\{0, 1, \ldots, n - 1\}$.*

Proof Let n be a positive integer. We show that $\equiv_n$ is an equivalence relation. At first glance, one might be tempted to try an inductive proof of this statement. However it is not clear how in general to relate divisibility by an integer n (by which $\equiv_n$ is defined) to divisibility by integers that are less than n. As it happens, a direct proof of the given statement is very straightforward.

(i) Reflexivity: Let $a \in \mathbf{Z}$. Then $a - a = 0 = n \cdot 0$; hence $a \equiv_n a$.

(ii) Symmetry: Suppose $a \equiv_n b$. We show $b \equiv_n a$. Since $a \equiv_n b$, there exists $k \in \mathbf{Z}$ such that $a - b = nk$. Thus $b - a = -n \cdot k = n \cdot (-k)$ and $b \equiv_n a$.

(iii) Transitivity: Suppose $a \equiv_n b$ and $b \equiv_n c$. Then $a - b = n \cdot k$ and $b - c = n \cdot m$ for some integers k and m. Hence $a - c = a - b + b - c = n \cdot k + n \cdot m = n \cdot (k + m)$, and $a \equiv_n c$.

We conclude that $\equiv_n$ is an equivalence relation.

We now prove the second statement. We show that any $a \in \mathbf{Z}$ is congruent modulo n to one and only one element from the set $\{0, 1, \ldots, n-1\}$.

By the Division Theorem (Theorem 1.3.3), there exist $q, r \in \mathbf{Z}$ such that $a = n \cdot q + r$ with $0 \leq r \leq n - 1$. Thus $a - r = n \cdot q$, which implies that $a \equiv_n r$ and $r \in \{0, 1, \ldots, n-1\}$. Therefore, a is congruent modulo n to at least one element from $\{0, 1, \ldots, n-1\}$.

Suppose there exist $r, s \in \{0, 1, \ldots, n-1\}$ such that $a \equiv_n r$ and $a \equiv_n s$. We show $r = s$. Since $\equiv_n$ is an equivalence relation, $r \equiv_n s$ and $r - s = nk$ for some $k \in \mathbf{Z}$. However $-(n-1) \leq r - s \leq n - 1$, since $0 \leq r, s \leq n - 1$. Now, if $k = 0$, then $r - s = 0$ or $r = s$. If $k \neq 0$, then by mathematical induction it follows that $|n \cdot k| \geq n$. Hence either $n \cdot k \leq -n$ or $n \cdot k \geq n$. Thus, if $k \neq 0$, then it is not the case that $-(n-1) \leq r - s = n \cdot k \leq n - 1$. Therefore $k = 0$ and $r = s$. (Note the use of an indirect argument to establish the uniqueness portion of the statement.) ∎

A final word on terminology and notation. In all likelihood, Gauss chose the word "congruence" because congruence modulo n is an equivalence relation just as is congruence of triangles. Hence, because of the similarity between the relations $\equiv_n$ and Cong, he borrowed the word "congruence" from geometry to use as a name for $\equiv_n$. As for notation, the symbol $\equiv_n$ has several advantages. First, the subscript reminds us of the modulus, thus preventing possible ambiguity on this score. Second, the symbol $\equiv_n$ suggests the analogy between congruence and equality. Nonetheless, the notation used by Gauss and most mathematicians after him is $\equiv \pmod{n}$. Henceforth, we follow the customs of the mathematical world and write $a \equiv b \pmod{n}$ to mean a is congruent to b modulo n.

Equivalence Classes

We have emphasized that an equivalence relation allows us to broaden the notion of equality from identity to similarity relative to a given property. Two elements need not be identical to be equivalent; they need only share a specified property. Nevertheless, looking closely at the set $\mathscr{F}$ of fractions, we realize that we usually regard equivalent fractions, such as $1/2$ and $2/4$, as representing the same quantity. Conceptually, we lump together all fractions equivalent to $1/2$ (namely the set $\{a/2a \mid a \in \mathbf{Z}$ and $a \neq 0\}$) and consider them to be a single entity. With the equivalence relation E on $\mathbf{Z}$ defined in Example 3.3.8, we identify any two even integers as being equivalent and any two odd integers as being equivalent. Thus the set $\mathbf{Z}$ of integers is split into two subsets (the evens and the odds) and any two elements in the same subset are equivalent. The practice followed in these examples can be generalized and applied to an arbitrary equivalence relation.

Definition 3.3.3 *Equivalence Class*

Let $\sim$ be an equivalence relation on a set A. For each $a \in A$, the *equivalence class of a* is the subset, denoted by $[a]_{\sim}$, consisting of all elements of A that are equivalent to a. In other words,

$$[a]_{\sim} = \{x \in A \mid x \sim a\}.$$

When there is no ambiguity, we write $[a]$ in place of $[a]_{\sim}$.

Let us look back at some of our principal examples.

1. $\mathscr{F} = \{a/b \mid a, b \in \mathbf{Z} \text{ and } b \neq 0\}$ with $a/b \equiv c/d$ if $ad = bc$. Then $[1/2] = \{x \in \mathscr{F} \mid x \equiv 1/2\} = \{a/b \in \mathscr{F} \mid 2a = b\} = \{a/2a \mid a \in \mathbf{Z} \text{ and } a \neq 0\}$, and $[2/3] = \{a/b \in \mathscr{F} \mid 3a = 2b\} = \{2a/3a \mid a \in \mathbf{Z} \text{ and } a \neq 0\}$. Notice that $[1/2] = [2/4]$ ($a/b \equiv 2/4$ iff $4a = 2b$ iff $2a = b$ iff $a/b \equiv 1/2$). Therefore, a given equivalence class can be described or represented by several distinct elements.

2. $\mathbf{Z}$ with $a \equiv b \pmod{n}$ if $n \mid a - b$. Consider first the case $n = 2$. Then, writing $[a]_2$ for $[a]_{\equiv_2}$,

$$[0]_2 = \{x \in \mathbf{Z} \mid x \equiv 0 \pmod 2\} = \{x \in \mathbf{Z} \mid x \text{ is even}\}$$

while

$$[1]_2 = \{x \in \mathbf{Z} \mid x \equiv 1 \pmod 2\} = \{x \in \mathbf{Z} \mid x \text{ is odd}\};$$

these are the only equivalence classes. Observe that $[0]_2 \cup [1]_2 = \mathbf{Z}$ and $[0]_2 \cap [1]_2 = \varnothing$. These are depicted in Figure 3.3.6. Again, each equivalence class can be described in several ways: $[0]_2 = [2]_2 = [4]_2$. The point we emphasize is that the *sets* $[0]_2$, $[2]_2$, and $[4]_2$ are equal; however, their descriptions, as all elements congruent to 0 mod 2 in the first case and all elements congruent to 2 mod 2 in the second case, etc., are different.

Next consider an arbitrary integer n. We write $[a]_{\equiv_n} = [a]_n$. By Theorem 3.3.1, each $a \in \mathbf{Z}$ is congruent to exactly one of the integers $0, 1, 2, \ldots, n-1$. If $0 \leq r \leq n-1$ and $a \equiv r \pmod{n}$, then $a \in [r]_n$.

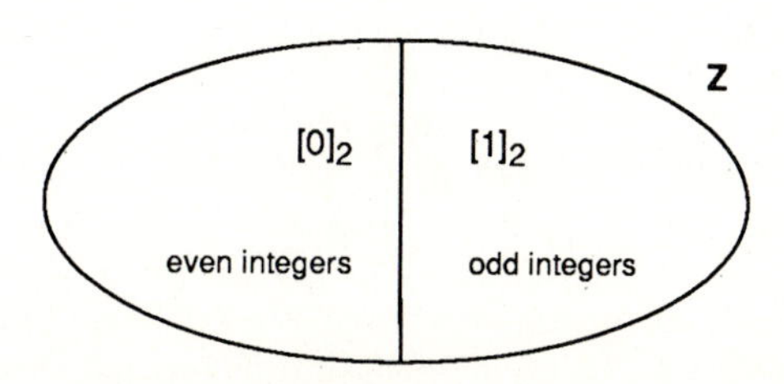

Figure 3.3.6

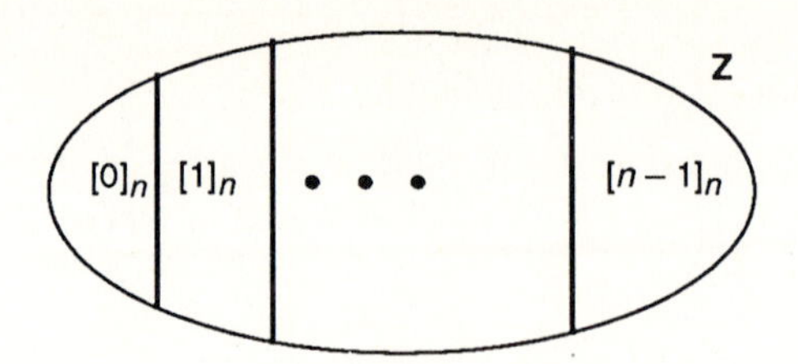

Figure 3.3.7. Equivalence classes of **Z** with respect to congruence modulo n

Also, if $0 \leq r, s \leq n - 1$ and $r \neq s$, then $[r]_n \cap [s]_n = \varnothing$. Thus there are exactly n equivalence classes of **Z** with respect to the congruence modulo n. They are $[0]_n, [1]_n, \ldots, [n-1]_n$ and $\mathbf{Z} = [0]_n \cup [1]_n \cup \cdots \cup [n-1]_n$ (Figure 3.3.7).

Not surprisingly, the phenomena visible in these examples are quite general.

Theorem 3.3.2 *Let* $\sim$ *be an equivalence relation on a set* A *and let* $x, y \in A$.

(i) *If* $x \sim y$, *then* $[x] = [y]$.

(ii) *If* $x \nsim y$, *then* $[x] \cap [y] = \varnothing$.

(iii) *The union of the equivalence classes of* $\sim$ *is* A: $A = \bigcup_{x \in A}[x]$.

Proof (i) Suppose $x \sim y$. We show $[x] = [y]$ by proving that $[x] \subseteq [y]$ and $[y] \subseteq [x]$. Let $u \in [x]$. Then $u \sim x$. Since $x \sim y$ and $\sim$ is transitive, $u \sim y$; hence $u \in [y]$. Therefore, $[x] \subseteq [y]$. A "symmetric" argument shows that $[y] \subseteq [x]$. (Just interchange x and y throughout the argument.)

(ii) We argue by contraposition. We assume $[x] \cap [y] \neq \varnothing$ and show that $x \sim y$. Let $u \in [x] \cap [y]$. Then $x \sim u$ and $y \sim u$. Therefore, $x \sim y$, which contradicts the assumption that $x \nsim y$.

(iii) The third statement is almost transparent. Each equivalence class is a subset of A. Thus the union of the equivalence classes is a subset of A. On the other hand, each $x \in A$ is in the equivalence class $[x]$; thus A is contained in the union of the equivalence classes. ■

Definition 3.3.4 *A Set Modulo an Equivalence Relation*

Let $\sim$ be an equivalence relation on A. The set of all equivalence classes is called *A modulo* $\sim$ and is written $A/\sim$.

We emphasize that $A/\sim$ is a set that is formed from the set A and the equivalence relation $\sim$. The elements of $A/\sim$ are certain subsets of

A, namely, the equivalence classes of A with respect to the equivalence relation $\sim$. Thus $A/\sim\ \subset P(A)$. The notation $A/\sim$ suggests the process of division, and in a sense this suggestion mirrors what is actually happening, for the set A is cut up or divided into equivalence classes and each of these sets becomes an element of $A/\sim$.

For the relation $\equiv \pmod{n}$ on $\mathbf{Z}$, it is customary to write either $\mathbf{Z}_n$ or $\mathbf{Z}/n\mathbf{Z}$ to denote the set of equivalence classes. We shall use $\mathbf{Z}/n\mathbf{Z}$. By Theorem 3.3.1, $\mathbf{Z}/n\mathbf{Z}$ has n elements:

$$\mathbf{Z}/n\mathbf{Z} = \{[0]_n, [1]_n, \ldots, [n-1]_n\}.$$

The interplay between the set A on which the equivalence relation $\sim$ is defined and the set of equivalence classes $A/\sim$ is important. Whenever one defines an equivalence relation on a set, one is usually interested in working with the set of equivalence classes. Nevertheless, a given equivalence class has to be described or represented by an element of the set A. Thus, if $\alpha \in A/\sim$, then we have $\alpha = [x]$ for some $x \in A$ and we call $x \in \alpha$ a *representative of* α. As we have seen, the choice of representative is not unique in general. But, in some cases, representatives of the equivalence classes can be chosen in a fairly reasonable way. An example is $\mathbf{Z}/n\mathbf{Z}$, where the classes are $[r]$ with $0 \le r \le n-1$. We mention two others: If x is an equivalence class of $\mathscr{F}$ relative to the relation $\equiv$ such that $x \neq [0/1]$, then we can write $x = [a/b]$ where a/b is a fraction in lowest terms. The set $\mathscr{F}/\equiv$ is precisely the set $\mathbf{Q}$ of rational numbers. Thus $\mathscr{F}/\equiv\ = \{[a/b] \| a = 0 \text{ and } b = 1 \text{ or } a/b \text{ is in lowest terms}\}$. Finally, if $\mathscr{L}$ and $\|$ are as defined in Example 3.3.6, then

$$\mathscr{L}/\| = \{[\ell] \| \ell \text{ is a line passing through } (0,0)\}.$$

Notice that in each case the set of equivalence classes is described by specifying a single element from each class. The following definition helps to provide some terminology for this kind of situation.

Definition 3.3.5 *Complete Set of Representatives*

Let $\sim$ be an equivalence relation on a set A. A subset of A containing exactly one element from each equivalence class is called a *complete set of representatives* of $A/\sim$.

For example, $\{0, 1, \ldots, n-1\}$ and $\{1, 2, \ldots, n\}$ are both complete sets of representatives of $\mathbf{Z}$ modulo n. In general, if B is a complete set of representatives of $A/\sim$, then $A/\sim\ = \{[x] \| x \in B\}$.

Example 3.3.9 On $\mathbf{R}^2$ define a relation R as follows: For $v, w \in \mathbf{R}^2$ with $v = (x, y)$ and $w = (u, v)$, $v R w$ iff $x + y = u + v$. It is easy to check

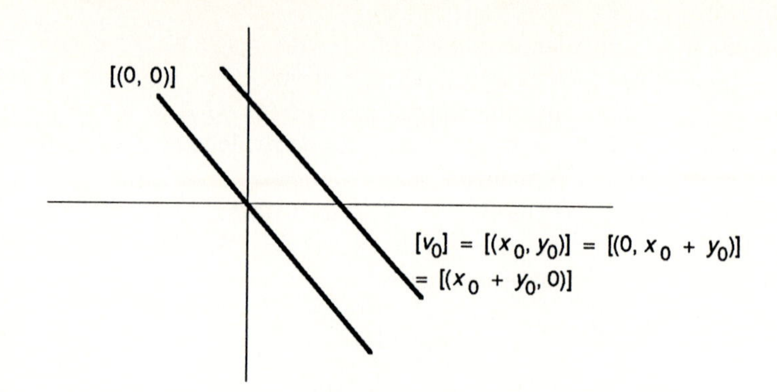

Figure 3.3.8

that R is an equivalence relation. Let $v_0 = (x_0, y_0)$ be fixed. What is $[v_0]$? By definition $[v_0] = \{v | v \, R \, v_0\} = \{(x, y) | x + y = x_0 + y_0\}$. This set is a straight line with slope -1 and y-intercept $(0, x_0 + y_0)$ as shown in Figure 3.3.8.

To obtain a complete set of representatives for this equivalence relation, we must choose one element from each equivalence class, i.e., one point from each line in $\mathbf{R}^2$ having slope -1. Since the y-axis intersects each line with slope -1 in exactly one point, the y-axis forms a complete set of representatives for the equivalence classes of $\mathbf{R}^2$ modulo R. The set of equivalence classes, $\mathbf{R}^2/R$, is

$$\mathbf{R}^2/R = \{[(0, y)] | y \in \mathbf{R}\}.$$

Another complete set of representatives consists of points along the x-axis:

$$\mathbf{R}^2/R = \{[(x, 0)] | x \in \mathbf{R}\}.$$

Partitions

Looking again at the picture of the equivalence classes of $\mathbf{Z}$ modulo n (Figure 3.3.9), we see that the distinct equivalence classes are disjoint and the union of these classes is all of $\mathbf{Z}$. This kind of division of a set has a name.

Definition 3.3.6 *Partition*

A *partition* of a set A is a collection P of subsets of A which are pairwise disjoint and whose union is A.

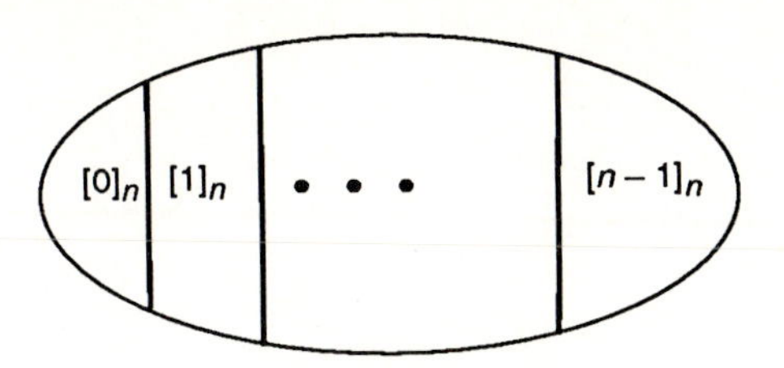

Figure 3.3.9

Note that P is just a collection of subsets of A and is not all of $P(A)$. Also we usually exclude $\varnothing$ from any partition. To say that the sets in P are pairwise disjoint is to say that if $B, C \in P$ with $B \neq C$, then $B \cap C = \varnothing$. The condition that the union of the sets of P is A translates into the equality:

$$A = \bigcup_{B \in P} B$$

Example 3.3.10 (i) $P_2 = \{[0]_2, [1]_2\}$ is a partition of $\mathbf{Z}$. The partition P_2 is a set that contains two elements, one of which is the set of even integers and the other of which is the set of odd integers. Note that P_2 is the set of equivalence classes of the equivalence relation $\equiv_2$ on $\mathbf{Z}$.

(ii) $P_n = \{[0]_n, [1]_n, \ldots, [n-1]_n\}$
$= \{[r]_n | r \in \mathbf{Z} \text{ and } 0 \leq r \leq n-1\}$
is a partition of $\mathbf{Z}$.

These examples suggest a close relationship between equivalence relations and partitions. In fact, Theorem 3.3.2 tells us that for any equivalence relation $\sim$ on a set A, the set of *distinct* equivalence classes of A modulo $\sim$ constitutes a partition of A. Thus, to each equivalence relation on a set A, there corresponds a partition of A. The next theorem assures us that the correspondence runs in the other direction as well.

Theorem 3.3.3 *Let P be a partition of a set A. Define a relation $\sim_P$ on A as follows: For $x, y \in A$, $x \sim_P y$ if there is a set B in the partition P such that $x \in B$ and $y \in B$. Then $\sim_P$ is an equivalence relation on A.*

Figure 3.3.10 depicts a partition of A and illustrates a pair of related elements and two pairs of unrelated elements.

Proof We show that $\sim_P$ is reflexive, symmetric, and transitive.

(1) Reflexivity. If $x \in A$, then $x \in B$ for some $B \in P$ (why?), and hence $x \in B$ and $x \in B$; thus $x \sim_P x$ and $\sim_P$ is reflexive.

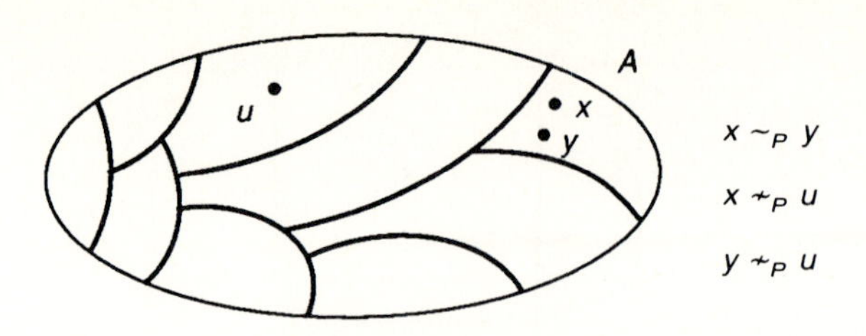

Figure 3.3.10

(2) Symmetry. If, for $x, y \in A$, $x \sim_P y$, then there is $B \in P$ such that $x \in B$ and $y \in B$, and hence $y \in B$ and $x \in B$; thus $y \sim_P x$ and $\sim_P$ is symmetric.

(3) Transitivity. Suppose for $x, y, z \in A$, $x \sim_P y$ and $y \sim_P z$. Then there exist $B, C \in P$ such that $x \in B$ and $y \in B$ and $y \in C$ and $z \in C$. Therefore, $y \in B \cap C$, which means that, since P is a partition of A, $B = C$. Thus there is a set $B \in P$ such that $x \in B$ and $z \in B$. Therefore, $x \sim_P z$ and $\sim_P$ is transitive. ■

Example 3.3.11 Consider the set $\mathbf{R}$ of real numbers. A partition of $\mathbf{R}$ can be obtained by defining for each $i \in \mathbf{Z}$, $A_i = \{x \in \mathbf{R} | i \leq x < i + 1\}$. Thus $A_0 = \{x \in \mathbf{R} | 0 \leq x < 1\}$ and $A_{-5} = \{x | -5 \leq x < -4\}$. (See Figure 3.3.11.) Then $P = \{A_i | i \in \mathbf{Z}\}$ is a partition of $\mathbf{R}$. What is the corresponding equivalence relation? If $x, y \in \mathbf{R}$, then $x \sim_P y$ if and only if there exist $A_i \in P$ such that $x \in A_i$ and $y \in A_i$ if and only if there exists $i \in \mathbf{Z}$ such that $i \leq x, y < i + 1$. We can express this relation using a commonly used function: For $a \in \mathbf{R}$, define $\lfloor a \rfloor$ to be the largest integer $\leq a$. Then $\lfloor 1/2 \rfloor = 0$, $\lfloor -1/2 \rfloor = -1$, and $\lfloor n \rfloor = n$ if $n \in \mathbf{Z}$. a is called either the *greatest integer function* evaluated at a or the *floor function* evaluated at a. Using this concept, we can state that $x \sim_P y$ iff $[x] = [y]$. This example illustrates how one can begin with a partition, use it to define an equivalence relation, and then find another way of describing the equivalence relation.

Example 3.3.12 Consider the set of all lines $\mathscr{L}_1$ in $\mathbf{R}^2$ parallel to the line with equation $3x + 2y = 0$. The set $\mathscr{L}_1$ constitutes a partition of $\mathbf{R}^2$.

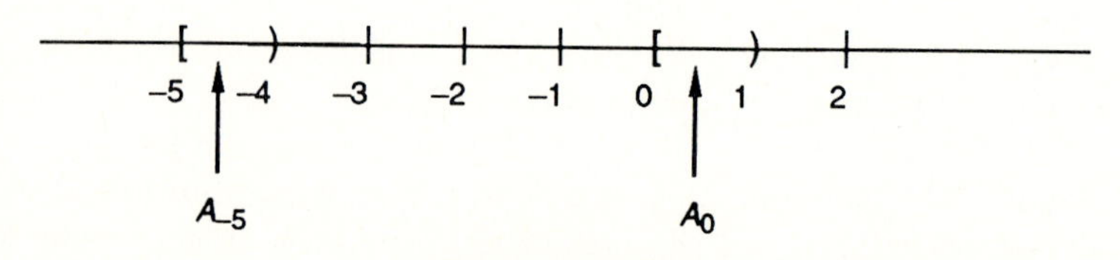

Figure 3.3.11

(Proof?) What is the corresponding equivalence relation on $\mathbf{R}^2$? If (x_1, y_2), $(x_2, y_2) \in \mathbf{R}^2$, then $(x_1, y_2) \sim_1 (x_2, y_2)$ iff (x_1, y_1) and (x_2, y_2) lie on the same line in $\mathscr{L}_1$, which happens iff $3x_1 + 2y_1 = 3x_2 + 2y_2$. Thus the relation $\sim_1$ can be described algebraically: $(x_1, y_1) \sim_1 (x_2, y_2)$ if $3x_1 + 2y_1 = 3x_2 + 2y_2$.

A summary of Theorems 3.3.2 and 3.3.3 might be in order. The thrust of Theorem 3.3.2 is that each equivalence relation $\sim$ on a set A determines a partition, let's call it $P(\sim)$, of A: $P(\sim)$ is just the set of distinct equivalence classes of A modulo $\sim$. (A fancier description $P(\sim) = \{[x] | x$ ranges over a complete set of representatives of A modulo $\sim\}$.) On the other hand, by Theorem 3.3.3, each partition P of A determines an equivalence relation $\sim_P$ on A.

What happens if we start with an equivalence relation $\sim$ on A, form the partition $P(\sim)$, and then from $P(\sim)$ construct the corresponding equivalence relation, which we shall call $\approx$ to simplify notation. A moment's thought should reveal that $\approx$ and $\sim$ are the same relation. Similarly, given a partition P, we can form the equivalence relation determined by P, $\sim_P$. Then the partition $P(\sim_P)$, determined by $\sim_P$, is the original partition P: $P(\sim_P) = P$. Thus the process of forming a partition from an equivalence relation and the process of constructing an equivalence relation from a partition are inverses to each other. Here is a convenient representation of the events:

$$\underset{\text{partition}}{P} \xrightarrow[]{\text{form eq. rel}} \underset{\substack{\text{equivalence}\\ \text{relation}}}{\sim_P} \xrightarrow[]{\text{form partition}} \underset{\text{partition}}{P(\sim_P) =}$$

$$\underset{\text{eq. rel.}}{\sim} \xrightarrow[]{\text{form partition}} \underset{\text{partition}}{P(\sim)} \xrightarrow[]{\text{form eq. rel.}} \underset{\text{eq. rel.}}{\approx}$$

Other examples of inverse processes have cropped up in our prior mathematical experience. The first examples we cite are inverse functions that we represent schematically as follows:

$$\underset{\text{real number}}{x} \xrightarrow[]{\text{multiply by 2 and add 1}} \underset{\text{real number}}{2x + 1} \xrightarrow[]{\text{subtract 1 and divide by 2}} \underset{\text{real number}}{[(2x + 1) - 1]/2 = x}$$

$$\underset{\text{real number}}{x} \xrightarrow[]{\text{exponentiate}} \underset{\substack{\text{pos. real}\\ \text{number}}}{e^x} \xrightarrow[]{\text{take natural logarithm}} \underset{\text{real number}}{\ln(e^x) = x}$$

Finally, we recall the Fundamental Theorem of Calculus, which states roughly that the process of differentiation is inverse to the process of integration:

$$\underset{\substack{\text{continuous}\\ \text{function in } [a,b]}}{f(x)} \xrightarrow[\text{}]{\substack{\text{integrate}\\ \text{from } a \text{ to } x}} \underset{\substack{\text{differentiable}\\ \text{function in } [a,b]}}{\int_a^x f(t)\,dt} \xrightarrow{\text{differentiate}} \underset{\substack{\text{continuous}\\ \text{function in } [a,b].}}{d/dx\left(\int_a^x f(t)\,dt\right) = f(x)}$$

Exercises §3.3

3.3.1. (a) How many equivalence relations are there on $\{a, b, c\}$?
(b) Give four examples of equivalence relations on $\{a, b, c, d\}$.

3.3.2. Let $\mathscr{L}$ be the set of lines in the Euclidean plane. For $\ell_1, \ell_2 \in \mathscr{L}$ define $\ell_1 \perp \ell_2$ to mean ℓ_1 and ℓ_2 are perpendicular. Is $\perp$ an equivalence relation?

3.3.3. (a) Let A be any set. Which equivalence relation(s) on A deserve the label "trivial"? Why?
(b) Characterize those sets A that have only one equivalence relation.

3.3.4. Describe two equivalence relations on $\mathscr{T}$, the set of triangles in the Euclidean plane, which are distinct from Cong and Sim.

3.3.5. Which of the following relations R are equivalence relations on $\mathbf{Z} - \{0\}$?
(a) aRb if $2|(a + b)$.
(b) aRb if $3|(a + b)$.
(c) aRb if $a|b$ and $b|a$.
(d) aRb if $a|b$ or $b|a$.
(e) aRb if $a|b$.

3.3.6. Let P be the set of all propositions in English. For $p, q \in P$ define $p \sim q$ if $p \Leftrightarrow q$ is true. Show: $\sim$ is an equivalence relation. In other words, logical equivalence is an equivalence relation on the set of English propositions.

3.3.7. Prove, or disprove and salvage: Let $[a, b] = \{x \in \mathbf{R} | a \le x \le b\}$. Let $C[a, b]$ denote the set of continuous functions on $[a, b]$. If $f, g \in C[a, b]$, then define $f \sim g$ if $\int_a^b f(x)\,dx = \int_a^b g(x)\,dx$. Then $\sim$ is an equivalence relation on $C[a, b]$.

3.3.8. Let $A = \mathbf{N} \times \mathbf{N}$. For $x, y \in A$ with $x = (a, b)$ and $y = (c, d)$, define $x \sim y$ if $a + d = b + c$.
(a) Prove: $\sim$ is an equivalence relation on A.
(b) Show: $\{(n, 0) | n \in \mathbf{N}\} \cup (0, m) | m \in \mathbf{N}$ and $m \neq 0\}$ is a complete set of representatives for $A/\sim$.

3.3.9. Let $M_n(\mathbf{R})$ denote the set of $n \times n$ matrices with real entries. For $A, B \in M_n(\mathbf{R})$, define $A \sim B$ if there exists $C \in M_n(\mathbf{R})$ such that $A = CBC^{-1}$. Is $\sim$ an equivalence relation on $M_n(\mathbf{R})$?

3.3.10. Prove, or disprove and salvage: Let U be an infinite set. For $A, B \in P(U)$, suppose $A \sim B$ if A and B are finite. Is $\sim$ an equivalence relation?

3.3.11. Let R and S be equivalence relations on a set A.
(a) Is $R \cap S$ necessarily an equivalence relation on A?
(b) Is $R \cup S$ necessarily an equivalence relation on A?

3.3.12. Let R and S be equivalence relations on A and B, respectively. Define a relation T on $A \times B$ as follows: $(a, b)T(a_1, b_1)$ iff aRa_1 and bSb_1. Prove or disprove: T is an equivalence relation on $A \times B$.

3.3.13. (a) Can a function (which is, of course, a relation) be an equivalence relation? Explain.
(b) Can a partial ordering also be an equivalence relation? Explain.

3.3.14. (a) Devise an equivalence relation on $\mathbf{R}$ with exactly two equivalence classes.
(b) Devise an equivalence relation on $\mathbf{R}$ with exactly three equivalence classes.

3.3.15. Let A be any nonempty set.
(a) Describe the partition of A given by the identity relation I_A.
(b) Describe the partition of A given by the equivalence relation $A \times A$.

3.3.16. Define the relation $\equiv_1$ on $\mathbf{R}$ by the rule: If $x, y \in \mathbf{R}$, then $x \equiv_1 y$ if $x - y$ is an integer.
(a) Show that $\equiv_1$ is an equivalence relation on $\mathbf{R}$.
(b) What is $[1/2]_{\equiv_1}$?
(c) Describe the set of equivalence classes.

3.3.17. Define a relation R on $\mathbf{R}^2$ as follows: For $v, w \in \mathbf{R}^2$ with $v = (x, y)$ and $w = (a, b)$, vRw if $|x| + |y| = |a| + |b|$.
(a) Prove that R is an equivalence relation on $\mathbf{R}^2$.
(b) Describe the equivalence classes geometrically.
(c) Find a complete set of representatives for $\mathbf{R}^2/R$.

3.3.18. Find a complete set of representatives of $(\mathbf{R}^2 - \{(0,0)\})/R$ where R is the equivalence relation defined in Example 3.3.7.

3.3.19. Let $A = \mathbf{Z}^2 = \mathbf{Z} \times \mathbf{Z}$. For $(a, b), (c, d) \in A$, we write $(a, b) \equiv_2 (c, d)$ if $a - c$ is even and $b - d$ is even.
(a) Show $\equiv_2$ is an equivalence relation on A.
(b) Find a complete set of representatives for $\equiv_2$.
(c) How many equivalence classes of $\equiv_2$ are there?
(d) Generalize this example.

3.3.20. (a) Let $a, b, c \in \mathbf{Z}$. Prove that if $a|b$ and $a|c$ then $a|b \pm c$.
(b) Prove that if $a, b, c, d, n \in \mathbf{Z}$ and $a \equiv b \pmod{n}$ and $c \equiv d \pmod{n}$, then $a + c \equiv b + d \pmod{n}$ and $ac \equiv bd \pmod{n}$.

3.3.21. Addition and Multiplication of Fractions. Let $a/b, c/d \in \mathscr{F}$. Define $a/b + c/d = (ad + bc)/bd$ and $(a/b) \cdot (c/d) = (a \cdot c)/(b \cdot d)$. Suppose $a/b \equiv a'/b'$ and $c/d \equiv c'/d'$.
(a) Prove: $a/b + c/d \equiv a'/b' + c'/d'$.
(b) Prove: $(a/b) \cdot (c/d) \equiv (a'/b') \cdot (c'/d')$.

3.3.22. Let A be a set. A relation R on A is called *circular* if xRy and yRz implies that zRx.

(a) Prove: A relation R on A is an equivalence relation if and only if R is reflexive and circular.

(b) Give an example of a set A and a relation R on A such that R is circular but R is not an equivalence relation.

3.3.23. Let R be a relation on a set A. Prove that R is an equivalence relation on A if and only if (a) R is reflexive, (b) for all $x, y, z \in A$ if xRz and yRz, then xRy.

3.3.24. Let $S = \{1, 2, 3, 4, 5\}$ and let P be the partition of S, $P = \{\{1, 2\}, \{3\}, \{4, 5\}\}$. List all the ordered pairs of the equivalence relation corresponding to P.

3.3.25. $A = \{1, 2, 3, 4\}$. Let R be the equivalence relation corresponding to the partition given by $P = \{\{1\}, \{2, 3\}, \{4\}\}$. Let S be the equivalence relation corresponding to the partition given by $P' = \{\{1, 3\}, \{2, 4\}\}$. Write down the ordered pairs in each of the following relations:

(a) $R \cup S$.

(b) $R \cap S$.

(c) $R \circ S$.

3.3.26. Let $|A| = k \in \mathbf{N}^+$. Suppose P is a partition of A into n (disjoint) subsets: $A = A_1 \cup A_2 \cup \cdots \cup A_n$ and $|A_i| = k_i$, $1 \leq i \leq n$ (hence $k = k_1 + k_2 + \cdots + k_n$). What is the size of the equivalence relation on A corresponding to P? (i.e., how many ordered pairs?)

§3.4 Functions

We observed in Section 3.1 that a function is actually a special kind of relation. The purpose of this section is to record some general properties of functions and to discuss various ways of representing functions. First we recall the basic definition.

Definition 3.4.1 *Function*

Let A and B be sets. A *function* from A to B is a relation, f, from A to B such that if for $a \in A$ and $b, c \in B$, $(a, b) \in f$ and $(a, c) \in f$, then $b = c$. If $(a, b) \in f$, then we write $b = f(a)$. A function from A to B is also called a *mapping* from A to B.

Several technical terms and concepts are associated with the function concept. These names and ideas provide the vocabulary in which discussions involving functions take place. Most of the following ideas are familiar from calculus, although some of the labels for them might be new.

Definition 3.4.2 *Domain and Range*

If f is a function from A to B, then

(i) the *domain of* f, written $\text{Dom}(f)$, is the set: $\text{Dom}(f) = \{a \in A \mid \text{There exists } b \in B \text{ such that } b = f(a)\}$.

(ii) the *range of* f, written $\text{Ran}(f)$, is the set: $\text{Ran}(f) = \{b \in B \mid \text{There exists } a \in A \text{ such that } b = f(a)\}$.

Usually, when we consider a function from A to B, we assume that $A = \text{Dom}(f)$. In this case we write $f\colon A \to B$ to denote the function f. Note, however, that when we use this notation, we are not assuming that $\text{Ran}(f) = B$, merely that $\text{Ran}(f) \subseteq B$. (Note: Some authors call B the *codomain* of f.)

The symbol $f\colon A \to B$ is suggestive of the definition of a function that is used in less formal mathematics: A function from A to B is a rule or correspondence, f, that assigns to each $a \in A$ a unique element $f(a) \in B$. The "correspondence" definition of function has a special virtue: It suggests that a function is a dynamic, as opposed to a static, entity that transforms elements from one set into elements of another set. The view of a function as the embodiment of an active process assists in the visualization of functions and reflects the way in which functions actually arise in many applications.

We next consider a special set of functions called sequences. Sequences appear throughout mathematics and in a wide range of applications of mathematics. Sequences play a particularly important role in discrete mathematics, the discipline that is concerned with properties of finite and denumerable sets (a certain class of infinite sets). The mathematical problems that arise in computer science are often problems in discrete mathematics and lead in many cases to questions involving sequences. Often in these situations, the sequences that occur are approximated by continuous functions and hence are studied using techniques of calculus. We shall provide examples in Chapter 5, where we shall examine the interplay between discrete mathematics and continuous mathematics more closely.

Definition 3.4.3 *Sequences*

Let A be a set. A *sequence with values in A* is a function $s: \mathbf{N} \to A$. For $n \in \mathbf{N}$ one often writes s_n in place of $s(n)$ and denotes the sequence s by $\{s_n | n \in \mathbf{N}\}$.

Example 3.4.1 (i) Define $p: \mathbf{N} \to \mathbf{N}$ by $p(n) = p_n = 2^n$. Thus p is the sequential analog of the base 2 exponential function $f(x) = 2^x$ whose domain is all of $\mathbf{R}$.

(ii) Define $f: \mathbf{N} \to \mathbf{N}$ by $f_0 = 1 = f_1$ and for $n \geq 2$, $f_n = f_{n-1} + f_{n-2}$. The sequence $\{f_n | n \in \mathbf{N}\}$ is called the *Fibonacci sequence* and is one of the most famous sequences in mathematics. We shall encounter the Fibonacci sequence again later in the text. (The interested reader might wish to consult the *Fibonacci Quarterly*, a research journal that publishes papers that focus on the Fibonacci sequence and related matters.)

(iii) Define the *factorial* function fact: $\mathbf{N} \to \mathbf{N}$ by fact(0) = 1 and for $n \geq 1$ fact(n) = $n \cdot$ fact($n - 1$). Thus, for example, fact(3) = $3 \cdot$ fact(2) = $3 \cdot 2 \cdot$ fact(1) = $3 \cdot 2 \cdot 1 \cdot$ fact(0) = $3 \cdot 2 \cdot 1 \cdot 1 = 6$. It is customary to write fact(n) = $n!$ and to read $n!$ as "n factorial." The following table gives the first few values of the factorial sequence.

n	0	1	2	3	4	5	6	7	8	9	10
$n!$	1	1	2	6	24	120	720	5040	40320	362880	3628800

Properties of Functions

As we have just mentioned, a function is often regarded as a means of corresponding to each element of a certain set an element in another set.

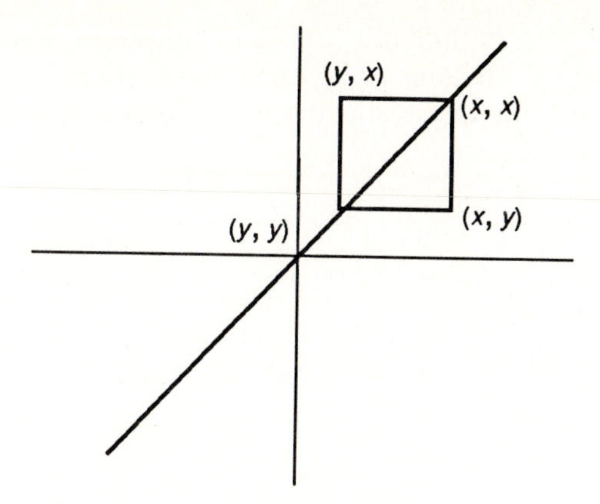

Figure 3.4.12

R. Suppose $f = f^{-1}$. (In other words, suppose $f(x) = f^{-1}(x)$ for all $x \in \mathbf{R}$. Indeed, this can happen, an example being provided by the function $f(x) = -x$.) What can be said about the graph of f? By the comments in the previous paragraph, we see that if $f = f^{-1}$, then the graph of f itself is symmetric about the line $y = x$. Conversely, if the graph of a function $f: \mathbf{R} \to \mathbf{R}$ is symmetric about the line $y = x$, then f^{-1} is a function since the graph of f^{-1} (which is the graph of f) is the graph of a function; moreover, $f = f^{-1}$. (See Figure 3.4.13.)

We summarize the results of the last two paragraphs: Suppose $f: \mathbf{R} \to \mathbf{R}$ is a bijection. Then $f^{-1}: \mathbf{R} \to \mathbf{R}$ is also a bijection and $f = f^{-1}$ if and only if the graph of f is symmetric about the line $y = x$. From our discussion in

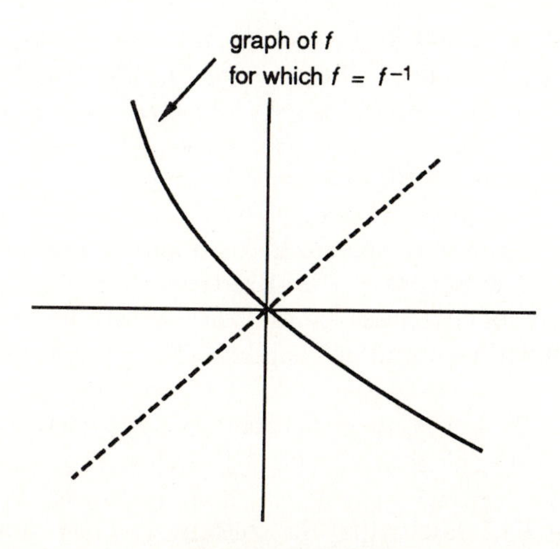

Figure 3.4.13

Section 3.1 we know that the last condition holds if and only if the function f when regarded as a relation is a symmetric relation.

We complete our discussion of properties of functions with an important technical idea.

Definition 3.4.7 *Inverse Image*

Let $f: A \to B$ be any function and let $C \subseteq B$. The *inverse image of C under f* or the *preimage of C under f* is the set

$$f^{-1}(C) = \{x \in A \mid f(x) \in C\}.$$

When $C = \{b\}$ where $b \in B$, we write $f^{-1}(b)$ in place of $f^{-1}(C) = f^{-1}(\{b\})$.

The inverse image of C under f is simply the set of elements of A carried into C by f. For example $f^{-1}(B) = A$. The use of the symbol f^{-1} in $f^{-1}(C)$ might suggest that f is an injective function. However, such need not be the case. For any function $f: A \to B$ and for any $C \subseteq B$, the subset of A, $f^{-1}(C)$, is defined.

We describe the behavior of the inverse image relative to the set operations of $\cap$ and $\cup$ in our next result.

Theorem 3.4.4 *Let $f: A \to B$ be a function and let $B_1, B_2 \subseteq B$.*
(i) $f^{-1}(B_1 \cap B_2) = f^{-1}(B_1) \cap f^{-1}(B_2)$.
(ii) $f^{-1}(B_1 \cup B_2) = f^{-1}(B_1) \cup f^{-1}(B_2)$.

Proof We give a quick proof of (i). We show that $x \in f^{-1}(B_1 \cap B_2)$ if and only if $x \in f^{-1}(B_1) \cap f^{-1}(B_2)$: $x \in f^{-1}(B_1 \cap B_2)$ iff $f(x) \in B_1 \cap B_2$ iff $f(x) \in B_1$ and $f(x) \in B_2$ iff $x \in f^{-1}(B_1)$ and $x \in f^{-1}(B_2)$ iff $x \in f^{-1}(B_1) \cap f^{-1}(B_2)$.

We leave the proof of (ii) as an exercise. ■

These results can be generalized to intersections and unions of an arbitrary family of subsets of B. (See Exercise 3.4.15.) The next proposition is a special case of Exercise 3.4.15. We record this result separately, however, since it will be useful in Chapter 4. The proof is left as an exercise.

Proposition 3.4.5 Let $f: A \to B$. Then $\bigcup_{b \in B} f^{-1}(b) = A$. Moreover, if $b, b' \in B$ with $b \neq b'$, then $f^{-1}(b) \cap f^{-1}(b') = \varnothing$.

Proposition 3.4.5 asserts that the collection $\{f^{-1}(b) \mid b \in B\}$ partitions A. Thus this collection determines an equivalence relation on A. See Miscellaneous Exercise 1 at the end of this chapter for details.

Finally we pick up a loose end left from Section 2.3. On a set theoretic basis how can we define the concept of a binary operation? The following definition answers this question.

> Definition 2.2.1 (revised) *Binary Operation*
>
> Let A be a set. A *binary operation* on A is a function, denoted by $*$, from $A \times A$ to A.

Representation of Functions

With functions, as with any mathematical object, the problem of representation arises. How can we usefully represent or describe a given function? Experience teaches us rather quickly that the answer depends on the circumstance. On certain occasions a particular kind of representation possesses advantages over others. In situations of even moderate complexity, insight usually occurs when the function (or whatever the mathematical object of concern) is regarded from several points of view. The following discussion presents a variety of viewpoints and serves to summarize a number of the ideas presented in this section.

1. *Ordered Pairs*. To represent a function, we can revert to the definition: a function, f, from A to B is a subset of $A \times B$ with the property that for each $a \in A$, there exists exactly one $b \in B$ such that $(a, b) \in f$. Even though it is given to us in the definition, the ordered pair representation of a function is difficult to work with. First, the dynamic nature of a function, as a transformation of elements of A to elements of B, is not readily apparent in this representation. Second, the really interesting member of the ordered pair (a, b), namely b, is not strongly emphasized. The primary virtue of the ordered pair approach is in its offering of a clean, set theoretic definition of the function concept. Most of the time we can operate with descriptions of functions that enable us to manipulate function easily and accurately.

2. *Equations or Formulas*. When asked to imagine a function $f: \mathbf{R} \to \mathbf{R}$, most of us conjure up an equation that allows us to compute the value of $f(x)$ corresponding to the input x. This computation usually involves a combination of the arithmetic operations of addition, subtraction, multiplication, and division, the algebraic operation of taking roots, and converging infinite series. Examples include:

$$f(x) = x^2 + 2x + 3 \qquad f(x) = \frac{x+1}{x^2+1}$$

$$f(x) = \sum_{n=0}^{\infty} \frac{x^n}{n!} (= e^x) \qquad f(x) = \sum_{n=0}^{\infty} x^n (= 1/(1-x) \text{ if } |x| < 1).$$

The general classification of functions described here is:

1. Polynomial Functions. $a_0 + a_1x + \cdots + a_nx^n$ where $n \in \mathbf{N}$ and $a_i \in \mathbf{R}$ for $0 \le i \le n$.
2. Rational Functions. $P(x)/Q(x)$ where P and Q are polynomials.
3. Algebraic Functions. Functions that are solutions of algebraic equations whose coefficients are rational functions. For example, if $R(x)$ is a rational function, then $\sqrt[n]{R(x)}$ is an algebraic function since it is a solution of the equation $z^n - R(x) = 0$.
4. Analytic Functions. $\sum_{n=0}^{\infty} a_n x^n$ where $a_i \in \mathbf{R}$.

Given our repeated exposure to the arithmetic, algebraic, and analytic operations, it is natural for us to think of functions in these terms. Looking at the history of science, we can see the function concept arising in the late Middle Ages and early Renaissance in the work of mathematicians and physicists. From the time of Newton, around 1660, until about 1840, functions were defined to be quantities expressible in one of the forms 1–4 given above. In the mid-nineteenth century, the function concept was enlarged to include the "correspondence" definition. The ordered pair definition, which had to await the birth of set theory at the turn of the twentieth century, was apparently first formulated in 1939 by N. Bourbaki, who is not a real person but which is actually a pseudonym for a group of French mathematicians.

5. Fourier Series. For a certain class of functions, another type of infinite series representation is important. Let $f: \mathbf{R} \to \mathbf{R}$ be a periodic function with period 1. This means that for all $x \in \mathbf{R}$, $f(x + 1) = f(x)$. Examples of such functions are $\sin(2\pi x)$, $\cos(2\pi x)$, and $\sin(4\pi x)$. Periodic functions arise frequently in the study of natural phenomena. Mathematicians and physicists have long been interested in the problem of expressing an arbitrary periodic function of period 1 in terms of the trigonometric functions $\sin(2\pi x)$, $\cos(2\pi x)$, $\sin(4\pi x)$, etc. Specifically, if f has period 1, there are these constants $a_0, a_1, b_1, a_2, b_2, \ldots$ such that:

$$\begin{aligned} f(x) &= a_0 + a_1 \sin(2\pi x) + b_1 \cos(2\pi x) \\ &\quad + a_2 \sin(4\pi x) + b_2 \cos(4\pi x) + \cdots \\ &= a_0 + \sum_{n=1}^{\infty} (a_n \sin(2\pi n x) + b_n \cos(2\pi n x)). \end{aligned}$$

Such a representation for f is called a *Fourier series* for f. The theory of Fourier series is concerned with the existence of and properties of Fourier series representation of periodic functions.

3. *Tables*. Suppose $f: A \to B$ is a function whose domain is a finite set. Thus $A = \{a_1, \ldots, a_n\}$ where n is a positive integer. Let $b_1 = f(a_1)$, $b_2 = f(a_2), \ldots, b_n = f(a_n)$. Then we can represent f in tabular form:

x	a_1	a_2	$\cdots$	a_n
$f(x)$	b_1	b_2	$\cdots$	b_n

A sequence of real numbers, i.e., a function $s: \mathbf{N} \to \mathbf{R}$, can also be described in a table. Of course, this representation can never be complete. However, a tabular recording of the first few values of the sequence can help in the recognition of a pattern that may govern the values of the sequence.

To illustrate, let us recall Example 1.3.10. For $n \geq 1$ let $s_n = 1 + 3 + \cdots + (2n - 1)$ be the sum of the first n odd positive integers. The table for s_n, $n \leq 5$, is

n	1	2	3	4	5
s_n	1	4	9	16	25

On the basis of this table, it appears that $s_n = n^2$ for $n \geq 1$. We can check that indeed for $s = 6, 7, 8, 9, 10$, $s_n = n^2$. Armed with this evidence, we can now proceed to prove by mathematical induction that $s_n = n^2$ for all $n \geq 1$.

4. *Graphs*. If $f: \mathbf{R} \to \mathbf{R}$, then we have the standard pictorial representation of f, the graph of f in the Cartesian plane. The graph of f is the picture in the Cartesian plane of the set $\{(x, f(x)) | x \in \text{Dom}(f)\}$. The graph provides a representation of f that is immediately derived from the definition. The close proximity between the definition of and the graph of a function, together with the apparent enhancement that pictures provide to our thinking, makes the graph one of the most useful ways of representing a function. Recall that in one-variable calculus, the graph of a function is used to motivate the concepts of derivative (as the slope of a tangent line) and integral (as area under the graph).

5. *Recursive Definitions of Sequences*. Our final example illustrates a way of defining sequences. A sequence $s: \mathbf{N} \to \mathbf{R}$ of real numbers is called *recursive* if there is a number n_0 such that for each $n \geq n_0$, s_{n+1} can be expressed in terms of $s_0, s_1, \ldots, s_n$. In other words, a sequence is recursive if any given element in the sequence, beyond a certain point, can be defined in terms of previous members of the sequence.

Example 3.4.1 (revisited) (i) The Fibonacci sequence is also recursive for, if $n \geq 2$, then $f_n = f_{n-1} + f_{n-2}$.

(ii) The sequence fact: $\mathbf{N} \to \mathbf{R}$ defined by $\text{fact}(n) = n!$ is recursive, for whenever $n \geq 1$, $\text{fact}(n + 1) = (n + 1) \cdot \text{fact}(n)$.

Let $s: \mathbf{N} \to \mathbf{R}$ be a recursive sequence and let n_0 be the integer given in the definition of recursivity. Thus $n_0 = 1$ for the factorial sequence and $n_0 = 2$ for the Fibonacci sequence. The numbers $s(0), s(1), \ldots, s(n_0 - 1)$ are called the *initial values* or *initial conditions*. The initial values can be prescribed arbitrarily; thus many distinct sequences exist with the same "recurrence relation." For example, if we define $L: \mathbf{N} \to \mathbf{R}$ by $L(0) = 1$, $L(1) = 3$ and $L(n) = L(n-1) + L(n-2)$ for $n \geq 2$, we obtain the sequence $1, 3, 4, 7, 11, 18, \ldots$. This close relative of the Fibonacci sequence is called the *Lucas sequence* and is also well known within mathematics.

Recursive sequences arise frequently in problems that involve the counting of objects. Such problems occur frequently in discrete mathematics including combinatorics (see Chapter 5) and computer science. The idea of a recursive sequence is actually a special case of the more general idea of recursion. Another example is that of a recursive computer program, which is roughly speaking a program that operates by calling "earlier" instances of itself. As you might suspect, recursion is closely related to mathematical induction. A thorough discussion of recursion would take us deep into the realm of set theory, far beyond the scope of this text. Suffice it to say that recursive sequences and recursion in general are ubiquitous themes in mathematics, and as far as sequences are concerned, recursion provides a way of defining and thereby representing sequences.

Exercises §3.4

3.4.1. Find all functions $f: A \to B$ when:
(a) $A = \{1, 2\}$ and $B = \{1\}$.
(b) $A = \{1, 2, 3\}$ and $B = \{1\}$.
(c) $A = \{a_1, \ldots, a_n\}$ and $B = \{b\}$.
(d) $A = \{1\}$ and $B = \{1, 2\}$.
(e) $A = \{1\}$ and $B = \{1, 2, 3\}$.
(f) $A = \{a\}$ and $B = \{b_1, \ldots, b_n\}$.

3.4.2. In each case state whether the given function is injective, surjective, and/or bijective.
(a) $f: \mathbf{R} \to \mathbf{R}$, $f(x) = 2x$ for $x \in \mathbf{R}$.
(b) $f: \mathbf{R} \to \mathbf{R}$, $f(x) = 3 - x$ for $x \in \mathbf{R}$.
(c) $f: \mathbf{R} \to \mathbf{R}$, $f(x) = x^2 + 2x + 3$ for $x \in \mathbf{R}$.
(d) $f: [0, \pi) \to [0, 1]$, $f(x) = \sin(x)$ for $x \in [0, \pi)$.
(e) $f: \mathbf{R} \to \mathbf{R}^+$, $f(x) = e^{(x^2)}$ for $x \in \mathbf{R}$.

3.4.3. Define $g: \mathbf{Z} \to \mathbf{N}$ as follows:

$$g(x) = \begin{cases} 2x & \text{if } x \geq 0. \\ -2x - 1 & \text{if } x < 0. \end{cases}$$

(a) Evaluate $g(x)$ for $-5 \leq x \leq 5$.
(b) Describe the definition of g in a sentence.
(c) Prove that if $x \in \mathbf{Z}$, then $g(x) \in \mathbf{N}$.
(d) Prove that g is injective.
(e) Prove that g is surjective.
(f) Find g^{-1}.

3.4.4. Let $a, b \in \mathbf{R}$ with $a \neq 0$. Define $f: \mathbf{R} \to \mathbf{R}$ by $f(x) = a \cdot x + b$.
(a) Show f is injective.
(b) What is $\text{Ran}(f)$?
(c) Find $f^{-1}(x)$ for $x \in \text{Ran}(f)$.
(d) Sketch the graphs of f and f^{-1}.
(e) As a check of (c) show that if $f(x) = -x$, then $f = f^{-1}$.

3.4.5. For $x \in (-\pi/2, \pi/2)$, let $f(x) = \tan(x)$. Sketch the graph of f^{-1}.

3.4.6. Show that $\cos \circ \sin: [0, \pi/2] \to \mathbf{R}$ is injective. What is $\text{Ran}(\cos \circ \sin)$?

3.4.7. (a) For $n \geq 1$, $n \in \mathbf{N}$, let $S_n = 1 + 2 + \cdots + n$. Give a recursive definition of the sequence S_n.
(b) Give a recursive definition of the sequence t_n where $t_n = 1^2 + 2^2 + \cdots + n^2$.

3.4.8. (a) Let $A = \{a, b\}$. How many distinct bijective mappings are there from A to A?
(b) Let $A = \{a, b, c\}$. How many distinct bijective mappings are there from A to A?

3.4.9. Let X be any set. Define a relation $\approx$ on $P(X)$ by the rule: $A \approx B$ if there exists a bijective mapping $f: A \to B$. Prove: $\approx$ is an equivalence relation.

3.4.10. Let $f: A \to B$ with $A_1 \subseteq A$ and $B_1 \subseteq B$.
(a) Prove, or disprove and salvage: $f(f^{-1}(B_1)) = B_1$.
(b) Prove, or disprove and salvage: $f^{-1}(f(A_1)) = A_1$.

3.4.11. Let $f: A \to B$ and $g: B \to C$.
(a) Suppose f and g are bijective. Express $(g \circ f)^{-1}$ in terms of f^{-1} and g^{-1}. Prove that your expression is correct.
(b) Give an example to show that $g \circ f$ can be bijective yet neither f nor g need be bijective.
(c) If $g \circ f$ and f are bijective, then is g necessarily bijective?
(d) Prove: If $g \circ f$ and g are bijective, then f is bijective.

3.4.12. Prove the second statement in Theorem 3.4.2.

3.4.13. Prove Theorem 3.4.4 (ii).

3.4.14. Prove Proposition 3.4.5.

3.4.15. Suppose $f: A \to B$. Let $\{B_i | i \in X\}$ be an arbitrary family of subsets of B.
(a) Prove: $f^{-1}(\bigcup_{i \in X} B_i) = \bigcup_{i \in X} f^{-1}(B_i)$.
(b) Prove: $f^{-1}(\bigcap_{i \in X} B_i) = \bigcap_{i \in X} f^{-1}(B_i)$.

3.4.16. The graph of a function f in the figure below is symmetric about the line $y = x$. What is $f(x)$ for $x \in \mathbf{R}$? What is $f^{-1}(x)$ for $x \in \mathbf{R}$?

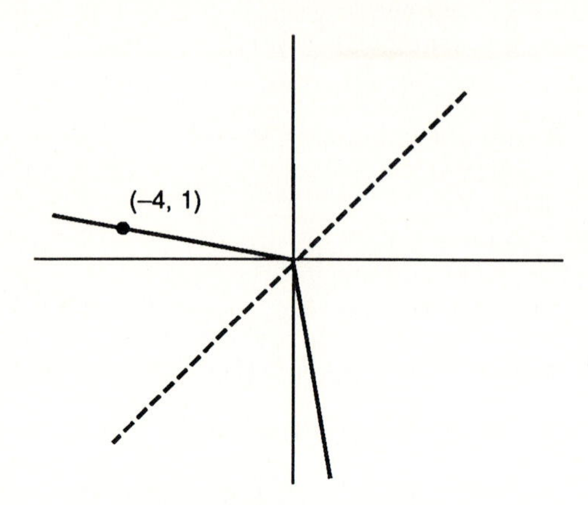

3.4.17. (a) Let $f: \mathbf{R}^+ \rightarrow \mathbf{R}^+$ be defined by $f(x) = 1/x$. Show f is bijective. What is f^{-1}?

(b) Let $g: \mathbf{R}^+ \rightarrow \mathbf{R}^+$ be defined by $g(x) = 1/x^2$. Show g is bijective. What is g^{-1}?

(c) Generalize the statements in (a) and (b) and prove that your generalizations are correct.

3.4.18. Let a be a finite set with n elements. Let B_n denote the set of binary sequences of length n. Fill in the blank: The function from $P(A)$ to B_n that assigns each subset of A to its corresponding binary sequence is a ______ function.

Notes

The function concept began to emerge in the late Middle Ages and early Renaissance. The idea is certainly implicit in the work of Galileo on the motion of objects. In the seventeenth century, with the rise of calculus, functions became the central focus of mathematical research and remained so for at least two centuries. While various definitions of a function were proposed, no clear definition appeared until the nineteenth century. Until then, mathematicians thought of a function as something defined by a single analytic expression. Examples of such expressions include polynomials, rational and algebraic functions, and power series. In the 1830s, P. Dirichlet defined a function to be a rule by which each number x in an interval is corresponded to a unique value y. Dirichlet's definition essentially captures the generality of the function concept as it is used by modern mathematicians. An interesting account of the history of the function concept is sprinkled throughout Kline's valuable history [1].

The origins of the relation concept are quite murky. As noted in Section 3.4, special cases of equivalence relations go back to Gauss and even to the Greeks. Other examples of equivalence relations appeared throughout the nineteenth century and evidently the concept was formalized around 1900. The idea of an order relation seems to have followed a similar development, although the possibility of well-ordering an arbitrary set (a well-ordering is a special kind of total ordering) was studied extensively in the early 1900s.

Bibliography

1. M. Kline, *Mathematical Thought from Ancient to Modern Times*, Oxford University Press, 1972.

Miscellaneous Exercises

1. Let $f: A \to B$ be a function. For $x, y \in A$, define $x \sim_f y$ if $f(x) = f(y)$.
   (a) Prove: $\sim_f$ is an equivalence relation on A.
   (b) If f is injective, how large is each equivalence class?
   (c) Which equivalence relations presented in this chapter are special cases of $\sim_f$?

2. Let $f: A \to B$ be a function and let R be an equivalence relation on B. Define a relation S on A as $x S y$ if $f(x) R f(y)$. Prove: S is an equivalence relation on A. Give an example of an equivalence relation S on a set A that arises in this way.

3. Show that the function $f: \mathbf{N} \times \mathbf{N} \to \mathbf{Z}^+$ defined by $f(a, b) = 2^a(2b + 1)$ is bijective.

4. Let $F(\mathbf{R})$ denote the set of functions from $\mathbf{R}$ to $\mathbf{R}$. For $f, g \in F(\mathbf{R})$, we say that $f \sim g$ if $f - g$ is a differentiable function (i.e., $f - g$ has a derivative at each $x \in \mathbf{R}$). Show: $\sim$ is an equivalence relation on $F(\mathbf{R})$.

5. Suppose $f: A \to B$ and $g: B \to A$ are functions such that $g \circ f = I_A$ and $f \circ g = I_B$. Show: f and g are both bijections and $f = g^{-1}$. What is f^{-1}?

6. (a) Sketch the graph of a function $f: \mathbf{R} \to \mathbf{R}$ with the property that for each $y \in \mathbf{R}$, the set $f^{-1}(y) = \{x | f(x) = y\}$ is infinite.
   (b) Give an explicit formula for a function that has the property described in (a).

7. Let A and B be any sets. Find a one-to-one correspondence $f: A \times B \to B \times A$.

8. Let A, B, C be any sets. Find a bijection $f: (A \times B) \times C \to A \times (B \times C)$. Note: $(A \times B) \times C = \{((a, b), c) | a \in A, b \in B, c \in C\}$.

9. Let $f: \mathbf{R} \to \mathbf{R}$ and $g: \mathbf{R} \to \mathbf{R}$. Define $f + g: \mathbf{R} \to \mathbf{R}$ by $(f + g)(x) = f(x) + g(x)$ for all $x \in \mathbf{R}$ and $fg: \mathbf{R} \to \mathbf{R}$ by $(fg)(x) = f(x) \cdot g(x)$.
   (a) Prove or disprove: If f and g are bijective, then $f + g$ is bijective.
   (b) Prove or disprove: If f and g are bijective, then fg is bijective.

10. Let R be a relation on a set A. Let $R^1 = R$, $R^2 = R \circ R$, and for $n \geq 3$ $R^n = R \circ R^{n-1}$. Finally, let $R^+ = \bigcup_{n=1}^{\infty} R^n$.
    (a) Prove: R^+ is a transitive relation. (Hint: Problem 3.1.25 might be useful.)
    (b) Prove: If S is any transitive relation on A that contains R, then S contains R^+. (Because of this fact, R^+ is called the *transitive closure of* R. R^+ is the smallest (in terms of containment) transitive relation on A containing R.)
    (c) Let $A = \{a, b, c\}$ and let $R = \{(a, b), (b, a), (a, c)\}$. Find R^+.
    (d) Let $R^* = R^+ \cup I_A$. Prove: R^* is reflexive and transitive.
    (e) Prove that any reflexive, transitive relation on A containing R contains R^*. (R^* is called the *reflexive, transitive closure of* R.)

11. Let R_1 and R_2 be relations on a set A.
    (a) Prove or disprove: If R_1 and R_2 are reflexive, then $R_1 \circ R_2$ is reflexive.
    (b) Prove or disprove: If R_1 and R_2 are symmetric, then $R_1 \circ R_2$ is symmetric.
    (c) Prove or disprove: If R_1 and R_2 are transitive, then $R_1 \circ R_2$ is transitive.

12. Congruence Relations. Let A be a set on which a binary operation $*$ is defined. For example on the set $\mathbf{Z}$, addition and multiplication are both binary operations. Let $\sim$ be an equivalence relation on A. The relation $\sim$ is called a *congruence relation with respect to* $*$ if for each $a, a', b, b' \in A$ such that $a \sim a'$ and $b \sim b'$, it follows that $a * b \sim a' * b'$.
    (a) Prove that $\equiv_n$ is a congruence relation on $\mathbf{Z}$ with respect to each of the operations of addition and multiplication. (See Exercise 3.3.20.)
    (b) Prove that on the set $\mathscr{F}$ of fractions of integers, the relation $\equiv$ is a congruence relation with respect to both addition and multiplication. (See Exercise 3.3.21.)

13. Give an example of a set A, a binary operation $*$ on A, and an equivalence relation $\sim$ on A such that $\sim$ is not a congruence relation on A with respect to $*$.

14. Let A be a set and let $\Pi(A)$ denote the set of partitions of A. For $P_1, P_2 \in \Pi(A)$, we say that $P_1 \leq P_2$ if for every $B_1 \in P_1$, there exists $B_2 \in P_2$ such that $B_1 \subseteq B_2$.
    (a) Show that $\leq$ is a partial ordering on $\Pi(A)$.
    (b) Show that in $\Pi(\mathbf{Z})$, $P(\equiv_4) \leq P(\equiv_2)$.
    (c) Show that in $\Pi(\mathbf{Z})$, $P(\equiv_3) \nleq P(\equiv_2)$.
    (d) Let $n, m \in \mathbf{Z}^+$. Conjecture a necessary and sufficient condition on n and m for $P(\equiv_n) \leq P(\Pi_m)$ to hold in $\Pi(\mathbf{Z})$.

15. Prove: If R is a reflexive and transitive relation, then $R \cap R^{-1}$ is an equivalence relation.

16. Let f: $A \to A$ be a function.
    (a) Can f be a partial ordering? Explain.
    (b) Can f be a strict partial ordering? Explain.

17. Let A and B be finite sets with m and n elements, respectively, where $m, n \in \mathbf{Z}^+$. Investigate the following question: How many functions f: $A \to B$ are there?

18. Suppose f: $A \to B$ is a function. Let $\{B_i | i \in I\}$ be a partition of B. Prove: $\{f^{-1}(B_i) | i \in I\}$ is a partition of A.

19. (See Miscellaneous Exercise 1 of Chapter 2.) Suppose that $A, B \subseteq \mathbf{R}$ and that $f\colon A \to B$ is a bijection. Let $*$ be a self-distributive binary operation on B. Define a binary operation $\circ$ on A as follows: For $a, b \in A$, $a \circ b = f^{-1}(f(a) * f(b))$.
    (a) Prove that $\circ$ is self-distributive on A.
    (b) By choosing f appropriately, use the fact that $a * b = (a+b)/2$ is self-distributive on $\mathbf{R}$ to show that the geometric mean, $a \circ b = \sqrt{ab}$, is self-distributive on $\mathbf{R}^+$.
    (c) Show that the harmonic mean, $a * b = 1/2(ab/(a+b))$ is self-distributive on $\mathbf{R}^+$.
    (d) Find at least one more self-distributive operation on $\mathbf{R}^+$.

20. Let A be a set and let $f\colon A \to \mathbf{R}$ be a function. Define a relation $\leq_1$ on A by $x \leq_1 y$ if $f(x) \leq f(y)$. Find a necessary and sufficient condition on f for $\leq_1$ to be a partial ordering on A.

21. Let m and n be positive integers. Let R and S, respectively, be the relations $\equiv_m$ and $\equiv_n$ on $\mathbf{Z}$.
    (a) Find $R \circ S$ when $m = 2$ and $n = 6$.
    (b) Find $R \circ S$ when $m = 9$ and $n = 6$.
    (c) Make a conjecture about $R \circ S$ in general.

22. Define a relation R on $\mathbf{R}^2$ as follows: For $(x_1, y_1), (x_2, y_2) \in \mathbf{R}^2$, $(x_1, y_2)\ R\ (x_2, y_2)$ if and only if $|y_1 - x_1| = |y_2 - x_2|$.
    (a) Show that R is an equivalence relation on $\mathbf{R}^2$.
    (b) Describe the equivalence classes geometrically. In particular, describe the equivalence classes of $(2, 5)$ and $(1, -1)$.
    (c) Find a complete set of representatives for $\mathbf{R}^2/R$.

23. A function f whose domain includes a subset A of $\mathbf{R}$ is called *subadditive* if $f(a+b) \leq f(a) + f(b)$ for all $a, b \in A$.
    (a) Show that if f is subadditive, then for all positive integers n and for all $a_1, \ldots, a_n \in A$,

$$f(a_1 + \cdots + a_n) \leq f(a_1) + \cdots + f(a_n).$$

    (b) Show that if $a_1, \ldots, a_n$ are nonnegative real numbers, then $\sqrt{a_1 + \cdots + a_n} \leq \sqrt{a_1} + \cdots + \sqrt{a_n}$.

24. Let $\{f_n \mid n \geq 0\}$ denote the Fibonacci sequence.
    (a) Compute $f_1 \cdot f_4 - f_2 \cdot f_3$, $f_2 \cdot f_5 - f_3 \cdot f_4$, $f_3 \cdot f_6 - f_4 \cdot f_5$, and $f_4 \cdot f_7 - f_5 \cdot f_6$.
    (b) Make a conjecture on the basis of your data in (a).
    (c) Prove your conjecture.

25. Let $f\colon A \to B$ be a function and let X and Y be subsets of B.
    (a) Suppose f is surjective. Show that if $f^{-1}(X) \subseteq f^{-1}(Y)$, then $X \subseteq Y$.
    (b) Give an example to show that if f is not surjective, then $f^{-1}(X) \subseteq f^{-1}(Y)$ does not necessarily imply that $X \subseteq Y$.

CHAPTER FOUR

Cardinality

One of the most basic of all mathematical concepts is that of *size* or *measure*. For several kinds of mathematical objects, mathematicians have proposed definitions for size. For example, let a and b be real numbers with $a < b$ and let I be the interval of numbers, $I = (a, b) = \{x \in \mathbf{R} | a < x < b\}$. Then the *length of* I is defined to be $b - a$. We think of $b - a$ as the size or the measure of I. (See Figure 4.1.1.) As another example consider the Cartesian plane $\mathbf{R}^2$. For a point $\mathbf{v} = (x, y) \in \mathbf{R}^2$, we define $|\mathbf{v}| = \sqrt{x^2 + y^2}$. Geometrically, $|\mathbf{v}|$ is the Euclidean distance from $\mathbf{v}$ to $(0, 0)$ and is the length of the line segment joining $\mathbf{v}$ to $(0, 0)$. Thus $|\mathbf{v}|$ can be thought of as the size of $\mathbf{v}$. (See Figure 4.1.2.)

In the past two hundred years the types of systems considered by mathematicians have grown significantly in number. Simultaneously, the mathematical ideas have been applied to explain the behavior of phenomena and to solve problems in a variety of other areas. A consequence of this expansion in theoretical and applied mathematics has been the extension of the notion of size or measure to many new situations. Here are two examples:

1. Let f be a continuous function over an interval $I = [a, b] = \{x \in \mathbf{R} | a \leq x \leq b\}$. Then the *norm of* f on I, written $\|f\|$, is defined to be: $\|f\| = (\int_a^b f^2(x)\, dx)^{1/2}$. The norm of f is analogous to the length of a vector in $\mathbf{R}^2$ defined above (think of the sum of a set of numbers and the integral of a function as being analogous) and provides a way of measuring the size of any continuous function on I.

2. Let A be an $n \times n$ matrix. Define the *norm* of A, written $|A|$, to be the maximum absolute value of the entries of A: Thus, if $A = (a_{ij})$, where $1 \leq i \leq m$ and $1 \leq j \leq n$, then $|A| = \max\{|a_{ij}| | 1 \leq i \leq m, \text{ and } 1 \leq j \leq n\}$. This measure of the size of a matrix plays a useful role in matrix theory.

In this chapter we focus on the concept of size. The setting in which we work is both the most primitive and the most general in all of mathematics. We consider an arbitrary set A and attempt to define a way of measuring A.

When A is finite, a simple method of assessing the size of A is apparent: Count the elements of A. Counting, perhaps the most basic of all

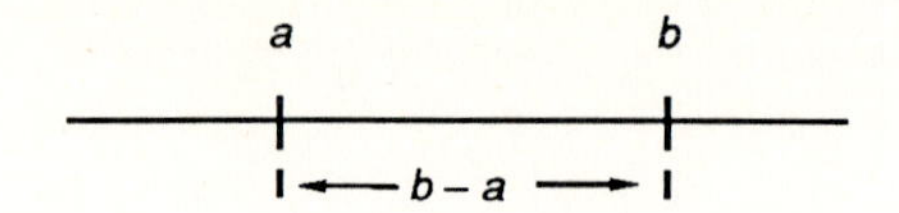

Figure 4.1.1. Size or measure of $I = b - a$.

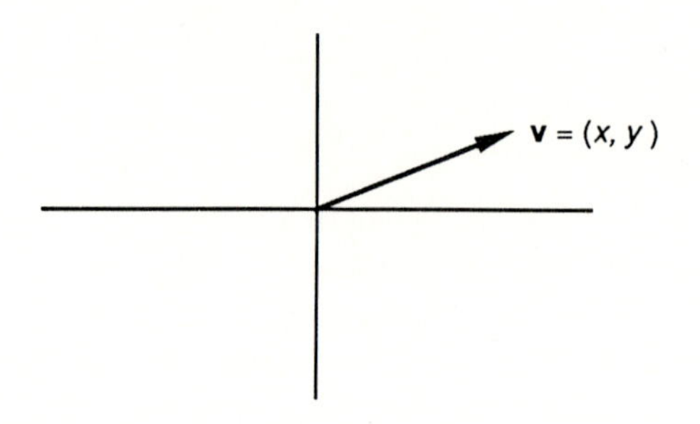

Figure 4.1.2. Size or measure of $\mathbf{v} = \sqrt{x^2 + y^2}$.

human mathematical activities, provides a measure of the size of a finite set. What happens when A is infinite? In this case, clearly, counting will not do. To attain a measure of the size of an infinite set A, we take an indirect route. In Section 4.1, we propose a way of comparing any two sets for size. Our method of comparison is suggested by the properties of finite sets. Once we have a way of comparing sets for size, several interesting questions arise: Is $\mathbf{Z}$ "larger than" $\mathbf{N}$? Is $\mathbf{Q}$ larger than $\mathbf{Z}$? Is $\mathbf{R}$ larger than $\mathbf{Q}$? Given any infinite set A, can one find an infinite set B that is larger than A? Then we approach the problem of assigning a size measure to each infinite set. In doing so we are led to the notion of cardinal number and to some fascinating problems in the foundations of set theory.

§4.1 Equinumerous Sets

The question we consider is the following: For any set A, how should we define the size of A? Looking first at finite sets, we can see a clear way of defining size. Let A be a nonempty set. Then A is finite if its elements can be matched via a one-to-one correspondence with the set $\{1, \ldots, n\}$ for some positive integer n. In this case we can reasonably call the integer n the size of A. Thus $\{1, 2, 3\}$ and $\{a, b, c\}$ have size 3 while $\{1, 2, a, b\}$ has size 4.

With infinite sets, however, we face a difficulty. If A is infinite, how can we assign a number to the size of A? For example, since the set $\mathbf{Z}^+ = \{m \in \mathbf{Z} | m > 0\}$ contains the set $\{1, \ldots, n\}$, the size of $\mathbf{Z}^+$ is larger than n for each positive n. Should we simply say that $\mathbf{Z}^+$ is infinite and leave matters at that? Should we, in addition, regard all infinite sets as having the same size? Such a measure of size would indeed be very crude since we could not distinguish among infinite sets according to size.

Let us return briefly to finite sets. The sets $A = \{1, 2, 3\}$ and $B = \{a, b, c\}$ each contain exactly three elements. There is, however, another way of seeing that A and B have the same size. The function $f: A \to B$ defined by $f(1) = a$, $f(2) = b$, and $f(3) = c$ is a one-to-one correspondence between A and B. Notice, however, that the set A, which has three elements, cannot be put in one-to-one correspondence with the set $C = \{1, 2, a, b\}$, which has four elements, since any function $g: A \to C$ must "miss" at least one element of C. These examples suggest a way of determining if two finite sets have the same size: Two finite sets have the same size if a one-to-one correspondence between the sets exists. The following definition, which applies to an arbitrary pair of sets, formalizes this approach to comparing sets.

Definition 4.1.1 *Equinumerous Sets*

Two sets A and B are *equinumerous* if there exists a bijection $f: A \to B$. We write $A \approx B$ if A and B are equinumerous and write $A \not\approx B$ if A and B are not equinumerous.

In general then, we regard two equinumerous sets A and B as having the same size since the elements of A and B can be matched in a one-to-one and onto fashion. In particular, using this definition, we can reformulate the definition of finite set. For each positive integer n, let $\mathbf{N}_n = \{1, \ldots, n\} = \{k \in \mathbf{N} | 1 \leq k \leq n\}$.

Definition 4.1.2 *Finite Set*

A set A is *finite* if $A = \varnothing$ or if there is an integer n such that $A \approx \mathbf{N}_n$. If $A \approx \mathbf{N}_n$, then we say that A has n elements and write $|A| = n$. If $A = \varnothing$, then we say that A has 0 elements and write $|A| = 0$.

Definition 4.1.3 *Infinite Set*

A set A is *infinite* if A is not finite.

Thus a nonempty set A is infinite if for each positive integer n, $A \not\approx \mathbf{N}_n$. We now look at several examples.

Example 4.1.1 (i) Let $A = \{x\}$. Then $A \approx \mathbf{N}_1$ and $|A| = 1$.

(ii) Let $B = \{a, b\}$ and $C = \{d, e, f\}$. Then $|B| = 2$, $|C| = 3$, and $B \not\approx C$, since no function $f: A \to B$ can be surjective and hence bijective.

Our next example involves infinite sets. The proofs of many of the following assertions were given in §3.4.

Example 4.1.2 Let E denote the set of even integers. Thus $E^c = \mathbf{Z} - E$ is the set of odd integers.

(i) $E \approx E^c$. The function $f: E \to E^c$ defined by $f(x) = x + 1$ is a bijection from E to E^c. (See Exercise 3.4.2.)

(ii) $\mathbf{Z} \approx E$. The function $f: \mathbf{Z} \to E$ given by $f(n) = 2n$ is a one-to-one correspondence from $\mathbf{Z}$ to E. (See Example 3.4.2.)

(iii) Let $\mathbf{N}^+ = \{n \in \mathbf{N} | n > 0\} = \mathbf{N} - \{0\}$. Then $\mathbf{N} \approx \mathbf{N}^+$, since $s: \mathbf{N} \to \mathbf{N}^+$ defined by $s(n) = n + 1$ is a one-to-one correspondence from $\mathbf{N}$ to $\mathbf{N}^+$.

(iv) $\mathbf{Z} \approx \mathbf{N}$. Define $g: \mathbf{Z} \to \mathbf{N}$ by

$$g(x) = \begin{cases} 2x & \text{if } x \geq 0 \\ -(2x+1) & \text{if } x < 0 \end{cases}.$$

In other words, g sends a nonnegative integer to an even nonnegative integer and sends a negative integer to an odd positive integer. To see that g is a bijection from $\mathbf{Z}$ to $\mathbf{N}$, first note that for $x \in \mathbf{Z}$, if $x \geq 0$, then $2x \in \mathbf{N}$, while if $x < 0$, then $2x + 1 \in \mathbf{Z}$, $2x + 1 < 0$, and hence $-(2x + 1) \in \mathbf{N}$. Thus g is indeed a function from $\mathbf{Z}$ to $\mathbf{N}$. Next we show that g is injective. Suppose $x, y \in \mathbf{Z}$ and $g(x) = g(y)$. We consider two cases. If $g(x)$ is even, then $2x = g(x) = g(y) = 2y$; hence $x = y$. If $g(x)$ is odd, then $-(2x+1) = g(x) = g(y) = -(2y+1)$; thus $x = y$. Therefore, g is injective. Finally, we argue that g is surjective. Let $y \in \mathbf{N}$. If y is even, then $y = 2n$ for some $n \in \mathbf{N}$ and $y = g(n)$. If y is odd, then $y = 2n - 1$ for some $n \in \mathbf{N}^+$; therefore, $y = 2n - 1 = -[2(-n) + 1] = g(-n)$. We now see that g is surjective. Thus g is bijective.

(v) Let $S = \{n \in \mathbf{N} | n = k^2 \text{ for some } k \in \mathbf{N}\} = \{0, 1, 4, 9, \dots\}$. Then $\mathbf{N} \approx S$ since the function $\text{sq}: \mathbf{N} \to S$ defined by $\text{sq}(k) = k^2$ is a bijection from $\mathbf{N}$ to S.

These examples illustrate an unusual feature of the concept of equinumerous sets. Both E and $\mathbf{N}$ are proper subsets of $\mathbf{Z}$, yet each set is in one-to-one correspondence with $\mathbf{Z}$. Thus a set (like $\mathbf{Z}$) can be equinumerous with one of its proper subsets (like $\mathbf{N}$ or E). In this sense of measuring the size of a set, a given set can be equally as large as one of its proper subsets. A question to ponder: Is this property true of every infinite set? In other words, if A is an infinite set, is there a proper subset B of A such that $A \approx B$?

Example 4.1.3 Let us consider intervals on the real number line. For $a, b \in \mathbf{R}$ with $a < b$, recall that $(a, b) = \{x \in \mathbf{R} | a < x < b\}$.

(i) $(0,1) \approx (1,2)$. Define $f: (0,1) \to (1,2)$ by $f(x) = x + 1$. As an exercise, check that f is a bijection. (To do so, first check that f is a function from $(0,1)$ to $(1,2)$, then check that f is one-to-one and onto.)

(ii) $(0,1) \approx (0,2)$. In this case a bijection is given by $g(x) = 2x$.

(iii) $(0,1) \approx (1,3)$. For $x \in (0,1)$ define $h(x) = 2x + 1$. Then h is a bijection from $(0,1)$ to $(1,3)$. Notice that h can be represented as a composition of two functions: $h(x) = h_2 \circ h_1(x)$ where $h_1(x) = 2x$ and $h_2(x) = x + 1$. The function h_1 stretches the interval $(0,1)$ into $(0,2)$ and the function h_2 translates $(0,2)$ into the interval $(1,3)$. By composing these functions we obtain $h = h_2 \circ h_1$, which is, by Theorem 3.4.1, a bijection from $(0,1)$ to $(1,3)$.

Example 4.1.4. Now let us expand our horizons somewhat.

(i) By Example 3.4.3(iii), the function $f: (-\pi/2, \pi/2) \to \mathbf{R}$ given by $f(x) = \tan(x)$ is a one-to-one correspondence. Thus the interval $(-\pi/2, \pi/2)$, which has length π, is equinumerous with the entire real line. This amazing result says in effect that a certain interval of finite length has as many points as the set of real numbers.

(ii) $\mathbf{R} \approx \mathbf{R}^+$ since the function $f(x) = e^x$ is a bijection from $\mathbf{R}$ to $\mathbf{R}^+$.

These examples might suggest that in general any two intervals of real numbers (be they finite in length like $(0,1)$ or infinite in length like $\mathbf{R}$ or $\mathbf{R}^+$) are equinumerous. We leave this issue as an exercise. The next result will be of assistance in dealing with this question.

Theorem 4.1.1 *Let A, B, and C be any sets.*
(i) $A \approx A$.
(ii) *If* $A \approx B$, *then* $B \approx A$.
(iii) *If* $A \approx B$ *and* $B \approx C$, *then* $A \approx C$.

Remark We might be tempted to assert that $\approx$ is an equivalence relation. Before doing so, however, we must specify the set on which the relation $\approx$ is defined. We could confine ourselves to the set of subsets of a given set U. Then Theorem 4.1.1 does assure us that $\approx$ is an equivalence relation on $\mathscr{P}(U)$. Nevertheless, since Theorem 4.1.1 holds for any sets A, B, and C, this restriction is unnecessary. On the other hand, as noted earlier, Russell's paradox (Section 2.1) warns us not to speak of the set of all sets, for the concept of the set of all sets leads to a contradiction. By stating Theorem 4.1.1 as we have, we avoid any set theoretic difficulties while conveying the full generality of the result.

Proof (1) The identity function $I_A: A \to A$ is a bijection. Therefore, $A \approx A$.

(2) If $A \approx B$, then there exists a bijection $f: A \to B$. By Theorem 3.4.3 (i), $f^{-1}: B \to A$ is also a bijection. Thus $B \approx A$.

(3) If $A \approx B$ and $B \approx C$, then there exist bijections $f: A \to B$ and $g: B \to C$. By Theorem 3.4.1(iii), $g \circ f: A \to C$ is a bijection. Therefore, $A \approx C$. ■

We now give some applications of this result.

Example 4.1.5 (i) By Example 4.1.2, we know that $\mathbf{Z} \approx \mathbf{N}$ and $\mathbf{N} \approx \mathbf{N}^+$. Hence $\mathbf{Z} \approx \mathbf{N}^+$.

(ii) One can show that $(0,1) \approx (-\pi/2, \pi/2)$. (See Exercise 4.1.2(f).) Since $(-\pi/2, \pi/2) \approx \mathbf{R}$, as noted in Example 4.1.4, it follows that $(0,1) \approx \mathbf{R}$.

As an aside we pause to outline a geometric argument that shows that $I = (0,1) \approx \mathbf{R}$. In the plane consider the line $y = x - 1/2$ for $0 < x < 1$. (See Figure 4.1.3.) Let P be the point $(1/2, 1/2)$ in $\mathbf{R}^2$ and let Q be the point $(1/2, -1/2)$. Define $f: (0,1) \to \mathbf{R}$ as follows:

(i) For $1/2 \leq x < 1$, consider the straight line L_1 joining P and $(x, x - 1/2)$. Define $f(x)$ to be the first coordinate of the point of intersection of L_1 with the x-axis.

(ii) For $0 < x < 1/2$, let L_2 be the line joining Q and $(x, x - 1/2)$. Define $f(x)$ to be the first coordinate of the point of intersection of L_2 with the x-axis.

It seems apparent geometrically that f is injective: If $x_1 \neq x_2$, then $f(x_1) \neq f(x_2)$. Also f is surjective. For example, if $y \in \mathbf{R}$ and $y \geq 1/2$, then take the line segment joining P and $(y, 0)$. This segment intersects the

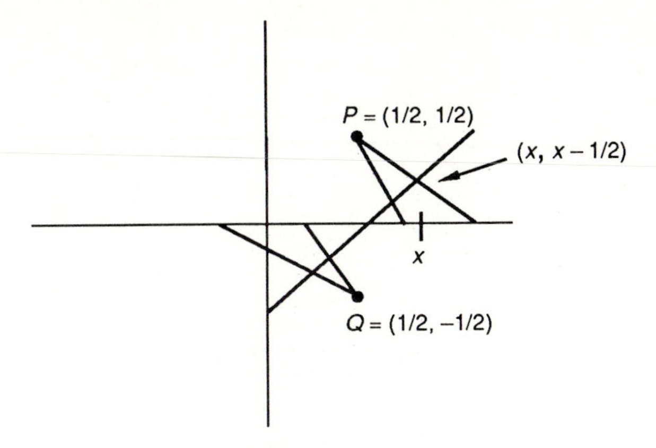

Figure 4.1.3

segment $y = x - 1/2$ for $1/2 \leq x < 1$. Calling this intersection point $(x_0, x_0 - 1/2)$, we see that $f(x_0) = y$.

We stress that despite its persuasiveness this geometric argument is not a rigorous proof. The principal drawback is that the function f is not explicitly and precisely defined. As a result we cannot be certain that f is indeed a function from $(0, 1)$ to **R** that is bijective. Nevertheless, the details can be supplied. We leave it to be solved in Exercise 4.1.11.

Set Operations

It is reasonable to ask how the "relation" $\approx$ behaves relative to the various set operations. For instance, if $A \approx A'$ and $B \approx B'$, then is it true that $A \cup B \approx A' \cup B'$? A simple example reveals that the answer is No: $\{1\} \approx \{1\}$ and $\{1\} \approx \{2\}$, yet $\{1\} \cup \{1\} = \{1\} \quad \{1, 2\} = \{1\} \cup \{2\}$. By adding a suitable hypothesis, however, a reasonable result can be salvaged.

Theorem 4.1.2 *Suppose A, B, A', and B' are sets such that $A \cap B = \varnothing = A' \cap B'$. If $A \approx A'$ and $B \approx B'$, then $A \cup B \approx A' \cup B'$.*

Proof To prove that $A \cup B \approx A' \cup B'$, we must define a bijection $f: A \cup B \to A' \cup B'$. Since $A \approx A'$ and $B \approx B'$, there exist bijections $g: A \to A'$ and $h: B \to B'$. (See Figure 4.1.4.) We use the functions g and h to define a function $f: A \cup B \to A' \cup B'$ as follows: If $x \in A \cup B$, then $x \in A$ or $x \in B$. If $x \in A$, then define $f(x) = g(x)$; if $x \in B$, then define $f(x) = h(x)$. Since $A \cap B = \varnothing$, x cannot be in both A and B, and hence $f(x)$ is an unambiguously defined element of $A' \cup B'$. Thus f is a function from $A \cup B$ to $A' \cup B'$.

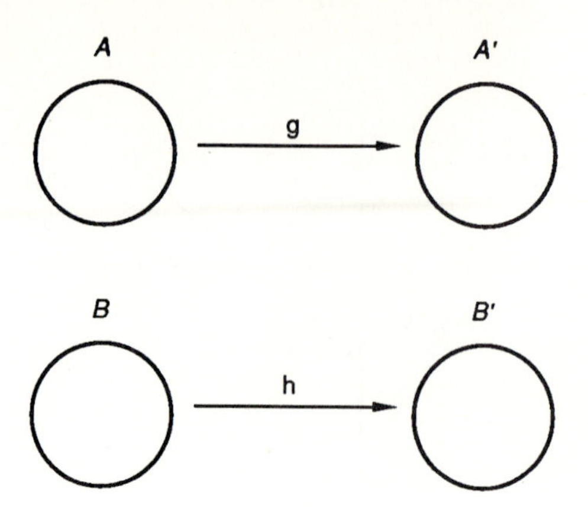

Figure 4.1.4

We show that f is bijective by showing that f is injective and surjective.

First, we establish the injectivity of f. Suppose $x, y \in A \cup B$ and $f(x) = f(y)$ in $A' \cup B'$. Since $A' \cap B' = \varnothing$, either $f(x) = f(y) \in A'$ or $f(x) = f(y) \in B'$. We consider the first case, $f(x) = f(y) \in A'$. Then $g(x) = f(x) = f(y) = g(y)$. Since g is a bijective, and hence an injective, function, and since $g(x) = g(y)$, we conclude that $x = y$. The second case, $f(x) = f(y) \in B'$, also leads to the conclusion that $x = y$. Thus f is injective.

Now for the surjectivity of f. Let y be an arbitrary element of $A' \cup B'$. We must find $x \in A \cup B$ such that $f(x) = y$. Since $y \in A' \cup B'$, either $y \in A'$ or $y \in B'$. If $y \in A'$, then since $g\colon A \to A'$ is surjective, there exists $x \in A \subseteq A \cup B$ such that $y = g(x)$, and hence, by the definition of f, $y = f(x)$. If $y \in B'$, then by the surjectivity of h, there exists $x \in B \subseteq A \cup B$ such that $y = h(x) = f(x)$. Therefore, f is surjective.

The proof that $A \cup B \approx A' \cup B'$ is now complete. ■

Corollary 4.1.3 *If A and B are disjoint finite sets, then $|A \cup B| = |A| + |B|$.*

Proof Suppose $|A| = m$ and $|B| = n$ where m and n are positive integers. Therefore, $A \approx \mathbf{N}_m$ and $B \approx \mathbf{N}_n$. But $\mathbf{N}_n = \{1, \ldots, n\} \approx \{m + 1, \ldots, m + n\} = \{k \in \mathbf{N} | m + 1 \leq k \leq m + n\}$. (Proof: Exercise 4.1.5.) Since by assumption $A \cap B = \varnothing$ and $\{1, \ldots, m\} \cap \{m + 1, \ldots, m + n\} = \varnothing$, $A \cup B \approx \{1, \ldots, m\} \cup \{m + 1, \ldots, m + n\} = \mathbf{N}_{m+n}$. Therefore, $|A \cup B| = m + n = |A| + |B|$. ■

As a few examples suggest, the question, "If $A \approx A'$ and $B \approx B'$, then does $A \cap B \approx A' \cap B'$?" does not have a reasonable answer. For Cartesian products, however, a very nice result holds.

Theorem 4.1.4 *If $A \approx A'$ and $B \approx B'$, then $A \times B \approx A' \times B'$.*

Proof Since $A \approx A'$ and $B \approx B'$, there exist one-to-one correspondences $f\colon A \to A'$ and $g\colon B \to B'$. We define a function $h\colon A \times B \to A' \times B'$ by $h((a, b)) = (f(a), g(b))$ for all $(a, b) \in A \times B$. We claim that h is a one-to-one correspondence.

We show that h is a one-to-one mapping and leave the proof that h is onto as an exercise. Suppose (a, b) and (a_1, b_1) are elements of $A \times B$ with the property that $h((a, b)) = h((a_1, b_1))$. We show that $(a, b) = (a_1, b_1)$. Since $h((a, b)) = h((a_1, b_1))$, $(f(a), g(b)) = (f(a_1), g(b_1))$ and hence $f(a) = f(a_1)$ and $g(b) = g(b_1)$. Since f and g are both one-to-one functions, $a = a_1$ and $b = b_1$. Therefore $(a, b) = (a_1, b_1)$. ■

Example 4.1.6 Let $I = (0,1)$. Since $I \approx \mathbf{R}$, $I \times I \approx \mathbf{R} \times \mathbf{R} = \mathbf{R}^2$. Geometrically, $I \times I$ is the interior of a unit square in $\mathbf{R}^2$; thus, the entire Cartesian plane can be put in one-to-one correspondence with the interior of the unit square.

Cardinal Numbers

Let us recap our approach to the problem of defining a measure of the size of an arbitrary set. We noted that if A is a finite set and $A \neq \varnothing$, then $A \approx \mathbf{N}_n$ for some positive integer n, and we said that A has n elements. For an infinite set A, we did *not* propose a definition of the size of A. We did, however, define when two sets A and B have the same size: A and B have the same size if $A \approx B$; that is, if there exists a bijection $f\colon A \to B$. But the question remains: Can we assign a "numerical" measure of size to an arbitrary set?

If we insist on using real numbers to measure the size of infinite sets, then the answer is certainly not apparent. But if we expand our notion of number, then perhaps we can. If we think of the number 2 as representing the size of all sets that can be put in one-to-one correspondence with the set $\{1, 2\} = \mathbf{N}_2$, then by analogy we can propose the following definition.

> Definition 4.1.4 *Cardinality*
>
> The *cardinal number* of a set A, written $|A|$, consists of all sets B such that $A \approx B$. We also call $|A|$ the *cardinality* of the set A.

We must stress that our definition of cardinal number is not rigorous. When we speak of all sets equinumerous with a given set, we are running the risk of encountering a Russell-type paradox. The usual way of avoiding such difficulties is to define the cardinal number, A, of a given set to be itself a set. This definition is achieved through the description of a clear, unambiguous procedure that will assign to each set, A, another set, called the cardinal number of A and written $|A|$, which is equinumerous to A. Using the Zermelo–Fraenkel axioms, we can indeed describe such a procedure. For instance, this procedure assigns to the set $\{a, b\}$, where $a \neq b$, the cardinal number $\{1, 2\}$: i.e., for any set A such that $A \approx \{a, b\}$, $|A| = \{1, 2\}$; moreover, in this case, mathematicians identify the cardinal number of the set $\{1, 2\}$ with the integer 2. Readers wishing to see a rigorous treatment of cardinal numbers can consult *Sets: Naive, Axiomatic, and Applied* by van Dalen et al.

Nevertheless, in spite of the logical pitfalls, we shall maintain an informal picture of a cardinal number as a kind of equivalence class. This class consists of sets all of which can be put in a one-to-one correspondence with a given set.

We are now able to phrase a number of definitions in terms of cardinal numbers.

Definition 4.1.5 *Finite, Denumerable, and Uncountable Sets*

(i) If $A \approx \mathbf{N}_n$, then we say A if finite and has *cardinality* n or that A has *cardinal number* n.
(ii) If $A \approx \mathbf{N}$, then we say that A is denumerable and that the *cardinality* or *cardinal number* of A is $\aleph_0$. ($\aleph_0$ is pronounced "aleph-null.")
(iii) A set A is *countable* if A is finite or if A is denumerable.
(iv) A set A is *uncountable* if A is not countable.

We should think of Definition 4.1.5 as an elaboration of Definition 4.1.4. According to Definition 4.1.1, a cardinal number is to be regarded as the aggregate of all sets equinumerous to a given set. If the given set happens to be equinumerous with $\mathbf{N}_n$, then we say that this cardinal number is n. If the given set is $\mathbf{N}$, then we say that the cardinal number is $\aleph_0$. An admittedly crude picture of the cardinality of sets is shown in Figure 4.1.5.

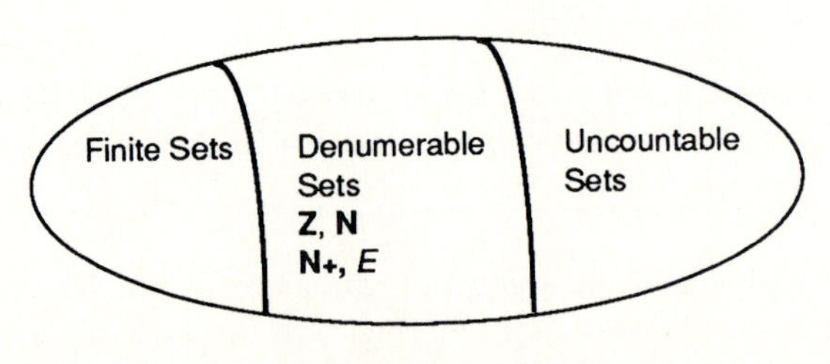

Figure 4.1.5

We can now extend the notion of ordering from real numbers to cardinal numbers. We shall let α and β denote cardinal numbers. Recall that a cardinal number consists of all sets D with the property that there exists a set A for which the $D \approx A$.

Definition 4.1.6 *Ordering of Cardinal Numbers*

Let α and β be cardinal numbers. We say that $\alpha < \beta$ if for any set A in α and any set B in β, A has the same cardinality as a subset of B but B does not have the same cardinality as any subset of A. We say that $\alpha \leq \beta$ if $\alpha < \beta$ or $\alpha = \beta$.

Let us restate this definition in somewhat different terms. Let α and β be cardinal numbers and let A and B be sets such that $|A| = \alpha$ and $|B| = \beta$. Then $\alpha < \beta$ if there exists a subset $B' \subset B$ such that $A \approx B'$ and for every subset $A' \subset A$, $B \not\approx A'$.

As cardinal numbers, therefore, $3 < 4$ since if $|A| = 3$ and $|B| = 4$ with $A = \{a, b, c\}$ and $B = \{d, e, f, g\}$, then $A \approx \{d, e, f\} \subset B$ while for all $A' \subseteq A$, $B \not\approx A'$. Also, for each positive integer n, $n < \aleph_0$ as cardinal numbers.

A whole host of interesting questions about cardinal numbers arises almost immediately. Here are two such questions that will be discussed later in this chapter:

1. Is $\aleph_0 < |\mathbf{R}|$? In other words, is the set of positive integers of smaller cardinality than the set of real numbers?
2. For any two cardinal numbers α and β, is is true that either $\alpha < \beta$, $\alpha = \beta$, or $\beta < \alpha$? In other words, does $<$ provide a total ordering of cardinal numbers?

Exercises §4.1

4.1.1. (a) Let $C = \{n \in \mathbf{N} | n = k^3 \text{ for some } k \in \mathbf{N}\}$. Show: $C \approx \mathbf{N}$ and $C \approx \mathbf{Z}$.
(b) Show $\mathbf{Z}^+ \approx \mathbf{Z}^-$.
(c) Show $\mathbf{Q}^+ \approx \mathbf{Q}^-$.
(d) Show $\mathbf{R}^+ \approx \mathbf{R}^-$.

4.1.2. (a) Show that $(0,1) \approx (0,3)$.
(b) Show that $(0,1) \approx (3,6)$.
(c) Show that for each $a \in \mathbf{R}$, $(0,1) \approx (a, a+1)$.
(d) Show that for each $a \in \mathbf{R}$, $(0,1) \approx (a, a+2)$.
(e) Show that for each $a \in \mathbf{R}$, $(0,1) \approx (a, a+3)$.
(f) Show that for each $a \in \mathbf{R}$ and each $b \in \mathbf{R}$ such that $b > 0$, $(0,1) \approx (a, a+b)$.

4.1.3. (a) Prove that $\mathbf{R}^+ = \{x \in \mathbf{R} \mid x > 0\} = (0, \infty) \approx (1, \infty)$.
(b) Prove that $(0,1) \approx (1, \infty)$.

4.1.4. (a) Show that $I \times I \times I \approx \mathbf{R}^3$. Describe this result in a sentence using no mathematical symbols.
(b) State and prove a generalization of the result proved in part (a).

4.1.5. (a) Show that for every $n, m \in \mathbf{N}^+$ $\mathbf{N}_n = \{1, \ldots, n\} \approx \{m+1, \ldots, m+n\}$.
(b) Prove that if $A \approx \mathbf{N}_k$ where $k \in \mathbf{N}^+$ and if $x \in A$, then $A - \{x\} \approx \mathbf{N}_{k-1}$.

4.1.6. Prove: If $|A| = n$ and $A \approx B$, then $|B| = n$.

4.1.7. Show: If $A \approx B$, then $P(A) \approx P(B)$.

4.1.8. Let A and B be finite sets. Conjecture a formula for $|A \cup B|$ in the case that $A \cap B \neq \varnothing$.

4.1.9. Prove that the function h defined in the proof of Theorem 4.1.4 is onto.

4.1.10. Give a geometric definition of a bijective function $g:(0,1) \times (0,1) \to \mathbf{R}^2$ analogous to the geometric description given after Example 4.1.5 of a bijective function from $(0,1)$ to $\mathbf{R}$.

4.1.11. Refer to the function $f:(0,1) \to \mathbf{R}$ defined in the paragraph after Example 4.1.5.
(a) For $1/2 \leq x < 1$ find an explicit formula for $f(x)$.
(b) For $0 < x < 1/2$ find an explicit formula for $f(x)$.
(c) Show that f is a bijection.

4.1.12. Let $\mathbf{Q}_1 = (0,1) \cap \mathbf{Q} = \{x \in \mathbf{Q} \mid 0 < x < 1\}$. Prove that $\mathbf{Q}_1 \approx \mathbf{Q}$.

4.1.13. As noted in Example 4.1.2, the set $\mathbf{Z}$ of integers is equinumerous with one of its proper subsets, the set E of even integers.
(a) Find a proper subset A of $\mathbf{Q}$ such that $A \approx \mathbf{Q}$.
(b) Find a proper subset B of $\mathbf{R}$ such that $B \approx \mathbf{R}$.

§4.2 *Finite Sets*

We now consider some basic properties of finite sets. All our results concern the behavior of finite sets relative to various set theoretic operations. Since, by definition, a nonempty set A is finite if $A \approx \mathbf{N}_k$ for some positive integer k, any statement involving finite sets can be phrased in terms of a positive integer, and hence is susceptible to a proof by mathematical induction. It is worth mentioning that although the theorems we prove are believable, perhaps even obvious, the precise proofs in some cases are somewhat subtle.

Our first result concerns the taking of subsets of a finite set.

Theorem 4.2.1 *If A is finite and $B \subseteq A$, then B is finite.*

Proof Let $S(k)$ be the statement: If $|A| = k$ and $B \subseteq A$, then B is a finite set. We prove that $S(k)$ is true for all $k \geq 0$.

Basis step If $k = 0$, then $A = \varnothing$, and hence $B = \varnothing$ and B is finite.

Inductive step We suppose $S(k)$ is true and prove $S(k + 1)$ is true.

Suppose $|A| = k + 1$ and $B \subseteq A$. We consider two (disjoint if you will) cases.

Case 1. $B = A$. Then $|B| = |A| = k + 1$ and B is finite.

Case 2. $B \subset A$. Then there exists $x \in A$ such that $x \notin B$. Thus $B \subseteq A - \{x\}$. Since $|A - \{x\}| = k$ by Lemma 4.2.2, which is proved below, statement $S(k)$ applies to the sets $A - \{x\}$ and $B \subseteq A - \{x\}$. Therefore, B is a finite set.

By PMI the proof is complete. ■

Lemma 4.2.2 *Let A be a nonempty finite set. If $x \in A$, then $A - \{x\}$ is a finite set and $|A - \{x\}| = |A| - 1$.*

Proof Suppose $|A| = n$ for some positive integer n. Let $x \in A$. We consider two cases.

Case 1. $n = 1$. Then $A = \{x\} = \varnothing$ and the statement of the theorem is valid.

Case 2. $n > 1$. Then $A \approx \mathbf{N}_n$, and hence there exists a bijection $f: A \to \mathbf{N}_n$. Suppose $f(x) = i$. We now define a function $g: A - \{x\} \to$

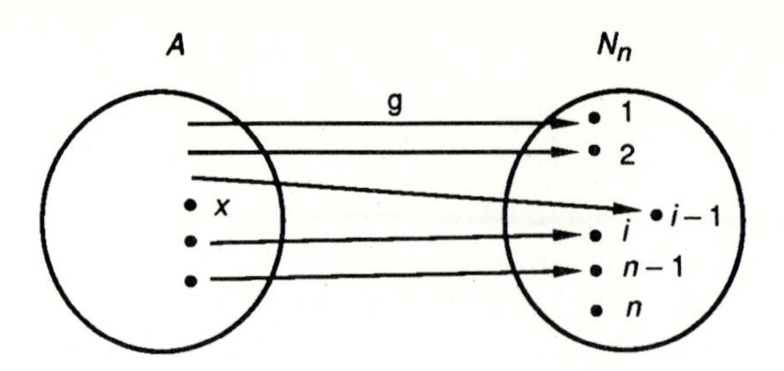

Figure 4.2.1

$\mathbf{N}_{n-1}$ as follows: For $y \in A - \{x\}$,

$$g(y) = \begin{cases} f(y) & \text{if } f(y) < i \\ f(y) - 1 & \text{if } f(y) > i \end{cases}.$$

(See Figure 4.2.1.) One can check that g is a function from $A - \{x\}$ to $\mathbf{N}_{n-1}$ and that g is a bijection. (We leave the proofs as an exercise.) Thus $|A - \{x\}| = n - 1 = |A| - 1$. ■

What about the union of two finite sets? If A and B are disjoint finite sets, then by Corollary 4.1.3 $A \cup B$ is finite and $|A \cup B| = |A| + |B|$. In general, it seems clear that if A and B are finite, then $A \cup B$ is finite and $|A \cup B| \leq |A| + |B|$. But is there a precise formula for $|A \cup B|$ in terms of $|A|, |B|$, and some other numbers determinable from A and B? Some experimentation with simple examples suggests the following theorem, which is a special case of the so-called Inclusion–Exclusion Principle.

Theorem 4.2.3 *Inclusion–Exclusion Principle. Let A and B be finite sets. Then $A \cup B$ is finite and $|A \cup B| = |A| + |B| - |A \cap B|$.*

Proof A direct inductive proof of Theorem 4.2.3 can be constructed. We have, however, proved a special case of the result in Corollary 4.1.3. Our approach is to reduce the proof of Theorem 4.2.3 back to the special case, Corollary 4.1.3. Thus we have another example of a bootstrap-type case analysis argument.

The general plan then is as follows: To prove $|A \cup B| = |A| + |B| - |A \cap B|$, we consider two cases:

Case 1. $A \cap B = \varnothing$. This is Corollary 4.1.3.

Case 2. $A \cap B \neq \varnothing$. In this case, our strategy is to reduce back to Case 1.

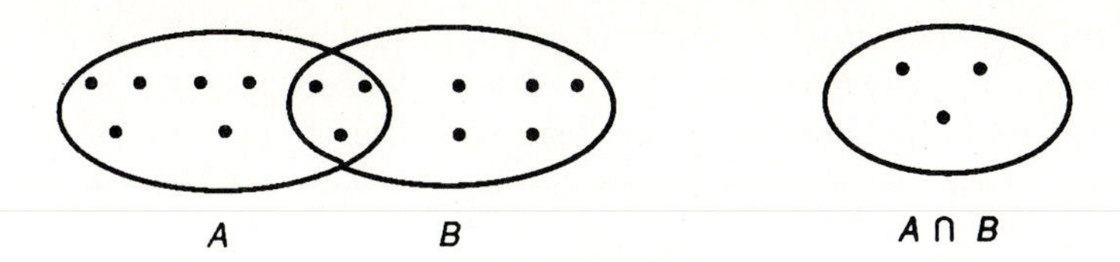

Figure 4.2.2

Since $A \cup B = A \cup (B - A)$ and $A \cap (B - A) = \varnothing$, it follows that $A \cup B$ is finite and $|A \cup B| = |A| + |B - A|$. Now $B = (A \cap B) \cup (B - A)$ where $(A \cap B) \cap (B - A) = \varnothing$; therefore, $|B| = |A \cap B| + |B - A|$. Thus $|A \cup B| = |A| + |B| - |A \cap B|$. ∎

Figure 4.2.2 illustrates Theorem 4.2.3. The point is that when the elements of A and of B are counted separately, the elements of $A \cap B$ are counted twice. We shall investigate other cases of the Inclusion–Exclusion Principle in Exercises 4.2.6 and 4.2.7.

Example 4.2.1 How many integers n, $1 \leq n \leq 100$ are divisible by either 2 or 5?

Let $A = \{n | 1 \leq n \leq 100 \text{ and } 2|n\}$ and $B = \{n | 1 \leq n \leq 100 \text{ and } 5|n\}$. Then $A \cup B = \{n | 1 \leq n \leq 100 \text{ and } 2|n \text{ or } 5|n\}$, while $A \cap B = \{n | 1 \leq n \leq 100 \text{ and } 10|n\}$. Thus $|A \cup B| = |A| + |B| - |A \cap B| = 50 + 20 - 10 = 60$.

Next we consider the set $A \times B$ when A and B are finite.

Theorem 4.2.4. *If A and B are finite sets, then $A \times B$ is finite and $|A \times B| = |A| \cdot |B|$.*

Before giving a formal proof, let us describe the basic idea behind the argument. To show $A \times B$ is finite, we shall try to list its elements in a systematic way. We begin by listing the elements of A and B: $A = \{a_1, \ldots, a_m\}$ and $B = \{b_1, \ldots, b_n\}$. Then we write down the elements of $A \times B$ whose first coordinate is a_1: $(a_1, b_1), \ldots, (a_1, b_n)$. We continue with the elements of $A \times B$ with first coordinate a_2: $(a_2, b_1), \ldots, (a_2, b_n)$. Repeating this procedure we eventually exhaust $A \times B$ when we write down the elements $(a_m, b_1), \ldots, (a_m, b_n)$. Thus the elements of $A \times B$ have been described in a sequence of m lists, each of which has n elements. To make this procedure precise, we use mathematical induction.

Proof For each positive integer m, let $S(m)$ be the statement: If A and B are finite sets with $|A| = m$, then $A \times B$ is finite and $|A \times B| = |A| \cdot |B|$. We prove that $S(m)$ holds for $m \geq 1$ by PMI.

Basis step Let $m = 1$. Then $A = \{a_1\}$ and $A \times B = \{a_1\} \times B = \{(a_1, b) | b \in B\}$. The function $f: B \to A \times B$ defined by $f(b) = (a_1, b)$ is easily seen to be a bijection. Therefore, $A \times B$ is finite and $|A \times B| = |B| = |A| \cdot |B|$.

Inductive step Suppose $S(m)$ is true. We show $S(m + 1)$ is true. Let $A = \{a_1, \ldots, a_m, a_{m+1}\}$ be any set with $m + 1$ elements and let B be any finite set. Let $A' = \{a_1, \ldots, a_m\}$. Then $A \times B = (A' \cup \{a_{m+1}\}) \times B = (A' \times B) \cup (\{a_{m+1}\} \times B)$ by Theorem 2.3.1. Since $|A'| = m$, $A' \times B$ is finite and $|A' \times B| = |A'| \cdot |B|$. By the basis step, $\{a_{n+1}\} \times B$ is finite with $|\{a_{n+1}\} \times B| = |B|$. Moreover, by Theorem 2.4.1, $(A' \times B) \cap (\{a_{n+1}\} \times B) = (A' \cap \{a_{n+1}\}) \times B = \varnothing \times B = \varnothing$. Therefore, by Theorem 4.2.3, $A \times B$ is a finite set containing exactly

$$\begin{aligned} |A \times B| &= |A' \times B| + |\{a_{m+1}\} \times B| \\ &= |A'| \cdot |B| + |B| = (m + 1)|B| = |A| \cdot |B| \end{aligned}$$

elements. ■

Example 4.2.2 Suppose two dice, one red and one blue, are rolled. Let R and B be the set of possible outcomes for the respective dice. Thus $R = B = \{1, 2, 3, 4, 5, 6\}$. Then the set of possible outcomes for the pair is $R \times B = \{(r, b) \in \mathbf{N} \times \mathbf{N} | 1 \leq r, b \leq 6\}$. By Theorem 4.2.4 $|R \times B| = 36$. In how many rolls does at least one 6 appear? Let A_1 (resp. A_2) be the set of rolls in which a 6 appears on the red (resp. blue) die. Then, by the Inclusion–Exclusion Principle, there are $|A_1 \cup A_2| = |A_1| + |A_2| - |A_1 \cap A_2| = 6 + 6 - 1 = 11$ ways in which at least one 6 occurs.

We proved another result concerning the behavior of finite sets under set operation back in Chapter 2, namely, Theorem 2.2.4. We recall that result.

Theorem 2.2.4 (revisited) *If A is a finite set, then $\mathscr{P}(A)$ is a finite set with $|\mathscr{P}(A)| = 2^{|A|}$.*

The final theorem in this section concerns functions between finite sets. This theorem is known as the Pigeonhole Principle because it has the following vivid interpretation. Suppose you are stuffing m pigeons into n pigeonholes where $m > n$. Then at least one pigeonhole will contain two or more pigeons when you are finished stuffing.

Theorem 4.2.5 *The Pigeonhole Principle. Suppose $|A| = m$ and $|B| = n$ with $m > n$. If $f: A \to B$ is any function, then f is not injective.*

Remark The Pigeonhole Principle can be restated in terms of inverse images. If $f: A \to B$ where $|A| = m > n = |B|$, then there exists $b \in B$ such that $|f^{-1}(b)| = |\{a \in A | f(a) = b\}| > 1$, for there exists $b \in B$ such that $|f^{-1}(b)| > 1$ if and only if there exist $a_1, a_2 \in A$ such that $a_1 \neq a_2$ and $f(a_1) = f(a_2) = b$ if and only if f is not injective.

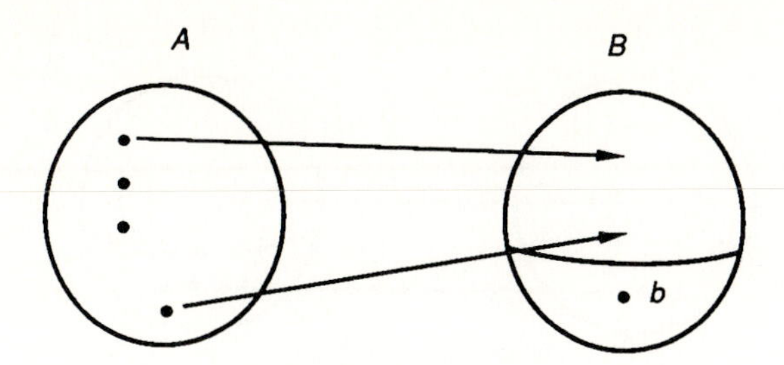

Figure 4.2.3

Proof We argue by induction on $|B|$. For each positive integer n, let $S(n)$: If $f: A \to B$ with $|A| = m > n = |B|$, then f is not injective.

Recall that to show f is not injective, we must find $x, y \in A$ with $x \neq y$ and $f(x) = f(y)$.

Basis step If $|B| = 1$ then $B = \{b\}$ and for all $x \in A$, $f(x) = b$. Since $m = |A| > 1$, f is not injective.

Inductive step We assume $S(n)$ holds and prove that $S(n + 1)$ is true. Suppose $f: A \to B$ where $|A| = m > n + 1 = |B|$. Let b be a fixed element of B. We consider three distinct cases:

Case 1. (See Figure 4.2.3.) For all $x \in A$, $f(x) \neq b$. Then, for all $x \in A$, $f(x) \in B - \{b\} = B_1$; thus $f: A \to B_1$. Since $|B_1| = n$ (Lemma 4.2.2) and since $S(n)$ holds, f is not injective.

Case 2. (See Figure 4.2.4.) There exist $x, y \in A$ such that $x \neq y$ and $f(x) = f(y) = b$. Then f is not injective.

Case 3. (See Figure 4.2.5.) There exists a unique $x_0 \in A$ such that $f(x_0) = b$. Let $A_1 = A - \{x_0\}$ and $B_1 = B - \{b\}$. For each $a \in A_1$, $f(a) \in B_1$. Thus we can define a function $g: A_1 \to B_1$ by $g(a) = f(a)$. (In technical terms, g is the *restriction* of f to A_1.) Since $|A_i| = m - 1 > n =$

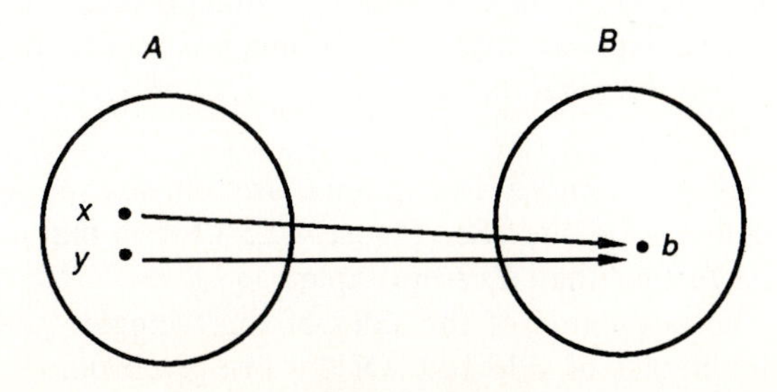

Figure 4.2.4

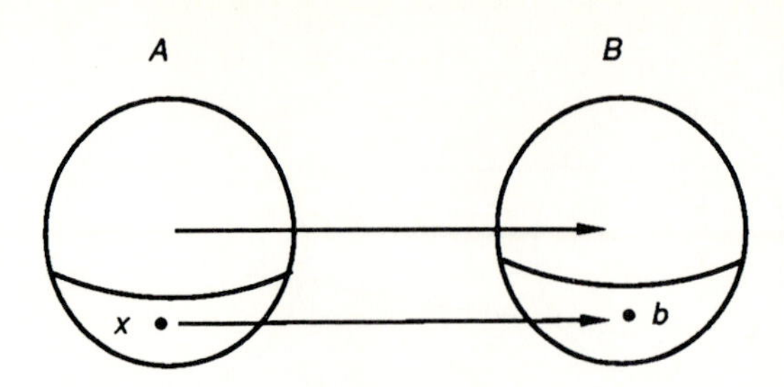

Figure 4.2.5

$|B_1|$, by inductive hypothesis g is not injective and thus there exist $x, y \in A_1 \subseteq A$ such that $x \neq y$ and $f(x) = g(x) = g(y) = f(y)$. Therefore, in Case 3, f is not injective.

This completes the proof. ■

The Pigeonhole Principle has a variety of applications. In fact, the pigeonhole argument can be applied in so many diverse situations that it is regarded by many as a standard problem-solving tool in mathematics. In most cases, the Pigeonhole Principle is applied in a subtle fashion: Neither the pigeons nor the pigeonholes are readily apparent in the statement of the problems and the problem solver must interpret the problem so as to make the Pigeonhole Principle applicable. In general, the Pigeonhole Principle is used to establish the existence of configurations or arrangements of a certain type. The next two examples illustrate the Pigeonhole Principle in action.

Example 4.2.3 Prove that at least two residents of Minneapolis have the same number of hairs on their heads. We need to know two facts, one of anatomy, the other of demography. Each human has at most 300,000 hairs on his or her head and the city of Minneapolis has over 400,000 residents.

Let M be the set of residents of Minneapolis. Define $f: M \to \mathbf{N}_{300,000} \cup \{0\}$ by letting, for $x \in M$, $f(x)$ be the number of hairs on x's head. Since $|M| > 300{,}001$, f is not injective. Thus there are x and $y \in M$, $x \neq y$, with $f(x) = f(y)$; hence x and y have the same number of hairs on their heads. Note that in our interpretation of this problem, the residents of Minneapolis are the pigeons while the nonnegative integers less than or equal to 300,000 are the pigeonholes.

Example 4.2.4 Suppose five points are chosen on or inside an equilateral triangle of side one. (See Figure 4.2.6.) Prove that at least two of the points are no farther than $1/2$ unit apart.

If we join the midpoints of the sides of the triangles, then we obtain four equilateral triangles of side $1/2$. Of the five given points, at least two

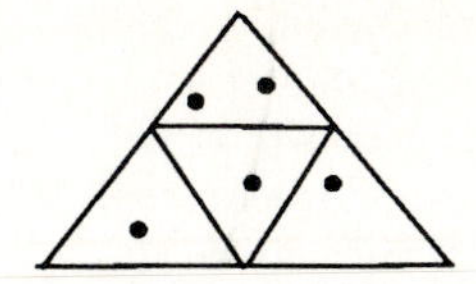

Figure 4.2.6

must lie on or inside one of the four smaller triangles. These points are no farther than $1/2$ unit apart. To apply Theorem 4.2.5 formally, we let P be the set of five given points and T be the set of four triangles of side $1/2$ and let $f: P \to T$ be the function that assigns to any point a triangle in T in which the point lies; in case a point lies on the boundary of two or three triangles, simply assign the point to any one of the triangles.

The Pigeonhole Principle has become an important tool in discrete or combinatorial mathematics, that portion of mathematics concerned with properties of finite and countable sets. The last two examples illustrate the usefulness of the Pigeonhole Principle in establishing the existence of a certain type of configuration or arrangement. More Pigeonhole applications are given in the exercises.

Exercises §4.2

4.2.1. Prove: If A is finite and B is any set, then $A \cap B$ is finite.

4.2.2. Prove: If A is finite, B is infinite and $A \subseteq B$, then $B - A$ is infinite.

4.2.3. Prove: If $\mathscr{P}(A)$ is finite, then A is finite.

4.2.4. Complete the proof of Lemma 4.2.2 by proving that g is a bijection.

4.2.5. (a) Prove that if $A_1, \ldots, A_n$ are pairwise disjoint finite sets (i.e., $A_i \cap A_j = \varnothing$ if $i \neq j$), then $|A_1 \cup \cdots \cup A_n| = |A_1| + \cdots + |A_n|$.
(b) Prove that if $A_1, \ldots, A_n$ are finite sets, then $A_1 \cup \cdots \cup A_n$ is finite.

4.2.6. The Inclusion–Exclusion Principle.
(a) Let A, B, and C be finite sets. Conjecture a formula for $|A \cup B \cup C|$ (in terms of $|A|$, $|B|$, $|C|$, and other integers determined by A, B, and C) and prove your conjecture. Note that as a check on your conjectured formula, it should reduce to Theorem 4.2.3 in case one of the sets is empty. Describe some other ways of checking your formula.
(b) How many positive integers n up to 120 are divisible by either 2, 3, or 5?

4.2.7. (a) Conjecture a formula for $|A \cup B \cup C \cup D|$ where A, B, C, and D are finite sets, then prove your conjecture.
(b) Generalize the results proved in Exercises 4.2.6 and 4.2.7(a).

4.2.8. Prove: If $A_1, \ldots, A_n$ are finite sets, then $A_1 \times \cdots \times A_n$ is a finite set.

4.2.9. Suppose A is infinite and B is finite. Suppose $f: A \to B$ is a function. Prove that for at least one $b \in B$, the set $f^{-1}(b) = \{x \in A | f(x) = b\}$ is infinite.

4.2.10. (a) Let A and B be sets with $|A| = |B| = 2$. Show that there exist exactly two one-to-one correspondences from A to B.
(b) Let A and B be sets with $|A| = |B| = 3$. Show that there exist exactly six one-to-one correspondences from A to B.
(c) Suppose $|A| = |B| = n$ where n is an arbitrary positive integer. State and prove a result about the number of one-to-one correspondences from A to B.

Exercises 4.2.11–4.2.16 all consist of applications of the Pigeonhole Principle.

4.2.11. (a) Ten points are chosen on or inside an equilateral triangle of side 1. Prove that at least two of the points are no more than $1/3$ unit apart.
(b) Find an integer $p(n)$ such that if $p(n)$ points are chosen on or inside an equilateral triangle of side 1, then there are at least two points whose distance apart is no more than $1/n$.

4.2.12. Suppose n people are at a party. Show that at least two of the people know the same number of people among those present. (Note: Assume that if A knows B, then B knows A.)

4.2.13. Suppose 20 distinct integers are chosen from the set $\{1, 4, 7, \ldots, 100\} = \{n \in \mathbf{Z} | n = 3k + 1 \text{ where } 0 \leq k \leq 33\}$. Prove that there exist two distinct integers among the chosen 20 whose sum is 104.

4.2.14. (a) Show that from any four integers it is possible to choose a pair whose difference is even.
(b) Show that from any three integers it is possible to choose a pair whose difference is divisible by 3.
(c) Generalize the results given in (a) and (b), then prove your generalization.

4.2.15. A point in $\mathbf{R}^n$ is called a *lattice point* if its coordinates are integers. Thus $(-1, 1)$ is a lattice point in $\mathbf{R}^2$, while $(-1, 1/2)$ is not a lattice point in $\mathbf{R}^2$.
(a) Prove that if A is a set of five lattice points in $\mathbf{R}^2$, then there exist $P, Q \in A$ with $P \neq Q$ such that the line segment PQ contains a lattice point (which is not necessarily in A) that is different from P and Q.
(b) Prove that if A is a set of nine lattice points in $\mathbf{R}^3$, then there exist $P, Q \in A$ with $P \neq Q$ such that the line segment PQ contains a lattice point that is different from P and Q.

4.2.16. Prove the following generalization of the Pigeonhole Principle: If A and B are finite sets with $|B| = n$ and $|A| \geq kn + 1$ where k and n are positive integers and $f: A \to B$ is any function, then there exists $b \in B$ such that $|f^{-1}(b)| \geq k + 1$. In less technical words, if $kn + 1$ or more objects are placed into n containers, then at least one container contains at least $k + 1$ objects.

4.2.17. Prove that if A and B are finite sets and $f: A \to B$ is injective, then $|A| \leq |B|$.

§4.3 Denumerable Sets

In this section we study properties of denumerable sets. We can put the elements of a finite set into a list that eventually ends. Thus, if A is finite and nonempty, we can write $A = \{a_1, \ldots, a_n\}$ for some integer n. Denumerable sets are the next best thing. A denumerable set is infinite, but, as we shall see, its elements can also be placed into an ordered list: $\{a_0, a_1, a_2, \ldots\}$. In this section we take a close look at denumerable sets. We study the behavior of denumerable sets under set operations. These theorems say essentially that when you put together denumerable sets using most set operations, you get denumerable sets. Many of these theorems are analogous to the theorems about finite sets proved in the last section. As an application of these results, we prove that the set of rational numbers is denumerable. We begin by recalling the definitions of denumerability and countability.

> Definition 4.3.1 A set A is *denumerable* if $A \approx \mathbf{N}$. A set A is *countable* if A is finite or denumerable.

If A is denumerable, then there exists a bijection $f: \mathbf{N} \to A$. Let $a_0 = f(0)$, $a_1 = f(1)$, and for each $n \in \mathbf{N}$, $a_n = f(n)$. Then $A = \{a_0, a_1, \ldots, a_n, \ldots\}$. In other words, the elements of a denumerable set form a sequence.

As proved in Section 3.4 and Section 4.1, the following sets are denumerable: $\mathbf{N} - \{0\}$, E (set of even integers), and E^c (set of odd integers). Our first result is the analogue of Theorem 4.2.1 for denumerable sets.

Theorem 4.3.1 (i) *Let A be a denumerable set and let B be an infinite subset of A. Then B is denumerable.*

(ii) *Let A be a countable set and let $B \subseteq A$. Then B is countable.*

Proof (i) We give a bootstrap-type case analysis proof. First we prove the theorem in case $A = \mathbf{N}$ and B is an infinite subset of $\mathbf{N}$. The general case can then be reduced to the case that $A = \mathbf{N}$. We leave the last portion of the proof to the reader.

Suppose B is an infinite subset of $\mathbf{N}$. We define a function $f: \mathbf{N} \to B$ in stages. (Actually, we are using mathematical induction to define f.) Let b_0 be the smallest element of B. (Here we use the fact that any nonempty subset of **N** has a smallest member. We can prove this property of **N**, known as the Well-Ordering Principle, by mathematical induction. For details, see Section 7.2, especially Theorem 7.2.9.) Then we define $f(0) = b_0$. Next let b_1 be the smallest element of $B - \{b_0\}$ and define $f(1) = b_1$. In general, for $n \geq 1$, define

$$f(n) = b_n = \text{smallest element of } B - \{b_0, \ldots, b_{n-1}\}.$$

It is clear that f is a function from **N** to B. We claim that f is one-to-one and onto. First, we show that if $n < m$, then $f(n) \neq f(m)$. If $n < m$, then $b_m \in \mathbf{N} - \{b_0, \ldots, b_n, \ldots, b_{m-1}\}$, hence $f(m) = b_m \neq b_n = f(n)$. Thus f is one-to-one. To see that f is onto, note that if $b \in B$, then there exists an integer k such that b is the smallest elements of $B - \{b_0, \ldots, b_k\}$. (We cannot specify the exact value of k in general except to say that $k \leq b - 1$.) Thus $b = f(k + 1)$ and f is onto.

(ii) Again we consider cases. First, if A is finite, then B is finite by Theorem 4.2.2, and hence B is countable. Next suppose A is infinite. Since A is countable, it follows that A is denumerable. Now B is clearly finite or infinite. If B is finite, then B is countable. If B is infinite, then by part (i), B is denumerable and hence countable. ■

Next we analyze the behavior of denumerable sets under Cartesian products.

Theorem 4.3.2 $\mathbf{N} \times \mathbf{N}$ *is denumerable.*

First Proof Define a function $g: \mathbf{N} \times \mathbf{N} \to \mathbf{N}$ by $g(a, b) = 2^a 3^b$.

We claim that g is one-to-one. Suppose $g(a, b) = g(a', b')$. We show $(a, b) = (a', b')$. Since $g(a, b) = g(a', b')$, $2^a 3^b = 2^{a'} 3^{b'}$. Suppose $a \leq a'$. Then, after canceling 2^a from both sides, we have $3^b = 2^{(a'-a)} 3^{b'}$. Since 3^b is odd, $2^{(a'-a)} 3^{b'}$ must be odd, and hence $a - a' = 0$ and $a = a'$. If $a' \leq a$, then a similar argument shows $a = a'$. Therefore, $3^b = 3^{b'}$, which implies that $b = b'$. Therefore, $g(a, b) = g(a', b')$ implies that $(a, b) = (a', b')$; hence g is one-to-one.

Note that g is not onto. Nonetheless, $\mathbf{N} \times \mathbf{N} \approx \operatorname{Ran}(g)$ where $\operatorname{Ran}(g) = \{n \in \mathbf{N} | n = 2^a 3^b \text{ for some } a, b \in \mathbf{N}\}$. Since $\operatorname{Ran}(g)$ is an infinite subset of **N**, $\operatorname{Ran}(g) \approx \mathbf{N}$. Therefore, $\mathbf{N} \times \mathbf{N} \approx \mathbf{N}$ by the transitivity of $\approx$.

Second Proof Define $f: \mathbf{N} \times \mathbf{N} \to \mathbf{N}^+ = \{n \in \mathbf{N} | n \geq 1\}$ by $f(a, b) = 2^a(2b + 1)$. As an exercise (see Miscellaneous Exercise 3 of Chapter 3), show that f is a one-to-one correspondence. Hence $\mathbf{N} \times \mathbf{N} \approx \mathbf{N}^+ \approx \mathbf{N}$.

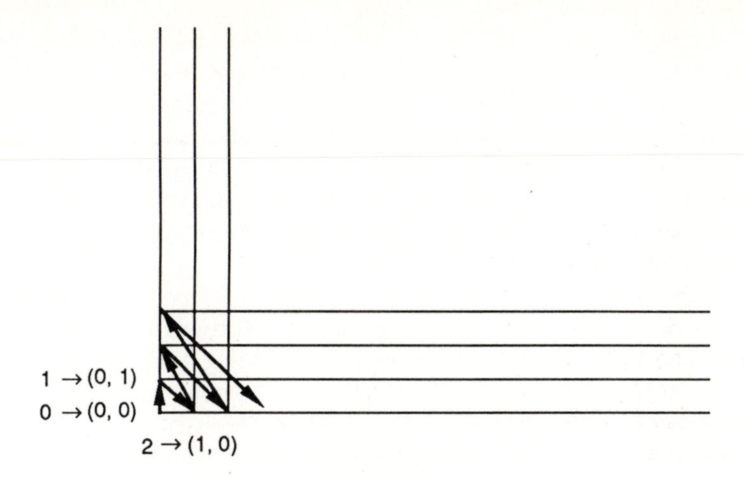

Figure 4.3.1

Third Proof Our third proof is based on the geometric representation of $\mathbf{N} \times \mathbf{N}$ as certain points in the first quadrant in $\mathbf{R}^2$. (See Figure 4.3.1.) Given this representation we can list the elements of $\mathbf{N} \times \mathbf{N}$ by moving along the diagonal line $y + x = k$ where $k = 0, 1, 2, \ldots$. An explicit formula for the labeling is $h(a, b) = (a + b)(a + b + 1)/2 + a$. We leave as an exercise the proof that $h\colon \mathbf{N} \times \mathbf{N} \to \mathbf{N}$ is a one-to-one correspondence. ∎

Corollary 4.3.3 *If A and B are denumerable sets, then $A \times B$ is denumerable.*

Proof If A and B are denumerable, then $A \approx \mathbf{N}$ and $B \approx \mathbf{N}$. Therefore, by Theorem 4.1.4, $A \times B \approx \mathbf{N} \times \mathbf{N}$. But $\mathbf{N} \times \mathbf{N} \approx \mathbf{N}$, and by Theorem 4.1.1 $A \times B \approx \mathbf{N}$. Therefore, $A \times B$ is denumerable. ∎

Notice that Theorem 4.3.2 and Corollary 4.3.3 provide, in effect, a bootstrap proof that $A \times B$ is denumerable if A and B are denumerable. Two cases are considered:

1. $A = B = \mathbf{N}$.
2. A and B are arbitrary denumerable sets.

To prove Case 1 we exploit the fact that $A \times B = \mathbf{N} \times \mathbf{N}$ consists of all ordered pairs of nonnegative integers. Having an explicit form for the elements of $A \times B$, we can construct a one-to-one function from $A \times B$ to $\mathbf{N}$, thereby proving that $A \times B$ is denumerable. We then reduce Case 2 back to Case 1 using Theorem 4.1.4.

In our next theorem we consider the set of rational numbers $\mathbf{Q}$. When pictured geometrically along the real number line, the set of rationals appears to be distinctly larger than the set of integers. Thus one might suspect that $\mathbf{Q} \not\approx \mathbf{N}$ and that $\mathbf{Q}$ is not denumerable and therefore is uncountable. However, as Georg Cantor proved in an epoch-making paper published in 1874, $\mathbf{Q}$ is indeed denumerable. Cantor, in fact, gave several proofs that $\mathbf{Q}$ is denumerable, the most famous being quite similar to our third proof of Theorem 4.3.2.

Throughout the 1870s, 1880s, and 1890s, Cantor proved a number of remarkable and surprising results about sets. Although certain famous mathematicians of his day, the most notable being Leopold Kronecker, refused to accept Cantor's methods (Kronecker objected to the use of "nonconstructive" arguments, in particular to the use of proof by contradiction), Cantor's work was responsible for establishing set theory as a separate discipline within mathematics. Interestingly enough, Cantor first began to investigate properties of sets of real numbers while studying the convergence of Fourier series. The theory of Fourier series is an important subject within theoretical mathematics, which originally arose in the eighteenth century from physical studies of heat flow and which has since its inception found many practical applications. In Cantor's work we can see that the study of Fourier series led to a new branch of pure mathematics. This episode in the history of mathematics provides a vivid example of how apparently separate areas of mathematics are intertwined.

Theorem 4.3.4 $\mathbf{Q}$ *is denumerable.*

Proof We first prove that $\mathbf{Q}^+$, the set of positive rational numbers, is denumerable. The key idea is to represent the elements of $\mathbf{Q}^+$ in such a way that the denumerability of $\mathbf{Q}^+$ becomes apparent. Following Cantor's idea, we correspond each positive rational $x = a/b$ to the point (a, b) in $\mathbf{N} \times \mathbf{N}$. Since a given rational can have several representations as a fraction, we must be careful when formulating this correspondence. Here is a precise definition:

For $x \in \mathbf{Q}^+$, write $x = a/b$ where a and b are positive integers having no common positive integer factor except 1. (In other words, write x as a fraction in lowest terms.) Define $f: \mathbf{Q}^+ \to \mathbf{N} \times \mathbf{N}$ by $f(x) = (a, b)$. We see that f is one-to-one, since if for $x, y \in \mathbf{Q}^+$, $f(x) = (a, b) = f(y)$, then $x = a/b = y$. Therefore, $\mathbf{Q}^+ \approx f(\mathbf{Q}^+)$, the latter being an infinite subset of $\mathbf{N} \times \mathbf{N}$. Since $\mathbf{N} \times \mathbf{N}$ is denumerable, Theorem 4.3.1 assures us that $f(\mathbf{Q})^+$ is denumerable. Since $\mathbf{Q}^+ \approx f(\mathbf{Q}^+)$, $\mathbf{Q}^+$ is denumerable.

Next let $\mathbf{Q}^-$ be the set of negative rational numbers. Since $\mathbf{Q}^- \approx \mathbf{Q}^+$, $\mathbf{Q}^-$ is also denumerable.

Last, we show that $\mathbf{Q}$ itself is denumerable. (See Figure 4.3.2.) Since $\mathbf{Q}^+$ is denumerable, $\mathbf{Q}^+ \approx \mathbf{N}$. But $\mathbf{N} \approx \mathbf{N}^+ = \mathbf{N} - \{0\}$. Thus $\mathbf{Q}^+ \approx \mathbf{N}^+$. Also $\mathbf{Q}^- \approx \mathbf{N}^- = \{-1, -2, -3, \dots\} = \{n \in \mathbf{Z} | n < 0\}$. Now we apply Theorem

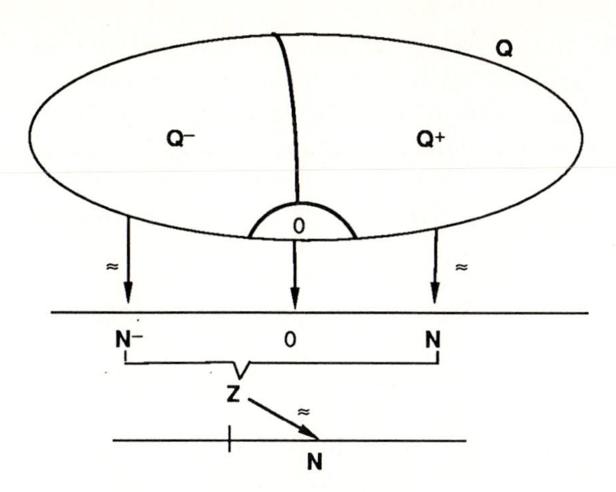

Figure 4.3.2

4.1.2 twice, first to conclude that $\mathbf{Q}^+ \cup \{0\} \approx \mathbf{N}^+ \cup \{0\} = \mathbf{N}$, then to conclude that $\mathbf{Q} = (\mathbf{Q}^+ \cup \{0\}) \cup \mathbf{Q}^- \approx \mathbf{N} \cup \mathbf{N}^- = \mathbf{Z}$. But by Example 4.1.2(iv) we know that $\mathbf{Z} \approx \mathbf{N}$. Therefore, $\mathbf{Q}$ is denumerable. ∎

Returning now to the behavior of denumerable sets under set operations, we ask if the union of two denumerable sets is denumerable. We can, in fact, answer this question if we look back at the end of the proof of Theorem 4.3.4. In that proof we showed that $\mathbf{Q}^+ \cup \{0\} \approx \mathbf{N}$ and $\mathbf{Q}^- \approx \mathbf{N}^- \approx \mathbf{N}$; in other words, $\mathbf{Q}^+ \cup \{0\}$ and $\mathbf{Q}^-$ are denumerable. Thus $\mathbf{Q} = \mathbf{Q}^+ \cup \{0\} \cup \mathbf{Q}^- \approx \mathbf{N} \cup \mathbf{N}^- = \mathbf{Z} \approx \mathbf{N}$, and hence $\mathbf{Q}$ is denumerable.

What are the general characteristics of the preceding argument? The sets $\mathbf{Q}^+ \cup \{0\}$ and $\mathbf{Q}^-$ are disjoint denumerable sets with $\mathbf{Q}^+ \cup \{0\} \approx \mathbf{N}$ and $\mathbf{Q}^- \approx \mathbf{N}^-$. Thus, in general, if A and B are disjoint denumerable sets, then $A \approx \mathbf{N}$ and $B \approx \mathbf{N}$. But $\mathbf{N} \approx \mathbf{N}^-$; hence $B \approx \mathbf{N}^-$. Thus, by Theorem 4.1.2, $A \cup B \approx \mathbf{N} \cup \mathbf{N}^- = \mathbf{Z} \approx \mathbf{N}$. This argument proves one case (the case in which $A \cap B = \varnothing$) of the following theorem. In order to provide a different way of representing the union of denumerable sets, we give yet another proof of the case just established.

Theorem 4.3.5 (i) *If A and B are denumerable sets, then $A \cup B$ is denumerable.*
(ii) *If A and B are countable sets, then $A \cup B$ is countable.*

Proof (i) We give a bootstrap-type case analysis proof. We consider two cases:

1. $A \cap B = \varnothing$.
2. $A \cap B \neq \varnothing$.

Case 1. $A \cap B = \varnothing$. Let $A_1 = \{(n, m) \in \mathbf{N} \times \mathbf{N} \mid n = 1\} = \{(1, m) \mid m \in \mathbf{N}\}$. It is easy to show that $A_1 \approx \mathbf{N}$. Similarly, if $A_2 = \{(n, m) \in \mathbf{N} \times \mathbf{N} \mid n = 2\} = \{(2, m) \mid m \in \mathbf{N}\}$, then $A_2 \approx \mathbf{N}$. Since $A \cap B = \varnothing = A_1 \cap A_2$, and $A_1 \approx A$ and $B_1 \approx B$, it follows from Theorem 4.1.2 that $A \cup B \approx A_1 \cup A_2$. But $A_1 \cup A_2$ is an infinite subset of $\mathbf{N} \times \mathbf{N}$, and hence $A_1 \cup A_2$ is denumerable. Therefore, $A \cup B$ is denumerable.

Case 2. $A \cap B \neq \varnothing$. Then $A \cup B = A \cup (B - A)$ and $A \cap (B - A) = \varnothing$. Since $B - A \subseteq B$, $B - A$ is countable by Theorem 4.3.1. We have then two subcases:

Subcase 1. $B - A$ is denumerable. Then by Case 1 of this proof, $A \cup B = A \cup (B - A)$ is denumerable.

Subcase 2. $B - A$ is finite. If $B - A = \varnothing$, then $A \cup B = A$ is denumerable. Otherwise, $B - A \approx \mathbf{N}_k$ for some positive integer k. Since $A \approx \mathbf{N} \approx \{n \in \mathbf{N} \mid n \geq k + 1\}$, Theorem 4.1.2 implies that $A \cup B = A \cup (B - A) \approx \{n \in \mathbf{N} \mid n \geq k + 1\} \cup \mathbf{N}_k = \mathbf{N}^+ \approx \mathbf{N}$.

Therefore, in each possible case, we conclude that $A \cup B$ is denumerable.

The proof of part (ii) is left as an exercise. ■

Theorem 4.3.5 provides both empirical evidence for and the start of an inductive proof of the following theorem. Once again we are pursuing the general question of how properties of sets are preserved under set operations.

Theorem 4.3.6 (i) *If $A_1, \ldots, A_n$ are denumerable sets, then $A_1 \cup \cdots \cup A_n$ is denumerable.*

(ii) *If $A_1, \ldots, A_n$ are countable sets, then $A_1 \cup \cdots \cup A_n$ is countable.*

We leave the (inductive) proof of Theorem 4.3.6 as an exercise.

A question arises in the wake of Theorem 4.3.6. Suppose that, instead of taking a finite set of denumerable sets, we take an infinite set of denumerable sets, specifically a denumerable set of denumerable sets. Is the union of these sets also denumerable? In symbolic terms we ask: If $A_1, A_2, \ldots, A_n, \ldots$ is a denumerable collection of denumerable sets, then is $A_1 \cup A_2 \cup \cdots \cup A_n \cup \cdots = \bigcup_{n=1}^{\infty} A_n$ a denumerable set?

To develop a feel for the question, we look for an example of a denumerable collection of denumerable sets. A natural place to look is within the real number system.

Consider first the set of all fractions, D_1, with denominator 1. This set is precisely the set $\mathbf{Z}$ of integers and hence is denumerable. Next consider all fractions with denominator 2: $D_2 = \{0/2, 1/2, 2/2, \ldots\} = \{n/2 \mid n \in \mathbf{Z}\}$. D_2 is clearly a denumerable set. For each positive integer k we have the set D_k of all fractions with denominator k: $D_k = \{n/k \mid n \in \mathbf{Z}\}$. For each k, D_k is denumerable. (The rule $f(n) = n/k$ defines a one-to-one correspondence $f: \mathbf{Z} \to D_k$.) Now what about the set $\bigcup_{k=1}^{\infty} D_k$? Clearly, $\bigcup_{k=1}^{\infty} D_k = \mathbf{Q}$, a denumerable set. While this example does not quite constitute a general proof, it does lead us to the next theorem.

Theorem 4.3.7 (i) *If* $\{A_n | n \in \mathbf{N}^+\}$ *is a denumerable collection of denumerable sets, then* $\bigcup_{n=1}^{\infty} A_n$ *is a denumerable set.*
(ii) *If* $\{A_n | n \in \mathbf{N}^+\}$ *is a countable collection of countable sets, then* $\bigcup_{n=1}^{\infty} A_n$ *is a countable set.*

Proof (i) Let $A = \bigcup_{n=1}^{\infty} A_n$. Since A_n is denumerable for each $n \geq 1$, there exists, for each $n \geq 1$, a one-to-one correspondence $f_n\colon A_n \to \mathbf{N}$. We use the collection $\{f_n | n \geq 1\}$ to show that A is equinumerous to an infinite subset of a denumerable set.

We first assume that the sets $\{A_n | n \geq 1\}$ are pairwise disjoint. We now define a function $F: A \to \mathbf{N} \times \mathbf{N}$. Let $a \in A$. Then there exists a unique $k \in \mathbf{N}^+$ such that $a \in A_k$, unique since if $n \neq m$, then $A_n \cap A_m = \varnothing$. Define $F(a) = (f_k(a), k)$. Note that since $f_k(a) \in \mathbf{N}$, $F(a) \in \mathbf{N} \times \mathbf{N}$. In other words, we send A into $\mathbf{N} \times \mathbf{N}$ by sending A_k into the points of $\mathbf{N} \times \mathbf{N}$ with second coordinate k.

We show that F is a one-to-one function from A onto an infinite subset of $\mathbf{N} \times \mathbf{N}$. First suppose there exist $a, b \in A$ with $F(a) = F(b)$. We show $a = b$, thereby establishing that F is one-to-one. If $F(a) = (f_k(a), k) = (f_n(b), n) = F(b)$, then $k = n$, and hence $f_k(a) = f_k(b)$. But since $f_k\colon A_k \to \mathbf{N}$ is a one-to-one function, we know that $a = b$. Thus F is a one-to-one function from A onto an infinite subset of $\mathbf{N} \times \mathbf{N}$. Therefore, A is denumerable.

If the sets $\{A_n | n \geq 1\}$ are not pairwise disjoint, then the relation F as defined above is not a function. In this case, we modify F as follows: For $a \in A$, find the least positive integer k such that $a \in A_k$, and define

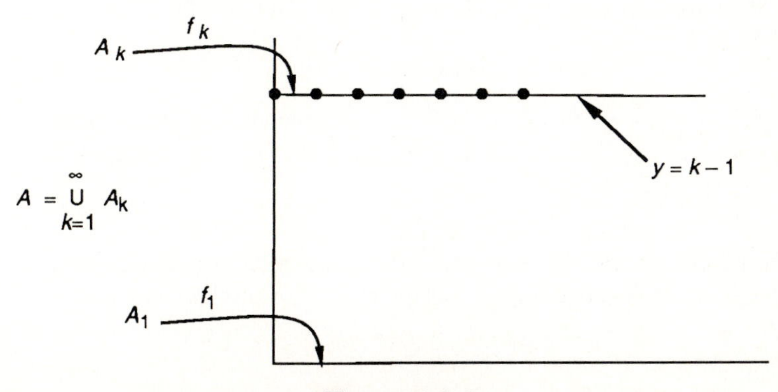

Figure 4.3.3

$F(a) = (f_k(a), k)$. Then, as an exercise, the reader can check that F is a one-to-one function from A onto a subset of $\mathbf{N} \times \mathbf{N}$. Thus A is denumerable in case the sets $\{A_n | n \geq 1\}$ are not pairwise disjoint.

We leave the proof of (ii) as an exercise. ■

The final result of this section asserts that any infinite set contains a denumerable set. In the language of cardinal numbers, this statement can be phrased as follows: For any infinite cardinal number α, $\aleph_0 \leq \alpha$. We are stating Theorem 4.3.8 at this point because of its reference to denumerable sets; its proof is deferred until Section 4.5.

Theorem 4.3.8. *If A is an infinite set, then there exists a subset $B \subseteq A$ such that B is denumerable.*

Let us close this section by reviewing the methods that have been used to prove that a given set A is denumerable. The first method is to verify the definition of denumerability by finding a one-to-one correspondence between A and $\mathbf{N}$. The second approach, based on Theorem 4.3.1, is to show that A is equinumerous with an infinite subset of a denumerable set. The third method, which relies on Theorems 4.3.6 and 4.3.7, is to represent A as a union of a finite or denumerable number of denumerable sets.

We illustrate the second and third methods by giving two proofs that the set, which we denote by $\mathbf{N}(2)$, of all two-element subsets of $\mathbf{N}$ is denumerable.

To show that $\mathbf{N}(2)$ is denumerable using Theorem 4.3.1, we show that $\mathbf{N}(2)$ is equinumerous with a subset of $\mathbf{N} \times \mathbf{N}$. To do so, we define a function $f: \mathbf{N}(2) \to \mathbf{N} \times \mathbf{N}$ as follows: If $A \in \mathbf{N}(2)$, then A can be written as $A = \{a, b\}$ where $a, b \in \mathbf{N}$ and $a < b$; define $f(A)$ to be the ordered pair (a, b). It is easy to see that if $A \neq B$, then $f(A) \neq f(B)$. Hence f is an injective function from $\mathbf{N}(2)$ to $\mathbf{N} \times \mathbf{N}$. Thus $\mathbf{N}(2)$ is equinumerous with an infinite subset of $\mathbf{N} \times \mathbf{N}$ and is therefore denumerable.

To illustrate the third method, we argue as follows: Let A_0 be the set of two-element subsets of $\mathbf{N}$ whose smallest element is 0. Then

$$A_0 = \{\{0, a\} \in \mathscr{P}(\mathbf{N}) | a \geq 1\}.$$

Clearly, $A_0 \approx \{n \in \mathbf{N} | n \geq 1\} \approx \mathbf{N}$, and hence A_0 is denumerable. Let A_1 be the set of two-element subsets of $\mathbf{N}$ whose smallest element is 1. Then

$$A_1 = \{\{1, a\} \in \mathscr{P}(\mathbf{N}) | a \geq 2\}.$$

As above, one can show that A_1 is denumerable. In general, for $k \geq 0$ let A_k be the set of two-element subsets of $\mathbf{N}$ whose smallest element is k. Then

$$A_k = \{\{k, a\} \in \mathscr{P}(\mathbf{N}) | a \geq k + 1\}$$

and A_k is denumerable.

Thus $\{A_k | k \geq 0\}$ is a denumerable collection of denumerable sets. Notice that $\mathbf{N}(2) = \bigcup_{k=0}^{\infty} A_k$ since if $\{n, m\}$ is a two-element subset of $\mathbf{N}$ with $n < m$, then $\{n, m\} \in A_n$. By Theorem 4.3.7, $\mathbf{N}(2)$ is denumerable.

Exercises §4.3

4.3.1. (a) Prove that $\mathbf{Z} \times \mathbf{N}$ is denumerable.
(b) Prove that $\mathbf{Z} \times \mathbf{Z}$ is denumerable.

4.3.2. (a) Let $A = \{a_0, a_1, a_2, \ldots\}$ be denumerable. Show $A - \{a_0\} \approx A$.
(b) Let A be any denumerable set and let $a \in A$. Show $A \approx A - \{a\}$.

4.3.3. (a) Show there exists a subset $A \subseteq \mathbf{N}$ such that $\mathbf{N} \times \mathbf{N} \times \mathbf{N} \approx A$.
(b) Show $\mathbf{N} \times \mathbf{N} \times \mathbf{N}$ is denumerable.
(c) Show that if $A_1, \ldots, A_n$ are denumerable, then $A_1 \times \cdots \times A_n$ is denumerable.

4.3.4. (a) For $k \in \mathbf{N}$, let $A_k = \{(k, n) \in \mathbf{N} \times \mathbf{N} | n \in \mathbf{N}\}$. Show A_k is denumerable.
(b) For $k \in \mathbf{N}^+$, let $D_k = \{r \in \mathbf{Q} | r = n/k$ for some $n \in \mathbf{Z}\}$. Show D_k is denumerable.

4.3.5. (a) Prove Theorem 4.3.6.
(b) Prove Theorem 4.3.7(ii).

4.3.6. Let A and B be sets. Define $A \sim B$ to mean $(A - B) \cup (B - A)$ is countable. Show that for all sets A, B, and C,
(a) $A \sim A$,
(b) if $A \sim B$, then $B \sim A$, and
(c) if $A \sim B$ and $B \sim C$, then $A \sim C$.

4.3.7. Show that if A is denumerable and B is uncountable, then $B - A$ is uncountable.

4.3.8. Let A be denumerable. Let B be any set such that there exists a surjective function $f: A \to B$. Prove B is countable.

4.3.9. (a) Let $A_2 = \{x \in \mathbf{R} | x \in \mathbf{Q}$ or $x^2 \in \mathbf{Q}\}$. Show A_2 is denumerable. (Hint: Write A_2 as a union of a finite number (i.e., two or three) of denumerable sets.)
(b) Let $A_3 = \{x \in \mathbf{R} | x \in \mathbf{Q}$ or $x^3 \in \mathbf{Q}\}$. Show A_3 is denumerable.
(c) Generalize the results of parts (a) and (b).

4.3.10. (a) Let $\mathscr{P}_1(\mathbf{Q}) = \{P(x) | P(x) = ax + b$ where $a, b \in \mathbf{Q}\}$. Thus $\mathscr{P}_1(\mathbf{Q})$ is the set of polynomials of degree at most 1 with rational coefficients. Prove $\mathscr{P}_1(\mathbf{Q})$ is denumerable.
(b) Let $\mathscr{P}_2(\mathbf{Q}) = \{P(x) | P(x) = ax^2 + bx + c$ where $a, b, c \in \mathbf{Q}\}$. Show $\mathscr{P}_2(\mathbf{Q})$ is denumerable.
(c) For each positive integer n let $\mathscr{P}_n(\mathbf{Q}) = \{P(x) | P(x) = a_n x^n + \cdots + a_1 x + a_0$ where $a_i \in \mathbf{Q}\}$ for $0 \leq i \leq n$. Show $\mathscr{P}_n$ is denumerable.
(d) Let Poly($\mathbf{Q}$) $= \{P(x) | P \in \mathscr{P}_n$ for some $n\}$. Thus Poly($\mathbf{Q}$) is the set of polynomials with rational coefficients. Show Poly($\mathbf{Q}$) is denumerable.

4.3.11. For each positive integer k, let $\mathbf{N}(k)$ denote the set of subsets of $\mathbf{N}$ having exactly k elements.
(a) Show that $\mathbf{N}(1) \approx \mathbf{N}$; hence $\mathbf{N}(1)$ is denumerable.
(b) Show that $\mathbf{N}(3)$ is denumerable.
(c) Show that for all $k \geq 1$, $\mathbf{N}(k)$ is denumerable.
(d) Let $\mathbf{N}_f$ be the set of all finite subsets of $\mathbf{N}$. Show that $\mathbf{N}_f$ is denumerable.

4.3.12. We outline another proof that $\mathbf{Z} \times \mathbf{Z}$ is denumerable. For $n \geq 0$, let $R(n) = \{(a, b) \in \mathbf{Z} \times \mathbf{Z} | \max\{|a|, |b|\} = n\}$. Thus $R(0) = \{(0, 0)\}$ and $R(1) = \{(1,0), (1,1), (0,1), (-1,1), (-1,0), (-1,-1), (0,-1), (1,-1)\}$.
(a) In $\mathbf{Z} \times \mathbf{Z}$ draw a picture of the sets $R(1)$, $R(2)$, $R(3)$.
(b) Show that $\bigcup_{n=0}^{\infty} R(n) = \mathbf{Z} \times \mathbf{Z}$.
(c) Show that $R(n) \cap R(m) = \varnothing$ if $n \neq m$.
(d) Show that $|R(0)| = 1$ and $|R(n)| = 8n$ for $n \geq 1$.
(e) Show that $\mathbf{Z} \times \mathbf{Z}$ is denumerable.

§4.4 Uncountable Sets

Thus far we have constructed a hierarchy of sets—the empty set, finite sets, and denumerable sets. Any set contains the empty set; any denumerable set contains finite sets of arbitrary size; and any infinite set contains a denumerable set. Along the way we have proved many theorems about the behavior under the various set operations of sets in each of these categories. But what happens beyond denumerable sets?

We have already introduced a name for such a set.

> **Definition 4.4.1** *Uncountable Set*
>
> A set A is *uncountable* if A is not countable.

The purpose of this section is to show that uncountable sets do indeed exist. Specifically, we prove that *set of real numbers is uncountable*. This remarkable theorem was proved by Cantor in his 1874 paper on set theory. We prove that **R** is uncountable by showing that the open interval $(0, 1)$ is uncountable. Since $\mathbf{R} \approx (0, 1)$ (Example 3.1.5), **R** is also uncountable.

To prove that $(0, 1)$ is uncountable, it is necessary to have a workable representation of the elements of the real number system. For our purposes, the most useful representation of a real number is the decimal representation. Let x be a real number in $(0, 1)$. Then x can be represented in the form

$$x = .a_1 a_2 a_3 \cdots a_n \cdots$$

where, for $i \geq 1$, the digit a_i is an integer in the set $\{0, 1, 2, 3, 4, 5, 6, 7, 8, 9\}$. This representation can be interpreted geometrically as in Chapter 2 or in the following analytic fashion: To say that $x = .a_1 a_2 a_3 \cdots a_n \cdots$ is to say that the infinite series

$$\sum_{i=1}^{\infty} a_i/10^i = a_i/10 + a_2/10^2 + \cdots + a_n/10^n + \cdots$$

converges to x. Thus $.12333\cdots = 1/10 + 2/10^2 + 3/10^3 + 3/10^4 + 3/10^5 + \cdots$. While each real number has a decimal expansion, certain real numbers have more than one such expansion. For example, the real number

$$\begin{aligned} .1\overline{9} &= .1999 = 1/10 + 9/10^2 + 9/10^3 + \cdots \\ &= 1/10 + \sum_{i=2}^{\infty} 9/10^i = 1/10 + (9/10^2)/(1 - 1/10) \\ &= 1/10 + 1/10 = 1/5. \end{aligned}$$

(We are using the well-known formula for the sum of a geometric series.) Thus $.2 = .1999 \cdots$. In general, any real number x in $(0, 1)$ has a unique decimal representation unless

(1) $x = .a_1 \cdots a_n\overline{9}$ where $a_n \neq 9$, in which case
(2) $x = .a_1 \cdots (a_n + 1)\overline{0} \cdots$.

The decimal representation of a real number will play a crucial role in our proof that $(0, 1)$ is uncountable.

How do we show that a set is uncountable? By definition, a set is uncountable if it is not countable. Since the concept of uncountability is defined via negation, we should consider an indirect proof whenever we are trying to prove that a given set A is uncountable: Assume A is countable and derive a contradiction. Proof by contradiction is an often-used technique to demonstrate the uncountability of a given set. Of course, once a set A has been proved to be uncountable, then a second approach to demonstrating uncountability is available: If A is uncountable, then a given set B is uncountable provided that $B \approx A$. Now to the principal result of this section.

Theorem 4.4.1 *The interval* $(0, 1)$ *is uncountable.*

First Proof Suppose $(0, 1)$ is countable. Then, since $(0, 1)$ is infinite, $(0, 1)$ is denumerable and $(0, 1) \approx \mathbf{N}^+$. Then there exists a one-to-one correspondence $f: \mathbf{N}^+ \to (0, 1)$. Letting, for $n \in \mathbf{N}^+$, $a_n = f(n)$, we have $(0, 1) = \{a_1, a_2, \ldots, a_n, \ldots\}$.

For each $n \in \mathbf{N}$, we write a_n in decimal form:

$$\begin{aligned} a_1 &= .a_{11}a_{12} \cdots \\ a_2 &= .a_{21}a_{22} \cdots \\ &\vdots \\ a_n &= .a_{n1}a_{n2} \cdots a_{nn} \cdots \end{aligned}$$

where $a_{ij} \in \{0, 1, 2, 3, 4, 5, 6, 7, 8, 9\}$. Also, for those a_m that have two decimal representations, we choose the terminating form as in equation (2) above.

Our strategy is to produce a number $x \in (0, 1)$ that is distinct from a_n for all $n \geq 1$. But since $(0, 1) = \{a_n | n \geq 1\}$ and $x \neq a_n$ for all $n \geq 1$, $x \notin (0, 1)$. This contradiction means that the assumption that $(0, 1)$ is countable is false.

Now how can we produce a number $x \in (0, 1)$ such that $x \neq a_n$ for all $n \geq 1$? Here is the basic idea: We shall define x by prescribing the digits in its decimal expansion. To be certain that $x \neq a_1$, we shall choose the first digit of x to be unequal to the first digit of a_1. To ensure that $x \neq a_2$, we chose the second digit of x to be unequal to the second digit of a_2, and so on. Now the details.

We define the number x by giving its decimal expansion; $x = .b_1b_2 \cdots b_n \cdots$ where the digits b_n for $n \geq 1$ are chosen as follows:

1. b_1 is an integer, $1 \leq b_1 \leq 8$, and $b_1 \neq a_{11}$.
2. b_2 is an integer, $1 \leq b_2 \leq 8$, and $b_2 \neq a_{22}$.
3. In general for $n \geq 1$, b_n is an integer, $1 \leq b_n \leq 8$, and $b_n \neq a_{nn}$. (In other words, the digit b_n differs from the nth digit of a_n and is different from 0 and 9.)

We claim that $x \in (0, 1)$ and $x \neq a_n$ for all $n \geq 1$. By its very construction, $x \in (0, 1)$. We now show $x \neq a_n$ for all $n \geq 1$. Suppose to the contrary that $x = a_m$ for some integer $m \geq 1$. Then $.b_1b_2 \cdots b_m \cdots = x = a_m = .a_{m1}a_{m2} \cdots a_{mm} \cdots$; hence $b_i = a_{mi}$ for all $i \geq 1$. But by the definition of x, $b_m \neq a_{mm}$. This contradiction shows that $x \neq a_m$. Therefore, $x \neq a_n$ for all n and $x \notin \{a_n | n \geq 1\} = (0, 1)$.

Therefore, we have reached the contradiction that $x \in (0, 1)$ and $x \notin (0, 1)$. Thus $(0, 1)$ must indeed be uncountable. ∎

Second Proof This proof is a slight variation on the preceding argument. The change comes in the definition of x: Define $x = .b_1b_2 \cdots b_n \cdots$ where for $n \geq 1$, $b_n = 4$ if $a_{nn} \neq 4$ and $b_n = 5$ if $a_{nn} = 4$. (The choice of the pair 4 and 5 is irrelevant; any pair of integers between 1 and 8 will do.) Just as in the first proof it follows that $x \in (0, 1)$, yet $x \neq a_n$ for all n.

Remarks Some further comments on the choice of the b's: First, we want to be certain that $x \neq 0$, so we choose each $b_n \neq 0$. This is a bit of overkill since we only need one of the $b_m \neq 0$ to ensure that $x \neq 0$. Second, we take $b_m \neq 9$ for all m so as to prevent x from having the form $x = .b_1 \cdots b_n\overline{9}$, for in that case $x = .b_1 \cdots (b_n + 1)\overline{0}$ and x might equal one of the a_k.

Corollary 4.4.2 **R** *is uncountable.*

Proof By Example 4.1.5, $(0, 1) \approx \mathbf{R}$. If $\mathbf{R}$ is countable, then $(0, 1)$ is countable, which contradicts Theorem 4.4.1. ∎

Back in Section 4.1 we asked: Is it true that $\aleph_0 = |\mathbf{N}| < |\mathbf{R}|$? We now see that the answer is yes, for **N** is equinumerous with a subset of **R**, i.e., with **N** itself; on the other hand, **R** is not equinumerous with a subset of **N**, as otherwise **R** would be denumerable. As a result, we know that more than one infinite cardinal number exists: The cardinal numbers $\aleph_0$ and $|\mathbf{R}|$ are distinct.

Since the cardinal number $|\mathbf{R}|$ occurs so frequently in mathematics, it has acquired a name and a special symbol.

Definition 4.4.2 *Cardinality of the Continuum*

The cardinal number $|\mathbf{R}|$ is called the *cardinal number of the continuum* and is denoted by c.

Corollary 4.4.3 *Let* Irr $= \mathbf{R} - \mathbf{Q}$ *be the set of irrational real numbers. Then* Irr *is uncountable.*

Proof Again we argue by contradiction. Suppose Irr is countable. Since **Q** is countable, $\mathbf{Q} \cup \text{Irr}$ is countable by Theorem 4.3.5. But $\mathbf{Q} \cup \text{Irr} = \mathbf{R}$, which means that **R** is countable, a contradiction of Corollary 4.4.2. Thus Irr is uncountable. ■

The argument used in the proof of Theorem 4.4.1 is known, for obvious reasons, as the *diagonal* argument or *Cantor's diagonal argument*. As mentioned earlier, Cantor first proved **R** is uncountable in 1874. In 1891 he gave another proof in which the diagonal argument is introduced. Since Cantor's work, the diagonal argument has taken on a life of its own, becoming a significant proof technique in mathematical logic and theoretical computer science.

To master any technique we must of course know how to use it; at the same time, however, it is important for us to understand the limitations of the method. What are the limitations of the diagonal method? In particular, suppose we attempt to use the diagonal argument to prove that **Q** is uncountable. Since **Q** is countable, the argument must have a hole in it. Exercise: Find the hole.

Cantor's work with infinite sets did not end with his theorem on the uncountability of **R**. In fact, that theorem marks the beginning of a flood of significant theorems and conjectures about sets of real numbers and set theory in general. For example, Cantor proved an important theorem about algebraic and transcendental numbers. A real number α is *algebraic* if there exists a nonzero polynomial function $f(x) = a_0 + a_1x + \cdots + a_nx^n$ with rational coefficients (i.e., each a_i is rational) such that $f(\alpha) = 0$. For example, $\sqrt{2}$ is an algebraic number since if $f(x) = x^2 - 2$, then $f(\sqrt{2}) = 0$.

Also, any rational number is algebraic: If $r \in \mathbf{Q}$, then $f(r) = 0$ where $f(x) = x - r$. A real number that is not algebraic is called *transcendental*. It is difficult to find explicit examples of transcendental numbers. For instance, it was not until 1873 that C. Hermite proved that e, the base of the natural logarithms, is transcendental. Nine years later, F. Lindemann proved that π is transcendental. The first known example of a transcendental number was provided by the French mathematician J. Liouville in 1844 when he proved a general theorem that implies, for instance, that the number

$$\sum_{1}^{\infty} \frac{1}{10^{n!}} = \frac{1}{10} + \frac{1}{10^{2!}} + \frac{1}{10^{3!}} + \frac{1}{10^{4!}} + \cdots$$

$$= .1100010 \cdots$$

is transcendental.

Given the difficulty of the results of Liouville and Hermite, Cantor's proof, published in his 1874 paper, that the set of transcendental numbers is uncountable stunned the mathematicians of his day. Cantor's proof follows the lines of Corollary 4.4.3. He first proved that the set of algebraic numbers is countable. Since the union of the set of algebraic and transcendental real numbers is all of $\mathbf{R}$, the set of transcendental numbers must be uncountable. Thus, in a sense, "most" real numbers are transcendental since the set of algebraic numbers is merely a countable subset of the uncountable set of real numbers. This theorem convinced many hitherto skeptical mathematicians (among them the well-known German mathematician Richard Dedekind) of the usefulness of Cantor's notions of countability and uncountability.

Shortly thereafter, Cantor turned to the consideration of higher-dimensional Euclidian spaces. Recall that for each positive integer n, $\mathbf{R}^n$ is defined to be

$$\mathbf{R}^n = \{(x_1, \ldots, x_n) | x_i \in \mathbf{R}\}.$$

We can prove rather straightforwardly that $\mathbf{R}^n$ is uncountable, but a more subtle question arises: Can $\mathbf{R}^n$ be put into one-to-one correspondence with $\mathbf{R}$? In other words, is $\mathbf{R}^n \approx \mathbf{R}$? Even for the case $n = 2$ the answer to the question is not obvious. Some mathematicians believed that the answer must be in the negative because surely two independent variables (which are needed to describe $\mathbf{R}^2$) cannot be reduced to one variable. Cantor, however, proved otherwise: For each positive integer n, $\mathbf{R}^n \approx \mathbf{R}$. This theorem led to other interesting theorems and eventually, in the twentieth century, to the development of a new area of specialization within mathematics—dimension theory.

We now outline a proof of a special case of the theorem mentioned in the previous paragraph by showing that the Euclidean plane $\mathbf{R}^2$ has cardinality c. We let $I_1 = \{x \in \mathbf{R} | 0 < x \leq 1\} = (0, 1]$. Thus $I_1 \times I_1 \approx (0, 1] \times (0, 1] \approx \mathbf{R} \times \mathbf{R}$. Therefore, the next theorem implies that $\mathbf{R} \approx \mathbf{R} \times \mathbf{R}$.

Theorem 4.4.4 $I_1 \times I_1 \approx I_1$.

Proof (Outline) For each $x \in (0, 1]$ we take the decimal expansion of x, $x = .a_1 a_2 \cdots a_n \cdots$; this time, however, we consider only those expansions that do not terminate in 0's. Thus, for example, we write $1/2 = .4\bar{9}$ and not $1/2 = .5\bar{0}$.

We divide the expansion of $x = .a_1 a_2 \cdots a_n \cdots$ into blocks where each block consists of either a single nonzero digit or a string of 0's and a nonzero digit. For example, if $x = .1201003806 \cdots$, the first six blocks are $1, 2, 01, 003, 8, 06$.

We now define a function $\phi: I_1 \times I_1 \to I_1$. If $(x, y) \in I_1 \times I_1$, then we write the decimal expansions of x and y, divide each expansion into blocks according to the recipe described in the previous paragraph: We form a real number z by alternately juxtaposing blocks from x and y. For instance, if

$$x = .1201003806 \cdots$$

and

$$y = .0870311 \cdots$$

then

$$z = .10827010300318106 \cdots .$$

This process defines ϕ as a function from $I_1 \times I_1$ to I_1. We can prove that ϕ is a one-to-one correspondence; we leave this step as an exercise. ∎

Cantor's announcement in 1877 that $\mathbf{R}^2 \approx \mathbf{R}$ stunned the mathematical world. (Moreover, the result apparently stunned Cantor himself, for he wrote to Dedekind: "I see it but I don't believe it.") Although still a young man, Cantor had by this time come to be recognized as one of the leading mathematicians of his day. Nevertheless, as indicated earlier, his work in set theory was far from finished.

Cantor considered a number of other problems, among them the question of how two sets can be compared with respect to size. In particular, he conjectured for sets an analog of the trichotomy property of real numbers. Cantor conjectured that for any two sets A and B either

1. $A \approx B$,
2. $A \approx C$ where $C \subset B$, and for all $D \subseteq A$, $D \not\approx B$, or
3. $B \approx D$ where $D \subset A$, and for all $C \subseteq B$, $A \not\approx C$.

In terms of cardinal numbers this conjecture asserts that if α and β are cardinal numbers, then either $\alpha = \beta$ $\alpha < \beta$, or $\beta < \alpha$. In other words, any two cardinal numbers are comparable with respect to $<$. A result that is logically equivalent to the comparability property of cardinal numbers is the Cantor–Bernstein (also known as the Schroeder–Bernstein) Theorem. As it happens, the assertion that any two sets are comparable is also equivalent to the Axiom of Choice, one of the most important and controversial statements in set theory. These three topics, the comparability property for sets, the Cantor–Bernstein Theorem, and the Axiom of Choice, constitute the focus of the next section.

Exercises §4.4

4.4.1. Prove that if A is uncountable and $B \supseteq A$, then B is uncountable.

4.4.2. Prove: A is uncountable if and only if $A \times A$ is uncountable.

4.4.3. (a) Prove: $(0, 1]$ is uncountable.
(b) Prove: $[0, 1]$ is uncountable.
(c) Prove: If $a < b$ where $a, b \in \mathbf{R}$, then (a, b) is uncountable.

4.4.4. Let $I' = \{x \mid 0 < x < 1 \text{ and } x \text{ is irrational}\}$. Is I' countable or uncountable?

4.4.5. Prove: $\{x \in (0,1) \mid x = .a_1 a_2 \cdots a_n \cdots \text{ where each } a_i = 3 \text{ or } 8\}$ is uncountable.

4.4.6. Use Cantor's diagonal argument to show that $\{x \mid 0 < x < 1$ and each digit in the decimal expansion of x is either 0 or 1$\} = \{x = .a_1 a_2 \cdots a_n \cdots \mid x \neq 0 \text{ and each } a_i = 0 \text{ or } 1\}$ is uncountable.

4.4.7. Prove or disprove: $\{x \in (0,1) \mid x = .a_1 a_2 \cdots a_n \cdots$ where each $a_i = 0$ or 1 and only finitely many $a_i = 0\}$ are uncountable.

4.4.8. Prove or disprove: $\{x \in (0,1) \mid x = .a_1 a_2 \cdots a_n \cdots$ where $0 \leq a_i \leq 9$ and only finitely many $a_i = 0\}$ are uncountable.

4.4.9. (a) Prove that $\mathbf{R}^3 \approx \mathbf{R}$.
(b) Prove that for every positive integer n, $\mathbf{R}^n \approx \mathbf{R}$.
(c) Prove that if $A_i \approx \mathbf{R}$ for $1 \leq i \leq n$, then $A_1 \times \cdots \times A_n \approx \mathbf{R}$.

4.4.10. Define a relation $\sim$ on $\mathbf{R}$ as follows: For $x, y \in \mathbf{R}$, $x \sim y$ if $x - y \in \mathbf{Q}$.
(a) Show: $\sim$ is an equivalence relation on $\mathbf{R}$.
(b) Show: For every $x \in \mathbf{R}$, $[x]_\sim$ is denumerable.
(c) What can be said about the cardinality of the set of equivalence classes?

4.4.11. Let $A = \{x \in (0,1) \mid x = .a_1 a_2 \cdots a_n \cdots$ such that $a_i = 0$ or 1 and no two consecutive $a_i = 0\}$. (Thus $.\overline{10} \in A$ while $.100\overline{1} \notin A$.) Prove or disprove: A is uncountable.

4.4.12. Let $A = \{x \in (0,1) \mid x = .a_1 a_2 \cdots a_n \cdots$ such that each $a_i = 1$, 2, or 3 and for $i \geq 1$, $a_i \neq a_{i+1}\}$. Prove or disprove: A is uncountable.

§4.5 Cardinal Numbers

We close this chapter by taking a closer look at cardinal numbers. Recall that a cardinal number is defined to be all sets equinumerous with a given set. As we noted earlier, this definition is informal, if not imprecise. We usually insist that any mathematical object be defined either as a set or as a member of a set; because we are hesitant to speak of the set of all sets equinumerous with a given set, the exact nature of a cardinal number remains unspecified. Thus we simply say that if A is a set, then the *cardinal number of A* or the *cardinality* of A, written $|A|$, comprises all sets B such that $B \approx A$.

Earlier in this chapter we saw that to each positive integer n, there corresponds a cardinal number, $|\mathbf{N}_n| = |\{1, \ldots, n\}|$. Moreover, by the Pigeonhole Principle, if $n, m \in \mathbf{N}^+$ with $n \neq m$, then $\mathbf{N}_n \not\approx \mathbf{N}_m$ and hence the cardinal numbers $|\mathbf{N}_n|$ and $|\mathbf{N}_m|$ are distinct. If A is an arbitrary nonempty finite set, then, by definition, $A \approx \mathbf{N}_n$ for some $n \in \mathbf{N}^+$, and in this case we write $|A| = |\mathbf{N}_n| = n$.

We also defined an ordering, $<$, on cardinal numbers by saying $|A| < |B|$ if A is equinumerous with some subset of B but B is not equinumerous with any subset of A. We then investigated the cardinality of infinite sets. In particular, we asked if any two infinite sets are equinumerous and answered this question in the negative by showing that $|\mathbf{N}| = \aleph_0 < c = |\mathbf{R}|$.

In this section we consider the following two questions about infinite cardinal numbers:

1. For each infinite set A, is there an infinite set B such that $|A| < |B|$?

If not, then there exists an infinite set A_0 such that for all sets B, $|B| \leq |A_0|$. Thus a negative response to the question asserts that there exists a largest infinite set. Clearly an affirmative answer implies that there exists no largest infinite set. We can therefore restate the original question as follows: Is there a largest infinite set?

2. Are two cardinal numbers comparable? Specifically if A and B are infinite sets, then is it true that either $|A| < |B|$, $|A| = |B|$ (i.e., $A \approx B$), or $|B| < |A|$?

Power Sets and the Cantor – Bernstein Theorem

We begin by considering the first question. How can we attack this problem? One method is that of analogy. Recall a question often asked by children: Is there a biggest number? The most common response is to say: Take any number, add 1 to that number, and you will get a bigger number. Thus there is no biggest number.

We can try the same idea with sets: Given an infinite set A, try to obtain a set that is larger than A by joining to A an element not already in A. As a specific example, take $A = \mathbf{N} = \{0, 1, 2, \cdots\}$ and consider $\mathbf{N} \cup \{-1\} = \{-1, 0, 1, 2, \cdots\}$. The function $f: \mathbf{N} \to \mathbf{N} \cup \{-1\}$ given by $f(x) = x - 1$ is clearly a bijection; hence $\mathbf{N} \approx \mathbf{N} \cup \{-1\}$. In general, if A is a denumerable set and x is any object, then $A \approx A \cup \{x\}$. (See Exercise 4.5.1.) Thus the process of adjoining a new element to a denumerable set produces a denumerable set, i.e., a set of the same cardinality as A.

What happens when A is an arbitrary infinite set? Is $A \approx A \cup \{x\}$ for any x? Certainly, $A \approx A \cup \{x\}$ if $x \in A$. Suppose $x \notin A$. If we assume A contains a denumerable set B, then we can reason as follows: Since B is denumerable, $B \approx B \cup \{x\}$. Also, since $(A - B) \cap B = \varnothing$, $(A - B) \cup B \approx (A - B) \cup B \cup \{x\}$ by Theorem 4.1.2. Therefore, $A \approx A \cup \{x\}$. Thus, if it is true that every infinite set contains a denumerable subset, then for each infinite set A, $A \approx A \cup \{x\}$. We return to the question of whether every infinite set contains a denumerable subset later in this section.

Let us now return to the challenge of trying to find for each infinite subset A a set B such that $|A| < |B|$. As we have just seen, the set $A \cup \{x\}$ has the same cardinality as A (at least if A is denumerable). A simple modification of this idea might be fruitful. Perhaps the set $\{x\}$ is just not large enough. Suppose instead we take a set A', which is disjoint from A yet has the same cardinality as A, and then let $B = A \cup A'$. Perhaps $|A| < |B|$.

First, how do we find such a set A'? One way is to let $A' = A \times \{1\} = \{(a, 1) | a \in A\}$. Then $A \approx A'$ and $A \cap A' = \varnothing$. Now let $B = A \cup A'$. Is $|A| < |B|$?

We can answer this question in at least one specific case. Let $A = \mathbf{N}$. Then, since $A \approx A'$, both A and A' are denumerable and by Theorem 4.3.5(i), $B = A \cup A'$ is also denumerable.

The moral of these attempts seems to be: Given an infinite set, it does not appear to be possible to produce a set of larger cardinality using the operation of $\cup$. It is natural to wonder if we can use any of the other set operations to produce sets of larger cardinality. The other basic set operations are intersection, complement, and power set. Let us look closely at the power set of a given set.

The only result that we have about the size of a power set is Theorem 2.2.4: If $|A| = n$, then $|\mathscr{P}(A)| = 2^n$. As noted in Section 2.2, 2^n is significantly larger than n in the sense that $2^n/n \to \infty$ as $n \to \infty$. From this interpretation of Theorem 2.2.4, one might indeed believe that for any set A, $|A| < |\mathscr{P}(A)|$.

This suspicion is essentially confirmed in Theorem 4.5.1. Let us pause to gather some evidence that supports the statement of Theorem 4.5.1. The argument that follows will be sketchy and may be skipped by readers eager to see the proof of the next theorem.

We consider the special case that $A = \mathbf{N}$. We want to show that $|\mathbf{N}| < |\mathscr{P}(\mathbf{N})|$. In fact, we outline a proof of a stronger statement: $[0, 1] \approx$ a subset of $\mathscr{P}(\mathbf{N})$. Since $|\mathbf{N}| < |[0, 1]|$, we have $|\mathbf{N}| < |\mathscr{P}(\mathbf{N})|$.

1. As in Section 2.1 we present each subset of $\mathbf{N}$, i.e., each element of $\mathscr{P}(\mathbf{N})$ as a binary sequence; in this case, the sequence representing a subset $B \subseteq \mathbf{N}$ has denumerably many entries: $b_0 b_1 b_2 \cdots b_n \cdots$ where for each $i \in \mathbf{N}$,

$$b_i = \begin{cases} 1 & \text{if } i \in B \\ 0 & \text{if } i \notin B \end{cases}.$$

Thus there is a one-to-one correspondence between $\mathscr{P}(\mathbf{N})$ and binary sequences that are denumerably long.

2. To each binary sequence of denumerable length, we associate a real number in $[0, 1]$ as follows: Assign to the sequence $b_0 b_1 b_2 \cdots$ the real number $x = b_0/2 + b_1/2^2 + b_2/2^3 + \cdots$ (in other words, $b_0, b_1, b_2, \cdots$ are the digits in the binary expansion of x). For instance, the set $E = \{n \in \mathbf{N} \mid n \text{ is even}\}$ has the sequence $101010 \cdots$, which corresponds to

$$1/2 + 1/2^3 + 1/2^5 + \cdots = (1/2)/(1 - 1/2^2) = 2/3.$$

3. Conversely to each $x \in [0, 1]$ we can assign a binary sequence of denumerable length consisting of the digits in the binary expansion of x. In fact, nearly every $x \in [0, 1]$ corresponds to a unique binary sequence of denumerable length, the exceptions being certain rational numbers which correspond to two different sequences. For example, $1/2$ corresponds to $1000 \cdots$ and $01111 \cdots$ since

4. $1/2 = 1/4 + 1/8 + \cdots + 1/2^n + \cdots$.

To summarize, $\mathscr{P}(\mathbf{N})$ is equinumerous with the set of binary sequences of denumerable length and the latter set contains a subset equinumerous with $[0, 1]$. Thus

5. $|\mathbf{R}| = |(0, 1)| \leq |[0, 1]| \leq \mathscr{P}(\mathbf{N})|$.

Since $|\mathbf{N}| < |\mathbf{R}|$, $|\mathbf{N}| < |\mathscr{P}(\mathbf{N})|$.

Remark We can in fact show that $|\mathbf{R}| = |\mathscr{P}(\mathbf{N})|$.

Now let us turn to Theorem 4.5.1.

Theorem 4.5.1 *For any set A, $A \not\approx \mathscr{P}(A)$.*

Proof We argue by contradiction. Suppose $A \approx \mathscr{P}(A)$. Then there exists a one-to-one correspondence $f\colon A \to \mathscr{P}(A)$. Hence, for each $a \in A$, $f(a)$ is a subset of A. Define $B = \{a \in A \mid a \notin f(a)\}$.

(Before continuing with the proof, let us record a few observations. First, with respect to the definition of B, since $f(a) \subseteq A$, it makes sense to ask if $a \notin f(a)$. Thus, the set B is a legitimately defined subset of A. To develop a feel for B, consider the element $a_1 \in A$ such that $f(a_1) = \varnothing$. Since $f\colon A \to \mathscr{P}(A)$ is a one-to-one correspondence, such an element a_1 exists. Since $a_1 \notin \varnothing = f(a_1)$, $a_1 \in B$. On the other hand, consider the element $a_2 \in A$ such that $f(a_2) = A$. Since $a_2 \in A = f(a_2)$, it follows that $a_2 \notin B$. Thus B is a nonempty subset of A which is not equal to A.)

Since f is a one-to-one correspondence, there exists $a_0 \in A$ such that $B = f(a_0)$. We ask: Is $a_0 \in B$?

If $a_0 \in B$, then by the defining property of B, $a_0 \notin f(a_0) = B$. Thus if $a_0 \in B$, then $a_0 \notin B$.

If $a_0 \notin B = f(a_0)$, then by the definition of B, $a_0 \in B$. Thus if $a_0 \notin B$, then $a_0 \in B$.

In either case we have obtained a contradiction, and hence no one-to-one correspondence exists between A and $\mathscr{P}(A)$. ∎

Remark Note the similarity between the proof of Theorem 4.5.1 and the argument in Russell's paradox.

What does Theorem 4.5.1 say in terms of cardinal numbers? First, since $A \not\approx \mathscr{P}(A)$, $|A| \neq |\mathscr{P}(A)|$. In addition, there exists an injective function $f\colon A \to \mathscr{P}(A)$, namely $f(a) = \{a\}$, which tells us that A is equinumerous with a subset of $\mathscr{P}(A)$. Thus it is possible that $|A| < |\mathscr{P}(A)|$. To justify this assertion, however, we have to show that there exists no one-to-one correspondence between $\mathscr{P}(A)$ and a subset of A. The next theorem allows us to demonstrate this fact.

Theorem 4.5.2 *Cantor–Bernstein. Let A and B be sets. Suppose $A \approx B'$ where $B' \subseteq B$ and $B \approx A'$ where $A' \subseteq A$. Then $A \approx B$.*

Remarks (1) In words, Theorem 4.5.2 states that if A and B are each equinumerous with a subset of the other, then A and B are themselves equinumerous.

(2) Let us show how Theorem 4.5.2 applies to the cardinality of A and $\mathscr{P}(A)$. We know A is equinumerous with a subset of $\mathscr{P}(A)$. If $\mathscr{P}(A) \approx B$ where $B \subseteq A$, then by the Cantor–Bernstein Theorem, $A \approx$

$\mathcal{P}(A)$. But this conclusion contradicts Theorem 4.5.1. Therefore, for each $B \subseteq A$, $\mathcal{P}(A) \not\approx B$. Hence $|A| < |\mathcal{P}(A)|$.

(3) The Cantor–Bernstein Theorem is often referred to as the Schroeder–Bernstein Theorem in the mathematical literature. As van Dalen [3] points out, this theorem was first proved in the 1890s by Cantor in a special case and conjectured by Cantor in the general case. Previously (in 1887), however, Dedekind had already given a proof of the theorem. In 1896 Schroeder and Bernstein independently gave proofs of Theorem 4.5.2, but only Bernstein's was correct. Much later (in 1908) a proof was published by Zermelo. We follow van Dalen in calling this result the Cantor–Bernstein Theorem.

Proof of Theorem 4.5.2 Since $A \approx B'$ and $B \approx A'$, there exist bijections $f\colon A \to B' \subseteq B$ and $g\colon B \to A' \subseteq A$. We shall construct a bijection $h\colon A \to B$.

Let $f^{-1}\colon B' \to A$ and $g^{-1}\colon A' \to B$ be the inverses of f and g, respectively. Let $x \in A$. If $x \in A'$, then we can apply g^{-1} to x to obtain $g^{-1}(x) \in B$. We call $g^{-1}(x)$ the first ancestor of x. If $g^{-1}(x) \in B'$, then we can obtain $f^{-1}(g^{-1}(x)) \in A$, which we call the second ancestor of x. If $f^{-1}(g^{-1}(x)) \in A'$, then we can obtain $g^{-1}(f^{-1}(g^{-1}(x))) \in B$, which we call the third ancestor of x. Notice that each element of $A - A'$ has zero ancestors.

For each $x \in A$, one of the following three possibilities holds:

1. x has infinitely many ancestors.
2. x has a last ancestor that is a member of A.
3. x has a last ancestor that is a member of B.

Define
$$\begin{aligned} A_i &= \{x \in A \mid x \text{ has infinitely many ancestors}\}, \\ A_0 &= \{x \in A \mid x \text{ has an even number of ancestors}\}, \\ A_1 &= \{x \in A \mid x \text{ has an odd number of ancestors}\}. \end{aligned}$$

(Note that $A - A' \subseteq A_0$.) Clearly, $A = A_i \cup A_0 \cup A_1$ and $A_i \cap A_0 = A_i \cap A_1 = A_0 \cap A_1 = \varnothing$. Define subsets B_i, B_0, and B_1 in analogous fashion. We leave it for the reader to check that f sends A_i onto B_i and A_0 onto B_1 while g^{-1} sends A_1 onto B_0. Thus $A_i \approx B_i$, $A_0 \approx B_1$, and $A_1 \approx B_0$. (See Figure 4.5.1.)

We are now ready to define $h\colon A \to B$: Let $x \in A$. Then

$$h(x) = \begin{cases} f(x) & x \in A_i \cup A_0 \\ g^{-1}(x) & x \in A_1 \end{cases}.$$

Since $f\colon B_i \to A_i$, $f\colon A_0 \to B_1$, and $g^{-1}\colon A_1 \to B_0$ are bijections, h is a bijection. ∎

We now give an application of Cantor–Bernstein.

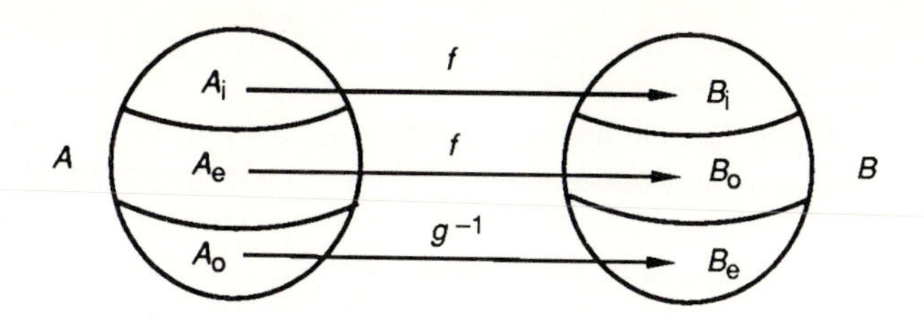

Figure 4.5.1

Corollary 4.5.3 $(0, 1) \approx [0, 1]$.

Proof First note $(0, 1) \subseteq [0, 1]$. Next define $f: [0, 1] \to (0, 1)$ by $f(x) = (1/2)x + 1/4$. It follows easily that f is a one-to-one correspondence from $[0, 1]$ onto $[1/4, 3/4] \subseteq (0, 1)$. Since each of $(0, 1)$ and $[0, 1]$ is equinumerous with a subset of the other, $(0, 1) \approx [0, 1]$. ∎

Comparability and the Axiom of Choice

We now consider the second question raised at the start of this section: Are any two cardinal numbers comparable? Given any two sets A and B, can we show that either $|A| < |B|$, $|A| = |B|$, or $|B| < |A|$?

Let us attack this question in a straightforward fashion. We shall try to construct a function $f: A \to B$ that is injective. If we succeed, then either $|A| < |B|$ or $|A| = |B|$. If we fail, then perhaps we can show that $|B| < |A|$.

How can we define a function $f: A \to B$? The most direct way is that we choose $x_1 \in A$ and $y_1 \in B$ and define $f(x_1) = y_1$. Next, assuming that $A - \{x_1\} \neq \varnothing$ and $B - \{y_1\} \neq \varnothing$, we choose $x_2 \in A - \{x_1\}$ and $y_2 \in B - \{y_1\}$ and define $f(x_2) = y_2$. Continuing in this manner, we construct at each stage a set $A' \subseteq A$ and a set $B' \subseteq B$ such that f is a bijection from A' to B'. At each step there are several possibilities:

1. $A' = A$ and $B' = B$. In this case, $A \approx B$ and $|A| = |B|$.
2. $A' = A$ and $B' \neq B$. Then A is equinumerous with a subset of B and either $|A| < |B|$ or $|A| = |B|$, the latter possibility occurring if B is equinumerous with a subset of A.
3. $A' \neq A$ and $B' = B$. In this case, the function $f^{-1}: B' \to A'$ is a bijection from $B' = B$ to a subset of A, and hence either $|B| < |A|$ or $|B| = |A|$.
4. $A' \neq A$ and $B' \neq B$. In this situation, we choose $x \in A - A'$ and $y \in B - B'$ and extend f to a function $f: A' \cup \{x\} \to B' \cup \{y\}$ by defining $f(x) = y$.

A natural question arises: Does this process ever end in the sense that at some stage $A' = A$ or $B' = B$? It would appear that since A' and B' are obtained by adding one element at a time, both A' and B' are finite sets. If A and B happen to be infinite, then A' and B' will never equal A and B, respectively. Thus the approach indicated above seems to be inadequate for the task of proving the comparability property for cardinal numbers.

Nevertheless, an argument based on this approach can be formulated. This argument, however, uses a powerful and controversial assertion about sets, called the Axiom of Choice (abbreviated AC). The Axiom of Choice can be stated informally as follows: Given an arbitrary family of nonempty sets, an element can be chosen from each set. Technically AC has the following form:

Axiom of Choice Let A be any set. Let $\{X_a | a \in A\}$ be a family of nonempty sets indexed by A. Then there exists a function $F: A \to \bigcup_{a \in A} X_a$ such that for each $a \in A$, $F(a) \in X_a$.

It is interesting to note that the Axiom of Choice was used in the first proof of the uncountability of $(0, 1)$. In that case, the set A is $\mathbf{N}$, the set of natural numbers. For each $n \in \mathbf{N}$, the set X_n is $\{1, 2, \ldots, 8\} - \{a_{nn}\}$. The function $F: \mathbf{N} \to \bigcup_{n=0}^{\infty} X_n$ selects for each $i \in \mathbf{N}$ an element $F(i) = b_i \in X_i = \{1, \ldots, 8\} - \{a_{ii}\}$. Notice that the second proof of Theorem 4.4.1 does not use the Axiom of Choice.

The function F is often called a *choice function* since for each $a \in A$, F *chooses* an element $F(a) \in X_a$. On the one hand, AC seems to be a plausible assertion: Since each X_a is nonempty, we can surely choose an element from each X_a. On the other hand, AC gives no prescription for how the element $F(a)$ is to be chosen from X_a. AC simply states that the function F exists. This lack of precision about the nature of F has led some mathematicians to doubt AC. Moreover, in spite of its apparent plausibility, AC is logically equivalent to many statements that some mathematicians find highly implausible. One example is the so-called Banach–Tarski paradox, which implies that a ball in 3-space of radius 1 can be dissected into finitely many disjoint pieces and then reassembled to form a ball of radius 2.

Among the consequences of the Axiom of Choice is an interesting generalization of some results we encountered earlier in this chapter. In Sections 4.3 and 4.4, respectively, we proved that $\mathbf{N} \times \mathbf{N} \approx \mathbf{N}$ and $\mathbf{R} \times \mathbf{R} \approx \mathbf{R}$. Using the Axiom of Choice, we can show that for any infinite set A, $A \times A \approx A$.

As mentioned earlier, AC has aroused controversy since its introduction by Zermelo. Some mathematicians such as Cantor and Bernstein readily accepted AC; others, such as Peano, strongly rejected it. The status of AC in the Zermelo–Fraenkel axiomatic development of sets became a major problem in set theory. This problem was resolved by K. Gödel in the

1930s and P. Cohen in the 1960s. The results of Gödel and Cohen imply that neither AC nor its negation can be derived from the axioms of ZF. Thus mathematicians working with ZF have the option of assuming AC or the negation of AC and deriving the theorems that follow from each of these assumptions.

Among the many statements that are equivalent to AC is Zorn's Lemma. In order to state Zorn's Lemma, we need to introduce two new concepts. Let $(X, \leq)$ be a poset. A *chain* in X is a subset $B \subseteq X$ such that B is totally ordered by the partial ordering $\leq$. An element $x_0 \in X$ is called *maximal* if for all $x \in X$, $x \leq x_0$.

Zorn's Lemma *Let $(X, \leq)$ be a poset with the property that every chain in X has an upper bound in X. Then X contains at least one maximal element.*

Since Zorn's Lemma is equivalent to AC, Zorn's Lemma is not really a lemma in ZF. ZL (our abbreviation for Zorn's Lemma) cannot be derived from the ZF axioms and similarly the negation of ZL cannot be proved in ZF. ZL has proved to be a very important tool in set theory. For example, by using ZL we can answer question 2 posed at the start of this section.

Comparability of Cardinal Numbers

If α and β are cardinal numbers, then exactly one of the following holds: $\alpha < \beta$, $\alpha = \beta$, or $\beta < \alpha$. Equivalently, if A and B are sets, then exactly one of the following holds: $|A| < |B|$, $|A| = |B|$, or $|B| < |A|$.

The proof that ZL implies the comparability of cardinals is beyond the scope of this text. Readers interested in seeing a proof of this implication and of the equivalence of AC and ZL can consult Hamilton's book [5]. One last note on these matters. In 1915 F. Hartogs proved that the comparability property of cardinal numbers implies AC. Thus the comparability property is equivalent to both AC and ZL. Therefore, the second question raised at the start of the chapter cannot be answered using the ZF axioms. Neither the comparability property nor its negation can be proved in ZF!

To give a glimpse of the powers of AC, we use it to show that every infinite set contains a denumerable subset. As we noted earlier, one consequence of this result is that for each infinite set A, $A \approx A \cup \{x\}$.

Theorem 4.5.4 *Every infinite set contains a denumerable subset.*

Proof Let A be an infinite set. For each positive integer n let A_n denote the set of all subsets of A having cardinality exactly n. Note that since A is infinite, there exists at least one subset of A of cardinality n. Thus $A_n \neq \varnothing$ for all $n \in \mathbf{N}^+$. For each $n \in \mathbf{N}^+$, choose an element of A_n, call it α_n. In other words, for each $n \in \mathbf{N}^+$, we choose a subset of A that contains exactly n elements. (At this point we are using AC. Formally, we are asserting the existence of a function $f: \mathbf{N}^+ \to \bigcup_{n \in \mathbf{N}^+} A_n$ such that $f(n) \in A_n$, i.e., $f(n)$ is a subset of A of cardinality n.)

Let $B = \bigcup_{n \in \mathbf{N}^+} \alpha_n$. (Thus B consists of all the elements in the chosen subsets.) Then $B \subseteq A$ and B is infinite, since B contains subsets of arbitrary finite size. Also, B is countable by Theorem 4.3.7(ii). Since B is an infinite, countable set, B is denumerable. ■

Exercises §4.5

4.5.1. Show that if A is any denumerable set and x is any object, then $A \approx A \cup \{x\}$.

4.5.2. Let A be any set. Show $A \approx A \times \{1\}$.

4.5.3. Complete the proof of the Cantor–Bernstein Theorem by showing that:
(a) f sends A_i onto B_i and A_0 onto B_1, and that
(b) g^{-1} sends A_1 onto B_0.

4.5.4. (a) Prove: $(0,1) \approx [0,1)$.
(b) Prove: $(0,1) \approx (0,1]$.
(c) Prove: If $a < b$ and $c < d$, then $(a, b) \approx [c, d]$.

4.5.5. Prove: $(0,1) \times (0,1) \approx [0,1] \times [0,1]$.

4.5.6. Use AC to show that if $f: X \to Y$ is surjective, then there exists $g: Y \to X$ such that g is injective.

4.5.7. Use AC to show that if $\{X_a | a \in A\}$ is a family of pairwise disjoint sets indexed by A, then there exists a set B such that for each $a \in A$, $B \cap X_a$ is a singleton set.

4.5.8. Prove, using an argument similar to the proof of Theorem 4.5.1 but without using the statement of Theorem 4.5.1, that for any set A, there does not exist an injection $f: P(A) \to A$.

Notes

Most of the ideas presented in this chapter originated in the work of Cantor. For example, the denumerability of **Q**, the uncountability of **R**, the diagonal procedure, cardinal numbers, and the statements of the Cantor–Bernstein Theorem and the comparability property are all found in Cantor's work. The controversies generated by Cantor's methods and results were heated and in some cases personal. Accounts of the early history of set theory are found in Dauben's article [1], van Dalen's essay [4], and Kline's book [6]. For discussions on the work of Gödel and Cohen, see Nagel and Newman [7] and Cohen and Hersh [2]. References [4] and [5] are good sources for the history and status for the Axiom of Choice. The results on finite sets are part of the mathematical folklore, although the Pigeonhole Principle is sometimes attributed to P. Dirichlet.

Bibliography

1. J. W. Dauben, "Georg Cantor and the origins of transfinite set theory," *Scientific American*, 248 (June 1983), 122–131, 154.
2. P. Cohen and R. Hersh, "Non-Cantorian set theory," *Scientific American*, 217 (December 1967), 104–116, 160.
3. D. van Dalen, H. C. Doets, and H. de Swart, *Sets: Naive, Axiomatic, and Applied*, Pergamon, 1978.
4. D. van Dalen and A. Monna, *Sets and Integration: An Outline of the Development*, Walters-Noordhoff, 1972.
5. A. G. Hamilton, *Numbers, Sets, and Axioms*, Cambridge University Press, 1982.
6. M. Kline, *Mathematical Thought from Ancient to Modern Times*, Oxford University Press, 1972.
7. E. Nagel and J. Newman, "Gödel's proof," *Scientific American*, 194 (June 1956), 71–86, 168, 170.

Miscellaneous Exercises

1. Let A and B be sets. Show that if $A - B \approx B - A$, then $A \approx B$. Is the converse true?

2. Let A and B be sets with $a \notin A$ and $b \notin B$. Show that if $A \cup \{a\} \approx B \cup \{b\}$, then $A \approx B$.

3. Is the set of all relations on $\mathbf{N}$ denumerable?

4. Let A be an uncountable subset of $\mathbf{R}$. For each $n \in \mathbf{Z}$ let $A_n = A \cap [n, n+1]$. Show that there exists $m \in \mathbf{Z}$ such that A_m is uncountable.

5. (a) Find a function $f: \mathbf{N} \to \mathbf{N}$ that is onto but is not one-to-one.
   (b) Find a function $f: \mathbf{N} \to \mathbf{N}$ such that for each $n \in \mathbf{N}$, $|f^{-1}(n)| = 2$.

6. Let $A_1, \ldots, A_n \cdots$ be pairwise disjoint sets each of cardinality c. Determine the cardinality of $\bigcup_{n=1}^{\infty} A_n$.

7. Show $\mathbf{R} \approx \mathbf{R} - \mathbf{N}$.

8. Let Irr denote the set of irrational numbers. Show $\mathbf{R} \approx \text{Irr}$.

9. Prove, or disprove and salvage: Let A, B, and C be sets. If $|A| < |B|$, then $|A \times C| < |B \times C|$.

10. Investigate the following questions: Let A and B be finite sets with $|A| = m$ and $|B| = n$. Let $F = \{f | f \text{ is a function from } A \text{ to } B\}$.
    (a) What is $|F|$?
    (b) How many functions in F are one-to-one?
    (c) How many are onto?
    (d) How many are both one-to-one and onto?

11. (a) Let U be a finite set and let $A, B \subseteq U$. Show that $|A^c \cap B^c| = |U| - |A| - |B| + |A \cap B|$.
    (b) How many positive integers ≤ 100 are divisible by neither 2 nor 3?

12. Which of these sets are equinumerous? Explain.
    (a) $\mathbf{R}$. (d) $\mathbf{Q}$. (g) $\mathbf{N} \times \mathbf{R}$.
    (b) $\mathbf{N}$. (e) $\mathbf{R} \times \mathbf{R}$.
    (c) Irr. (f) $\mathbf{N} \times \mathbf{Q}$.

13. Investigate the following question: Does there exist a function $f\colon \mathbf{R} \to \mathbf{R}$ such that for each $y \in \mathbf{R}$ the set $f^{-1}(y)$ is uncountable?

14. Prove that
    (a) $\mathbf{Q} \times \mathbf{Q} \approx \mathbf{Q}$.
    (b) $[0,1] \cup [2,3] \approx (0,1)$.
    (c) $\mathbf{R} \times \mathbf{Q} \approx \mathbf{R}$.

15. Prove that if $f\colon A \to B$ is an injective function, then there exists a surjective function $g\colon B \to A$.

16. For a set A, define $\mathbf{Q}(A)$ to be the set of all functions $f\colon A \to \{0,1\}$. Show that for any set A, $\mathscr{P}(A)$ and $\mathbf{Q}(A)$ are equinumerous.

17. Prove, or disprove and salvage: If A and B are sets and $A \times B$ is finite, then A and B are finite.

CHAPTER FIVE

Discrete Structures

§5.1 An Overview of Discrete Structures

In Chapter 4 we focused on the concept of cardinality of a set. We saw that in studying cardinality, it is natural to divide sets into two categories—finite sets and infinite sets. In certain ways, finite sets and infinite sets behave similarly with respect to cardinality: If A and B are finite (resp., infinite) sets, then $A \cup B$ and $A \times B$ are finite (resp., infinite) sets. In other ways, finite and infinite sets behave quite differently. For instance, suppose A and B are nonempty, disjoint sets. If A and B are nonempty, disjoint finite sets, then $|A \cup B| \neq |A|$ and $|A \cup B| \neq |B|$. If A and B are infinite, however, then $|A \cup B|$ might equal $|A|$ or $|B|$. For another example, consider sets A and B with $A \subset B$. If B is finite, then $B \not\approx A$, i.e., $|B| \neq |A|$; but if B is infinite, then it might be the case that $B \approx A$. (In fact, using the Axiom of Choice we can show that for each infinite set B, there exists a subset $A \subset B$ such that $A \approx B$.) The moral of this story is that while the theory of finite sets and the theory of infinite sets are not identical, they are similar or analogous. Thus we can use either theory as a guide or model in an investigation of the other without expecting the two theories to coincide completely.

Another example of a significant analogy in mathematics is that between the theory of discrete structures and the theory of continuous structures. Continuous mathematics is concerned with properties of the real number system and spaces, such as $\mathbf{R}^n$, Euclidean n-space, that can be built from the real numbers. An important aspect of continuous mathematics is the study of functions, especially continuous functions, through the use of derivatives and integrals.

In discrete mathematics we are concerned with finite and denumerable sets rather than with the real numbers or with intervals of real numbers. In studying finite and denumerable sets, we encounter the discrete analog of a function, namely, the concept of sequence. Instead of considering a function $f: \mathbf{R} \to \mathbf{R}$, we consider a sequence, s, which is a function $s: \mathbf{N} \to \mathbf{R}$. The terms of the sequence are the numbers $a_0 = s(0)$, $a_1 = s(1), \ldots, a_n = s(n), \ldots$.

A large portion of discrete mathematics is concerned with *combinatorial* properties of countable sets, i.e., with properties related to the arrangement of elements of the set into an orderly pattern. When investigating combinatorial aspects of a set, we would usually focus on three types of problems:

1. The existence of an arrangement of the desired type.
2. The listing or counting of all arrangements of the given type.
3. Procedures or algorithms to produce an arrangement of the desired type.

The purpose of this chapter is to provide an introduction to discrete structures. We begin with a survey of several useful combinatorial principles. We then apply these principles to the study of subsets of a finite set. A corollary of this investigation is the Binomial Theorem, which in turn leads to a general study of sequences. Finally, we close the chapter with a quick introduction to graph theory, that portion of combinatorics that describes relations on finite sets.

To close this section, let us consider some problems that illustrate the kinds of questions arising in discrete mathematics. The first problem asks for the existence of a certain kind of arrangement.

Problem 1 Show that in any collection of six people there are either three mutual friends or three mutual strangers. The problem requires us to find three people among the six who are mutual friends (i.e., each person knows the other two) or three people who are mutual strangers (i.e., each person knows neither of the other two).

Problem 2 Consider an 8×8 chessboard. In how many ways can eight rooks be placed on the board so that none of the rooks attacks any of the others? Recall that a rook controls all squares in the row and the column in which it is placed. Also for now, we assume that the eight rooks are indistinguishable. Thus the question amounts to asking in how many ways eight rooks can be placed on an 8×8 chessboard so that no two rooks lie in the same row or same column. Clearly, the existence of such an arrangement is not in question; for example, placing eight rooks along either of the two diagonals yields an arrangement of eight nonattacking rooks. What we wish to know is exactly how many such arrangements exist?

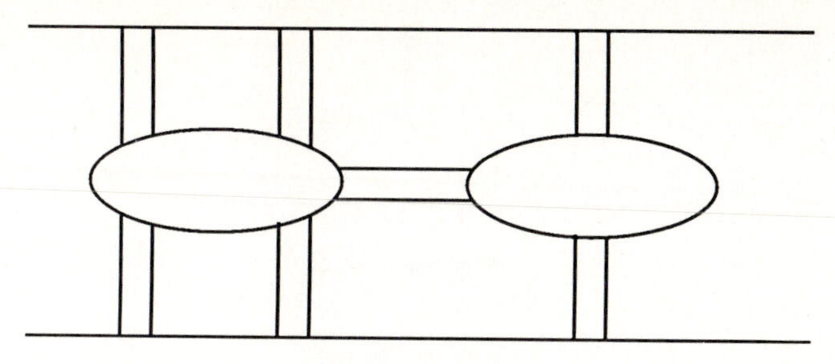

Figure 5.1.1

Problem 3 Back in the early 1700s, in the Prussian town of Koenigsberg (now Kaliningrad located in the Soviet Union), a series of seven bridges connected the shores and two islands in the Pregel River. Figure 5.1.1 depicts the bridges and islands. Since the citizens of Koenigsberg were fond of walking over the various bridges, the following question arose: Can a person walk over the bridges in such a way that he or she crosses each bridge exactly once? The Koenigsberg Bridge Problem was solved by Leonhard Euler, the most eminent mathematician of the eighteenth century, who in devising his solution created the mathematical discipline of graph theory. In Section 5.6 we present Euler's analysis of the bridge problem. For now we leave the problem as a challenge for the reader.

Problem 4 Suppose a highway inspector must inspect a system of highways such as the one shown in Figure 5.1.2. Because the inspector wishes to perform his job efficiently, he wants to traverse the highway system in such a way that he travels over each section of highway (represented by line segments joining the small circles that represent the towns) exactly once. First, can the inspector actually traverse the highway system in this way? Second, if so, then describe a procedure that the inspector can follow so as to obtain the desired method of travel. We return to this problem in Section 5.6. For now, we invite the reader to inspect highway systems similar to the one pictured in Figure 5.1.2, and to formulate a conjecture as to when efficient traversals of the highway system are possible. For example, for the highway system shown in Figure 5.1.3, the inspector

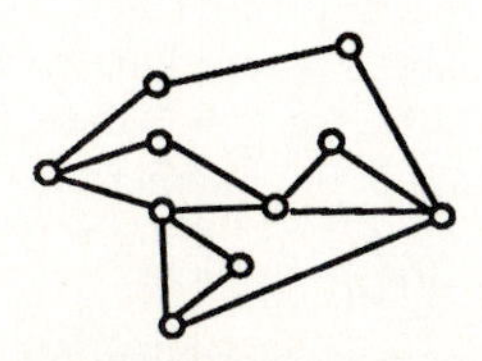

Figure 5.1.2

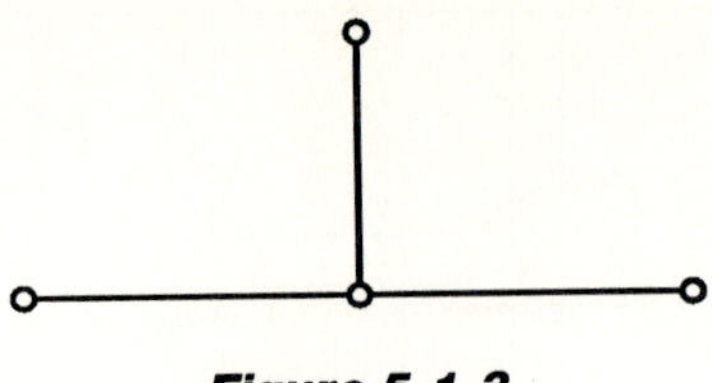

Figure 5.1.3

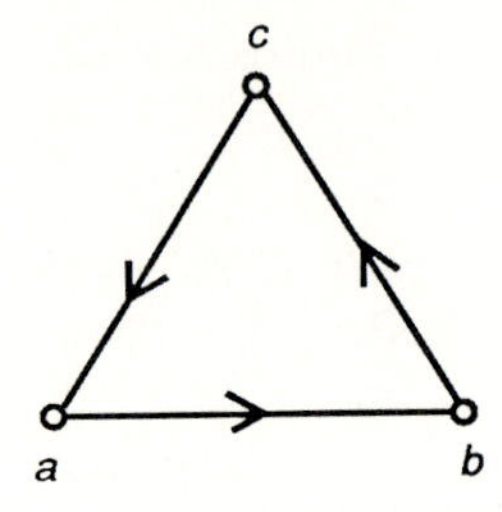

Figure 5.1.4

cannot inspect all the roads without passing over some road at least twice, while for the system in Figure 5.1.4, the inspector can inspect all the roads efficiently.

The four problems presented in this section convey some of the principal themes of discrete mathematics. The problems are all concerned with specific types of arrangements or patterns on countable sets. The questions ask for either the existence, number, or generation of arrangements of a given type.

Exercises §5.1

5.1.1. Represent the elements of Problem 1 as follows: Let each person be represented by a small circle in the plane, ∘, and join the circles representing two people by a solid line if they are friends, ∘—∘, and by a dotted line if they are strangers, ∘ · · · · ∘. Restate Problem 1 in terms of this representation of the problem.

5.1.2. Try the following approach to Problem 2: Consider the number of ways of placing k indistinguishable nonattacking rooks on a $k \times k$ chessboard as k varies from 1 to 7. Clearly, for $k = 1$ there is only one way, and for $k = 2$ there are two ways. What is the number of ways when $k = 3$? When $k = 4$?

5.1.3. Solve the Koenigsberg Bridge problem.

5.1.4. (a) Solve the highway inspector problem for the highway system depicted in Figure 5.1.2.

(b) Draw a number of other highway systems and solve the highway inspector problem for each such system.

(c) Formulate a conjecture about those highway systems in which it is possible to inspect all the roads by passing over each road exactly once.

5.1.5. Consider an 8×8 chessboard with two opposite corners removed. Can such a board be covered by 31 dominos, each of which covers exactly two squares on the chessboard?

5.1.6. Let $n \in \mathbf{Z}^+$. Suppose n lines are given in the plane in such a way that no two of the lines are parallel and no three meet at a point. Let p_n denote the number of regions into which the n lines divide the plane. Calculate p_n for $1 \leq n \leq 5$. Conjecture a general formula for p_n.

5.1.7. Prove that in any group of eight people, at least two people have the same number of acquaintances within the group.

5.1.8. Let A be a set with n elements where $n \in \mathbf{Z}^+$. How many subsets of A have exactly one element? Exactly $n - 1$ elements? Exactly 2 elements?

§5.2 Basic Combinatorial Principles

In this section we present several basic principles of combinatorics. Most of these ideas and results are familiar to us from Section 4.2. Our purpose here is to summarize and to illustrate these principles.

We begin with the Addition Principle and the Multiplication Principle, two tools that facilitate the calculation of the number of elements of certain types of finite sets. Then we move on to one of the guiding heuristic principles in combinatorics, the Fubini Principle. We close with the Pigeonhole Principle, which is invaluable in proving the existence of configurations of a given type. As usual, for a finite set A, $|A|$ denotes the number of elements of A.

The Addition Principle

The Addition Principle can be stated in varying degrees of generality. We start with a particular form of the Addition Principle and then extend this result to the general Inclusion–Exclusion Principle. Our first version is an extension of Corollary 4.1.3. (See Exercise 4.2.5(a).)

Throughout this section n denotes a positive integer.

The Addition Principle Let $A_1, \ldots, A_n$ be any collection of pairwise disjoint finite sets, i.e., $A_i \cap A_j = \varnothing$ for all i, j such that $1 \leq i, j \leq n$, and $i \neq j$. Then $|A_1 \cup \cdots \cup A_n| = |A_1| + \cdots + |A_n|$.

Proof We can prove the Addition Principle using mathematical induction. For a natural number $n \geq 2$, let $S(n)$ be the statement: If $A_1, \ldots, A_n$ are pairwise disjoint finite sets, then $|A_1 \cup \cdots \cup A_n| = |A_1| + \cdots + |A_n| = \sum_{i=1}^{n} |A_i|$.

The basis step is the case $n = 2$, which is covered in Corollary 4.1.3. Now for the inductive step.

We assume that statement $S(n)$ holds and prove that $S(n + 1)$ holds. To this end, let $A_1, \ldots, A_{n+1}$ be any collection of $n + 1$ pairwise disjoint finite sets. We must show that $|A_1 \cup \cdots \cup A_{n+1}| = \sum_{i=1}^{n+1} |A_i|$.

Since the sets $A_1, \ldots, A_{n+1}$ are pairwise disjoint, $(A_1 \cup \cdots \cup A_n) \cap A_{n+1} = (A_1 \cap A_{n+1}) \cup \cdots \cup (A_n \cap A_{n+1}) = \varnothing$. Thus $|A_1 \cup \cdots \cup A_{n+1}| = |(A_1 \cup \cdots \cup A_n) \cup A_{n+1}| = |A_1 \cup \cdots \cup A_n| + |A_{n+1}|$ by Corollary 4.1.3 (which is the basis step in this inductive proof). Now, by inductive hypothesis,

$$|A_1 \cup \cdots \cup A_n| = \sum_{i=1}^{n} |A_i|.$$

Therefore,

$$|A_1 \cup \cdots \cup A_{n+1}| = \sum_{i=1}^{n} |A_i| + |A_{n+1}| = \sum_{i=1}^{n+1} |A_i|.$$

Thus, by PMI, the Addition Principle holds. ■

As we shall see in the next few sections, the Addition Principle is often used to calculate the number of elements in a finite set A by writing A as a union of a finite number of pairwise disjoint subsets, say $A = A_1 \cup \cdots \cup A_n$, and then determining the number of elements in each of the sets $A_1, \ldots, A_n$. For example, suppose A is a finite set such that $A = A_1 \cup A_2 \cup A_3 \cup A_4 \cup A_5$ where $A_1, \ldots, A_5$ are pairwise disjoint subsets of A such that $|A_k| = k^2$ for $1 \leq k \leq 5$. Then $|A| = 1^2 + 2^2 + 3^2 + 4^2 + 5^2 = 55$.

Example 5.2.1 Let $A = \{n \in \mathbf{Z} | 1 \leq n \leq 100$ and n has at least one 4 in its base 10 expansion$\}$. Write $A = A_1 \cup A_2 \cup \cdots \cup A_{10}$ where $A_1 = \{n \in A | 1 \leq n \leq 9\}$, $A_2 = \{n \in A | 10 \leq n \leq 19\}, \ldots, A_{10} = \{n \in A | 90 \leq n \leq 100\}$. The sets A_i are pairwise disjoint and $|A_5| = |\{n \in A | 40 \leq n \leq 49\}| = 10$, while $|A_1| = |A_2| = |A_3| = |A_4| = |A_6| = \cdots = |A_{10}| = 1$. Thus $|A| = |A_1| + \cdots + |A_{10}| = 19$.

The Addition Principle can be generalized to cover the cases in which the sets $A_1, \ldots, A_n$ are not necessarily pairwise disjoint. In such cases the Addition Principle goes under the name of the Inclusion–Exclusion Principle which we first met back in Section 4.2. Let us restate the cases of the Inclusion–Exclusion Principle that were presented in Section 4.2. (See Theorem 4.2.3 and Exercise 4.2.6.)

The Inclusion – Exclusion Principle

IE1. (For two sets.) Let A and B be finite sets. Then $|A \cup B| = |A| + |B| - |A \cap B|$.

IE2. (For three sets.) Let A, B, and C be finite sets. Then $|A \cup B \cup C| = |A| + |B| + |C| - |A \cap B| - |A \cap C| - |B \cap C| + |A \cap B \cap C|$.

Upon first encountering the Inclusion–Exclusion Principle, we might reasonably wonder about the rationale for its name. Looking at IE1, we see that to count the number of elements of $A \cup B$, we *include* in our count all the elements of both A and B to obtain $|A| + |B|$, and then we *exclude* the elements that have been counted twice by subtracting $|A \cap B|$. Thus the name Inclusion–Exclusion Principle.

Proof of IE2 The proof of IE1 is given in Section 4.2. We can prove IE2 by repeatedly applying IE1:

$$
\begin{aligned}
|A \cup B \cup C| &= |A \cup B| + |C| - |(A \cup B) \cap C| \\
&= |A| + |B| + |C| - |A \cap B| - |(A \cup B) \cap C| \\
&= |A| + |B| + |C| - |A \cap B| - |(A \cap C) \cup (B \cap C)| \\
&= |A| + |B| + |C| - |A \cap B| \\
&\quad - (|A \cap C| + |B \cap C| - |A \cap C \cap B \cap C|) \\
&= |A| + |B| + |C| - |A \cap B| - |A \cap C| \\
&\quad - |B \cap C| + |A \cap B \cap C|.
\end{aligned}
$$

■

Example 5.2.2 At a meeting of 100 people, 55 are male, 30 are American, and 15 of the Americans are male. How many people at the meeting are either male or American? Let A be the set of males at the meeting and B be the set of Americans at the meeting. We want to determine $|A \cup B|$. By Inclusion–Exclusion, $|A \cup B| = |A| + |B| - |A \cap B| = 50 + 30 - 15 = 65$.

Example 5.2.3 How many positive integers less than or equal to 100 are divisible by 2, 3, or 5? Let $A = \{n \in \mathbf{N} | 1 \leq n \leq 100 \text{ and } 2|n\}$, $B = \{n \in \mathbf{N} | 1 \leq n \leq 100 \text{ and } 3|n\}$, and $C = \{n \in \mathbf{N} | 1 \leq n \leq 100 \text{ and } 5|n\}$. We want to determine $|A \cup B \cup C|$. It is easy to check that $|A| = 50$, $|B| = 33$, $|C| = 20$, $|A \cap B| = |\{n | 1 \leq n \leq 100 \text{ and } 6|n\}| = 16$, $|A \cap C| = |\{n | 1 \leq n \leq 100 \text{ and } 10|n\}| = 10$, $|B \cap C| = |\{n | 1 \leq n \leq 100 \text{ and } 15|n\}| = 6$, and $|A \cap B \cap C| = |\{n | 1 \leq n \leq 100 \text{ and } 30|n\}| = 3$. Thus

$$
\begin{aligned}
|A \cup B \cup C| &= 50 + 33 + 20 - 16 - 10 - 6 + 3 \\
&= 74.
\end{aligned}
$$

IE1 and IE2 provide formulas for the number of elements in the union of two or three sets in terms of the number of elements in the given sets and all possible intersections of the sets. Naturally, one would suspect that a general version of the Inclusion–Exclusion Principle holds. Specifically, if $A_1, \ldots, A_n$ are finite sets, then $|A_1 \cup \cdots \cup A_n|$ should be expressible in terms of $|A_1|, \ldots, |A_n|$, $|A_1 \cap A_2|, \ldots, |A_1 \cap A_2 \cap A_3|, \ldots, |A_1 \cap \cdots \cap A_n|$. In other words, the cardinality of $A_1 \cup \cdots \cup A_n$ is a function of the cardinalities of the A's, the cardinalities of all the possible intersections of two of the A's, the cardinalities of all the possible intersections of three of the A's, and ..., and the cardinality of the intersection of all the A's. What should this precise expression be? Looking at the pattern visible in IE1 and IE2, we naturally conjecture that

$$\begin{aligned}
|A_1 \cup \cdots \cup A_n| &= |A_1| + \cdots + |A_n| - \big(|A_1 \cap A_2| + |A_1 \cap A_3|\big) \\
&\quad + \cdots + |(A_{n-1} \cap A_n|) \\
&\quad + \big(|A_1 \cap A_2 \cap A_3| + \cdots + |A_{n-2} \cap A_{n-1} \cap A_n|\big) \\
&\quad - \cdots \pm |A_1 \cap \cdots \cap A_n| \\
&= \sum_{i=1}^{n} |A_i| - \sum_{i,j} |A_i \cap A_j| + \sum_{i,j,k} |A_i \cap A_j \cap A_k| \\
&\quad - \cdots \pm |A_1 \cap \cdots \cap A_n|,
\end{aligned}$$

where the sums have the following interpretations: $\sum_{i,j}|A_i \cap A_j|$ represents the sum of the cardinalities of all sets obtained by intersecting exactly two of the sets among $A_1, \ldots, A_n$ and can be written as $\sum_{1 \le i < j \le n}|A_i \cap A_j|$; similarly, the next sum is the total of the cardinalities of all the sets obtained by intersecting three of the sets among $A_i, \ldots, A_n$ and can be written as $\sum_{1 \le i < j < k \le n}|A_i \cap A_j \cap A_k|$. Notice that the term $\sum_{i,j}|A_i \cap A_j|$ is multiplied by (-1) while $\sum|A_i \cap A_j \cap A_k|$ is multiplied by $+1$. In general, the term that is the sum of the cardinalities of the intersection of exactly m of the sets among $A_1, \ldots, A_n$ is multiplied by $(-1)^{m+1}$. Thus the last term in the expression for $|A_1 \cup \cdots \cup A_n|$ is $(-1)^{n+1}|A_1 \cap \cdots \cap A_n|$.

With these comments in mind, we have the following formulation of the general Inclusion–Exclusion Principle. For a proof, consult Brualdi [3].

Theorem 5.2.1 *The Inclusion–Exclusion Principle. Let $A_1, \ldots, A_n$ be finite sets. Then*

$$\begin{aligned}
|A_1 \cup \cdots \cup A_n| &= \sum_{i=1}^{n} |A_i| - \sum_{1 \le i < j \le n} |A_i \cap A_j| + \cdots \\
&\quad + (-1)^{m+1} \sum_{1 \le i_1 < \cdots < i_m \le n} |A_{i_1} \cap \cdots \cap A_{i_m}| \\
&\quad + \cdots + (-1)^{n+1} |A_1 \cap \cdots \cap A_n|.
\end{aligned}$$

The Multiplication Principle

Next we consider a second important tool for counting the number of elements in a finite set. In fact, we present two versions of the so-called Multiplication Principle.

The Multiplication Principle MP1. If $A_1, \ldots, A_n$ are finite sets, then $|A_1 \times \cdots \times A_n| = |A_1| \cdot \cdots \cdot |A_n|$.

MP2. Suppose $B_1, \ldots, B_n$ is a sequence of tasks or operations that are to be performed in order. Suppose that for each k, $2 \leq k \leq n$, the number of ways in which task B_k can be performed is independent of the number of ways that $B_1, \ldots, B_{k-1}$ have been performed. Let b_1 be the number of ways of performing operation B_1, and for $2 \leq k \leq n$, let b_k be the number of ways of performing operation B_k after operations $B_1, \ldots, B_{k-1}$ have been performed. Then the total number of ways of performing the tasks $B_1, \ldots, B_n$ in order is $b_1 \cdot \cdots \cdot b_n$.

The most useful version of the Multiplication Principle is MP2. We shall provide several examples of MP2 in action after discussing briefly the proofs of MP1 and MP2.

First recall that in Theorem 4.2.4, we proved that if A and B are finite sets, then $|A \times B| = |A| \cdot |B|$. Thus MP1 is a generalization of Theorem 4.2.4. Naturally, MP1 can be proved by induction with Theorem 4.2.4 constituting the basis step in the proof. We leave the proof of the inductive step to the reader.

We shall outline briefly the proof of MP2 for the case $n = 2$. Thus we have two tasks, B_1 and B_2, to be performed in order. Task B_1 can be performed in b_1 ways, and for each way in which B_1 is performed, B_2 can be carried out in b_2 ways. (Our assumption is that the number of ways of performing B_2 is the same for all outcomes of B_1, even though the actual outcomes of task B_2 may vary with outcomes of B_1.) Let $a_1, \ldots, a_{b_1}$ be the outcomes of performing task B_1. Given outcome a_i of task B_1, let $c_{i,1}, \ldots, c_{i,b_2}$ be the possible outcomes of task B_2. Thus we can list the possible outcomes of performing B_1 followed by B_2 as follows:

$$\begin{array}{llcl}
(a_1, c_{1,1}) & (a_1, c_{1,2}) & \cdots & (a_1, c_{1,b_2}) \\
(a_2, c_{2,1}) & (a_2, c_{2,2}) & \cdots & (a_2, c_{2,b_2}) \\
\vdots & & & \\
(a_{b_1}, c_{b_1,1}) & (a_{b_1}, c_{b_1,2}) & \cdots & (a_{b_1}, c_{b_1,b_2})
\end{array}$$

Clearly, the number of ways of performing B_1 followed by B_2 is $b_1 \cdot b_2$.

Example 5.2.4 (i) How many three-letter strings can be formed using the English alphabet? We are looking for the number of sequences of length 3 such as *aab*, *can*, *zzz* from English letters. One creates such a sequence by giving the first letter, which can be done in 26 ways, then giving the second letter, which can be done in 26 ways, and finally giving the third letter, which can also be done in 26 ways. Therefore, the total number of three-letter strings is $26 \cdot 26 \cdot 26 = 26^3$.

(ii) How many three-letter strings contain no letter more than once? The three tasks, B_1, B_2, and B_3, consist of choosing the first, second, and third letters, respectively. B_1 can be accomplished in 26 ways. Once B_1 is performed, B_2 can be carried out in 25 ways, since the second letter must differ from the first. Given that B_1 and B_2 have been performed, B_3 can be carried out in 24 ways. Thus the number of three-letter strings in which no letter appears more than once is $26 \cdot 25 \cdot 24$.

Example 5.2.5 In how many ways can a sequence of five cards be dealt from a standard fifty-two card deck? In this question we are distinguishing the order in which the cards are dealt, and thus dealing ace-king-queen-jack-ten of spades is different from dealing ten-jack-queen-king-ace of spades. Letting the tasks of dealing the first through fifth cards be B_1, B_2, B_3, B_4, B_5, respectively, we see that B_1 can be performed in 52 ways; after B_1, B_2 can be performed in 51 ways; given B_1 and B_2, B_3 can be done in 50 ways; once B_1, B_2, and B_3 are performed, B_4 can be carried out in 49 ways; and after the first four tasks are completed, B_5 can be done in 48 ways. Thus the five cards can be dealt in $52 \cdot 51 \cdot 50 \cdot 49 \cdot 48$ ways.

Example 5.2.6 Let A be a set having n elements. How many subsets of A contain exactly two elements? We use the binary sequence representation of subsets of A: Each two-element subset of A corresponds to a sequence of length n of 0's and 1's containing exactly two 1's. Thus counting the number of two-element subsets of A amounts to counting the number of binary sequences of length n having exactly two 1's. To form such a sequence, imagine a sequence of n boxes, numbered 1 to n, in which a 0 or a 1 is to be placed.

1	2	3	• • •	n

We have n places in which to put the first 1 and after that choice is made, $n - 1$ places in which to stick the second 1. The sequence is then completed if we put a 0 in each of the remaining places. Thus there are $n \cdot (n - 1)$ ways of placing two 1's into the n boxes. Notice, however, that each sequence of two 1's and $n - 2$ 0's is obtained in exactly two ways. (If we put a 1 in slot 3 first followed by a 1 in slot 5, then we obtain the same sequence as if we place a 1 in slot 5 followed by a 1 in slot 3.) Thus there are $n \cdot (n - 1)/2$ binary sequences having exactly two 1's. Therefore, A contains exactly $n \cdot (n - 1)/2$ two-element subsets.

The Fubini Principle

Our last counting principle does not actually yield a counting formula, but is instead a valuable heuristic rule. As stated, this rule, called the Fubini Principle, is completely obvious; nonetheless, the Fubini Principle is a powerful tool with which many surprising results can be derived.

The Fubini Principle Let A be a finite set. Suppose the number of elements of A is determined by two distinct counting methods. Let a_1 and a_2 be the numbers obtained via these methods. Then $a_1 = a_2$.

Even though the numbers a_1 and a_2 are equal (since each equals $|A|$), their forms might well be distinct. Usually, the numbers a_1 and a_2 will be functions of some positive integer parameter n. Therefore, the resulting formula $a_1 = a_2$ will be an identity involving n. An example of such an identity is the formula $1 + \cdots + n = n(n + 1)/2$. Momentarily, we shall show how to derive this identity using the Fubini Principle. First, let us consider a very simple illustration of the Fubini Principle.

Example 5.2.7 Let us count the number of points in the following array.

$$\begin{matrix} \cdot & \cdot & \cdot \\ \cdot & \cdot & \cdot \end{matrix}$$

If we count by rows (first method), then we obtain a total of $3 + 3 = 2 \cdot 3$ points. Counting by columns (second method) yields a total of $2 + 2 + 2 = 3 \cdot 2$ points. Thus $2 \cdot 3 = 3 \cdot 2$.

Example 5.2.8 To derive $1 + \cdots + n = n(n + 1)/2$ using Fubini, consider the $n \times (n + 1)$ grid of squares shown in Figure 5.2.1. This grid contains $n \cdot (n + 1)$ squares of size 1×1. Let us count the number of 1×1 squares in another way. The stair-like grid of squares lying below the heavy line is congruent to the grid of squares lying above the heavy line. Thus $n \cdot (n + 1) = 2 \cdot S_n$ where S_n is the number of squares lying below the heavy line. But, by counting the number of squares in the columns from left to right, we find that

$$S_n = 1 + \cdots + n.$$

Thus

$$n \cdot (n + 1) = 2 \cdot (1 + \cdots + n)$$

Figure 5.2.1

or

$$1 + \cdots + n = n \cdot (n + 1)/2.$$

The Pigeonhole Principle

The final important combinatorial principle is the Pigeonhole Principle. We emphasize that the Pigeonhole Principle does not provide a counting technique, but, as noted in Chapter 4, it is used to establish the existence of a specified combinatorial configuration.

Informally, the Pigeonhole Principle asserts that if m objects are placed into n containers where $m > n$, then at least two objects must be placed into the same container. We recall the formal version of the Pigeonhole Principle (Theorem 4.2.5). For a proof, see Section 4.2.

The Pigeonhole Principle Suppose $|A| = m$ and $|B| = n$ with $m > n$. If $f\colon A \to B$ is any function, then f is not injective.

Example 5.2.9 Suppose n people are at a party. Show that at least two people at the party know the same number of people among those present.

Suppose we assign to each person at the party the number of people at the party whom that person knows. Now, a given person at the party can know between 0 and $n - 1$ other people. (We are assuming that if x and y are people at the party and x knows y, then y knows x. Also we do not say that x knows x.) Formally then, we have a function $f\colon P \to A$ where P is the set of people at the party and $A = \{0, 1, \ldots, n - 1\}$ such that for $x \in P$, $f(x) =$ number of people in P whom x knows. Note that $|P| = n = |A|$ and thus we are not yet in a position to use the Pigeonhole Principle.

Let us first assume that there exists $x \in P$ such that x knows 0 people in P (i.e., x is a stranger to everyone at the party). Then there exists no

$y \in P$ such that y knows $n-1$ people in P, since y does not know x for all $y \in P$. Therefore, in this case, $f\colon P \to \{0, 1, \ldots, n-2\} = A'$. Since $|P| = n > n - 1 = |A'|$, f is not injective and there exist $u, v \in P$ such that $f(u) = f(v)$. Hence u and v know the same number of people in P.

If there exists no $x \in P$ such that x knows 0 people, then for all $y \in P$, $f(y) \geq 1$ and $f\colon P \to \{1, \ldots, n-1\} = A''$. Since $|P| = n > n - 1 = |A''|$, f is not injective and again there exist $u, v \in P$ such that $f(u) = f(v)$.

The ideas presented in this section will be used repeatedly in the remainder of this chapter. As we shall see, they are fundamental in carrying out combinatorial analyses.

Exercises §5.2

5.2.1. (a) If $(2n)!/n! = x(1 \cdot 3 \cdot 5 \cdot \cdots \cdot (2n-1))$, then find x.
(b) Find y so that the following identity holds:

$$2 \cdot 6 \cdot 10 \cdot \cdots \cdot (y) = (n+1)(n+2) \cdot \cdots \cdot (2n).$$

5.2.2. (a) Find $|A|$ if $A = \{n \in \mathbf{Z} | 1 \leq n \leq 100$ and the base 10 expansion of n contains a 7$\}$.
(b) Find $|A|$ if $A = \{n \in \mathbf{Z} | 1 \leq n \leq 100$ and the base 10 expansion of n contains a 4$\}$.
(c) Find $|A|$ if $A = \{n \in \mathbf{Z} | 1 \leq n \leq 100$ and the base 10 expansion of n contains a 4 or a 7$\}$.

5.2.3. (a) Find $|A_1 \cup A_2|$ if $|A_1| = 8$, $|A_2| = 7$, and $|A_1 \cap A_2| = 3$.
(b) Suppose $|A_1 \cup A_2| = 7$, $|A_1| = 5$, and $|A_2| = 4$. Find $|A_1 \cap A_2|$.

5.2.4. State and prove the case $n = 4$ of the Inclusion–Exclusion Principle.

5.2.5. From a club with 20 members, a president, vice-president, and secretary are to be chosen. In how many ways can this be done?

5.2.6. (a) Find $|A|$ if $A = \{n \in \mathbf{Z} | 100 \leq n \leq 999$ and n has distinct digits in its base 10 expansion$\}$.
(b) Let A be as in part (a). Find $|\{n \in A | n \text{ is odd}\}|$ and $|\{n \in A | n \text{ is even}\}|$.
(c) How many integers between 1 and 1000 have distinct digits?

5.2.7. A *palindrome* is an integer n that equals the integer obtained by reversing its base 10 digits. For example, 12321 is a palindrome while 12312 is not.
(a) How many 4-digit palindromes are there?
(b) How many 5-digit palindromes are there?
(c) Generalize the results of parts (a) and (b).

5.2.8. In a certain state, a license number typically consists of three uppercase English letters followed by three digits from 0 to 9. How many license plates can this state produce?

5.2.9. How many three-letter strings that have exactly two letters the same can be formed using the English alphabet?

5.2.10. (a) How many four-letter strings can be formed in English?
(b) How many such strings have no two letters the same?

5.2.11. Fourteen horses enter the Kentucky Derby. In how many ways can horses finish "in the money"? That is, how many ways can the top three finishers be determined?

5.2.12. (a) In how many ways can 8 indistinguishable rooks be placed on an 8×8 chessboard?
(b) In how many ways can 8 indistinguishable rooks be placed on an 8×8 chessboard so that no rook attacks any other rook? Note: Two rooks attack each other if they are placed in either the same row or the same column.

5.2.13. A deci-domino is a 2×1 rectangle $\boxed{x|y}$ where x and y are integers with $0 \leq x, y \leq 9$.
(a) How many different deci-dominos are there provided that the deci-domino $\boxed{x|y}$ is considered to be the same as the deci-domino $\boxed{y|x}$?
(b) In how many ways can you select a pair of deci-dominos that match—that is, that share a common number?

5.2.14. Prove the inductive step in the proof that $|A_1 \times \cdots \times A_n| = |A_1| \cdot \cdots \cdot |A_n|$.

5.2.15. Let A be a finite set. Suppose $f\colon A \to A$ is an injective function. Show f is also surjective.

5.2.16. (a) Suppose 367 people are at a gathering. Show that at least two of the people at the gathering have the same birthday.
(b) Suppose 2000 are at a gathering. Fill in the blank and prove the resulting statement: At least ______ people at the gathering have the same birthday.

5.2.17. State and prove a general result which is suggested by Exercise 5.2.16(b).

5.2.18. Let A be a set having n elements.
(a) Show that A contains $n(n-1)(n-2)/6$ subsets with exactly three elements.
(b) How many four-element subsets does A contain?

§5.3 Permutations and Combinations

In this section, we apply the counting principles described in the preceding section to determine the number of elements in certain finite sets. The objects that we are considering arise frequently in mathematics, statistics, and computer science and hence form an important part of the repertoire of any student of mathematics. We begin with the concept of permutation.

Definition 5.3.1 *Permutation of a Set*

Let $A = \{a_1, \ldots, a_n\}$ be a set with n elements where $n \in \mathbf{Z}^+$. A *permutation of A* is an ordered listing of the elements of A.

Example 5.3.1 (i) Let $A = \{a_1, a_2\}$. The elements of A can be listed in two ways, (a) a_1, a_2 and (b) a_2, a_1, and hence there exist exactly two permutations of A.

(ii) Let $A = \{a_1, a_2, a_3\}$. The permutations of A are (a) a_1, a_2, a_3, (b) a_1, a_3, a_2, (c) a_2, a_1, a_3, (d) a_2, a_3, a_1, (e) a_3, a_1, a_2, and (f) a_3, a_2, a_1.

A permutation of a set with n elements is sometimes called a *permutation of n objects*. Given a set of n objects $A = \{a_1, \ldots, a_n\}$, how can an arbitrary permutation of A be represented? As in Example 5.3.1, we can adhere strictly to the definition and list the elements in order: $a_{m_1}, a_{m_2}, \ldots, a_{m_n}$ where for $1 \leq i \leq n$, m_i is an integer, $1 \leq m_i \leq n$, and for $i \neq j$, $m_i \neq m_j$. Thus, if $A = \{a_1, a_2, a_3\}$, then for the permutation a_3, a_1, a_2, $m_1 = 3$, $m_2 = 1$, $m_3 = 2$. With this notation in mind, we can devise another way of representing a permutation of A: Consider as fixed the original listing of $A = \{a_1, \ldots, a_n\}$. Then the permutation $a_{m_1}, \ldots, a_{m_n}$ is completely determined by the subscripts $m_1, \ldots, m_n$, and thus we can represent the permutation as

$$\begin{pmatrix} 1 & 2 & \cdots & n \\ m_1 & m_2 & \cdots & m_n \end{pmatrix},$$

or simply as $(m_1, m_2, \ldots, m_n)$. If we insist on working with the elements of

A rather than the subscripts, then we can represent this permutation of A as

$$\begin{pmatrix} a_1 & a_2 & \cdots & a_n \\ a_{m_1} & a_{m_2} & \cdots & a_{m_n} \end{pmatrix},$$

or as $(a_{m_1}, a_{m_2}, \ldots, a_{m_n})$. In any case, these representations are very reminiscent of tabular representations of functions and suggest an alternative way of defining the idea of a permutation.

Definition 5.3.2 *Permutation of a Set*

A *permutation of a set A* is a bijection $\pi: A \to A$.

Using the subscript representation just introduced, we arrive at the six permutations of the set $\{1,2,3\}$: $(1,2,3), (1,3,2), (2,1,3), (2,3,1), (3,1,2), (3,2,1)$.

Let A be a set with n elements. How many permutations are there on the set A? We have seen that if A has two elements, then there exist two permutations on A and that if A has three elements, then there are six permutations on A. The next theorem answers the general question.

Theorem 5.3.1 *There are $n!$ permutations on a set with n elements.*

Proof Let $A = \{a_1, \ldots, a_n\}$. We shall use the tabular representation of permutations to count the number of permutations on A:

$$\begin{pmatrix} a_1 & \cdots & a_n \\ a_{m_1} & \cdots & a_{m_n} \end{pmatrix}.$$

The entry a_{m_1} can be any of the n elements of A. Once a_{m_1} is chosen, there are $n - 1$ choices for the entry a_{m_2}. After a_{m_1} and a_{m_2} are chosen, there are $n - 2$ ways of choosing entry a_{m_3}. Continuing in this fashion, it is clear that after the entries $a_1, \ldots, a_{m_{n-1}}$ have been taken, there is exactly one choice for entry a_{mn}. Therefore, by the Multiplication Principle, there are $n \cdot (n-1)(n-2) \cdot \cdots \cdot 2 \cdot 1 = n!$ permutations of A. ∎

Example 5.3.2 Five candidates are running for mayor of Mudville. In how many ways can their names be listed on the ballot? Since each listing of the names is a permutation of the set of names, there are $5! = 120$ ways of listing the candidates' names on the ballot.

The next example suggests an extension of the idea of permutation.

Example 5.3.3 Ten horses compete in a race. There are, of course, $10!$ distinct ways in which these horses can finish the race. For reasons of finance, however, observers of horse races are interested only in the top three finishers. In how many possible ways can the first three finishers be determined in a ten-horse race? Evidently, there are 10 possibilities for the first finisher. Once first place is determined, there are 9 choices for the second place, and after first and second are fixed, there are 8 choices for third. Therefore, there are $10 \cdot 9 \cdot 8$ possible orders of finish for the top three horses in a field of ten horses. More generally, we have determined that there are $10 \cdot 9 \cdot 8$ lists of length three consisting of objects from a set of ten elements.

The next definition generalizes the idea implicit in the previous example.

Definition 5.3.2 *Permutation of Size k of n Objects*

Let A be a set with n elements. Let $k \in \mathbf{Z}$ with $1 \leq k \leq n$. A *permutation of size* k of the n objects in A is an ordered list of length k of elements of A. Let $P(n, k)$ denote the number of permutations of length k of a set with n elements.

Note that by Theorem 5.3.1, $P(n, n) = n!$. It is easy to see that $P(n, 1) = n$. Also Example 5.3.3 shows that $P(10, 3) = 10 \cdot 9 \cdot 8 = 720$.

Theorem 5.3.2 $P(n, k) = n(n-1) \cdot \cdots \cdot (n-k+1)$.

Proof The number $P(n, k)$ is the number of ordered lists of length k from a set with n elements. Such a list can be formed by filling in the first, second, third, $\ldots$, up to the kth entries in order. There are n ways of filling in the first entry, $n-1$ ways of filling in the second entry (once the first is chosen), $n-2$ ways of filling in the third entry (after the first two are chosen), etc. After the first $k-1$ entries on the list are selected, there are $n-(k-1) = n-k+1$ ways of filling in the kth entries. Thus, by the Multiplication Rule,

$$P(n, k) = n \cdot (n-1) \cdot \cdots \cdot (n-k+1). \qquad \blacksquare$$

By definition, a permutation of size k from a set with n elements is an ordered list of length k chosen from among the n objects. Suppose we relax the requirement that an ordered list be formed and consider only collections of k objects chosen from among the n objects. We have then the idea of a combination, without question one of the most important in mathematics.

Definition 5.3.3 *Combination*

Let A be a set with n elements. Let $k \in \mathbf{Z}$ with $1 \leq k \leq n$. A collection of k elements chosen from A is called a *combination of size* k chosen from A. Let $C(n, k)$ denote the number of combinations of size k taken from a set of size n.

In other words, a combination of size k chosen from A is just a subset of A having k elements.

Example 5.3.4 Let us determine $C(5, 2)$. Let $A = \{1, 2, 3, 4, 5\}$. We list all collections of two elements from A: $\{1, 2\}$, $\{1, 3\}$, $\{1, 4\}$, $\{1, 5\}$, $\{2, 3\}$, $\{2, 4\}$, $\{2, 5\}$, $\{3, 4\}$, $\{3, 5\}$, $\{4, 5\}$. Thus $C(5, 2) = 10$.

Example 5.3.5 Let $A = \{1, 2, 3\}$. Notice that $C(3, 3) = 1$ while $P(3, 3) = 3! = 6$. In fact, it might be worth comparing $C(3, 2)$ with $P(3, 2)$ and $C(3, 1)$ with $P(3, 1)$. We know that we can determine $P(3, 2) = 3 \cdot 2 = 6$; $C(3, 2)$ by calculating all the ways of choosing 2 elements from $\{1, 2, 3\}$: These are $\{1, 2\}$, $\{1, 3\}$, and $\{2, 3\}$. Thus $C(3, 2) = 3$. Finally, $P(3, 1) = 3$ while $C(3, 1) = 3$, as we can easily check. The following table summarizes our findings.

k	$P(3, k)$	$C(3, k)$
1	3	3
2	6	3
2	6	3

It is reasonable to ask if in general a connection exists between $P(n, k)$ and $C(n, k)$. For example, $P(3, 1) = C(3, 1)$. By Theorem 5.3.2, $P(n, 1) = n$ for all $n \in \mathbf{Z}^+$. It is easy to check that $C(n, 1) = n$, and hence $P(n, 1) = C(n, 1)$. At the other extreme, we have $P(n, n) = n!$ while $C(n, n) = 1$. What about other cases? Let us find $P(4, 2)$ and $C(4, 2)$. From our formula $P(4, 2) = 4 \cdot 3 = 12$. We can find $C(4, 2)$ by brute force: Let $A = \{1, 2, 3, 4\}$. The possible choices of two objects from A are: $\{1, 2\}$, $\{1, 3\}$, $\{1, 4\}$, $\{2, 3\}$, $\{2, 4\}$, and $\{3, 4\}$. Thus $C(4, 2) = 6 = P(4, 2)/2$. Note that $C(3, 2) = P(3, 2)/2$. Perhaps, in general, $C(n, 2) = P(n, 2)/2$.

Let us again summarize our calculations.

n	k	$P(n, k)$	$C(n, k)$
3	1	3	3
3	2	6	3
4	2	12	6
n	1	n	n
n	n	$n!$	1

Before reading on, conjecture a formula that relates $C(n, k)$ to $P(n, k)$ and test your conjecture in case $n = 4$ and $k = 3$. Note that $P(4, 3) = 4 \cdot 3 \cdot 2 = 24$ while $C(4, 3) = 4 = 24/6 = 24/3!$. The next theorem gives the general relationship between the number of combinations of size k from a set of n elements and the number of permutations of size k from a set of n elements.

Theorem 5.3.3 $C(n, k) = P(n, k)/k!$

Proof Let $A = \{1, 2, \ldots, n\}$ and let $1 \leq k \leq n$. Motivated by the Fubini Principle, we shall count the number of elements in the set, Perm(A, k), of all permutations of A of size k in two different ways. From the resulting identity, we shall derive the desired formula.

By definition, there are $P(n, k)$ permutations of A of size k, i.e., $|\text{Perm}(n, k)| = P(n, k)$. By counting in a different way, we shall show that $|\text{Perm}(n, k)| = C(n, k) \cdot k!$ from which the theorem follows.

The key idea is to divide the set Perm(n, k) into a collection of disjoint, nonempty subsets whose union is all of Perm(n, k). Each of the subsets will contain $k!$ elements and there will be a total of $C(n, k)$ subsets. Thus, by the Addition Principle, $P(n, k) = C(n, k) \cdot k!$.

How will the promised partition be formed? Perhaps it is not surprising that the partition will arise from an equivalence relation on Perm(n, k).

Let $p_1, p_2 \in \text{Perm}(n, k)$. Note that p_1 and p_2 are lists of length k of elements of A. Define a relation $\sim$ on Perm(n, k) by saying $p_1 \sim p_2$ if the elements in the list p_1 coincide (not necessarily in order) with the elements in p_2. (For example, if $A = \{1, 2, 3\}$ and $k = 2$, then $(1, 2) \sim (2, 1)$ and $(1, 2) \nsim (1, 3)$.) We can easily check that $\sim$ is an equivalence relation on Perm(n, k).

Let $p_1 \in \text{Perm}(n, k)$ with $p_1 = (a_{m_1}, \ldots, a_{m_k})$ where $a_{m_i} \in A$ for $1 \leq i \leq k$. Let $p_2 \in \text{Perm}(n, k)$. Then $p_1 \sim p_2$ if and only if the elements in the list p_2 are a permutation of the elements $a_{m_1}, \ldots, a_{m_k}$. Thus there are exactly $k!$ elements of Perm(n, k) that are equivalent to p_1. Since p_1 is an arbitrary element of Perm(n, k), it follows that each equivalence class has exactly $k!$ elements.

Next note that each $p \in \text{Perm}(n, k)$ determines a collection of k elements from A: Simply take the elements of A that make up the list p.

Distinct permutations can, however, determine the same collection of elements. (For example, the permutations $(1, 2)$ and $(2, 1)$ of $\{1, 2, 3\}$ each determine the collection $\{1, 2\}$.) In fact, if $p_1, p_2 \in \text{Perm}(n, k)$, then p_1 and p_2 determine the same collection of size k of elements of A if and only if $p_1 \sim p_2$. In other words, the elements of a given equivalence class of the relation $\sim$ determine a single collection of size k of elements of A, and elements from distinct equivalence classes determine distinct collections. Thus the number of equivalence classes equals the number of combinations of size k of elements of A, which is by definition $C(n, k)$.

The relation $\sim$ on $\text{Perm}(n, k)$ determines $C(n, k)$ equivalence classes, each containing $k!$ elements. Therefore, $P(n, k) = C(n, k) \cdot k!$. ∎

The number $C(n, k)$ can also be described as the number of subsets containing k elements of a set with n elements. This number is often represented by the symbol $\binom{n}{k}$ and is called the *binomial coefficient n over k*. Thus $C(n, k) = \binom{n}{k}$ and both symbols represent the number of combinations of size k from a set of n objects, the number of ways of choosing k objects from a set of n objects, and the number of k-element subsets contained in an n-element set. The symbols $\binom{n}{k}$ and $C(n, k)$ are often read as "n choose k." One other note: With the interpretation of $\binom{n}{k}$ as the number of k-element subsets of an n-element set, it makes sense to allow $k = 0$. Clearly, $\binom{n}{0} = 1$.

Given the condition that $k \in \mathbf{Z}$ and $0 \le k \le n$, we can have the following corollary of Theorem 5.3.3. We follow the usual practice and define $0! = 1$.

Corollary 5.3.4 *For $n \in \mathbf{Z}^+$ and $k \in \mathbf{Z}$ with $0 \le k \le n$,*

$$C(n, k) = \binom{n}{k} = \frac{n!}{k!(n-k)!}.$$

Henceforth, we shall use the symbol $\binom{n}{k}$ in place of $C(n, k)$ to represent the number of combinations of size k from a set of size n. Deeper properties of the binomial coefficients will be presented in the next section. For now, let us consider some elementary examples.

Example 5.3.6 (i) From a collection of 11 people, how many committees consisting of 5 people can be formed? The desired number is clearly

$$\binom{11}{5} = \frac{11 \cdot 10 \cdot 9 \cdot 8 \cdot 7}{5 \cdot 4 \cdot 3 \cdot 2 \cdot 1} = 11 \cdot 2 \cdot 3 \cdot 7 = 462.$$

(ii) Suppose that among the eleven people, there are five men and six women. How many committees of size five can be formed consisting of three men and two women? There are $\binom{5}{3}$ ways of choosing three men from among the five men present. For each such choice there are $\binom{6}{2}$ ways of choosing the two women. Thus there are

$$\binom{5}{3} \cdot \binom{6}{2} = \frac{5 \cdot 4 \cdot 3}{3!} \cdot \frac{6 \cdot 5}{2!} = 10 \cdot 15 = 150$$

committees of the stated type.

(iii) How many subsets of size 10 are there in a 20-element set? The number of 10-element subsets is

$$\binom{20}{10} = 19 \cdot 17 \cdot 13 \cdot 11 \cdot 4 = 323 \cdot 143 \cdot 4 = 184756.$$

(iv) (See Example 5.2.5.) How many 5-card hands can be dealt from a standard 52-card deck? The number of 5-card hands is the number of combinations of size 5 from a 52-element set, which is

$$\binom{52}{5} = \frac{52 \cdot 51 \cdot 50 \cdot 49 \cdot 48}{5 \cdot 4 \cdot 3 \cdot 2 \cdot 1} = 52 \cdot 51 \cdot 5 \cdot 49 \cdot 4 = 2598760.$$

Exercises §5.3

5.3.1. List all permutations of $\{1, 2, 3, 4\}$.

5.3.2. (a) In how many ways can the letters of the word *land* be ordered?
(b) In how many ways can the letters of the word *mara* be ordered?
(c) In how many ways can the letters of the word *llama* be ordered?

5.3.3. Determine the number of permutations of the letters of your first name.

5.3.4. In how many ways can 7 men and 7 women be lined up if a woman must be first in line and women and men alternate positions in line?

5.3.5. (a) In how many ways can 8 indistinguishable rooks be placed on a 9×9 chessboard so that no rook attacks another rook?
(b) Generalize the result in part (a).

5.3.6. (a) In how many ways can 4 people be seated around a round table? (Assume that the chairs are indistinguishable.)
(b) In how many ways can 6 people be seated around a round table?
(c) Generalize the results of parts (a) and (b).

5.3.7. Evaluate $C(n, k)$ for $n \leq 5$ and $0 \leq k \leq n$.

5.3.8. Show that $C(n, 1) = n$ for all $n \geq 1$.

5.3.9. Evaluate $C(n, 2)$ for $n \geq 2$.

5.3.10. Show that for all $n \geq 1$ and for all k, $0 \leq k \leq n$, $C(n, k) = C(n, n-k)$.

5.3.11. Use the Fubini Principle in counting the number of subsets of a set with n elements to show that for all $n \geq 1$, $C(n, 0) + C(n, 1) + \cdots + C(n, n) = 2^n$.

5.3.12. In how many ways can a hand of 13 cards be dealt from a 52-card deck?

5.3.13. Let A be the set $\{1, 2, \ldots, n\}$. How many two-element subsets of A do not contain consecutive integers?

5.3.14. (a) In how many ways can 6 men be paired off with exactly 6 women?
(b) In how many ways can 6 men be paired off with 6 women chosen from a collection of 10 women?

5.3.15. (a) Show: $k \cdot C(n, k) = n \cdot C(n-1, k-1)$.
(b) Show: $(n-k) \cdot C(n, k) = n \cdot C(n-1, k)$.

§5.4 Binomial Coefficients and the Binomial Theorem

The binomial coefficients $\binom{n}{k}$ where $n \in \mathbf{Z}^+$ and $0 \leq k \leq n$ are among the richest and most interesting objects in all of mathematics. The binomial coefficients have a double appeal: They are elementary in nature, requiring very little mathematical background for their definition. At the same time, the binomial coefficients can be interpreted in a variety of ways and they play an important role in diverse branches of mathematics.

Interpretation of the Binomial Coefficients

We consider now some of the interpretations and fundamental properties of the binomial coefficients. Our principal result is the Binomial Theorem, which we prove and from which we derive some interesting combinatorial identities. We shall, however, only touch the tip of the binomial coefficient iceberg. If you wish to learn more about the binomial coefficients, you can consult Brualdi [3].

Let $n \in \mathbf{Z}^+$ and $k \in \mathbf{Z}$ with $0 \leq k \leq n$. Then there are several interpretations or representations of the binomial coefficients $\binom{n}{k}$. We present three such representations beginning with the combinatorial approach that provided our original definition of the binomial coefficients.

1. *Combinatorial.* Define $\binom{n}{0} = 1$, and for $1 \leq k \leq n$ define $\binom{n}{k}$ to be the number of combinations or collections of size k that can be chosen from a set of n objects. Thus $\binom{n}{k}$ is the number of subsets that have k elements and that are contained in a set of n elements.

2. *Numerical.* From Corollary 5.3.4 we obtain a numerical expression for the binomial coefficients:

$$\binom{n}{k} = \frac{n!}{k!(n-k)!}.$$

3. *Geometric.* Consider a grid of squares as in Figure 5.4.1. This grid extends indefinitely downward. Starting with the vertex point 0, label the "rows" of the grid $0, 1, 2, 3, \ldots$. In row n of the grid, there are $n + 1$ intersection points of the grid; label these points $(n, 0), (n, 1), \ldots, (n, n)$. Consider the number of shortest zigzag paths from the point $(0, 0)$ (which is called the vertex) to the point (n, k). Denote this number by $z(n, k)$.

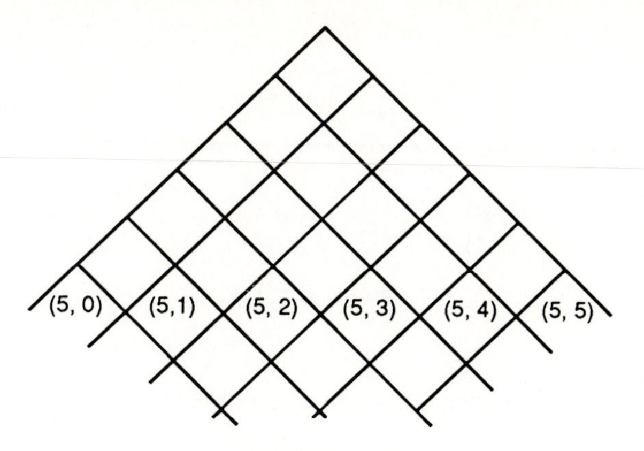

Figure 5.4.1

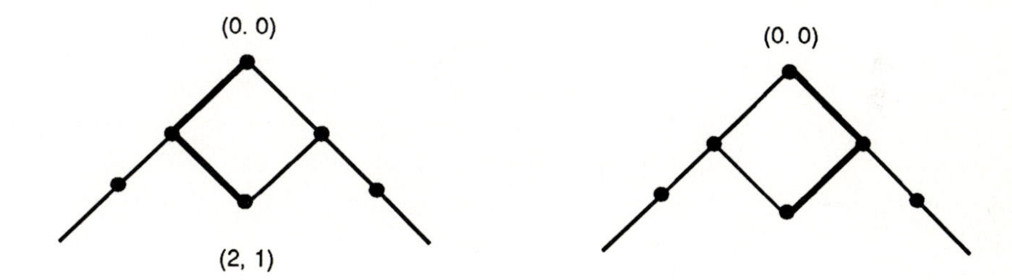

Figure 5.4.2

For example, $z(2,1) = 2$, as is shown by Figure 5.4.2. Also, $z(3,1) = 3 = z(3,2)$, as indicated by Figure 5.4.3. Finally, for each $n \geq 2$, $z(n,0) = 1$ and $z(n,n) = 1$. Note that it is also reasonable to assert that $z(0,0) = 1$.

Let us fill in the following table with the values of $z(n,k)$ for $n \leq 4$:

$$\begin{array}{ccccccccc} &&&& z(0,0) &&&& \\ &&& z(1,0) && z(1,1) &&& \\ && z(2,0) && z(2,1) && z(2,2) && \\ & z(3,0) && z(3,1) && z(3,2) && z(3,3) & \\ z(4,0) && z(4,1) && z(4,2) && z(4,3) && z(4,4). \end{array}$$

We obtain

$$\begin{array}{ccccccccc} &&&& 1 &&&& \\ &&& 1 && 1 &&& \\ && 1 && 2 && 1 && \\ & 1 && 3 && 3 && 1 & \\ 1 && 4 && 6 && 4 && 1, \end{array}$$

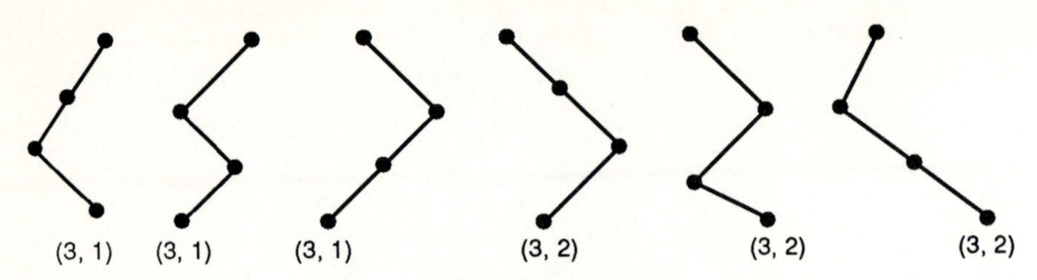

Figure 5.4.3

which happens to be identical with the corresponding table of binomial coefficients:

$$\binom{0}{0}$$

$$\binom{1}{0} \quad \binom{1}{1}$$

$$\binom{2}{0} \quad \binom{2}{1} \quad \binom{2}{2}$$

$$\binom{3}{0} \quad \binom{3}{1} \quad \binom{3}{2} \quad \binom{3}{3}$$

$$\binom{4}{0} \quad \binom{4}{1} \quad \binom{4}{2} \quad \binom{4}{3} \quad \binom{4}{4}.$$

Thus one might conjecture that $z(n, k) = \binom{n}{k}$ for all $n \in \mathbf{Z}$ with $n \geq 0$ and for all $k \in \mathbf{Z}$ $0 \leq k \leq n$. This conjecture is indeed true, as we shall show below. Once we have proved that $z(n, k) = \binom{n}{k}$, we have yet another interpretation of the binomial coefficients: $\binom{n}{k}$ is the number of shortest zigzag paths from the vertex $(0, 0)$ to the grid point labeled (n, k).

To prove that $z(n, k) = \binom{n}{k}$ we need the fundamental properties of the binomial coefficients that are summarized in the following theorem.

Theorem 5.4.1 *Let $n \in \mathbf{Z}$ with $n \geq 0$ and let $k \in \mathbf{Z}$ with $0 \leq k \leq n$. Then the binomial coefficients satisfy the following properties*:

(i) *Boundary Conditions.* $\binom{n}{0} = 1 = \binom{n}{n}$.

(ii) *Recurrence Relation.* $\binom{n+1}{k} = \binom{n}{k} + \binom{n}{k-1}$ *for* $1 \leq k \leq n$.

(iii) *Symmetry.* $\binom{n}{k} = \binom{n}{n-k}$.

Proof Statement (i) is immediate from the definition of $\binom{n}{k}$. We give two proofs of statement (ii) and leave statement (iii) as an exercise.

First Proof of (ii). Our first proof is based on the numerical formula for the binomial coefficients:

$$\begin{aligned}\binom{n}{k} + \binom{n}{k-1} &= \frac{n!}{k!\cdot(n-k)!} + \frac{n!}{(k-1)!\cdot(n-k+1)!}\\ &= \frac{n!}{(k-1)!\cdot(n-k)!}\cdot\left(\frac{1}{k} + \frac{1}{n-k+1}\right)\\ &= \frac{n!}{(k-1)!\cdot(n-k)!}\cdot\frac{n-k+1+k}{k(n-k+1)}\\ &= \frac{(n+1)\cdot n!}{k\cdot(k-1)!(n-k+1)\cdot(n-k)!} = \frac{(n+1)!}{k!(n-k+1)!}\\ &= \binom{n+1}{k}\end{aligned}$$

■

Second Proof of (ii). This time we use the combinatorial interpretation of $\binom{n}{k}$ as the number of k-element subsets of a set with n elements.

Let $A = \{a_1, \ldots, a_n, a_{n+1}\}$ be a set with $n + 1$ elements and let $\mathscr{S}$ denote the set of all k-element subsets of A. Then $|\mathscr{S}| = \binom{n+1}{k}$.

We count the number of elements of $\mathscr{S}$ in a different way. Let $B \in \mathscr{S}$. Then exactly one of the following possibilities holds: (1) $a_{n+1} \notin B$, in which case B is a k-element subset of $A' = \{a_1, \ldots, a_n\} = A - \{a_{n+1}\}$, or (2) $a_{n+1} \in B$ in which case $B = \{a_{n+1}\} \cup B'$ where $B' = B - \{a_{n+1}\}$ is a $(k-1)$-element subset of A'.

Let $\mathscr{S}_1$ be the sets in $\mathscr{S}$ of type (1) and let $\mathscr{S}_2$ be the sets in $\mathscr{S}$ of type (2). Then $\mathscr{S} = \mathscr{S}_1 \cup \mathscr{S}_2$ where $\mathscr{S}_1 \cap \mathscr{S}_2 = \varnothing$. By the Addition Principle, $|\mathscr{S}| = |\mathscr{S}_1| + |\mathscr{S}_2|$. But $|\mathscr{S}_1| = \binom{n}{k}$ and $|\mathscr{S}_2| = \binom{n}{k-1}$. Hence $\binom{n+1}{k} = \binom{n}{k} + \binom{n}{k-1}$. ■

We now use Theorem 5.4.1 to show that $z(n, k) = \binom{n}{k}$ for all $n \in \mathbf{Z}$ such that $0 \le k \le n$. We already know that for all $n \ge 0$, $z(n, 0) = 1 = \binom{n}{0}$ and $z(n, n) = 1 = \binom{n}{n}$.

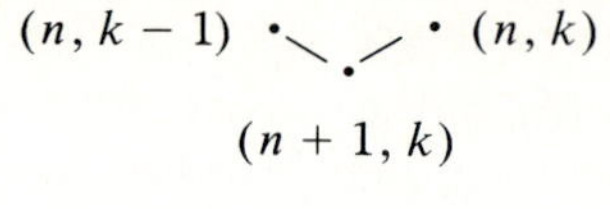

$$(n, k-1) \quad (n, k)$$
$$(n+1, k)$$

Figure 5.4.4

Our first goal is to show that for $1 \le k \le n$, $z(n+1, k) = z(n, k) + z(n, k-1)$. In other words, we show that the numbers $z(n, k)$ also satisfy relation (ii) of Theorem 5.4.1. To do so, we need only observe that any shortest zigzag path from vertex 0 to the point $(n+1, k)$ must pass through either the point $(n, k-1)$ or the point (n, k). (See Figure 5.4.4.) Conversely, any shortest path from 0 to $(n, k-1)$ (or (n, k)) extends to a shortest path from 0 to $(n+1, k)$. Thus the number of shortest zigzag paths from 0 to $(n+1, k)$ is the sum of the number of shortest zigzag paths to (n, k) and the number of shortest zigzag paths to $(n, k-1)$: $z(n+1, k) = z(n, k) + z(n, k-1)$.

Finally, we prove by induction on n that $z(n, k) = \binom{n}{k}$ for all $n \in \mathbf{Z}$ such that $n \ge 0$ and all $k \in \mathbf{Z}$ such that $0 \le k \le n$. For each $n \in \mathbf{Z}$, $n \ge 0$ let $S(n)$ be the statement: $z(n, k) = \binom{n}{k}$ for $k \in \mathbf{Z}$ such that $0 \le k \le n$.

Basis step For $n = 0$, we have $k = 0$ and $z(0, 0) = 1 = \binom{0}{0}$.

Inductive step Suppose $S(n)$ holds. We show $S(n+1)$ holds. First note that if $k = 0$ or $k = n+1$, then $z(n+1, k) = 1 = \binom{n+1}{k}$. Next assume that $1 \le k \le n$. Then, by the fact established in the preceding paragraph,

$$z(n+1, k) = z(n, k) + z(n, k-1).$$

But by inductive hypothesis, statement $S(n)$ holds and hence $z(n, k) = \binom{n}{k}$ and $z(n, k-1) = \binom{n}{k-1}$. Thus

$$\begin{aligned} z(n+1, k) &= z(n, k) + z(n, k-1) \\ &= \binom{n}{k} + \binom{n}{k-1} \\ &= \binom{n+1}{k}. \end{aligned}$$

Therefore, by PMI, for all $n \ge 0$ and for $0 \le k \le n$, $z(n, k) = \binom{n}{k}$. We have therefore established the geometric interpretation of the binomial coefficients.

Pascal's Triangle and the Binomial Theorem

The triangular representation of the binomial coefficients given below is known as Pascal's triangle.

$$\begin{array}{ccccccc} & & & \binom{0}{0} & & & \\ & & \binom{1}{0} & & \binom{1}{1} & & \\ & \binom{2}{0} & & \binom{2}{1} & & \binom{2}{2} & \\ \binom{3}{0} & & \binom{3}{1} & & \binom{3}{2} & & \binom{3}{3} \\ & & & \vdots & & & \end{array}$$

Once a given row of Pascal's triangle is known, the next row can be filled in using the boundary conditions, $\binom{n}{0} = 1 = \binom{n}{n}$, and the recurrence relation, $\binom{n+1}{k} = \binom{n}{k} + \binom{n}{k+1}$. Pascal's triangle through row 7 is given below.

$$\begin{array}{ccccccccccccccc} & & & & & & & 1 & & & & & & & \\ & & & & & & 1 & & 1 & & & & & & \\ & & & & & 1 & & 2 & & 1 & & & & & \\ & & & & 1 & & 3 & & 3 & & 1 & & & & \\ & & & 1 & & 4 & & 6 & & 4 & & 1 & & & \\ & & 1 & & 5 & & 10 & & 10 & & 5 & & 1 & & \\ & 1 & & 6 & & 15 & & 20 & & 15 & & 6 & & 1 & \\ 1 & & 7 & & 21 & & 35 & & 35 & & 21 & & 7 & & 1 \end{array}$$

Pascal's triangle is a goldmine of interesting identities involving the binomial coefficients. For example, summing the elements along the second diagonal running from upper right to lower left up to a certain row, e.g., $1, 1 + 2, 1 + 2 + 3, 1 + 2 + 3 + 4, 1 + 2 + 3 + 4 + 5$, etc., always seems to yield the element in the next row to the right of the final number in the sum: $1 = 1, 1 + 2 = 3, 1 + 2 + 3 = 6, 1 + 2 + 3 + 4 = 10, 1 + 2 + 3 + 4 + 5 = 15$.

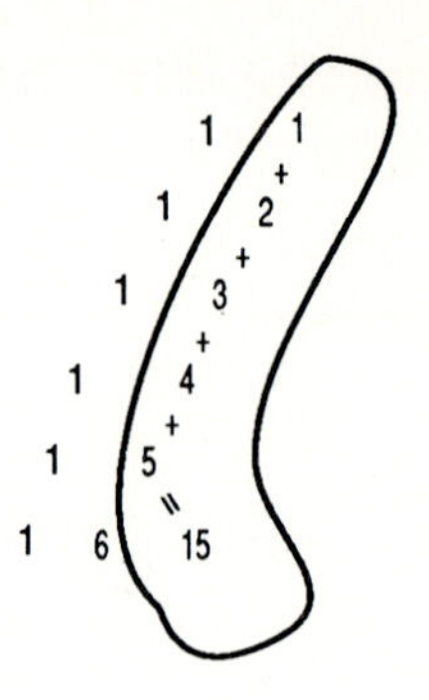

In general, it appears that

$$\binom{1}{1}+\binom{2}{1}+\cdots+\binom{n}{1}=\binom{n+1}{2}.$$

When both sides of this equation are calculated, we see that it becomes

$$1+2+\cdots+n=\frac{(n+1)n}{2},$$

a formula that we know to be true.

As a matter of fact, the binomial coefficient form of this identity, $\binom{1}{1}+\cdots+\binom{n}{1}=\binom{n+1}{2}$, provides yet another interpretation of the formula. (Remember the geometric representation presented in Sections 1.4 and 5.2.) To see this interpretation, let us count in two distinct ways the number of ways of choosing two numbers from the set $\{1, 2, \ldots, n, n+1\}$ (Fubini strikes again!).

By definition, this number is $\binom{n+1}{2}$. On the other hand, we can classify the ways of choosing two elements from $\{1, \ldots, n+1\}$ by considering the possible values of the largest of the two numbers. The largest number can vary from 2 to $n+1$ and by the Addition Principle the number of ways of choosing two elements from $\{1, \ldots, n+1\}$ is the number of ways of choosing two elements in which the largest is 2 plus the number of ways in which the largest is 3 plus $\cdots$ plus the number of ways in which the largest is $n+1$. For each $j \in \mathbf{Z}$ such that $2 \leq j \leq n+1$, the number of ways of choosing two elements from $\{1, \ldots, n+1\}$ such that the largest is j is precisely the number of ways of choosing one element from $\{1, \ldots, j-1\}$ that equals $\binom{j-1}{1}$. Thus the number of ways of choosing two elements from $\{1, \ldots, n+1\}$ is

$$\binom{n+1}{2}=\binom{1}{1}+\binom{2}{1}+\cdots+\binom{n}{1}.$$

We close this section by stating, proving, and applying the Binomial Theorem, one of the most important (yet elementary) results in mathematics.

Theorem 5.4.2 *The Binomial Theorem. For any real number x and positive integer n,*

$$(1+x)^n = \binom{n}{0} + \binom{n}{1}x + \binom{n}{2}x^2 + \cdots + \binom{n}{n}x^n = \sum_{k=0}^{n} \binom{n}{k}x^k.$$

Proof Since an integer parameter is involved in the statement of the theorem, it is reasonable to attempt an inductive proof.

For each $n \in \mathbf{Z}^+$, define $S(n)$: For each $x \in \mathbf{R}$, $(1+x)^n = \sum_{k=0}^{n}\binom{n}{k}x^k$.

Basis step $(1+x)^1 = 1 + x = \binom{1}{0}x^0 + \binom{1}{1}x^1.$

Inductive step Suppose $S(n)$ holds. To show that $S(n+1)$ holds, we must show that $(1+x)^{n+1} = \sum_{k=0}^{n+1}\binom{n+1}{k}x^k$. We work with the left side, $(1+x)^{n+1}$, using the assumption that $S(n)$ holds.

$$\begin{aligned}
(1+x)^{n+1} &= (1+x)(1+x)^n \\
&= (1+x)\cdot \sum_{k=0}^{n}\binom{n}{k}x^k \qquad \text{by inductive hypothesis} \\
&= \sum_{k=0}^{n}\binom{n}{k}x^k + \sum_{k=0}^{n}\binom{n}{k}x^{k+1} \qquad \text{by distributivity} \\
&= 1 + \sum_{k=1}^{n}\binom{n}{k}x^k + \sum_{k=1}^{n}\binom{n}{k-1}x^k + x^{n+1} \\
&\left(\text{since } \sum_{k=0}^{n}\binom{n}{k}x^{k+1} = \binom{n}{0}x + \binom{n}{1}x^2 + \cdots \right. \\
&\qquad \left. + \binom{n}{n-1}x^n + \binom{n}{n}x^{n+1} = \sum_{k=1}^{n+1}\binom{n}{k-1}x^k\right) \\
&= 1 + \sum_{k=1}^{n}\left[\binom{n}{k} + \binom{n}{k-1}\right]x^k + x^{n+1} \\
&= 1 + \sum_{k=1}^{n}\binom{n+1}{k}x^k + x^{n+1} \qquad \text{by Theorem 5.4.1} \\
&= \sum_{k=0}^{n+1}\binom{n+1}{k}x^k.
\end{aligned}$$

■

To illustrate the power of the Binomial Theorem, we derive some immediate corollaries.

Corollary 5.4.3 *For each $n \in \mathbf{Z}^+$, $2^n = \sum_{k=0}^n \binom{n}{k}$.*

Proof Set $x = 1$ in Theorem 5.4.2. ∎

The assertion $2^n = \sum_{k=0}^n \binom{n}{k}$ has a pretty combinatorial interpretation: Let $A = \{1, \ldots, n\}$. Then the left side counts the number of subsets of A while the right side counts the same number by totaling the number of subsets containing k elements as k ranges from 0 to n.

Corollary 5.4.4 *For each $n \in \mathbf{Z}^+$, $n \cdot 2^{n-1} = \sum_{k=1}^n k\binom{n}{k}$.*

Proof In the expression $(1 + x)^n = \sum_{k=0}^n \binom{n}{k} x^k$, regard x as a variable. From this viewpoint, the Binomial Theorem asserts the equality of the functions $(1 + x)^n$ and $\sum_{k=0}^n \binom{n}{k} x^k$.

These polynomial functions are clearly differentiable. Differentiating both sides of

$$(1 + x)^n = \sum_{k=0}^{n} \binom{n}{k} x^k,$$

we obtain

$$n(1 + x)^{n-1} = \sum_{k=1}^{n} k\binom{n}{k} x^{k-1}.$$

Setting $x = 1$ yields the desired conclusion. ∎

Corollary 5.4.5 *For each $n \in \mathbf{Z}^+$,*

$$\binom{2n}{n} = \binom{n}{0}^2 + \binom{n}{1}^2 + \cdots + \binom{n}{n}^2 = \sum_{k=0}^{n} \binom{n}{k}^2.$$

Proof We follow the spirit of Fubini by calculating the same quantity in two different ways. The coefficient of x^n in $(1 + x)^{2n}$ is, by the Binomial Theorem, $\binom{2n}{n}$.

On the other hand,

$$(1 + x)^{2n} = (1 + x)^n(1 + x)^n = \left(\sum_{k=0}^{n} \binom{n}{k} x^k\right)\left(\sum_{j=0}^{n} \binom{n}{j} x^j\right).$$

The term x^n in the product on the right is obtained by adding all terms of the form $\binom{n}{k}x^k\binom{n}{j}x^j$ where $k + j = n$. Since $j = n - k$, each such term has the form $\binom{n}{k}\binom{n}{n-k}x^n$. Thus the coefficient of x^n in the right-hand term is $\sum_{k=0}^{n}\binom{n}{k}\binom{n}{n-k}$. But, by symmetry, $\binom{n}{n-k} = \binom{n}{k}$. Hence

$$\binom{2n}{n} = \sum_{k=0}^{n}\binom{n}{k}^2. \qquad \blacksquare$$

Corollaries 5.4.4 and 5.4.5 can also be derived combinatorially. We challenge the reader to find combinatorial proofs of these results. Other applications of the Binomial Theorem are found in the exercises.

Exercises §5.4

5.4.1. Evaluate $\binom{11}{7}$.

5.4.2. Write out rows 8–10 of Pascal's triangle.

5.4.3. Find y if $\binom{n}{k} + 2\cdot\binom{n}{k-1} + \binom{n}{k-2} = \binom{y}{k}$.

5.4.4. Notice that

$$1 - 1 = 0$$

$$1 - 2 + 1 = 0$$

$$1 - 3 + 3 - 1 = 0$$

$$1 - 4 + 6 - 4 + 1 = 0.$$

Generalize these observations and prove your conjecture.

5.4.5. Prove that $\sum_{k=0}^{n} 2^k\binom{n}{k} = 3^n$ for all $n \geq 0$.

5.4.6. (a) Generalize the following observations, drawn from Pascal's triangle, and prove your conjecture.

$$1 = 1$$

$$1 + 3 = 4$$

$$1 + 3 + 6 = 10$$

$$1 + 3 + 6 + 10 = 20$$

$$1 + 3 + 6 + 10 + 15 = 35.$$

(b) Interpret your result combinatorially.

5.4.7. (a) Generalize the following observations, drawn from Pascal's triangle, and prove your conjecture.

$$1 = 1$$
$$1 + 4 = 5$$
$$1 + 4 + 10 = 15$$
$$1 + 4 + 10 + 20 = 35$$

(b) Interpret your result combinatorially.

5.4.8. Let $a_{n,j}$ denote the number of nonempty subsets of $\{1, 2, \ldots, n + 1\}$ of which j is the largest member. Find $a_{n,j}$ and explain why this verifies the identity $1 + 2 + \cdots + 2^n = 2^{n+1} - 1$.

5.4.9. Determine A in the following identity involving binomial coefficients:

$$1 + \frac{1}{2}\binom{n}{1} + \frac{1}{3}\binom{n}{2} + \cdots + \frac{1}{n+1}\binom{n}{n} = \frac{A}{n+1}.$$

(Hint: Use calculus and the Binomial Theorem.)

5.4.10. Expand each of the following expressions:
(a) $(2x + y)^3$
(b) $(2x + 3y)^4$.
(c) $(x + 7y)^5$.
(d) $(2x - y)^4$.

5.4.11. What is the sum of the coefficients of $(x + 1)^{53}$?

5.4.12. Let $n \in \mathbf{Z}$ with $n \geq 0$ and $k \in \mathbf{Z}$ with $0 \leq k \leq n$. Suppose $a_{n,k} \in \mathbf{R}$ and that $a_{n,0} = 1 = a_{n,n}$ and $a_{n+1,k} = a_{n,k} + a_{n,k-1}$ for $n > 0$ and $1 \leq k \leq n$. Show that $a_{n,k} = \binom{n}{k}$.

5.4.13. Find a combinatorial interpretation of Corollary 5.4.4.

5.4.14. Find a combinatorial or geometric interpretation of Corollary 5.4.5.

5.4.15. Use the Binomial Theorem to show that for all $x, y \in \mathbf{R}$ and all $n \in \mathbf{N}$, $(x + y)^n = \sum_{k=0}^{n}\binom{n}{k}x^{n-k}y^k$.

§5.5 Recurrence Relations

Let us introduce our study of recurrence relations by considering a familiar problem. This problem is indicative of a kind of question that arises frequently in the study of discrete structures and provides a simple example of a recurrence relation.

For each $n \in \mathbf{Z}$ with $n \geq 0$, let A_n be a finite set containing n elements. Let s_n denote the number of subsets of A_n. Our problem is to determine the number s_n. Of course, we already know that $s_n = 2^n$. We shall, however, ignore this fact and shall determine the number s_n, $n \geq 0$, in another way.

Our plan is to relate the number s_{n+1} to the number s_n. Consider the set A_{n+1} having $n + 1$ elements: $A_{n+1} = \{a_1, \ldots, a_n, a_{n+1}\}$. If $C \subseteq A_{n+1}$, then either (i) $C \subseteq \{a_1, \ldots, a_n\}$, or (ii) $C = B \cup \{a_{n+1}\}$ where $B \subseteq \{a_1, \ldots, a_n\}$ (see Lemma 2.2.5). There are exactly s_n subsets of A_{n+1} of type (i) and s_n subsets of A_{n+1} of type (ii) and no subset of A_{n+1} can be of both type (i) and type (ii). Thus, by the Addition Principle, $s_{n+1} = s_n + s_n = 2 \cdot s_n$.

The sequence of numbers $\{s_1, s_1, s_2, \ldots, s_n, \ldots\}$ satisfies the condition that for all $n \geq 0$, $s_{n+1} = 2 \cdot s_n$. Moreover, it is clear that $s_0 = 1$. It follows, therefore, that $s_1 = 2 \cdot s_0 = 2$, $s_2 = 2 \cdot s_1 = 2^2$, $s_3 = 2^3$, and by induction we can show that $s_n = 2^n$ for all $n \geq 0$. (Quick proof of the inductive step: Suppose $s_n = 2^n$. We show $s_{n+1} = 2^{n+1}$: By the basic relation and by inductive hypothesis, $s_{n+1} = 2 \cdot s_n = 2 \cdot 2^n = 2^{n+1}$.) Thus we have determined the value of s_n for all $n \geq 0$ by establishing the relation $s_{n+1} = 2 \cdot s_n$ (known as the *recurrence relation* for the sequence s_n), by knowing the value $s_0 = 1$ (known as the *initial condition* or *boundary condition* for the sequence), by calculating the first few terms of the sequence to develop a conjecture for the value for the general term, and by proving the conjecture using mathematical induction.

We shall demonstrate the procedure illustrated above with other examples after making some definition. First recall that a sequence with values in $\mathbf{R}$ is a function s: $\mathbf{N} \to \mathbf{R}$. For each $n \in \mathbf{N}$, we write $s(n) = s_n$ and denote the sequence s by $\{s_n | n \in \mathbf{N}\}$.

Definition 5.5.1 *Recursive Sequence and Recurrence Relation*

A sequence $\{s_n | n \in \mathbf{N}\}$ is called *recursive* if there exists $n_0 \in \mathbf{N}$ such that for all $n > n_0$, the term s_n can be expressed as a function of $s_0, s_1, \ldots, s_{n-1}$. A function that expresses s_n in terms of $s_0, \ldots, s_{n-1}$ is called a *recurrence relation* for the sequence.

Definition 5.5.2 *Initial Conditions*

Let $\{s_n | n \in \mathbf{N}\}$ be a recursive sequence and let n_0 be the least integer such that for all $n > n_0$, s_n can be expressed in terms of $s_0, \ldots, s_{n-1}$. Then the values $s_0, \ldots, s_{n_0}$ are called the *initial conditions* of the sequence $\{s_n | n \in \mathbf{N}\}$.

Example 5.5.1 The sequence $\{s_n | n \in \mathbf{N}\}$ where s_n is the number of subsets of a set with n elements is recursive. The recurrence relation is $s_n = 2 \cdot s_{n+1}$ for $n > 0$. The initial condition is $s_0 = 1$.

Example 5.5.2 The *Fibonacci sequence* $\{f_n | n \in \mathbf{N}\}$ is defined by the initial conditions $f_0 = f_1 = 1$ and the recurrence relation $f_n = f_{n-1} + f_{n-2}$. The *Lucas sequence*, $\{\ell_n | n \in \mathbf{N}\}$, is defined by the same recurrence relation as the Fibonacci sequence, $\ell_n = \ell_{n-1} + \ell_{n-2}$, but with initial conditions $\ell_0 = 1$, $\ell_1 = 3$.

Example 5.5.3 Let $n \in \mathbf{Z}^+$ and consider an $n \times n$ chessboard. Let r_n denote the number of ways of placing n indistinguishable rooks on the chessboard so that no rook attacks any other rook. (Recall that a rook on a chessboard controls the row and column in which it is placed and that two rooks attack each other if they lie in the same row or the same column.) Clearly, $r_1 = 1$. We can determine a recurrence relation for the sequence $\{r_n | n \in \mathbf{Z}^+\}$ as follows: Consider an $n \times n$ chessboard. We can place a rook in any of the n squares of the first column. (See Figure 5.5.1.) Once this is done, we can place a rook neither in the first column nor in the row in which the first rook is placed. If we delete this row and the first column, then we have in effect an $(n-1) \times (n-1)$ chessboard remaining. The number of ways of placing the remaining $n-1$ rooks so that no rook

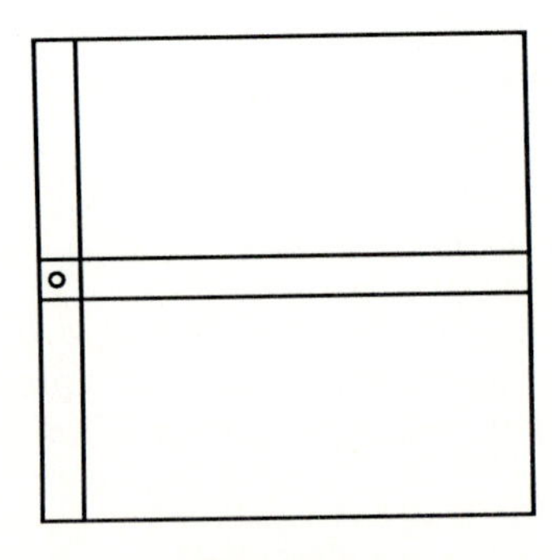

Figure 5.5.1

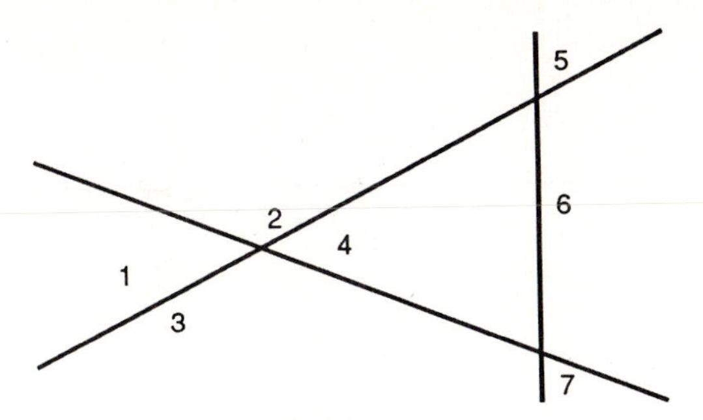

Figure 5.5.2

attacks any other is r_{n-1}. Thus, by the Multiplication Principle, $r_n = n \cdot r_{n-1}$. Notice that since $r_1 = 1$, $r_2 = 2 \cdot 1 = 2$, $r_3 = 3 \cdot r_2 = 3 \cdot 2 = 3!$, $r_4 = 4 \cdot r_3 = 4!$. It appears that for $n \geq 1$, $r_n = n!$, a conjecture that can be easily proved by induction.

Example 5.5.4 Let $n \in \mathbf{N}$. Consider any set of n straight lines in the plane with the property that no two of the lines are parallel and no three of the lines meet at a point. Such a collection of lines is said to be *in general position*. We want to calculate the number of regions in the plane determined by these lines. It is easy to see that 0 lines determine $p_0 = 1$ region, 1 line determines $p_1 = 2$ regions, and 2 lines determine $p_2 = 4$ regions. As can be seen in Figure 5.5.2, 3 lines determine $p_3 = 7$ regions.

Let us establish a recurrence relation for the sequence $\{p_n | n \in \mathbf{N}\}$. Suppose the number p_n is known and consider a set of $n + 1$ lines in the plane. Label the lines $\ell_1, \ldots, \ell_n, \ell_{n+1}$. The lines $\ell_1, \ldots, \ell_n$ divide the plane into p_n regions. Now the line ℓ_{n+1} intersects the lines $\ell_1, \ldots, \ell_n$ in n distinct points. (See Figure 5.5.3, in which lines $\ell_1, \ldots, \ell_5$ are solid and line ℓ_6 is dotted.) Label these points consecutively $x_1, \ldots, x_n$. The line ℓ_{n+1} passes through $n + 1$ of the p_n regions determined by lines $\ell_1, \ldots, \ell_n$. (These regions contain the $n - 1$ finite segments joining x_1 to x_2, x_2 to $x_3, \ldots, x_{n-1}$ to x_n, as well as the two infinite segments with endpoints x_1 and x_n.) Thus p_n, the number of regions determined by $\ell_1, \ldots, \ell_{n+1}$, is equal to p_n, the number of regions determined by $\ell_1, \ldots, \ell_n$ plus the number of regions determined by the intersections of ℓ_{n+1} with $\ell_1, \ldots, \ell_n$. Thus the recurrence relation for the sequence is $p_{n+1} = p_n + (n + 1)$.

Example 5.5.5 We list several examples of recurrence relations.

(i) $a_n = ka_{n-1}$ where $k \in \mathbf{R}$ is constant.
(ii) $a_n = ba_{n-1} + ca_{n-2}$ where $b, c \in \mathbf{R}$ are constant.
(iii) $a_n = a_{n-1} + a_{n-2} + a_{n-3}$.
(iv) $a_n = a_{n-1} \cdot a_{n-2}$.
(v) $a_n = a_{n-1} + n^2$.

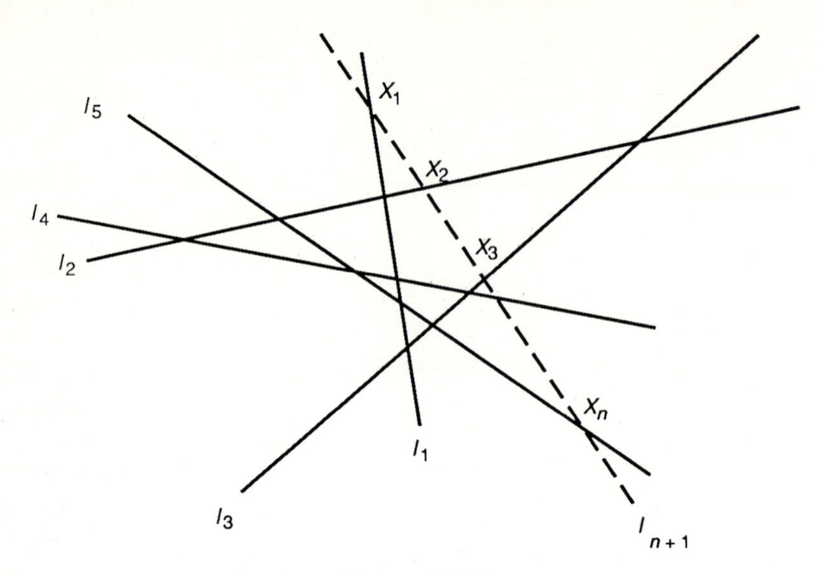

Figure 5.5.3

We devote the remainder of this section to a discussion of two techniques for solving recurrence relations. By this we mean the following: Given a recurrence relation with its initial conditions, we discuss methods of finding explicit formulas for the terms of the sequence. Our discussion will be quite limited. In fact, we shall avoid a general discourse on the theory behind the two methods and shall instead illustrate them in specific cases. In Examples 5.5.1 and 5.5.3 we were able to solve the given recurrence relations by conjecturing a formula based on the first few terms of the sequence and proving the conjecture by mathematical induction. Naturally, one would not expect the procedure to work in every case. The methods to be described now handle a number of general situations.

Linear Recurrence Relations

The first method applies to the so-called linear homogeneous recurrence relations of which Example 5.5.5(i)–(iii) are specific instances. In general, a *linear homogeneous recurrence relation* is a recurrence relation for which there exist $k, n_0 \in \mathbf{N}$ and $\alpha_1, \ldots, \alpha_k \in \mathbf{R}$ such that for all $n > n_0$, $a_n = \alpha_1 a_{n-1} + \alpha_2 a_{n-2} + \cdots + \alpha_k a_{n-k}$. The integer k is called the *order* of the recurrence relation. The recurrence relations $a_n = ka_{n-1}$, $a_n = ba_{n-1} + ca_{n-2}$, and $a_n = a_{n-1} + a_{n-2} + a_{n-3}$ are linear homogeneous recurrence relations of order 1, 2, and 3, respectively. Neither of the recurrence relations in (iv) and (v) is a linear homogeneous recurrence relation.

The technique used to solve linear homogeneous recurrence relations is called the method of the characteristic equation. We begin by solving the first order recurrence relation $a_n = ka_{n-1}$ by the guess-and-check method; we then illustrate the method of the characteristic equation in solving the special second order recurrence relation $f_n = f_{n-1} + f_{n-2}$ and the general second order linear homogeneous recurrence relation $a_n = ba_{n-1} + ca_{n-2}$ where $b, c \in \mathbf{R}$.

Before demonstrating the method of the characteristic equation, let us determine all solutions of the recurrence $a_n = ka_{n-1}$. (We use the experiment–conjecture–verify method of Examples 5.5.1 and 5.5.3.) For this recurrence the initial condition is simply the value a_0. Then $a_1 = k \cdot a_0$, $a_2 = k \cdot a_1 = k^2 \cdot a_0$, $a_3 = k \cdot a_2 = k^3 \cdot a_0$. In general, we conjecture that $a_n = a_0 \cdot k^n$. As before, this conjecture can be proved by induction.

Next we consider the specific second order linear homogeneous recurrence relation $f_n = f_{n-1} + f_{n-2}$. (When the initial conditions are $f_0 = f_1 = 1$, the resulting sequence is the Fibonacci sequence. The values f_0 and f_1 can, however, be arbitrary real numbers.) From the work in the previous paragraph, we know that any solutions to the first order linear homogeneous recurrence relation $a_n = ka_{n-1}$ has the form $a_n = a_0k^n$. Given this observation, let us look for solutions to $f_n = f_{n-1} + f_{n-2}$ which have the form $f_n = a \cdot r^n$ where $a, r \in \mathbf{R} - \{0\}$. At this time, we have no sound reason for believing that any solution of $f_n = f_{n-1} + f_{n-2}$ has the form $f_n = a \cdot r^n$. Nonetheless, we assume that a solution of this form exists and determine the possible values of r. We then check that these values of r do indeed yield solutions to the given recurrence relation.

If $f_n = a \cdot r^n$ for $n \geq 2$, then $f_n = a \cdot r^n = a \cdot r^{n-1} + a \cdot r^{n-2} = f_{n-1} + f_{n-2}$. Since $a \neq 0 \neq r$,

$$a \cdot r^n = a \cdot r^{n-1} + a \cdot r^{n-2}$$

if and only if $$r^2 = r + 1$$

if and only if $$r^2 - r - 1 = 0$$

if and only if $$r = \frac{1 \pm \sqrt{5}}{2}.$$

To summarize, we have proved that if $f_n = a \cdot r^n$ where $a, r \in \mathbf{R} - \{0\}$, and $f_n = f_{n-1} + f_{n-2}$, then $r = (1 \pm \sqrt{5})/2$; and conversely that if $r = (1 \pm \sqrt{5})/2$ and $f_n = a \cdot r^n$, then $f_n = f_{n-1} + f_{n-2}$.

(Notice also that if $a, b \in \mathbf{R}$ and $r_1 = (1 + \sqrt{5})/2$ and $r_2 = (1 - \sqrt{5})/2$, then $f_n = ar_1^n + br_2^n$ is a solution to $f_n = f_{n-1} + f_{n-2}$. In fact, one can prove that any solution to $f_n = f_{n-1} + f_{n-2}$ has the form $f_n = ar_1^n + br_2^n$ where $r_1 = (1 + \sqrt{5})/2$, $r_2 = (1 - \sqrt{5})/2$, and $a, b \in \mathbf{R}$. The numbers a and b are determined by the initial conditions: $f_0 = a + b$, $f_1 = ar_1$

$+ br_2 = (a + b)n/2 + (a - b)\sqrt{5}/2$. For instance, when $f_0 = f_1 = 1$, we find that $a = 1/\sqrt{5}$, $b = 1 - 1/\sqrt{5}$, and $f_n = (1/\sqrt{5})((1 + \sqrt{5})/2)^n + (1 - 1/\sqrt{5})((1 - \sqrt{5})/2)^n$.)

Recall that to solve the equation $a \cdot r^n = a \cdot r^{n-1} + a \cdot r^{n-2}$ for r, we had to solve the quadratic equation $r^2 - r - 1 = 0$. This equation is called the *characteristic equation* for the recurrence relation $f_n = f_{n-1} + f_{n-2}$. We now generalize this example to the case of an arbitrary second order linear homogeneous recurrence relation.

Let $a_n = b \cdot a_{n-1} + c \cdot a_{n-2}$ be an arbitrary second order linear homogeneous recurrence relation. Suppose that $a, r \in \mathbf{R}$ and that $a_n = a \cdot r^n$ is a solution to the given recurrence. Then $a \cdot r^n = b \cdot a \cdot r^{n-1} + c \cdot a \cdot r^{n-2}$, which occurs if and only if $r^2 - b \cdot r - c = 0$. This equation is called the *characteristic equation* of the recurrence relation $a_n = b \cdot a_{n-1} + c \cdot a_{n-2}$. The principal theorem in the theory of second order linear homogeneous recurrence relations expresses the solutions of the recurrence relation in terms of the roots of the characteristic equation. We state but do not prove this theorem.

Theorem 5.5.1 *Let $b, c \in \mathbf{R}$ and let $a_n = b \cdot a_{n-1} + c \cdot a_{n-2}$ define a second order recurrence relation with initial conditions a_0 and a_1. Let $r^2 - ar - b = 0$ be the characteristic equation of this recurrence relation.*

(i) If this equation has distinct real roots r_1 and r_2, then the sequence $\{a_n | n \in \mathbf{N}\}$ is a solution of the given recurrence if and only if there exist real numbers α_1 and α_2 such that for all $n \in \mathbf{N}$, $a_n = \alpha_1 \cdot r_1^n + \alpha_2 \cdot r_2^n$.

(ii) If the characteristic equation has only one real root, r_1 (i.e., if $r^2 - br - c = (r - r_1)^2$), then $\{a_n | n \in \mathbf{N}\}$ is a solution to the given recurrence if and only if there exist $\alpha_1, \alpha_2 \in \mathbf{R}$ such that $a_n = \alpha_1 \cdot r_1^n + \alpha_2 \cdot n \cdot r_1^n$.

The following examples illustrate Theorem 5.5.1 and show how to determine α_1 and α_2 from the initial conditions.

Example 5.5.6 Consider the recurrence relation $a_n = 3a_{n-1} - 2a_n$ with initial conditions $a_0 = 1$ and $a_1 = 3$. Then the characteristic equation of this recurrence is $r^2 - 3r + 2 = 0$, which has roots $r_1 = 2$ and $r_2 = 1$. Thus there exist $\alpha_1, \alpha_2 \in \mathbf{R}$ such that for all $n \in \mathbf{N}$ $a_n = \alpha_1 r_1^n + \alpha_2 r_2^n = \alpha_1 \cdot 2^n + \alpha_2$. When $n = 0$ and $n = 1$, we find that

$$a_0 = 1 = \alpha_1 \cdot 2^0 + \alpha_2 = \alpha_1 + \alpha_2$$

$$a_1 = 3 = \alpha_1 \cdot 2 + \alpha_2 = 2\alpha_1 + \alpha_2.$$

Solving for α_1 and α_2, we see that $\alpha_1 = 2$ and $\alpha_2 = -1$. Therefore, the general solution to the recurrence $a_n = 3a_{n-1} - 2a_{n-2}$ with initial conditions $a_0 = 1$ and $a_1 = 3$ is $a_n = 2^{n+1} - 1$.

Example 5.5.7 Let us solve the recurrence relation $a_n = 2a_{n-1} - a_{n-2}$ with initial conditions $a_0 = 3$ and $a_1 = 5$. The characteristic equation of this recurrence is $r^2 - 2r + 1 = (r - 1)^2 = 0$, which has only one real root, $r_1 = 1$. Thus there exist $\alpha_1, \alpha_2 \in \mathbf{R}$ such that for all $n \in \mathbf{N}$, $a_n = \alpha_1 \cdot 1^n + \alpha_2 \cdot n \cdot 1^n = \alpha_1 + n\alpha_2$. Setting $n = 0$ and $n = 1$, we find that

$$3 = \alpha_1 + 0 \cdot \alpha_2$$

$$5 = \alpha_1 + 1 \cdot \alpha_2,$$

which means that $\alpha_1 = 3$ and $\alpha_2 = 2$. Therefore, if $a_n = 2a_{n-1} - a_{n-2}$ with $a_0 = 3$ and $a_1 = 5$, then $a_n = 3 + 2n$.

In some cases, the characteristic equation of a given recurrence relation will have no real roots. In such a case, the roots will be nonreal, complex numbers. This situation is not covered by Theorem 5.5.1. Nevertheless, the general solution to a recurrence relation whose characteristic equation has no real roots can be described rather simply. For details see Brualdi [3].

Generating Functions

Our second technique for solving recurrence relations is the method of generating functions. The guiding idea behind the method is to associate to each sequence a certain function; by manipulating this function (perhaps using techniques of calculus), we can determine the terms of the sequence.

> Definition 5.5.3 *Generating Function*
>
> Let $\{a_n | n \in \mathbf{N}\}$ be a sequence. The expression $f(X) = a_0 + a_1 \cdot X + a_2 \cdot X^2 + \cdots + a_n \cdot X^n + \cdots = \sum_{n=0}^{\infty} a_n \cdot X^n$ is the *generating function* of the sequence $\{a_n | n \in \mathbf{N}\}$.

The expression $a_0 + a_1 \cdot X + \cdots + a_n \cdot X^n + \cdots$ is called a *power series* in the variable X. We regard this power series merely as a formal expression. A simple but effective way of thinking of a power series is as a polynomial-like expression with (denumerably) infinitely many terms. Two power series can be added and multiplied in exactly the same way that polynomials are added and multiplied.

Perhaps the most important example of a power series is the geometric series:

$$1 + X + X^2 + \cdots + X^n + \cdots = 1/(1 - X).$$

It is proved in calculus that the geometric series converges to the number $1/(1 - X)$ for $|X| < 1$. Since we are regarding a power series as a formal expression, we think of the above equality as an equality of formal expressions. In addition to taking sums and products of power series, we can also differentiate these series by differentiating each term. For example, differentiating each side of the equality

$$1/(1 - X) = 1 + X + X^2 + \cdots + X^n + \cdots$$

yields the equality

$$1/(1 - X)^2 = 1 + 2X + 3X^2 + \cdots + (n + 1)X^n + \cdots.$$

Let us now use the generating function method to solve certain recurrence relations.

Example 5.5.8 Let us reconsider the recurrence $s_n = 2s_{n-1}$ with $s_0 = 1$. The generating function of this sequence is

$$\begin{aligned} f(X) &= \sum_{n=0}^{\infty} s_n \cdot x^n = s_0 + s_1 \cdot X + s_2 \cdot X^2 + \cdots + s_n \cdot X^n + \cdots \\ &= s_0 + 2 \cdot s_0 \cdot X + 2 \cdot s_1 \cdot X^2 + \cdots + 2 \cdot s_{n-1} \cdot X^n + \cdots \\ &= s_0 + 2 \cdot X(s_0 + s_1 \cdot X + \cdots + x_{n-1} \cdot X^{n-1} + \cdots) \\ &= s_0 + 2 \cdot X \cdot f(X) \\ &= 1 + 2 \cdot X \cdot f(X). \end{aligned}$$

Therefore,

$$f(X) - 2 \cdot Xf(X) = f(X) \cdot (1 - 2X) = 1$$

or

$$f(X) = 1/(1 - 2X) = 1 + 2 \cdot X + 2^2 \cdot X^2 + \cdots + 2^n \cdot X^n + \cdots$$

by the formula for the geometric series. Therefore, for all $n \geq 0$, $s_n = 2^n$.

Example 5.5.9 Problem: Solve the recurrence from Example 5.5.4, $p_{n+1} = p_n + (n + 1)$ with initial condition $p_0 = 1$, using the generating function method. Let $f(X) = \Sigma_{n=0}^{\infty} p_n \cdot X^n$ be the generating function of the given sequence. Then

$$
\begin{aligned}
f(X) &= p_0 + p_1 \cdot X + p_2 X^2 + \cdots + p_{n+1} X^{n+1} + \cdots \\
&= p_0 + (p_0 + 1) \cdot X + (p_1 + 2) \cdot X^2 + \cdots \\
&\quad + (p_n + (n+1)) \cdot X^{n+1} + \cdots \\
&= p_0 + X + 2 \cdot X^2 + \cdots + (n+1) \cdot X^{n+1} + \cdots \\
&\quad + p_0 X + p_1 \cdot X^2 + \cdots + p_n \cdot X^{n+1} + \cdots \\
&= 1 + X + 2 \cdot X^2 + \cdots + n \cdot X^n + \cdots \\
&\quad + X \cdot (p_0 + p_1 \cdot X + \cdots + p_n \cdot X^n + \cdots) \\
&= 1 + X \cdot (1 + 2 \cdot X + \cdots + n \cdot X^{n-1} + \cdots) \\
&\quad + X \cdot (p_0 + p_1 \cdot X + \cdots + p_n \cdot X^n + \cdots) \\
&= 1 + X \cdot \frac{1}{(1-X)^2} + X \cdot f(X).
\end{aligned}
$$

Therefore,

$$f(X) - X \cdot f(X) = f(X)(1 - X) = 1 + \frac{X}{(1-X)^2}$$

and

$$f(x) = \frac{1}{(1-X)} + \frac{X}{(1-X)^3}.$$

To determine a power series expression for $1/(1-X)^3$, we can start with the power series expression for $1/(1-X)^2$ and differentiate both sides: Since

$$1/(1-X)^2 = 1 + 2 \cdot X + 3 \cdot X^2 + \cdots + (n+1) \cdot X^n + \cdots,$$

$$2/(1-X)^3 = 2 + 3 \cdot 2 \cdot X + 4 \cdot 3 \cdot X^2 + \cdots + (n+1) \cdot nX^{n-1} + \cdots$$

and thus

$$1/(1-X)^3 = 1 + \frac{3 \cdot 2}{2} X + \frac{4 \cdot 3}{2} X^2 + \cdots + \frac{(n+2)(n+1)}{2} X^n + \cdots,$$

therefore,

$$f(X) = \frac{1}{(1-X)} + \frac{X}{(1-X)^3}$$

$$= 1 + X + X^2 + \cdots + X^n + \cdots + X + \frac{3 \cdot 2}{2} \cdot X^2$$

$$+ \cdots + \frac{(n+1) \cdot n}{2} \cdot X^n + \cdots$$

$$= 1 + (1+1) \cdot X + \left(1 + \frac{3 \cdot 2}{2}\right) \cdot X^2 + \cdots$$

$$+ \left(1 + \frac{(n+1) \cdot n}{2}\right) \cdot X^n + \cdots.$$

It follows that

$$p_n = 1 + \frac{(n+1) \cdot n}{2}.$$

Exercises §5.5

5.5.1. Solve each of the following recurrence relations.
(a) $a_n = a_{n-1} + 2a_{n-2}$, $a_0 = 1$, and $a_1 = 2$.
(b) $a_n = 3a_{n-1} - 2a_{n-2}$, $a_0 = 3$, and $a_1 = 1$.
(c) $a_n = 3a_{n-1} + 4a_{n-2}$, $a_0 = 2$, and $a_1 = 7$.
(d) $a_n = 2a_{n-2}$, $a_0 = 1$, and $a_1 = 2$.
(e) $a_n = 5a_{n-1} + 6a_{n-2}$, $a_0 = 0$, and $a_1 = 1$.
(f) $2a_n = 3a_{n-1} - a_{n-2}$, $a_0 = 1$, and $a_1 = 2$.

5.5.2. Solve each of the following recurrence relations using generating functions.
(a) $a_n = 3a_{n-1}$, $a_0 = 1$.
(b) $a_n = 4a_{n-1}$, $a_0 = 2$.
(c) $a_n = 8a_{n-1} - 16a_{n-2}$, $a_0 = 1$, and $a_1 = 2$.

5.5.3. A robot can move forward in steps of size 1 meter or 2 meters. Let RB_n denote the number of ways the robot can walk n meters. Find RB_n for $n \leq 5$ and find a recurrence relation for RB_n. For example, $\mathrm{RB}_3 = 3$ since the robot can move 3 meters in three ways: (a) 1 m, 1 m, 1 m, (b) 1 m, 2 m, (c) 2 m, 1 m.

5.5.4. Let B_n be the number of binary strings (strings of 0's and 1's) of length n having no two consecutive 0's. Find a recurrence relation for B_n.

5.5.5. Let r_n^1 be the number of ways of placing n distinguishable rooks on an $n \times n$ chessboard. (For example, the rooks can be numbered $1, 2, \ldots, n$.) Find a formula for r_n^1.

5.5.6. Suppose n rooks are placed on an $n \times n$ chessboard in such a way that the arrangement is symmetric about the diagonal running from the lower left to the upper right end of the board.
(a) Let Q_n denote the number of such arrangements provided that the rooks are indistinguishable. Find a recurrence relation for Q_n.
(b) Let R_n denote the number of such arrangements provided that the rooks are distinguishable. Find a recurrence relation for R_n.

5.5.7. Suppose you have $\$n$ to spend and you buy juice for \$1, milk for \$2, and beer for \$2. Let A_n denote the number of ways of spending all $\$n$. (Assume that buying beer–milk–beer is different from buying milk–beer–beer.) Find a recurrence relation for A_n.

5.5.8. Consider n straight lines in the plane such that no two are parallel but exactly three are concurrent. Into how many regions is the plane separated?

5.5.9. (a) Suppose n points are given on a circle. How many line segments are there joining pairs of points?
(b) Suppose the n points are spaced so that no 3 line segments have a common intersection point inside the circle. Let I_n denote the number of intersection points inside the circle. (For instance, $I_4 = 1$ and $I_5 = 5$.) Find I_n for $n \geq 4$.

5.5.10. Given the situation as in the preceding problem, let R_n^1 denote the number of regions into which the interior of the circle is partitioned. For instance, $R_2^1 = 2$, $R_3^1 = 4$, $R_4^1 = 8$, and $R_5^1 = 16$. Find a recurrence relation for R_n^1.

5.5.11. (a) Let $r_{1,n}$ denote the number of regions (intervals) into which n points divide a line. Establish the recurrence relation $r_{1,n+1} = r_{1,n} + 1$ and prove that $r_{1,n} = \binom{n}{0} + \binom{n}{1}$.
(b) Let $r_{2,n}$ denote the number of regions into which n lines (no two parallel and no three concurrent) divide the plane. (Thus $r_{2,n} = p_n$ where p_n is as in Example 5.5.4.) Show that $r_{2,n+1} - r_{2,n} = r_{1,n}$ and prove that $r_{2,n} = \binom{n}{0} + \binom{n}{1} + \binom{n}{2}$.
(c) Let $r_{3,n}$ denote the number of regions into which n planes divide 3-space. (Assume no two planes are parallel and no three planes meet on a line.) Show that $r_{3,n+1} - r_{3,n} = r_{2,n}$ and find a formula for $r_{3,n}$.

5.5.12. Arrange the numbers $1, 2, \ldots, n$ consecutively about the circumference of a circle. Now, remove the number 2 and proceed in order removing every other number among those that remain until only one number is left. Let F_n denote the final number that remains. (For example, $F_2 = 1$, $F_3 = 3$, $F_4 = 1$, and $F_5 = 3$.) Show that $F_{2n} = 2 \cdot F_n - 1$ and $F_{2n+1} = 2 \cdot F_n + 1$.

§5.6 Graph Theory

As we saw in Chapter 3, any partial ordering on a finite set can be represented pictorially: Suppose $\leq$ is a partial ordering on a finite set A. To each point of A is associated a point in the plane and the points corresponding to the elements a and b of A are joined by a sequence of line segments if and only if $a \leq b$.

As it happens, this pictorial form of representing a relation is so useful and powerful that it has become a mathematical object of considerable interest in its own right. This object is called a *graph* and the study of graphs is known, therefore, as graph theory. (Note: Graphs as studied in graph theory are not to be confused with graphs of functions as studied in calculus and analysis.) Graph theory is both an old and a new mathematical discipline. Its origins can be traced to the early 1700s and the Koenigsberg Bridge Problem, which was presented at the start of this chapter. A major open problem, the Four-Color Conjecture, was proposed in 1879 and occupied the attention of graph theorists for nearly a century. Nevertheless, until the second half of the twentieth century, graph theory was not regarded as an area of central importance in mathematics. With the rise of computer science, discrete mathematics, and applications of these disciplines, opinions have changed. Graph theory is now recognized as a subject of significance both for mathematicians and for users of mathematics.

What is a graph? Informally, a graph is a finite set of points together with line segments joining certain pairs of the points. From a rigorous viewpoint, this definition might be considered deficient. For instance, what if the line segments joining two different pairs of points should intersect? We would want to exclude this intersection point from our graph, yet doing so requires extra stipulations and perhaps leads to ambiguity. To avoid these potential difficulties, we make the following formal definition of a graph. Immediately after doing so, we revert to our informal picture of a graph as a set of points, some pairs of which are joined by line segments.

Definition 5.6.1 *Simple Graph*

A *simple graph* G consists of a finite set $V(G)$ together with a collection $E(G)$ of two-element subsets of G. The elements of $V(G)$ are called *vertices* or *nodes*; the elements of $E(G)$ are called *edges*. We denote a simple graph by $G = (V(G), E(G))$.

We write $V = V(G)$ and $E = E(G)$ as long as this notation causes no ambiguity.

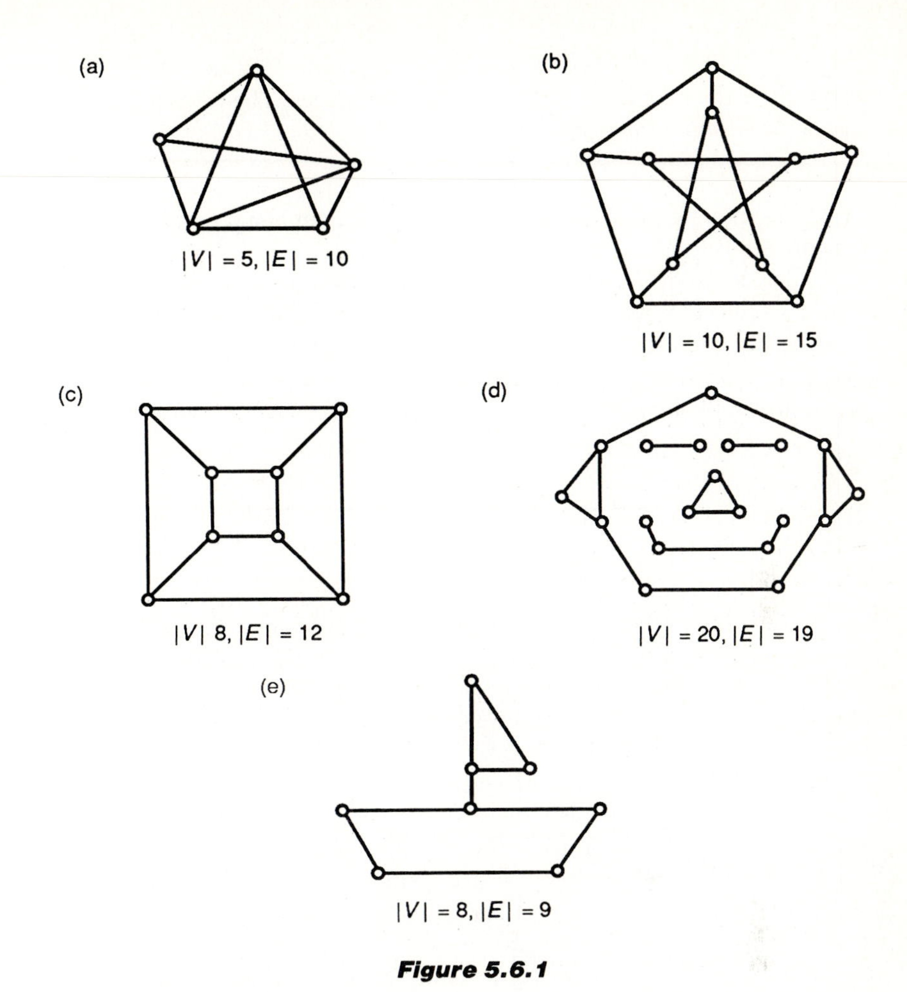

Figure 5.6.1

In our geometric picture of a simple graph, the elements of V are, of course, represented by points in the plane. The points corresponding to $a, b \in V$ are connected by a line segment if and only if $\{a, b\} \in E$. Thus Definition 5.6.1 does capture formally our intuitive idea of a graph.

Example 5.6.1 Figure 5.6.1 shows some examples of simple graphs. In each case, we indicate the number of vertices and edges.

Example 5.6.2 Graphs arise naturally in many everyday situations. Here are two examples.

(i) Let the nodes of G be cities in a given region (e.g., county or state) in the United States. For $v_1, v_2 \in V = V(G)$, join v_1 and v_2 by an edge of G if there is a highway joining v_1 to v_2 that does not pass through any other city (node) in $V(G)$.

(ii) Let V be a set of people. Each pair $v_1, v_2 \in V$ is joined by an edge if and only if the people v_1 and v_2 are acquaintances. The resulting graph is called a *friendship graph* for the set V.

Before moving on to more examples, a few comments on the definition might be helpful.

1. The edges of a simple graph are not considered to be directed: for if $a, b \in V$, then the sets $\{a, b\}$ and $\{b, a\}$ are equal. We can define a *directed graph* or *digraph* as a pair (V, E) where V is a set and E is a collection of *ordered* pairs of elements of V. In this case, if $(a, b) \in E$, then we think of the line segment joining a to b as having an arrow at b.

2. If $a, b \in V$, then at most one edge of E joins a to b. Mathematical structures with *multiple edges* can be defined and studied. We shall, however, only briefly mention graphs with multiple edges in this quick introduction to graph theory, and we do not consider graphs with loops at a vertex, although such graphs do arise naturally and are legitimate objects of study. (See Figure 5.6.2.) For the record, we have the following general definition: A *graph* G consists of a pair (V, E) where V is a finite set of points and E is a collection of subsets of V having one element or two elements in which a given one- or two-element subset might appear more than once. This definition allows for loops (each one-element subset of V defines a loop) and multiple edges.

Graph theory is filled with interesting problems that are easy to state and challenging to solve. In addition, with the increasing applications of graph theory in areas such as computer science, operations research, and engineering, it is important that *efficient* algorithms be devised to solve problems in graph theory. In other words, it is often not enough to know that a solution exists to a given problem. What is needed is an unambiguous procedure that solves the given problem in a finite number of steps. Moreover, for the algorithm to be regarded as efficient, the number of steps

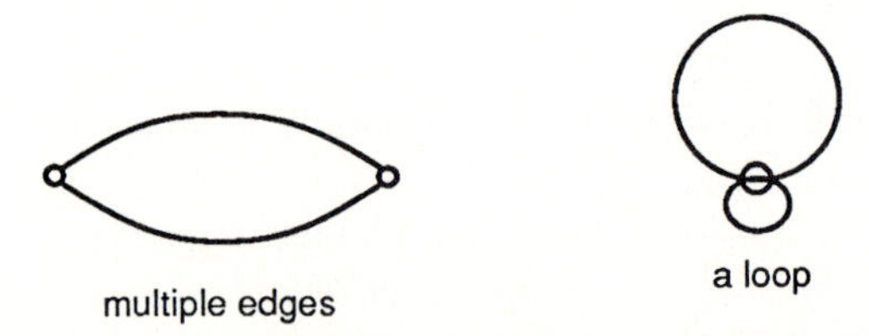

Figure 5.6.2

required by the algorithm to solve the problem should be a polynomial function (as opposed to an exponential function) of the parameter(s) that determine the size of the problem. In this section and the next, we describe some important problems in graph theory and algorithms that solve these problems.

Euler Circuits

Let us consider the Koenigsberg Bridge Problem and its solution by Euler. Recall that Euler's problem was to determine whether it is possible to start at either A, B, C, or D and traverse each of the seven bridges exactly once. (See Figure 5.6.3.)

First we can transform the configuration of bridges into a graph (which is not a simple graph). We form the graph by assigning a node to each body of land and an edge to each bridge. The resulting graph has 4 nodes and 7 edges. (See Figure 5.6.4.) We now introduce some terms that provide the vocabulary for stating Euler's problem for this graph.

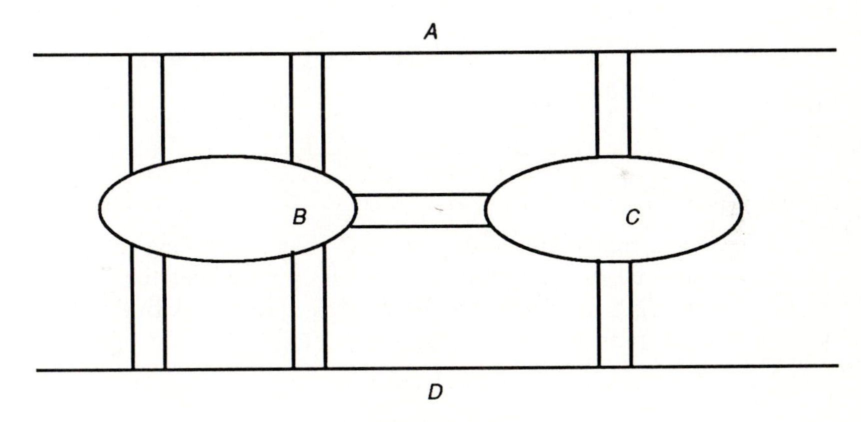

Figure 5.6.3

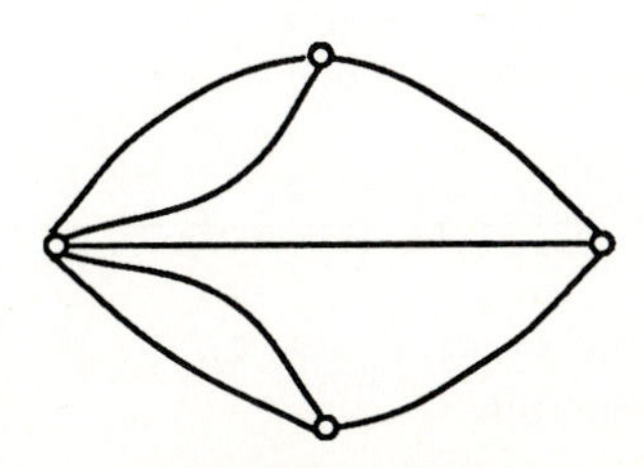

Figure 5.6.4

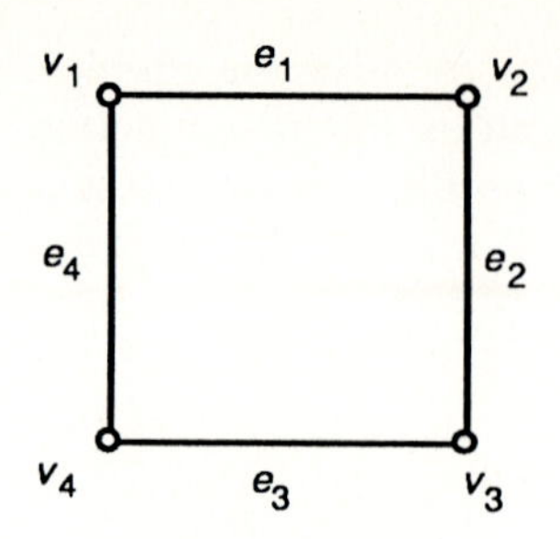

Figure 5.6.5

Definition 5.6.2 *Incidence and Adjacency*

Let $G = (V, E)$ be a simple graph. If $e = \{v, w\} \in E$, then e is said to be *incident* on v and w; v and w are called the *ends* of e and v and w are said to be *adjacent*.

Example 5.6.3 As shown in Figure 5.6.5, v_1 and v_2 are adjacent while v_1 and v_3 are not adjacent.

Definition 5.6.3 *Path*

Let G be a simple graph. A *path* in G is an alternating sequence of vertices and edges $v_1, e_1, v_2, e_2, \ldots, e_n, v_{n+1}$ such that for each i, $1 \leq i \leq n$, the ends of edge e_i are the vertices v_i and v_{i+1}. In this case we say that the path *begins* at v_1 and *ends* at v_{n+1}.

Definition 5.6.4 *Circuit*

Let G be a simple graph. A *circuit* in G is a path $v_1, e_1, v_2, e_2, \ldots, e_n, v_{n+1}$ such that $v_{n+1} = v_1$ and for $1 \leq i < j \leq n$, $e_i \neq e_j$.

In Example 5.6.3, $v_1, e_1, v_2, e_2, v_3, e_3$ is a path and $v_3, e_3, v_4, e_4, v_1, e_1, v_2, e_2, v_3$ is a circuit.

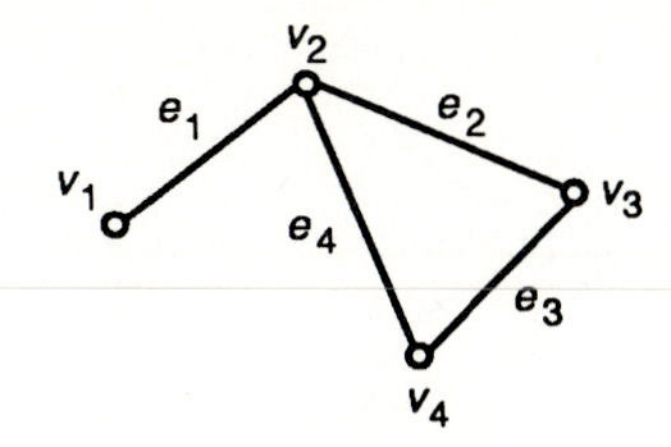

Figure 5.6.6

Remark Since in a simple graph, each pair of adjacent vertices determines a unique edge, a path (or circuit) in a simple graph can be specified by listing the sequence of adjacent vertices. Thus the path $v_1, e_1, v_2, e_2, \ldots, e_n, v_{n+1}$ in a simple graph G will be denoted as $v_1, v_2, \ldots, v_{n+1}$ or by $v_1 v_2 \cdots v_{n+1}$. Whenever we denote a path in this fashion, it is understood that vertices v_i and v_{i+1} are adjacent for $1 \leq i \leq n$. We denote a circuit by $v_1 v_2 \cdots v_n v_1$.

> **Definition 5.6.5** *Euler Path and Circuit*
>
> An *Euler path* in a simple graph G is a path in G that contains each edge of G exactly once. An *Euler circuit* is an Euler path that is also a circuit.

Thus, for the graph shown in Figure 5.6.6, $v_1 v_2 v_3 v_4 v_2$ (i.e., $v_1 e_1 v_2 e_2 v_3 e_3 v_4 e_4 v_2$) is an Euler path, and for the graph in Figure 5.6.7, $v_1 v_2 v_3 v_4 v_1$ is an Euler circuit.

A moment's thought reveals that the concepts of path, circuit, and Euler path can be formulated for general graphs as well as simple graphs. In fact, the Koenigsberg Bridge Problem, when stated in the language of

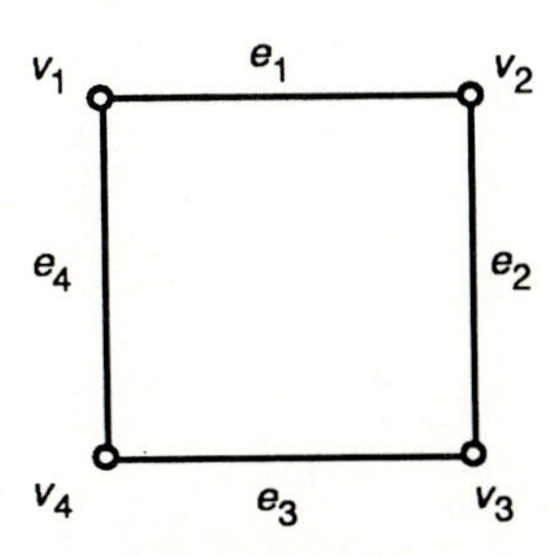

Figure 5.6.7

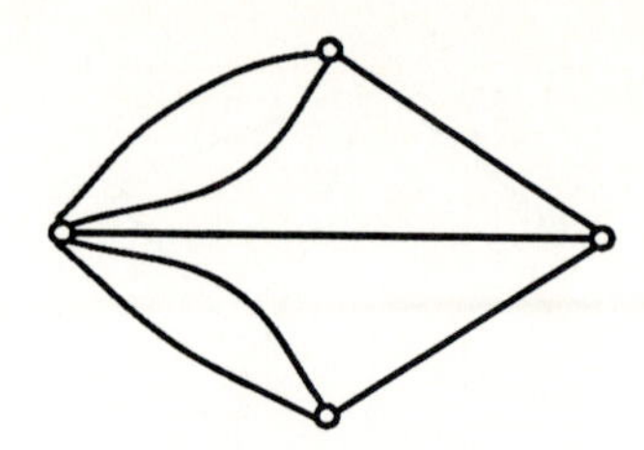

Figure 5.6.8

graph theory, becomes: Does the graph shown in Figure 5.6.8 have an Euler path?

Let us investigate the question of whether a simple graph G has an Euler circuit. (Actually, our remarks apply to an arbitrary graph G.) To carry out this investigation, we let G be a simple graph that possesses an Euler circuit, **C**. Then **C** consists of a sequence of vertices and edges $v_1e_1v_2e_2 \cdots e_nv_1$. Since every edge appears in **C**, every vertex must also appear in **C**; thus the graph G has the property that for any two of its vertices, v and w, there exists a path in G which begins at v and ends at w. For example, if $v = v_i$ and $w = v_j$ where $i < j$, then the path $v_ie_iv_{i+1} \cdots e_{j-1}v_j$ begins at v and ends at w. This property of a graph is important enough to deserve a name.

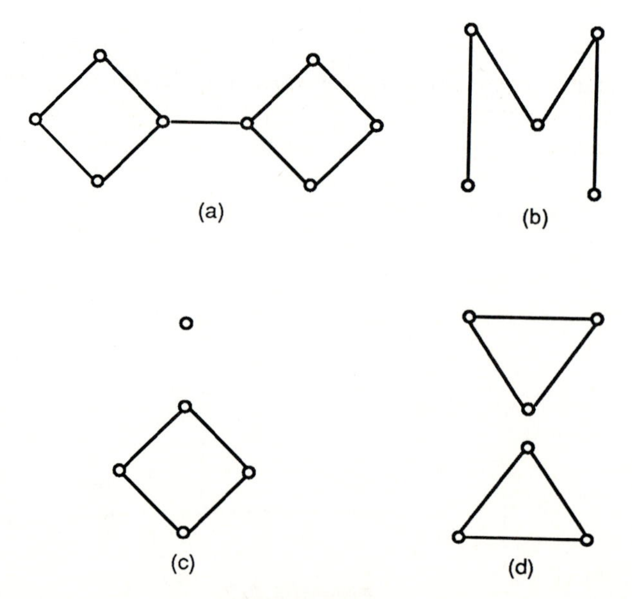

Figure 5.6.9

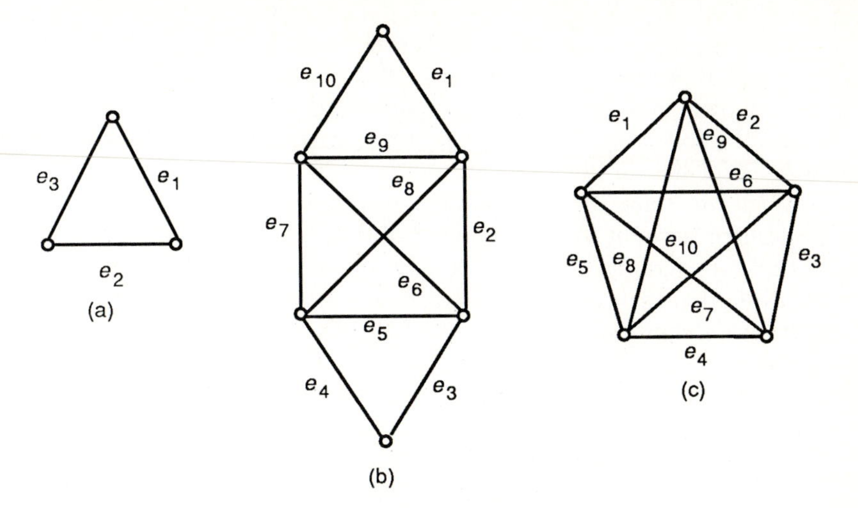

Figure 5.6.10

> **Definition 5.6.6** *Connected Graph*
>
> A graph g is *connected* if for all $v, w \in V$, there exists a path that begins at v and ends at w. A graph that is not connected is called *disconnected*.

Figure 5.6.9 illustrates both connected and disconnected graphs.

The argument in the previous paragraph shows that a necessary condition that G have an Euler circuit is that G is connected. Henceforth, we assume that G is a connected simple graph on which an Euler circuit **C** exists.

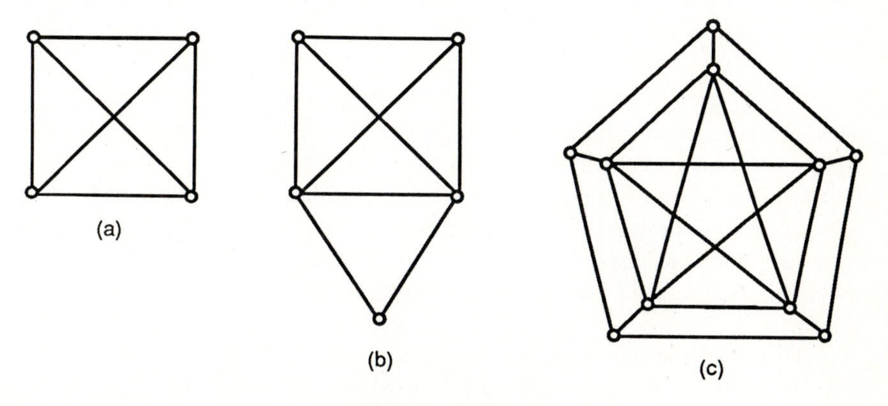

Figure 5.6.11

At this point, in order to develop a better feel for the question, we might find it worthwhile to consider some examples. Each of the simple graphs shown in Figure 5.6.10 has an Euler circuit (an Euler circuit is obtained by following the edges in sequence), whereas none of the graphs in Figure 5.6.11 has an Euler circuit.

What are some common features of graphs in Figure 5.6.10, features that perhaps account for the existence of an Euler circuit? One characteristic of all three graphs is that the number of edges incident to each vertex is even. Note that none of the graphs in Figure 5.6.11 has this property. Let us pause for a definition that is suggested by this remark.

Definition 5.6.7 *Degree of a Vertex*

Let G be a simple graph and let $v \in V$. The *degree of* v, written $\deg(v)$, is the number of edges incident on v.

At this juncture we might be tempted to conjecture that a simple graph G has an Euler circuit if and only if for each $v \in V$, $\deg(v)$ is even. We have several examples that support this statement. Before trying to prove this conjecture, let us test it by considering a specific type of graph: Let G be a simple graph having a vertex v of degree 1. Can G have an Euler circuit? If so, then this circuit, call it $\mathbf{C}$, cannot start (and end) at v; for if it does, then $\mathbf{C} = ve \cdots ev$ where e is the unique edge of $\mathbf{C}$ incident to v and edge e is used twice in $\mathbf{C}$. Similarly, if $\mathbf{C}$ starts elsewhere, then $\mathbf{C} = we_1 \cdots eve \cdots w$ and again edge e appears twice in $\mathbf{C}$. Thus, if G has an Euler circuit, then G cannot have a vertex of degree 1. With this additional evidence, we can formulate as a theorem the characterization of graphs that possess Euler circuits.

Theorem 5.6.1 *A connected simple graph G possesses an Euler circuit if and only if* $\deg(v)$ *is even for each vertex v of G.*

Proof Let G be a connected simple graph. First suppose that G has an Euler circuit $\mathbf{C} = v_1e_1v_2e_2 \cdots e_nv_1$. Let $v \in V$ with $v \neq v_1$. Then each time v appears in $\mathbf{C}$, it appears with two edges that are incident to v. Thus, if v appears k times in $\mathbf{C}$, then $2 \cdot k$ edges incident to v appear in $\mathbf{C}$. Since each edge of G appears exactly once in $\mathbf{C}$, there are $2 \cdot k$ edges of G having v as endpoint, i.e., $\deg(v) = 2 \cdot k$. A similar argument shows that $\deg(v_1)$ is even. Therefore, if G has an Euler circuit, then each vertex of G has even degree.

To prove the converse, we reason by contradiction. We want to show that if G is a connected simple graph in which each vertex has even degree, then G has an Euler circuit. Thus let us suppose that there exists a

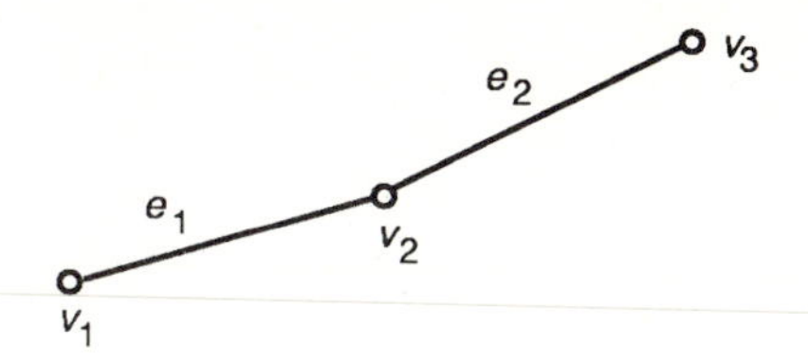

Figure 5.6.12

connected simple graph having every vertex of even degree and containing no Euler circuit. Among all such graphs, choose a graph G having the least number of edges. In other words, G is connected and simple, has every vertex of even degree, has no Euler circuit, and if G' is any other graph with these properties, then $|E(G')| \geq |E(G)|$.

Let $v_1 \in V(G)$. Take an edge e_1 incident to v_1 and let v_2 denote the other end of e_1. Since $\deg(v_2)$ is even, there exists an edge e_2 incident to v_2 such that $e_2 \neq e_1$. Let v_3 denote the other end of e_2. (See Figure 5.6.12.) We can continue in this fashion to form a path $P = v_1e_1v_2e_2v_3 \cdots v_ne_nv_{n+1}$. Since $V(G)$ is a finite set, we eventually repeat a vertex: There exist $j \in \mathbf{N}$ and $i < j$ such that $v_j = v_i$. Thus the path $v_ie_i \cdots e_{j-1}v_j$ is a circuit, since it has no repeated edges. Let **C** denote a circuit of G with no repeated edges having maximum length (i.e., maximum number of edges) among all circuits of G with no repeated edges. (See Figure 5.6.13.)

Since G has no Euler circuit, the circuit **C** does not contain all the edges of G. Let G_1 be the graph obtained by taking all the edges of G not in **C** together with the vertices incident to these edges. Figure 5.6.14 illustrates G_1 in a special case. In the graph G_1, each vertex has even degree, since each vertex of G has even degree and at each vertex of **C**, an even number of edges are removed to form G_1. Also note that G_1 has fewer edges than G since edges are removed from G to form G_1.

Since G is connected, there exists a vertex $v \in V(G)$ such that v is incident to edges of G that are in **C** and to edges that are in G_1. Now the graph G_1 might not be connected, but there exists a portion of G_1 that contains v and that is connected. Call this portion G_2. Since G_2 is a connected simple graph having every vertex of even degree and having

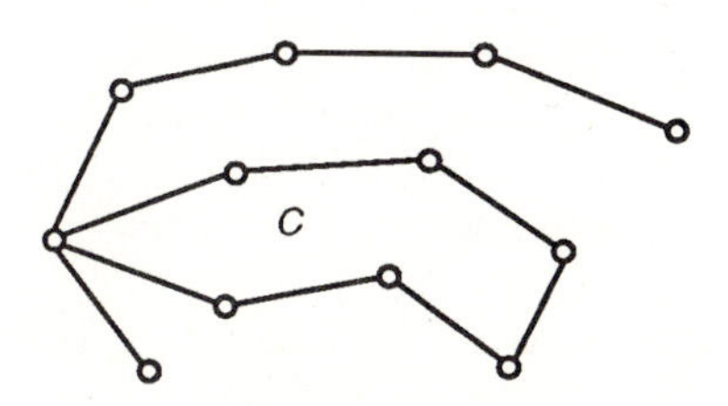

Figure 5.6.13

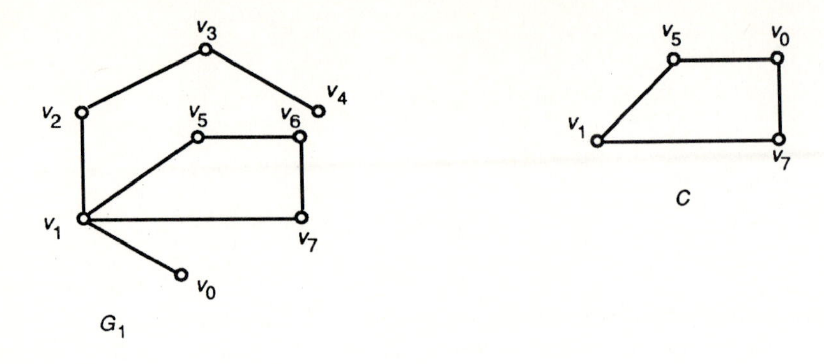

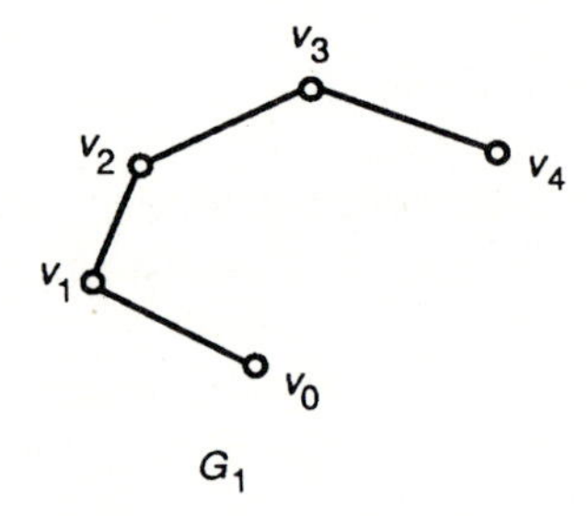

Figure 5.6.14

fewer edges than G, G_2 contains an Euler circuit $\mathbf{C}'$. (See Figure 5.6.15.) We can assume that the circuits $\mathbf{C}$ and $\mathbf{C}'$ both begin at v. Hence if we follow $\mathbf{C}$ by $\mathbf{C}'$, then we obtain a circuit of G, $\mathbf{CC}'$, that is longer than $\mathbf{C}$, which is a contradiction. This conclusion means that the graph G does indeed have an Euler circuit. ■

Theorem 5.6.1 provides a workable necessary and sufficient condition for the existence of an Euler circuit in a graph G. The existence of an Euler circuit can be considered a *global* property of G, namely, a property stated

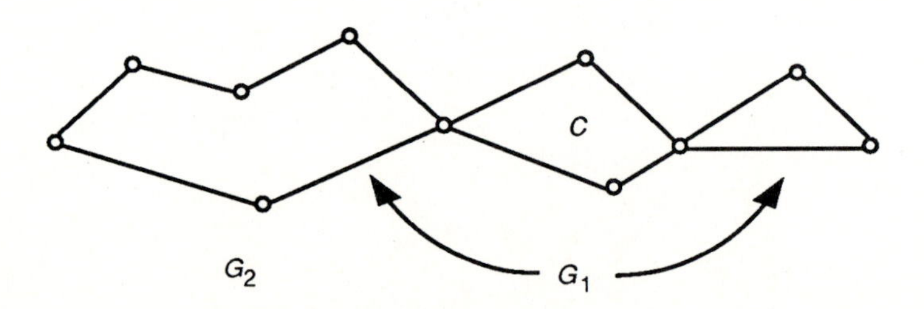

Figure 5.6.15

in terms of G as a whole. Theorem 5.6.1 relates this global property to a *local* property, namely, a property stated in terms of each vertex: G has an Euler circuit if and only if $\deg(v)$ is even for each vertex $v \in V$. In this sense, Theorem 5.6.1 provides a nontrivial and useful characterization of graphs possessing an Euler circuit.

We can relax the condition of Theorem 5.6.1 slightly and ask for conditions that guarantee the existence of an Euler path on a graph G. In other words, when does a graph G contain a path (that might be a circuit) that contains every edge exactly once? We find the answer in the next result, whose proof is left as an exercise.

Corollary 5.6.2 *A connected simple graph G contains an Euler path if and only if the number of vertices of odd degree in G is either* 0 *or* 2.

The proof of Theorem 5.6.1 is an example of an *existence proof*. The argument shows that if we assume that an Euler circuit does not exist in G, then we are led to a contradiction. Thus the existence of an Euler circuit is guaranteed even though the circuit itself is not constructed.

In many situations, of course, it is not enough for us to know that a configuration of a given type exists. We might also want to know how to construct or obtain such a configuration. In particular, it is natural to look for an efficient algorithm that constructs the desired object or configuration. We close our discussion of Euler circuits by outlining and illustrating an algorithm that constructs an Euler circuit in any connected simple graph all of whose vertices have even degree. This procedure is known as Fleury's algorithm.

The steps in Fleury's algorithm are:

1. Initial step. Choose any vertex v_0 and define $P_0 = v_0$. (P_0 is the start of a path that will become the Euler circuit.)
2. General step. Suppose that a path $P_i = v_0 e_1 v_1 e_2 \cdots e_i v_i$ has been formed. Choose an edge $e_{i+1} \in E - \{e_1, \ldots, e_i\}$ (i.e., an edge that has not yet been used) so that
   a. e_{i+1} is incident to v_i.
   b. Let G_i be the graph obtained from G by taking all the edges not in $\{e_1, \ldots, e_i\}$ and all vertices incident to these edges. (Clearly, e_{i+1} is an edge of G_i.) The edge e_{i+1} has the property that its removal from G_i does not create a disconnected graph. If no such edge exists, then choose any edge.
3. Stop if step 2 cannot be applied.

Let us illustrate Fleury's algorithm in action. For the graph G with vertex v_0 as labeled in Figure 5.6.16, suppose we follow Fleury's algorithm and choose e_1, v_1, e_2, v_2, e_3, $v_3 = v_0$, and e_4 and v_4, shown in Figure 5.6.17. At this point we must choose e_5. If we select $e_5 = e$, then the graph G_5

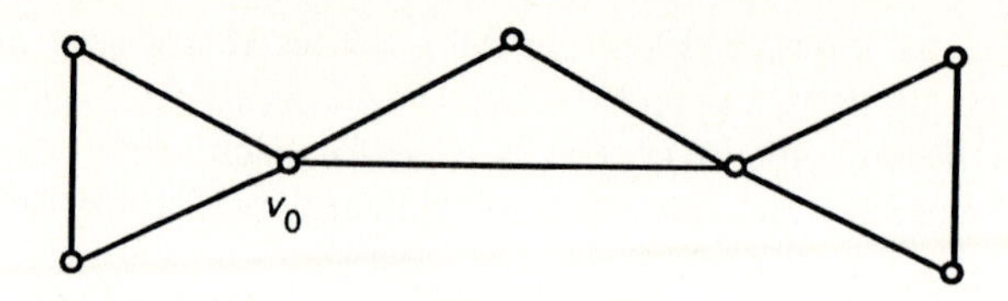

Figure 5.6.16

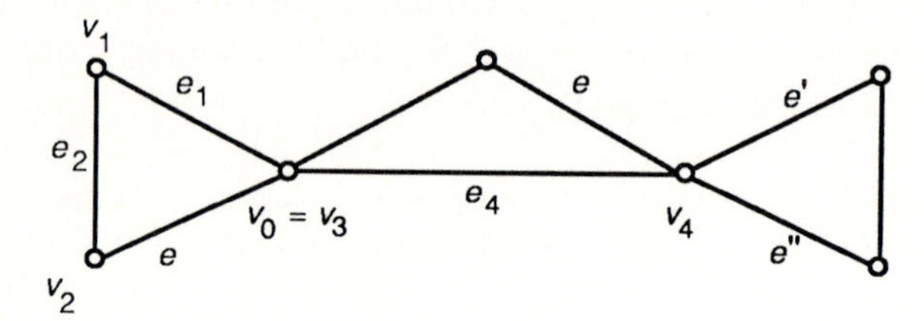

Figure 5.6.17

whose edge set is $E(G_5) = E(G) - \{e_1, e_2, e_3, e_4\}$ and whose vertex set is $V(G_5) = V(G) - \{v_1, v_2\}$ becomes disconnected upon the removal of $e_5 = e$. (See Figure 5.6.18.) Thus, according to Fleury's algorithm, we must choose $e_5 = e'$ or $e_5 = e''$. Once we make one of these choices, we can complete the Euler circuit easily. (See Figure 5.6.19.)

We shall not prove that Fleury's algorithm always works, i.e., that for any connected simple graph whose vertices all have even degree, Fleury's

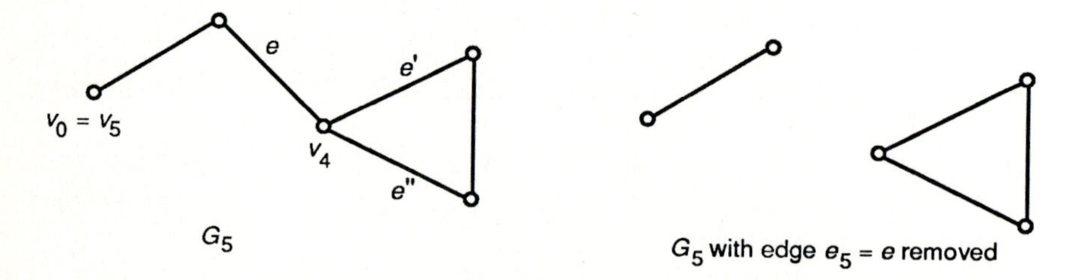

Figure 5.6.18

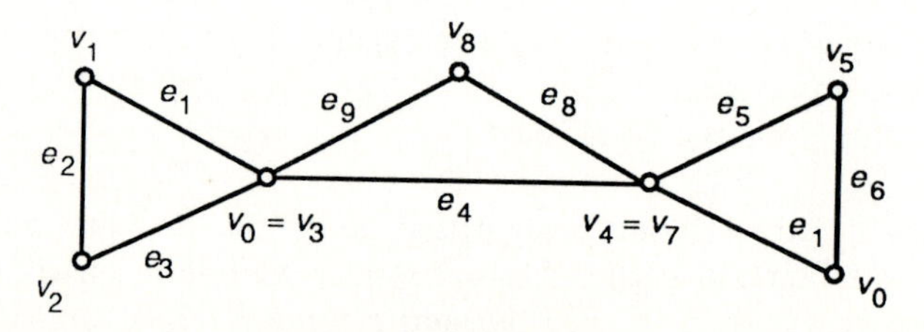

Figure 5.6.19

algorithm always yields an Euler circuit. A proof of this fact can be found in Bondi and Murty [2].

One can also prove that Fleury's algorithm is efficient in the sense that the number of steps required for the algorithm to produce the desired result is bounded by a polynomial function of the two numbers $\nu = |V(G)|$ and $\varepsilon = |E(G)|$ (such as $\nu^2\varepsilon^2 + 4\varepsilon^3\nu^6$). Algorithms that are inefficient include those for which the number of required steps is an exponential function of the input parameters (such as $2^\varepsilon + 3^\nu + \varepsilon^2\nu$).

Hamilton Circuits

Let G be a connected graph. Our primary concern thus far has been the existence of circuits that contain each edge of G exactly once. Let us now revise the role of edges and vertices and look for paths that contain each vertex exactly once.

Definition 5.6.8 *Hamilton Paths and Circuits*

A path in a graph is called a *Hamilton path* if the path contains every vertex of G exactly once. A circuit is called a *Hamilton circuit* if it contains every vertex of G exactly once except for the starting vertex, which appears twice.

A situation that requires such a path is the famous *traveling salesman problem*. Suppose that a salesman wishes to visit all the cities in his sales region. The cities in the region and the highways joining the cities form a simple graph such as in Figure 5.6.20. We assume that any two cities are joined by at most one highway that passes through none of the other cities. Also the mileage between two adjacent cities is indicated along the edge joining the cities. In the name of efficiency, the salesman wishes to devise a travel route that takes him through each city exactly once, brings him back

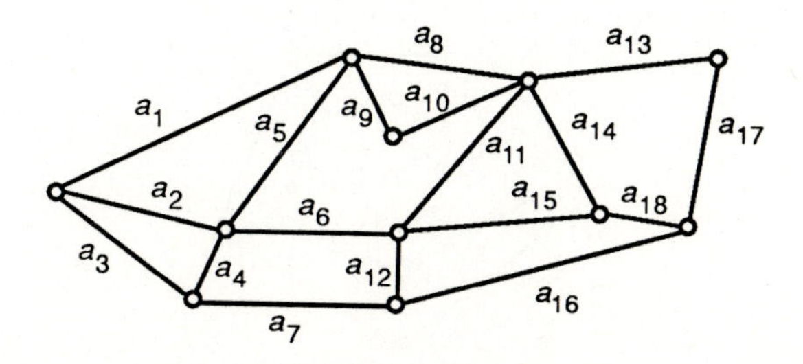

Figure 5.6.20

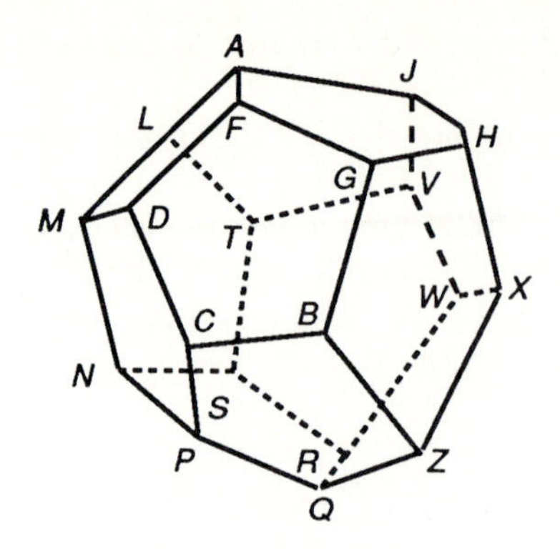

Figure 5.6.21

to his starting point, and requires the least total of miles traveled. A path that meets the requirements of the traveling salesman problem is called an *optimal circuit*. Thus an optimal circuit is a Hamilton circuit with the additional property that the sum of the mileages for that circuit is less than or equal to the sum of the mileages for any other Hamilton circuit.

The concept of a Hamilton circuit was first introduced by the Irish mathematician William Rowan Hamilton. Hamilton invented a game based on the dodecahedron, a solid consisting of 12 regular pentagons. (See Figure 5.6.21.) The game involves two players: player A places a pin in 5 adjacent vertices of the dodecahedron and player B must complete the path to a circuit that passes through every vertex exactly once. This game easily translates into a problem in graph theory. The vertices (corners) and edges of the dodecahedron can be projected into a graph (see Figure 5.6.22) and Hamilton's game corresponds to the following problem: Given a path consisting of five vertices and four edges on the dodecahedron graph, complete this path to a Hamilton circuit. We leave as an exercise for the reader the decision whether player B can always win Hamilton's game.

The question of whether every simple graph has a Hamilton circuit turns out to be very difficult. Currently, no workable necessary and sufficient condition (such as that for the existence of an Euler circuit given by Theorem 5.6.1) is known. Several sufficient conditions are known, but none

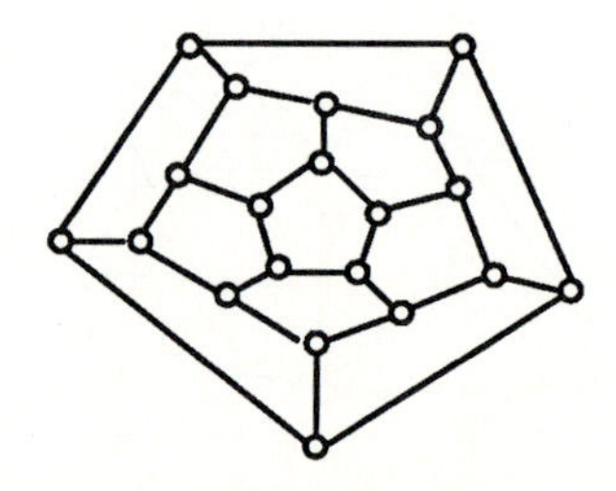

Figure 5.6.22

of these covers all cases in which Hamilton circuits are known to exist. An example of such a condition is given in the next theorem, put forth by G. A. Dirac in 1952, although proof is omitted. (For a proof see reference [2].)

Theorem 5.6.3 *Let G be a simple graph with $\nu \geq 3$ vertices. If $\deg(v) \geq \nu/2$ for all $v \in V(G)$, then G has a Hamilton circuit.*

Many graph theorists consider the finding of a nontrivial necessary and sufficient condition for the existence of a Hamilton circuit to be one of the major open problems in graph theory.

Returning to the traveling salesman problem, we can ask if there exists an efficient algorithm that solves this problem. In other words, is there an algorithm that constructs an optimal solution to the traveling salesman problem if one exists, or that shows that no solution exists, and that performs this in a number of steps that are bounded by a polynomial function of $v = |V(G)|$ and $e = |E(G)|$. At the present moment, the answer to this question is: No. In addition, many graph theorists believe that no such algorithm exists. Of course, such beliefs prove nothing. In any case, most graph theorists are certain that this problem is of the utmost difficulty.

In this section, we have discussed two of the classical topics in graph theory, Euler circuits and Hamilton circuits. There are many other important and interesting portions of graph theory that we shall not discuss, for example, matching theory, planarity, edge and vertex colorings (which includes the Four-Color Theorem), directed graphs, trees, and networks.

Exercises §5.6

In these exercises, for any graph G, $\nu = |V(G)|$ and $\varepsilon = |E(G)|$.

5.6.1. Let G be a simple graph. Show that $\varepsilon \leq \binom{\nu}{2}$.

5.6.2. Let G be the following graph.

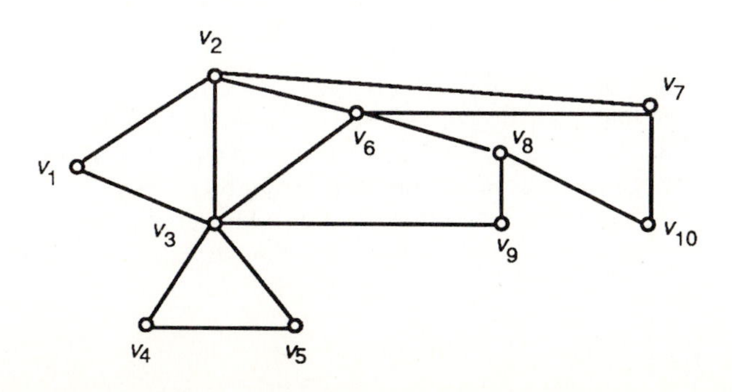

(a) Find the degree of each vertex of G.
(b) Does G contain an Euler circuit? If so, find at least one.

5.6.3. The *complete graph on n vertices* is the simple graph, K_n, having n vertices such that any two vertices are adjacent.
(a) Draw K_2, K_3, K_4, and K_5.
(b) What is $|E(K_n)|$?

5.6.4. Suppose G is a simple graph with vertex set V. If there exist nonempty disjoint subsets V_1, V_2 of V such that $V = V_1 \cup V_2$ with the property that every edge of G is incident to one vertex in V_1 and one vertex in V_2, then G is called a *bipartite graph*. The *complete bipartite on m and n vertices* is the bipartite graph $K_{m,n}$ for which $|V_1| = m$, $|V_2| = n$, and every vertex of V_1 is adjacent to every vertex of V_2.

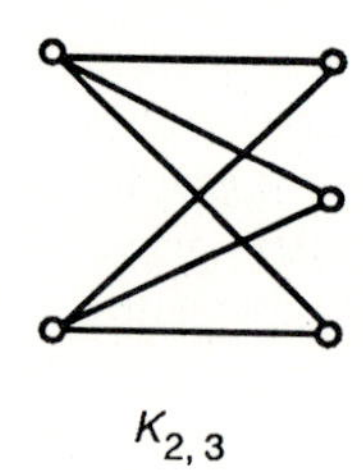

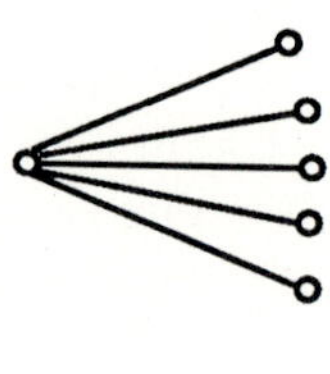

(a) Draw $K_{2,4}$, $K_{2,5}$, and $K_{3,5}$.
(b) What is $|E(K_{m,n})|$? Justify your answer.

5.6.5. Let K_n denote the complete graph on n vertices (see Exercise 5.6.3).
(a) For which n does K_n have an Euler circuit?
(b) For which n does K_n have a Hamilton circuit?

5.6.6. Let $K_{m,n}$ denote the complete bipartite graph on m and n vertices (see Exercise 5.6.4).
(a) For which m and n does $K_{m,n}$ have a Euler circuit?
(b) For which m and n does $K_{m,n}$ have a Hamilton circuit?

5.6.7. Translate the following problem into the language of graph theory: Suppose that in a group of people, each person knows at least half the people in the group. Prove that the people can be seated around a table so that each person knows both people on his or her left and right.

5.6.8. Let G be a simple graph.
(a) Prove that $\Sigma_{v \in V} \deg(v) = 2 \cdot \varepsilon$. (Let $S = \{(v, e) | v \in V, e \in E$ and e is incident to $v\}$. Count the number of elements of S in two ways, first by focusing on vertices, second by focusing on edges.)
(b) Prove that the number of vertices of odd degree is even.
(c) Suppose G is a simple graph such that $\deg(v) = d$ is constant for all $v \in V$. Show that $d \cdot \nu = 2 \cdot \varepsilon$.

5.6.9. Prove Corollary 5.6.2.

5.6.10. For each of the following graphs, find a Hamilton circuit.

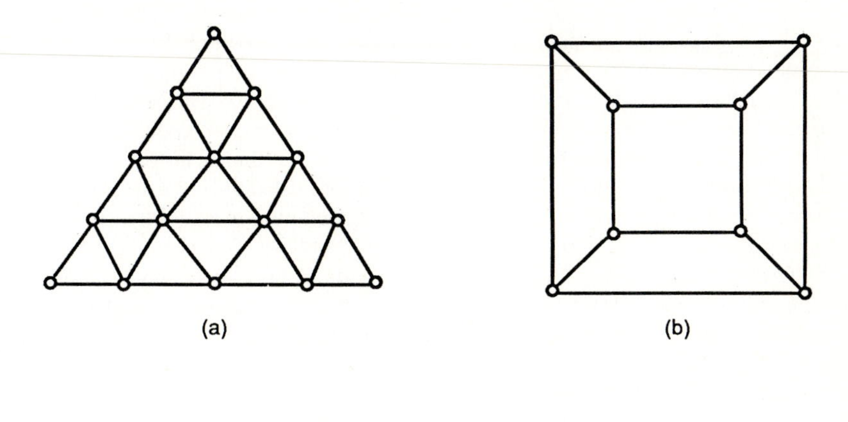
(a) (b)

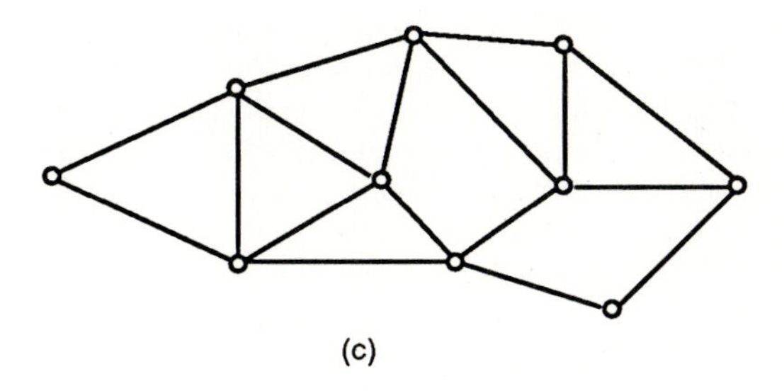
(c)

5.6.11. For each of the following graphs, prove that no Hamilton circuit exists.

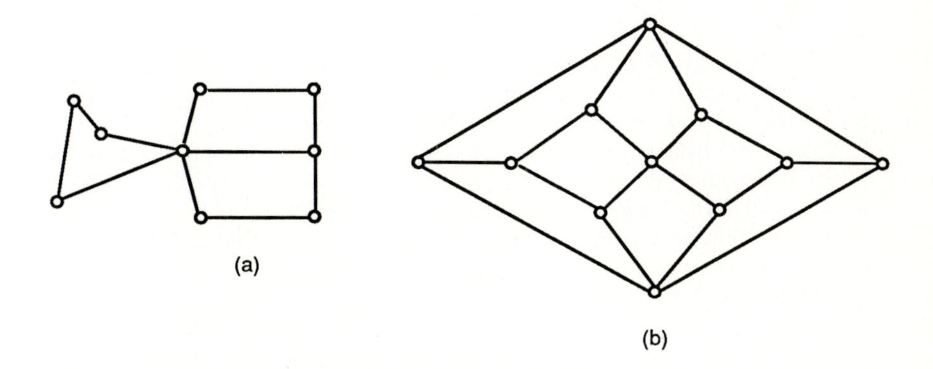
(a) (b)

5.6.12. (a) Suppose G is a simple graph such that for all $v \in V(G)$, $\deg(v) \geq (\nu - 1)/2$. Prove that G is connected.

(b) Suppose G is a simple graph such that $\varepsilon < \nu - 1$. Show that G is disconnected.

Notes to Chapter 5

Most of the results on combinatorics presented in this chapter go back centuries and can be considered part of mathematical folklore. For instance, the Binomial Theorem was familiar to Arab mathematicians of the thirteenth century. The binomial coefficients and Pascal's triangle were known and used by several sixteenth century mathematicians long before they were considered by Pascal himself in the 1650s. The important generalization of the Binomial Theorem to fractional exponents which is often presented in calculus courses was due to Newton. Similarly, the ideas of permutation and combination go back at least to the fourteenth century.

The origins of graph theory can be pinpointed quite precisely. In 1736 the Koenigsberg Bridge Problem was formally solved by Leonhard Euler. Euler's delightful solution is printed in [4]. In the same article, Euler characterized those graphs that contain an Euler circuit. As noted previously, Hamiltonian cycles arose from Hamilton's dodecahedron game in 1859.

Until its resolution in 1976, the Four-Color Problem was the most famous and tantalizing question in graph theory. Roughly stated, the question is whether in any possible map of countries on a sphere, the countries can be colored using only four colors in such a way that adjacent countries (i.e., those that share a common border) have different colors. The answer is yes and the solution by K. Appel and W. Haken was interesting and controversial because it used a computer to check several hundred cases.

A readable history of graph theory from 1736 to 1936 is found in a book by Biggs, Lloyd, and Wilson [1]. For an account of discrete mathematics and its role in computer science, many books are available; see, e.g., reference [5].

Bibliography

1. N. Biggs, E. Lloyd, and R. Wilson, *Graph Theory 1736–1936*, Oxford University Press, 1976.
2. J. A. Bondy and U. S. R. Murty, *Graph Theory with Applications*, North-Holland, 1976.
3. R. Brualdi, *Introductory Combinatorics*, North-Holland, 1977.
4. L. Euler, "The Konigsberg Bridges," in *Mathematics in the Modern World*, ed. by M. Kline, Freeman, 1968.
5. R. Johnsonbaugh, *Discrete Mathematics*, Macmillan, 1984.
6. D. Knuth, *The Art of Computer Programming*, I, Addison-Wesley, 1969.
7. G. Polya, *Mathematical Discovery*, I and II, J. Wiley, 1962, 1965.
8. A. Tucker, *Applied Combinatorics*, J. Wiley, 1980.

Miscellaneous Exercises

1. In how many ways can the numbers $1, 2, \ldots, n$ be listed such that each integer after the first on the list differs by one from an integer that is already listed? For example, $3, 4, 2, 1$ is such a list for $n = 4$.

2. There are $n + k$ lines in the plane situated so that exactly k lines are mutually parallel and no three of the $n + k$ lines are concurrent. Determine a formula for the number of regions into which the plane is divided by the $n + k$ lines.

3. In a recent survey of 400 women in the University of Minnesota class of 1980, one-half were married, one-half were employed, and one-half were blond. In addition, it was learned that there were 104 married women who were employed, 116 married women who were blond, 93 blonds who were employed, and 53 unemployed women who were single and not blond. Find the number of women surveyed who were simultaneously married, employed, and blond.

4. In how many ways can the vertices and the edges of a square be colored if four colors are used? (Each vertex and each edge are given one of the four colors.)

5. Can a simple graph have all its vertices of different degree? Justify your answer.

6. How many three-element subsets of $\{1, 2, \ldots, n\}$ are there which do not contain consecutive integers? For example, if $n = 4$, then there are no such subsets; if $n = 5$, then there exists exactly one such subset.

7. Let n be an odd positive integer and let $a_1, \ldots, a_n$ be any permutation of the set $\{1, \ldots, n\}$. Prove that the integer $(a_1 - 1) \cdot (a_2 - 1) \cdot \cdots \cdot (a_n - n)$ is even. Does the same conclusion hold if n is even?

8. Let $A_n = \{1, \ldots, n\}$. A *derangement* of A_n is a permutation of A_n, $a_1, \ldots, a_n$, such that for $1 \leq i \leq n$, $a_i \neq i$. Let D_n denote the number of derangements of A_n.
   (a) Calculate D_n for $2 \leq n \leq 4$.
   (b) Find a recurrence relation satisfied by $\{D_n | n \geq 2\}$.

9. Let G be a simple graph having an Euler circuit. Is it possible for $\nu = |V(G)|$ to be even and $\varepsilon = |E(G)|$ to be odd? (Hint: Use Exercise 5.6.8.)

10. For $n \in \mathbf{Z}^+$ and $1 \leq k \leq n$, let $S(n, k)$ denote the number of ways of partitioning the set $\{1, \ldots, n\}$ into k classes.
    (a) Evaluate $S(3, 1)$, $S(3, 2)$, $S(3, 3)$, and $S(4, 2)$.
    (b) For $n \geq 1$, evaluate $S(n, 1)$ and $S(n, n)$.
    (c) For $n \geq 2$ and $2 \leq k \leq n - 1$, prove the recurrence relation $S(n, k) = S(n - 1, k - 1) + k \cdot S(n - 1, k)$.

11. A *tree* is a connected graph having no circuits.
    (a) Give examples of trees having ≤ 5 vertices.
    (b) Prove: Any two vertices in a tree are connected by a unique path.
    (c) Prove: If T is a tree having ν vertices and ε edges, then $\varepsilon = \nu - 1$.

CHAPTER SIX

Algebraic Structures

§6.1 An Overview of Algebraic Structures

Roughly speaking, an algebraic structure is a set on which an operation (e.g., binary or unary) or several operations are defined. The mathematical discipline known as *algebra* or *abstract algebra* is the study of algebraic structures. The first algebraic structures encountered by mathematicians were number systems—the integers, the rationals, and the reals. With the increase in mathematical activity that occurred in the eighteenth and nineteenth centuries, new algebraic systems were created: vector spaces, groups, rings, fields, Boolean algebras, lattices, semigroups, Lie algebras, and many others.

In this section we provide a quick introduction to algebraic systems. Our primary purpose is to become familiar with some of the basic properties that an algebraic system might possess. Examples of such properties are associativity and commutativity. We also discuss a tabular form of representing binary operations defined on a finite set and close the section with a discussion of functions from one algebraic structure of a given type to another structure of that type which "preserve" the given algebraic structures. A typical example of such a function is a linear transformation from one vector space to another.

We begin by recalling some basic definitions.

Definition 6.1.1 *Binary and Unary Operations*

Let A be a set.

(i) A *binary operation on* A is a function $*: A \times A \to A$. If $(a, b) \in A \times A$, then the element of A assigned to (a, b) by the function $*$ is denoted by $a * b$.

(ii) A *unary operation on* A is a function $u: A \to A$.

One can, of course, define ternary operations on A (functions from $A \times A \times A$ to A), etc. But since such higher order operations occur rarely in practice, we restrict our attention to binary and unary operations.

In Chapter 2 we saw several examples of binary and unary operations. On $\mathbf{R}$ (or $\mathbf{Q}$ or $\mathbf{Z}$), addition, subtraction, and multiplication are binary operations. On $\mathbf{R}^* = \mathbf{R} - \{0\}$ division is a binary operation. If B is a set, then on $\mathscr{P}(B)$ intersection and union are binary operations and complementation is a unary operation.

Definition 6.1.2 *Algebraic Structure*

An *algebraic structure* is a set A together with a finite number of binary and unary operations defined on A. If $*_1, \ldots, *_r$ and $u_1, \ldots, u_s$ are the operations that define the algebraic structure on A, then we denote the structure by $(A;\ *_1, \ldots, *_r, u_1, \ldots, u_s)$.

In the notation of Definition 6.1.2, the following are examples of algebraic structures: $(\mathbf{Z};\ +, \cdot)$, $(\mathbf{Z};\ +)$, $(\mathbf{Z};\ \cdot)$, $(\mathbf{R};\ +, \cdot)$, $(\mathbf{R}^*;\ \cdot)$, $(P(A);\ \cap, \cup, {}^c)$, and $(\{0, 1\};\ \cdot)$.

Let us suppose that $(A;\ *)$ is an algebraic structure where A is a finite set and $*$ is a binary operation on A. Then $*$ can be conveniently represented by a table as follows: Let $A = \{a_1, \ldots, a_n\}$. Form an $n \times n$ table, called the *operation table* of $(A;\ *)$, whose rows and columns are labeled by the elements of A (written in the same order):

	a_1	a_2		a_j		a_n
a_1						
a_2						
$\vdots$			$\cdots$			
a_i				$a_i * a_j$		
a_n						

In the box corresponding to the row labeled a_i and the column a_j, insert the element $a_i * a_j$ of A. Clearly, any binary operation on A can be represented in this manner and any table of this form (i.e., one whose entries are elements of A) defines a binary operation on A. (The multiplication table that we memorized in our youth, giving all integer products $a \cdot b$ where $1 \leq a, b \leq 9$, is simply a portion of what would be the infinite operation table of $(\mathbf{Z}^+;\ \cdot)$.)

Algebraic structures are very general mathematical objects. In fact, it is difficult to analyze in detail an algebraic structure unless restrictions are imposed on the operations that define the structure. To illustrate our remarks, let us suppose that $(A; *)$ is an algebraic structure where $*$ is a binary operation on A. It is customary to assume that the operation $*$ satisfies certain properties. We can then deduce statements that are valid for all algebraic structures satisfying those properties. The properties that we introduce now were first observed in the operations of addition and multiplication on **R** and in the operations of intersection and union on the power set of a set.

Definition 6.1.3 *Associativity, Commutativity, and Idempotence*

Let A be a set on which a binary operation $*$ is defined.
(i) $*$ is *associative* if for all $a, b, c \in A$, $(a * b) * c = a * (b * c)$.
(ii) $*$ is *commutative* if for all $a, b \in A$, $a * b = b * a$.
(iii) $*$ is *idempotent* if for all $a \in A$, $a * a = a$.

Definition 6.1.4 *Identity and Inverse*

Let $(A; *)$ be an algebraic structure where $*$ is a binary operation.

(i) $*$ has an *identity element* or *identity* if there exists $e \in A$ such that for all $a \in A$, $a * e = e * a = a$.

(ii) If $*$ has an identity element e, then *inverses* exist for $*$ if for each $a \in A$, there exists $b \in A$ such that $a * b = b * a = e$. For each $a \in A$, an element b such that $a * b = b * a = e$ is called an *inverse* of a.

Definition 6.1.5 *Distributivity*

Let A be a set with two binary operations $*_1$ and $*_2$. Then $*_1$ *distributes* over $*_2$ if for all $a, b, c \in A$, $a *_1 (b *_2 c) = (a *_1 b) *_2 (a *_1 c)$.

Example 6.1.1 Addition on **Z** is associative, is commutative, has an identity, and possesses inverses. Multiplication is associative, is commutative, and has an identity on **Z**. Inverses, however, do not exist for multiplication on **Z**. Multiplication also distributes over addition, but addition does not distribute over multiplication. Notice that addition is an associative, commutative binary operation of $\mathbf{Z}^+$, but addition does not have an identity on $\mathbf{Z}^+$. Finally, observe that neither addition nor multiplication is idempotent on $\mathbf{Z}$.

Example 6.1.2 Let $A = \mathscr{P}(B)$ where B is any set. Then, on A, $\cap$ and $\cup$ are associative, commutative, idempotent operations and each possesses an identity element. (B is an identity for $\cap$ while $\varnothing$ is an identity for $\cup$.) Also each of the operations distributes over the other.

Example 6.1.3 We take some examples from linear algebra. Let $M_n(\mathbf{R})$ be the set of $n \times n$ matrices whose entries are in $\mathbf{R}$ and let $Gl_n(\mathbf{R})$ be the subset of $M_n(\mathbf{R})$ consisting of matrices of determinant $\neq 0$. Then addition and multiplication of matrices are associative operations on both $M_n(\mathbf{R})$ and $Gl_n(\mathbf{R})$, and multiplication distributes over addition in $M_n(\mathbf{R})$. For addition on $M_n(\mathbf{R})$, an identity element exists and inverses exist. The same remark holds for multiplication on $Gl_n(\mathbf{R})$. Addition is commutative on $M_n(\mathbf{R})$ while multiplication is not commutative on $Gl_n(\mathbf{R})$.

Modern algebra is that portion of mathematics in which algebraic structures are studied. As suggested above, an algebraist (i.e., a mathematician who specializes in algebra) proceeds by considering those algebraic structures that satisfy certain properties. For instance, a *semi-group* is an algebraic structure $(A; *)$ where $*$ is an associative binary operation; a *monoid* is a semi-group such that $*$ has an identity element; and a *group* is a monoid such that inverses exist for $*$.

In view of the discussion in Section 1.2, we can see that each algebraic structure, whether it be a semi-group, group, field, or vector space, etc., actually constitutes an axiomatic system, for an algebraic structure is defined to be a set A whose elements are not specified in advance to have a particular form and hence are among the undefined terms in the axiomatic system. The axioms include the statement that one or more binary or unary operations are defined on A and all the statements that give the properties (such as associativity) are satisfied by the operations. In the language of Section 1.2, an example of a particular algebraic structure is a model of the axiomatic system that defines the structure.

As mentioned at the start of this section, the purpose of this chapter is to introduce a few important algebraic structures: rings, fields, and groups. These structures were defined as axiomatic systems in the late nineteenth and early twentieth centuries. It is fair to say that a major thrust in the movement to axiomatize mathematics came from the study of algebraic structures. This development occurred more or less simultaneously with the effort to clean up the axiomatic presentation of geometry (spearheaded by David Hilbert in Germany) and the program to describe the real number system axiomatically carried out by several mathematicians in the late 1800s.

The latter topic will be the subject matter of Chapter 7. For now, we conclude this section with a discussion of homomorphisms—functions between two algebraic structures of a given type which preserve the structures.

Homomorphisms

Let $A = \{a, b\}$ and $B = \{c, d\}$ and consider the operations on A and B defined by the following tables:

$*_1$	a	b
a	a	b
b	b	a

$*_2$	c	d
c	c	d
d	d	c

The tables look remarkably similar, the only difference being the names of the elements involved. Specifically, the elements a and c seem to play the same role relative to the operations: For all $s \in A$, $s *_1 s = a$ and for all $t \in B$, $t *_2 t = c$. Similarly, b and d seem to play analogous roles: $a *_1 b = b$, $c *_2 d = d$, $b *_1 b = a$, $d *_2 d = c$. If we define a function $f: A \to B$ by $f(a) = c$ and $f(b) = d$, then f is a bijection and (as can be checked) for all $x, y \in A$, $f(x *_1 y) = f(x) *_2 f(y)$. The last property can be phrased as follows: If any two elements of A are combined under the operation in A and then carried to B by f, then the result is the same as if each of the elements of A were first carried to B by f and then combined under the operation in B. The next definition formalizes this property.

Definition 6.1.6 *Homomorphism and Isomorphism*

Let $(A_1; *_1)$ and $(A_2; *_2)$ be algebraic structures where $*_1$ and $*_2$ are binary operations. Let $f: A_1 \to A_2$ be a function from A_1 to A_2.

(i) f is a *homomorphism* if for all $x, y \in A_1$, $f(x *_1 y) = f(x) *_2 f(y)$.

(ii) f is an *isomorphism* if f is a bijective homomorphism. In this case, we write $(A_1; *_1) \approx (A_2; *_2)$.

Example 6.1.4 (i) Consider $(\mathbf{R}; +)$, the reals under addition. Define $f: \mathbf{R} \to \mathbf{R}$ by $f(x) = c \cdot x$ where $c \in \mathbf{R}$ is fixed. If $c \neq 0$, then f is an isomorphism of $(\mathbf{R}, +)$ with itself since f is bijective and for all $x, y \in \mathbf{R}$, $f(x + y) = c \cdot (x + y) = c \cdot x + c \cdot y = f(x) + f(y)$. If $c = 0$, then f is a homomorphism but not an isomorphism.

(ii) Let $(\mathbf{R}^+; \cdot)$ denote the positive reals under multiplication. Define $f:\mathbf{R}^+ \to \mathbf{R}^+$ by $f(x) = x^c$ where $c \in \mathbf{R}$ is fixed. If $c \neq 0$, then f is an isomorphism of $(\mathbf{R}^+, \cdot)$ with itself since f is bijective and for all $x, y \in \mathbf{R}^+$, $f(x \cdot y) = (x \cdot y)^c = x^c \cdot y^c = f(x) \cdot f(y)$. If $c = 0$, then f is a homomorphism but not an isomorphism.

(iii) Define f: $\mathbf{R}^+ \to \mathbf{R}$ by $f(x) = \ln(x) =$ the natural logarithm of x. Then, since f is bijective and $f(x \cdot y) = \ln(x \cdot y) = \ln(x) + \ln(y) = f(x) + f(y)$, f is an isomorphism. The same remark applies to the logarithm with respect to any positive base $\neq 1$.

Example 6.1.5 The function det: $Gl_n(\mathbf{R}) \to \mathbf{R}^*$ is a homomorphism since $\det(A \cdot B) = \det(A) \cdot \det(B)$ for all $A, B \in Gl_n(\mathbf{R})$. For $n \geq 2$, det is not an isomorphism, as the reader can easily check.

It is useful to think of a homomorphism as a structure-preserving function between two algebraic structures of a given type. An isomorphism enjoys the additional property of being a one-to-one correspondence, and hence it preserves the set theoretic properties of the two sets. For all practical purposes, two isomorphic algebraic structures are the same except possibly for the names of their elements and of the operations: The underlying sets are equinumerous, and under the associated correspondence, corresponding pairs of elements combine to yield corresponding elements. The next theorem describes basic facts about isomorphic structures.

Theorem 6.1.1 *Let $(A; *_1)$, $(B; *_2)$, and $(C; *_3)$ be algebraic structures defined by binary operations. Then*

(i) $(A; *_1) \approx (A; *_1)$.

(ii) *If* $(A; *_1) \approx (B; *_2)$ *then* $(B; *_2) \approx (A; *_1)$.

(iii) *If* $(A; *_1) \approx (B; *_2)$ *and* $(B; *_2) \approx (C; *_3)$, *then* $(A; *_1) \approx (C; *_3)$.

Proof We prove (ii) and leave (i) and (iii) as exercises. Suppose f: $A \to B$ is an isomorphism. We show that f^{-1}: $B \to A$ is an isomorphism.

Since f is bijective, f^{-1} is bijective by Theorem 3.4.3. We now show that for all $u, v \in B$, $f^{-1}(u *_2 v) = f^{-1}(u) *_1 f^{-1}(v)$. Since f is injective, $f^{-1}(u *_2 v) = f^{-1}(u) *_1 f^{-1}(v)$ as elements of A if and only if

$$f(f^{-1}(u *_2 v)) = f(f^{-1}(u) *_1 f^{-1}(v)).$$

But $f(f^{-1}(u *_2 v)) = u *_2 v$, while, since f is a homomorphism,

$$f(f^{-1}(u) *_1 f^{-1}(v)) = f(f^{-1}(u)) *_2 f(f^{-1}(v)) = u *_2 v.$$

Therefore, f^{-1} is a homomorphism. Since f^{-1} is a bijective homomorphism, f^{-1} is an isomorphism. ∎

We shall return to homomorphisms and isomorphisms in Section 6 of this chapter. Until then, we shall analyze a number of important algebraic structures. We begin with a concrete example, the set of integers under addition and multiplication. This set provides the motivating example for many of the ideas introduced in this chapter.

Exercises §6.1

6.1.1. In each case, state whether the given function defines binary operations. For those functions that do define binary operations, list those properties in Definition 6.1.3–6.1.4 satisfied by the operations. In each case, A is a given set and f is the alleged binary operation on A.
(a) $A = \mathbf{R}^+$, $f(x, y) = x + y$.
(b) $A = \mathbf{R}^*$, $f(x, y) = x/y$.
(c) $A = \mathbf{R}$, $f(x, y) = x - y$.
(d) $A = \mathbf{R}^+$, $f(x, y) = x \cdot y$.
(e) $A = \mathbf{R}^*$, $f(x, y) = x + y$.
(f) $A = \mathbf{R}^+$, $f(x, y) = x^y$.

6.1.2. (a) Prove that any binary operation on a set A has at most one identity.
(b) Prove that if a binary operation on a set is associative and has inverses, then each element in the set has a unique inverse.

6.1.3. (a) Give an example of a binary operation on a finite set A that is commutative but neither associative nor idempotent.
(b) Give an example of a binary operation on a finite set A that is idempotent but neither commutative nor associative.

6.1.4. (a) How many idempotent binary operations can be defined on $\{a, b\}$?
(b) How many idempotent binary operations can be defined on $\{a, b, c\}$?

6.1.5. Let A be a finite set and let $*$ be a binary operation on A.
(a) Suppose $*$ is commutative. How is this property reflected in the operation table of $(A; *)$?
(b) Suppose $*$ is idempotent. How is this property reflected in the operation table of $(A; *)$?

6.1.6. Let $(A, *_1)$ and $(B, *_2)$ be algebraic structures where $*_1$ and $*_2$ are binary operations.
(a) Show that if $|A| = |B| = 1$, then $(A; *_1) \approx (B; *_2)$.
(b) Suppose only that $|A| = 1$. Must there exist a homomorphism $f: A \to B$? State a necessary and sufficient condition.
(c) Suppose only that $|B| = 1$. Must there exist a homomorphism $f: A \to B$? State a necessary and sufficient condition.

6.1.7. Show (using results from calculus) that the exponential function exp: $(\mathbf{R}; +) \to (\mathbf{R}^+; \cdot)$ defined by $\exp(x) = e^x$ for $x \in \mathbf{R}$ is an isomorphism. What is the inverse of exp?

6.1.8. Show that the determinant function, det: $Gl_n(\mathbf{R}) \to \mathbf{R}^*$, is not an isomorphism for $n \geq 2$.

6.1.9. Prove, or disprove and salvage: Let $c \neq 0$ be an integer. The function $f: \mathbf{R}^* \to \mathbf{R}^*$ defined by $f(x) = x^c$ is an isomorphism of $(\mathbf{R}^*; \cdot)$ with itself.

6.1.10. Let $A = \{e, a, b, c\}$ and let $(A; *)$ be defined by the following operation table.

	e	a	b	c
e	e	a	b	c
a	a	e	c	b
b	b	c	e	a
c	c	b	a	e

(a) Given that $*$ is associative, list the other properties that $*$ possesses.
(b) Find an isomorphism f: $(A; *) \to (A; *)$ that is different from the identity.

6.1.11. Prove Theorem 6.1.1 (i) and (iii).

§6.2 *Algebraic Properties of the Integers*

We begin by studying the prototypical example of an algebraic structure, the set of integers. The set of integers constitutes the motivating example for the class of algebraic structures known as rings.

The major goal of this section is to prove a single theorem, the Unique Factorization Theorem for multiplication in $\mathbf{Z}$, and to give several applications of this result. All the algebraic properties of $\mathbf{Z}$ that we present were known to Euclid and most were proved by him. The proof of the Unique Factorization Theorem which we present is modeled upon the one given by C. F. Gauss in 1801 in his monumental treatise *Disquisitiones Arithmeticae*.

What are the basic algebraic properties of $\mathbf{Z}$? As a set, $\mathbf{Z} = \{0, \pm 1, \pm 2, \dots\}$ and this set is endowed with two binary operations, addition and multiplication. These operations have the following properties:

1. Associative Law for Addition. For all $x, y, z \in \mathbf{Z}$, $(x + y) + z = x + (y + z)$.
2. Additive Identity. There exists an element $0 \in \mathbf{Z}$ such that for all $x \in \mathbf{Z}$, $x + 0 = 0 + x = x$.
3. Additive Inverse. For all $x \in \mathbf{Z}$, there exists $y \in \mathbf{Z}$ such that $x + y = y + x = 0$.
4. Commutative Law for Addition. For all $x, y \in \mathbf{Z}$, $x + y = y + x$.
5. Associative Law for Multiplication. For all $x, y, z \in \mathbf{Z}$, $(x \cdot y) \cdot z = x \cdot (y \cdot z)$.
6. Multiplicative Identity. There exists an element $1 \in \mathbf{Z}$ such that for all $x \in \mathbf{Z}$, $x \cdot 1 = 1 \cdot x = x$.
7. Commutative Law for Multiplication. For all $x, y \in \mathbf{Z}$, $x \cdot y = y \cdot x$.
8. Distributive Law for Multiplication over Addition. For all $x, y, z \in \mathbf{Z}$, $x \cdot (y + z) = x \cdot y + x \cdot z$.

We assume the existence of the set $\mathbf{Z}$ satisfying properties 1–8. A rigorous development of $\mathbf{Z}$ would consist of a careful definition of $\mathbf{Z}$ as a set, a definition of each of the two operations, and proofs that properties 1–8 hold for the given set and the defined operations. Such a development is presented in Section 7.3. For the present, we work with our usual intuitive conception of the integers.

Recall that $\mathbf{Z} = \{0, \pm 1, \pm 2, \dots\}$ contains $\mathbf{N} = \{0, 1, 2, \dots\}$ and this set satisfies the Principle of Mathematical Induction. We can accomplish the proof of the Unique Factorization Theorem using mathematical induction. In order to state the theorem, we recall the concepts of divisibility in $\mathbf{Z}$ and of prime and composite integers.

Definition 6.2.1 *Divisibility in* $\mathbf{Z}$

Let $x, y \in \mathbf{Z}$. We say that x *divides* y, written $x|y$, if there exists $z \in \mathbf{Z}$ such that $x \cdot z = y$. If $x|y$, then we also say that x is a *factor* of y, x is a *divisor* of y, or y is a *multiple* of x. If x does not divide y, then we write $x \nmid y$.

We recall a basic fact about divisibility proved in Chapter 1 (see Lemma 1.3.3): If $a, b, c \in \mathbf{Z}$ and $a|b$ and $a|c$, then $a|b + c$. Also observe that for any $n \in \mathbf{Z}$, $\pm 1|n$; in other words, 1 and -1 are factors of each integer. Because of this fact, certain exceptions must be made when prime and composite integers are defined.

Definition 6.2.2 *Prime and Composite Integers*

An integer p is *prime* if (i) $p \neq 0, \pm 1$ and (ii) any factor of p is either $\pm p$ or ± 1. An integer n is *composite* if (i) $n \neq 0, \pm 1$ and (ii) n is not prime.

Notice that according to our definition, both 7 and -7 are prime integers. In fact, one can easily show that if p is prime, then $-p$ is prime.

We can now state and prove the principal result of this section.

Theorem 6.2.1 *Unique Factorization for Multiplication in* $\mathbf{Z}$. *Let n be an integer such that $n \neq 0, \pm 1$.*

(i) *There exists a positive integer r and prime integers $p_1, \ldots, p_r$ such that $n = p_1 \cdot \cdots \cdot p_r$.*

(ii) *If $n = p_1 \cdot \cdots \cdot p_r = q_1 \cdot \cdots \cdot q_s$ where $p_1, \ldots, p_r, q_1, \ldots, q_s$ are all prime, then $r = s$ and, after renumbering the q's, if necessary, $p_1 = \pm q_1, p_2 = \pm q_2, \ldots, p_r = \pm q_r$.*

Remark Statement (i) of Theorem 6.2.1 asserts that each integer $n \neq 0, \pm 1$ either is prime (in which case $r = 1$), or is expressible as or is decomposable into a product of primes. Statement (ii) claims that any two factorizations of n are unique in the sense that the p's and q's differ only in the order in which they appear and by factors of ± 1. Thus $60 = 2 \cdot 2 \cdot 3 \cdot 5 = 2 \cdot (-3)(-5) \cdot 2$ and in this case $r = s = 4$, $p_1 = p_2 = 2$, $p_3 = 3$, and $p_4 = 5$. We can rearrange the terms in the second factorization to find that $q_1 = q_2 = 2$, $q_3 = -3$, and $q_4 = -5$.

Proof of (i) See Chapter 1, Theorem 1.3.7.

We prove statement (i) first in the case that $n > 0$ and then reduce the case that $n < 0$ to the first case.

For an integer $n \geq 2$, let $S(n)$ be the statement: There exists a positive integer r and prime integers $p_1, \ldots, p_r$ such that $n = p_1 \cdot \cdots \cdot p_r$.

Using the strong form of mathematical induction, we prove that for $n \geq 2$, $S(n)$ is valid, thereby establishing statement (i).

Basis step $S(2)$ is true, since 2 is a prime integer.

Inductive step Suppose statements $S(2), \ldots, S(n)$ are all true. We show that $S(n+1)$ is true.

Clearly, if $n+1$ is prime, then $S(n+1)$ is true. If $n+1$ is not prime, then $n+1 = a \cdot b$ where a and b are integers such that $a \neq \pm 1$ and $b \neq \pm 1$. Now a and b are either both positive or both negative. If a and b are both positive, then $1 < a, b < n+1$. If a and b are both negative, then $n = (-a) \cdot (-b)$ where $1 < -a, -b < n+1$. Thus we can assume that $n = x \cdot y$ where $1 < x, y < n+1$.

By inductive hypothesis, $S(x)$ and $S(y)$ are both true. Thus there exist positive integers s and t and prime integers $p_1, \ldots, p_s$ and $q_1, \ldots, q_t$ such that $x = p_1 \cdot \cdots \cdot p_s$ and $y = q_1 \cdot \cdots \cdot q_t$. Therefore, since

$$n+1 = x \cdot y = p_1 \cdot \cdots \cdot p_s \cdot q_1 \cdot \cdots \cdot q_t,$$

$n+1$ can be expressed in the desired form.

By the Principle of Strong Induction, $S(n)$ is true for all $n \geq 2$.

Suppose now that $n < 0$ and $n \neq -1$. Then $-n \geq 2$ and by the previous argument $S(-n)$ is true. Thus $-n = p_1 \cdot \cdots \cdot p_r$ where $p_1, \ldots, p_r$ are prime integers. Therefore, $n = (-p_1) \cdot p_2 \cdot \cdots \cdot p_r$ is expressible as a product of prime integers. The proof of statement (i) is now complete.

Let us consider how we might prove statement (ii). We assume that $p_1, \ldots, p_r$ and $q_1, \ldots, q_s$ are primes and that $p_1 \cdot \cdots \cdot p_r = q_1 \cdot \cdots \cdot q_s$. We want to show that each p is $\pm$ one of the q's and vice versa. For example, we want to show that $p_1 = \pm q_i$ for some i between 1 and s. One way to reach this conclusion is to show that $p_1 | q_i$. Then, since q_i is prime, $q_2 = \pm p_1$. How can we show that $p_1 | q_i$ for some i? We know that $p_1 | p_1 \cdot \cdots \cdot p_r = q_1 \cdot \cdots \cdot q_s$. We would like to show that if $p_1 | q_1 \cdot \cdots \cdot q_s$, then $p_1 | q_i$ for some i. In other words, a key step in the proof of (ii) will be to show that whenever a prime divides a product of integers, then that prime divides at least one of the integers. This property of primes is established in Lemma 6.2.2, which is stated and proved below. We now use this lemma to prove statement (ii).

For $n \geq 2$, define the statement $S(n)$: If $n = p_1 \cdot \cdots \cdot p_r = q_1 \cdot \cdots \cdot q_s$ where $p_1, \ldots, p_r$ and $q_1, \ldots, q_s$ are prime integers, then $r = s$, and after renumbering the q's, if necessary, $p_1 = \pm q_1, p_2 = \pm q_2, \ldots, p_r = \pm q_r$.

We prove that $S(n)$ holds for $n \geq 2$ by PSI.

Basis step Exercise.

Inductive step Suppose $S(2), \ldots, S(n)$ are true. Suppose $n+1 = p_1 \cdot \cdots \cdot p_r = q_1 \cdot \cdots \cdot q_s$. Since $p_1 | n+1 = q_1 \cdot \cdots \cdot q_s$, it follows from repeated application of Lemma 6.2.2 (see Exercise 6.2.5) that $p_1 | q_1$, or $p_1 | q_2$, or ... or $p_1 | q_n$. Suppose $p_1 | q_i$ for some i such that $1 \leq i \leq s$. Then we relabel the q's by calling $q_i = q_1$, $q_1 = q_i$, and leaving all other subscripts unchanged. Thus $p_1 | q_1$. Since p_1 and q_1 are primes, $p_1 = \pm q_1$. Therefore, $n+1 = p_1 \cdot p_2 \cdot \cdots \cdot p_r = \pm p_1 \cdot q_2 \cdot \cdots \cdot q_s$ and $((n+1)/p_1) = p_2 \cdot \cdots \cdot p_r = \pm q_2 \cdot \cdots \cdot q_s$. By inductive hypothesis, $S((n$

$+ 1)/p_1)$ is true (note that $(n + 1)/p_1 \in \mathbf{N}$); hence $r - 1 = s - 1$ and, after renumbering the q's if necessary, $p_2 = \pm q_2, \ldots, p_r = \pm q_r$. The proof of statement (ii) is now complete. ■

Lemma 6.2.2 *Let p be a prime integer. If a and b are integers for which $p|a \cdot b$, then $p|a$ or $p|b$.*

Proof We prove the result in case p, a and b are positive integers, and leave the remaining cases as an exercise (see Exercise 6.2.5).

For $n \geq 2$ let $S(n)$ be the following statement: If a and b are positive integers such that $n = a + b$ and $p|a \cdot b$, then $p|a$ or $p|b$. We prove that $S(n)$ holds for $n \geq 2$ by PSI.

Basis step Suppose $n = 2$. Then $a + b = 2$ and $a = b = 1$. Since for each prime p, $p \nmid 1 \cdot 1 = 1$, $S(2)$ is true since its hypothesis is not satisfied.

Inductive step Suppose $S(2), \ldots, S(n)$ are true. We show $S(n + 1)$ also holds. Let a and b be positive integers such that $n + 1 = a + b$ and suppose $p|a \cdot b$. We consider three cases.

Case 1. $p = a$. Then $p|a$ and $S(n + 1)$ holds.

Case 2. $p < a$. Since $a \cdot b - p \cdot b = (a - p) \cdot b$ and since $p|a \cdot b$ and $p|p \cdot b$, p divides $(a - p) \cdot b$. Since $n + 1 = a + b > a + b - p = (a - p) + b$, it follows from the inductive hypothesis that $S(a + b - p)$ holds; hence since $p|(a - p) \cdot b$, either $p|a - p$ or $p|b$. But if $p|a - p$, then $p|a - p + p = a$. Therefore, if $p|a \cdot b$, $p|a$ or $p|b$.

Case 3. $p > a$. In this case, p cannot divide a; we shall show that $p|b$. By the Division Theorem (Theorem 1.3.6), there exist integers q and r such that $p = a \cdot q + r$ where $0 \leq r \leq a - 1$. It follows that $r \cdot b = (p - a \cdot q) \cdot b = p \cdot b - a \cdot b \cdot q$ is divisible by p. Since $r + b < a + b = n + 1$, $S(r + b)$ holds, and hence $p|r$ or $p|b$.

If $p|r$, then since $0 \leq r < a < p$, it must be the case that $r = 0$. Therefore, $p = a \cdot q$, which implies that p is divisible by a, which in turn can occur if and only if $a = 1$. But if $a = 1$, then $p|a \cdot b = b$. Therefore, $p|b$, which means that $S(n + 1)$ holds in Case 3.

Since $S(n + 1)$ is valid in each of the three cases and since these cases exhaust all possibilities, $S(n + 1)$ holds in every case. ■

Applications of the Unique Factorization Theorem

We now derive several interesting properties about integers from Theorem 6.2.1. Before doing so, we make an observation concerning the prime factorization of an integer. Let $n \in \mathbf{Z}^+$. Then we can factor n

uniquely as $n = p_1^{a_1} p_2^{a_2} \cdots p_r^{a_r}$ where $p_1, \ldots, p_r$ are distinct positive primes and a_i is a positive integer for $1 \leq i \leq r$. Thus $60 = 2^2 \cdot 3 \cdot 5$ and in this case $p_1 = 2$, $p_2 = 3$, $p_3 = 5$, $a_1 = 2$, $a_2 = 1$, $a_3 = 1$. This form of expressing the factorization of an integer is useful in many instances.

Let us consider some examples. Let $n = 420 = 2^2 \cdot 3 \cdot 5 \cdot 7$ and let $d = 22 = 2 \cdot 11$. Does d divide n? If so, then there exists an integer x such that $n = 420 = 2^2 \cdot 3 \cdot 5 \cdot 7 = d \cdot x = 2 \cdot 11 \cdot x$. We need not factor x into primes to realize that we have reached a contradiction, for 11 cannot appear in any factorization of $420 = 2^2 \cdot 3 \cdot 5 \cdot 7$. Similarly, we can ask if $d = 2 \cdot 3^2$ divides 420. If so, then we have $2^2 \cdot 3 \cdot 5 \cdot 7 = 2 \cdot 3^2 \cdot x$ for some $x \in \mathbf{Z}$. But again this situation is impossible, since 3 divides into 420 only once. On the other hand, $d = 2 \cdot 3 \cdot 5$ divides 420, since $420 = 2^2 \cdot 3 \cdot 5 \cdot 7 = d \cdot 2 \cdot 7 = 2 \cdot 3 \cdot 5 \cdot 2 \cdot 7$. These examples indicate a general pattern that is described in the next theorem.

Theorem 6.2.3 *Suppose $n = p_1^{a_1} \cdot \cdots \cdot p_r^{a_r} \in \mathbf{Z}^+$ where $p_1, \ldots, p_r$ are distinct positive primes and for $1 \leq i \leq r$, $a_i \in \mathbf{Z}^+$. Let $d \in \mathbf{Z}^+$. Then $d|n$ if and only if* (i) *for each positive prime p, if $p|d$, then $p = p_i$ for some i, and* (ii) *if $p_i^b|d$, then $b \leq a_i$.*

Proof We are given that $n = p_1^{a_1} \cdot \cdots \cdot p_r^{a_r}$. Write $d = q_q^{b_1} \cdot \cdots \cdot q_s^{b_s}$ where for $1 \leq i \leq s$, q_i is a positive prime and $b_i \in \mathbf{Z}^+$. Then $d|n$ if and only if there exists $x \in \mathbf{Z}$ such that $n = d \cdot x$, i.e.,

$$n = p_1^{a_1} \cdot \cdots \cdot p_r^{a_r} = q_1^{b_1} \cdot \cdots \cdot q_s^{b_s} \cdot x.$$

By Theorem 6.2.1, this equation holds if and only if for each j, $1 \leq j \leq s$, there exist $i \in \mathbf{Z}$ such that $1 \leq i \leq r$ and $q_j = p_i$, and $b_j \leq a_i$. ∎

In words, Theorem 6.2.3 asserts that d divides n if and only if the only primes dividing d are those dividing n and the highest power of any prime dividing d is less than or equal to the highest power of that prime dividing n. For example, if p is a positive prime and $a \in \mathbf{Z}^+$, then the only factors of $n = p^a$ are $1 = p^0, p, p^2, \ldots, p^a$. Thus p^a has $a + 1$ divisors. Here is another example. Let $n = 3^4 \cdot 5^3$. Then $d|n$ if and only if $d = 3^x \cdot 5^y$ where $x, y \in \mathbf{Z}$ and $0 \leq x \leq 4$ and $0 \leq y \leq 3$. Thus there are exactly as many divisors of $n = 3^4 \cdot 5^3$ as there are ordered pairs of integers (x, y) where $0 \leq x \leq 4$ and $0 \leq y \leq 3$. In other words, the number of divisors of n is $|\{(x, y) \in \mathbf{Z} \times \mathbf{Z} | 0 \leq x \leq 4 \text{ and } 0 \leq y \leq 3\}| = |\{0, 1, 2, 3, 4\} \times \{0, 1, 2, 3\}| = 5 \cdot 4 = 20$.

Theorem 6.2.4 *Let $n = p_1^{a_1} \cdot \cdots \cdot p_r^{a_r} \in \mathbf{Z}^+$ where the p_i are distinct positive primes. Then there exists $k \in \mathbf{Z}$ such that $n = k^2$ (i.e., n is a square in* $\mathbf{Z}$) *if and only if each a_i is an even integer.*

Proof If each a_i is even, then $a_i = 2 \cdot b_i$ for some $b_i \in \mathbf{Z}$. Thus $n = p_1^{2 \cdot b_1} \cdot \cdots \cdot p_r^{2 \cdot b_r} = (p_1^{b_1} \cdot \cdots \cdot p_r^{b_r})^2 = k^2$ where $k = p_1^{b_1} \cdot \cdots \cdot p_r^{b_r}$.

Conversely, if $n = k^2$ for some $k \in \mathbf{Z}$, then writing $k = q_1^{b_1} \cdot \cdots \cdot q_s^{b_s}$ where each q_i is a positive prime, we have $p_1^{a_1} \cdot \cdots \cdot p_r^{a_r} = n = k^2 =$

$(q_1^{b_1} \cdot \cdots \cdot q_s^{b_s})^2 = q_1^{2 \cdot b_1} \cdot \cdots \cdot q_s^{2 \cdot b_s}$. By Theorem 6.2.1 it follows that (after renumbering the q's if necessary), $p_1 = q_1, \ldots, p_r = q_r$ and $a_1 = 2b_1, \ldots, a_r = 2 \cdot b_r$. Therefore, each a_i is even. ■

Next we discuss one of the most important concepts in the study of algebraic properties of the integers, the greatest common divisor of two integers.

Definition 6.2.3 *Greatest Common Divisor*

Let $a, b \in \mathbf{Z}$. The *greatest common divisor* of a and b, written $\gcd(a, b)$ (or simply (a, b) by some authors), is the largest integer that divides both a and b.

Example 6.2.1 (i) One can easily check, by listing all the positive factors of both numbers, that $\gcd(6, 9) = 3$, $\gcd(8, -14) = 2$, and $\gcd(7, 22) = 1$.

(ii) If p is a positive prime, then $\gcd(p, a) = p$ if $p | a$ and $\gcd(p, a) = 1$ if $p \nmid a$.

(iii) Let $a = 2^2 \cdot 3 \cdot 5^3 \cdot 11^2 \cdot 13$ and $b = 2 \cdot 5^4 \cdot 13 \cdot 23$. What is $\gcd(a, b)$? Let $d = \gcd(a, b)$. Which positive primes appear in the factorization of d? By Theorem 6.2.3, the only possibilities are 2, 5, and 13, since these are the only primes dividing both a and b. In addition, 2 can appear only once in d since 2 appears only once in the factorization of b; similarly, 5 appears three times in d, and 13 appears once. Therefore $d = \gcd(a, b) = 2 \cdot 5^3 \cdot 13$. In general, for each pair of positive integers a and b, $\gcd(a, b)$ can be read off from the prime factorizations of a and b.

Theorem 6.2.5 *Let a and b be positive integers. Write $a = p_1^{a_1} \cdot \cdots \cdot p_r^{a_r}$ and $b = p_1^{b_1} \cdot \cdots \cdot p_r^{b_r}$ where $p_1, \ldots, p_r$ are positive primes and $a_1, \ldots, a_r, b_1, \ldots, b_r$ are nonnegative integers. Then $\gcd(a, b) = p_1^{c_1} \cdot \cdots \cdot p_r^{c_r}$ where for $1 \leq i \leq r$, $c_i = \min\{a_i, b_i\}$ = the least of the integers a_i and b_i.*

Proof Exercise. ■

Remark Our notation might be somewhat deceptive. The primes $p_1, \ldots, p_r$ are those dividing either a or b. For example, p_1 might divide a and not divide b. In this case, $b_1 = 0$. Thus if $a = 10$ and $b = 28$, then $p_1 = 2$, $p_2 = 5$, $p_3 = 7$, $a_1 = 1 = a_2$, $a_3 = 0$, $b_1 = 2$, $b_2 = 0$, $b_3 = 1$: $10 = 2^1 \cdot 5^1 \cdot 7^0$ and $28 = 2^2 \cdot 5^0 \cdot 7$.

By this point we have experienced at least some of the power of the Unique Factorization Theorem (UFT). We can use UFT to determine if one integer divides another or to find the greatest common divisor of two integers or the number of positive divisors of an integer (see Exercise 6.2.7), and, as we shall see below, to show that $\sqrt{2}$ is irrational. In general, any property of integers involving multiplication is susceptible to an attack in which unique factorization is used. If we consider all algebraic systems that satisfy properties 1–8, we obtain what are called commutative rings with unity. In the general theory of commutative rings with unity, the issue of whether unique factorization exists is of central importance. We shall study this matter in more detail in the next section.

We close this section by giving another proof of the fact that $\sqrt{2}$ is irrational. Suppose $\sqrt{2} = m/n \in \mathbf{Q}$ where $m, n \in \mathbf{Z}$. Then $2 \cdot n^2 = m^2$. Factoring each side into primes, we have

$$2 \cdot n^2 = 2p_1^{2a_1} \cdot \cdots \cdot p_r^{2a_r} = q_1^{2b_1} \cdot \cdots \cdot q^{2b_s} = m^2,$$

where $p_1, \ldots, p_r, q_1, \ldots, q_s$ are positive primes and $a_1, \ldots, a_r, b_1, \ldots, b_s$ are positive integers. In the prime decomposition on the left, the prime 2 appears an odd number of times (once in the factor 2 and an even number of times in $p_1^{2a_1} \cdot \cdots \cdot p_r^{2a_r}$). In the prime decomposition on the right, 2 appears an even number of times. Thus we have a contradiction of Theorem 6.2.1, since 2 must appear the same number of times in any two factorizations of any positive integer into positive primes. Therefore, $\sqrt{2} \notin \mathbf{Q}$. ■

Exercises §6.2

6.2.1. Let $a, b, c \in \mathbf{Z}$. (a) Prove $a \mid a$. (b) Prove: If $a \mid b$ and $b \mid c$, then $a \mid c$. (c) Prove: If $a \mid b$, then $a \mid b \cdot c$.

6.2.2. Prove: If p is prime, then $-p$ is prime.

6.2.3. List all positive primes less than 100.

6.2.4. Let $a \in \mathbf{Z}$, $n \in \mathbf{Z}^+$, and let p be a prime. Prove: If $p \mid a^n$, then $p \mid a$.

6.2.5. Prove by induction on r that if p is prime and $p \mid a_1 \cdot \cdots \cdot a_r$ where $a_i \in \mathbf{Z}$ for $1 \le i \le r$, then $p \mid a_i$ for at least one i.

6.2.6. Complete the proof of Lemma 6.2.2 by considering the cases in which p, a, and b are not necessarily all positive integers.

6.2.7. For each positive integer n, define $d(n)$ to be the number of positive divisors of n. As we have seen, $d(p^a) = a + 1$ if p is prime and $d(3^4 \cdot 5^3) = 20$.

(a) Let p and q be distinct positive primes and let $a, b \in \mathbf{Z}^+$. Conjecture and prove a formula for $d(p^a \cdot q^b)$. (First describe all divisors of $p^a \cdot q^b$.)

(b) Conjecture and prove a formula for $d(p_1^{a_1} \cdot \cdots \cdot p_r^{a_r})$ where for $1 \le i \le r$, $a_i \in \mathbf{Z}^+$ and p_i is a positive prime.

(c) For which $n \in \mathbf{Z}^+$ is $d(n)$ an odd integer? Prove your answer. (See Chapter 1, Miscellaneous Exercises 8 and 9.)

6.2.8. For $n \in \mathbf{Z}^+$, define $\sigma(n)$ to be the sum of the positive divisors of n. For example, $\sigma(2) = 1 + 2 = 3$ and $\sigma(6) = 1 + 2 + 3 + 6 = 12$.
(a) Let p be prime and $a \in \mathbf{Z}^+$. Find $\sigma(p)$, $\sigma(p^2)$, and $\sigma(p^a)$.
(b) A positive integer n is called *perfect* if $\sigma(n) = 2 \cdot n$. For example, 6 is perfect. Find all perfect integers less than 50.
(c) Let $n = p^a \cdot q^b$. Find a formula for $\sigma(n)$ and prove that your formula is correct. (Hint: Remember the geometric sum.)
(d) Let $n = p_1^{a_1} \cdot \cdots \cdot p_r^{a_r}$. Find a formula for $\sigma(n)$ and prove that your formula is correct.

6.2.9. In each case find $\gcd(a, b)$.
(a) $a = 24$, $b = 56$.
(b) $a = 14^3$, $b = 84$.
(c) $a = 10{,}000$, $b = 875$.
(d) $a = 91$, $b = 85$.

6.2.10. Find $\gcd(a, b)$ if $a|b$.

6.2.11. Prove Theorem 6.2.5.

6.2.12. Let $a, b \in \mathbf{Z}^+$. Define the *least common multiple* of a and b, written $\mathrm{lcm}(a, b)$, to be the smallest positive integer divisible by both a and b.
(a) Find lcm(6, 14).
(b) Find lcm(91, 85).
(c) Suppose $a = p_1^{a_1} \cdot \cdots \cdot p_r^{a_r}$ and $b = p_1^{b_1} \cdot \cdots \cdot p_r^{b_r}$ where $p_1, \ldots, p_r$ are primes and $a_i, b_i \geq 0$ for $1 \leq i \leq r$. Conjecture a formula for $\mathrm{lcm}(a, b)$ analogous to that for $\gcd(a, b)$ given in Theorem 6.2.5.
(d) Prove the conjecture you made in (c).
(e) What is $\gcd(a, b) \cdot \mathrm{lcm}(a, b)$?

6.2.13. Suppose $a, b \in \mathbf{Z}^+$ and $\gcd(a, b) = 1$. Prove: If $a \cdot b = k^2$ for some $k \in \mathbf{Z}$, then there exist $a_1, b_1 \in \mathbf{Z}$ such that $a = a_1^2$ and $b = b_1^2$.

6.2.14. Prove: $\sqrt{3} \notin \mathbf{Q}$.

6.2.15. Prove: $\sqrt{5} \notin \mathbf{Q}$.

6.2.16. Prove: $\sqrt{15} \notin \mathbf{Q}$.

6.2.17. Fill in the blank and prove the resulting statement: For $n \in \mathbf{Z}^+$, $\sqrt{n} \in \mathbf{Q}$ if and only if ______.

6.2.18. Let $n \in \mathbf{Z}^+$ and suppose $n = p_1^{a_1} \cdot \cdots \cdot p_r^{a_r}$ where $p_1, \ldots, p_r$ are positive primes. Find a necessary and sufficient condition on the a_i for n to be the cube of some integer.

6.2.19. Is $\sqrt[3]{3} \in \mathbf{Q}$?

6.2.20. Investigate the following question: For which $n \in \mathbf{Z}$ is $\sqrt[3]{n} \in \mathbf{Q}$?

6.2.21. Generalize the results you obtained in Exercises 6.2.17 and 6.2.19.

§6.3 Rings

In this section we follow a procedure that is used often in mathematics: We generalize from a particular example. As we observed in Section 6.2, the set of integers is endowed with an interesting algebraic structure. Two binary operations are defined on **Z** and these operations satisfy several properties, namely, statements 1–8 in Section 6.2. Moreover, the Unique Factorization Theorem holds for the operation of multiplication in **Z**. We generalize from this example by considering all algebraic structures that possess at least some of the properties satisfied by **Z**. Such structures are called rings. Our primary purpose in this section is to define and to give various examples of rings. A secondary purpose is to give a taste of general ring theory by discussing the concept of factorization in rings.

In order to keep our discussion from bogging down in technical matters, we shall confine ourselves primarily to a particular class of rings, commutative rings with unity. We formalize our approach by considering the axiomatic system formed by generalizing the properties 1–8 listed in Section 6.2. These axioms are listed in the next definition.

Definition 6.3.1 *Commutative Ring with Unity*

A *commutative ring with unity* is a set R on which two binary operations, denoted $+$ and $\cdot$, are defined satisfying the following conditions:

1. Associativity of $+$. For all $x, y, z \in R$, $(x + y) + z = x + (y + z)$.
2. Identity for $+$. There exists an element $0 \in R$ such that $1 \neq 0$ and for all $x \in R$, $x + 0 = 0 + x = x$.
3. Inverse for $+$. For each $x \in R$, there exists $y \in R$ such that $x + y = y + x = 0$.
4. Commutativity of $+$. For all $x, y \in R$, $x + y = y + x$.
5. Associativity of $\cdot$. For all $x, y, z \in R$, $(x \cdot y) \cdot z = x \cdot (y \cdot z)$.
6. Identity for $\cdot$. There exists an element $1 \in R$ such that for all $x \in R$, $x \cdot 1 = 1 \cdot x = x$.
7. Commutativity of $\cdot$. For all $x, y \in R$, $x \cdot y = y \cdot x$.
8. Distributivity of $\cdot$ over $+$. For all $x, y, z \in R$, $x \cdot (y + z) = x \cdot y + x \cdot z$.

Remarks (1) Just for the record, let us mention that a set R on which are defined two binary operations, denoted $+$ and $\cdot$, satisfying Axioms 1–5 and 8 of Definition 6.3.1, along with the additional distributive law

$(x + y) \cdot z = x \cdot z + y \cdot z$ for all $x, y, z \in R$, is called a *ring*; an algebraic system satisfying these properties and Axiom 6 is called a *ring with unity*; and an algebraic system satisfying Axioms 1–5, 7, and 8 is called a *commutative ring*.

(2) To specify a commutative ring with unity, we must define the set R and the two binary operations, $+$ and $\cdot$. Then we must check that Axioms 1–8 hold for R relative to these operations. In this case, we say that R is a commutative ring with unity under the given operations or with respect to the given operations.

(3) Let R be a ring under $+$ and $\cdot$. Then it is customary to call the operation $+$ *addition in* R, and the operation $\cdot$ *multiplication in* R. This usage is not intended to imply that the set R must consist of numbers and that $+$ and $\cdot$ are simply addition and multiplication of numbers. As we shall see from examples, such is not always the case. Rather, these terms are chosen because the concept of commutative ring with unity is based on the properties of number systems such as $\mathbf{Z}$ and $\mathbf{R}$ with respect to the usual operations of addition and multiplication of numbers.

(4) Note that since $0 \neq 1$, every commutative ring with unity has at least two elements.

Example 6.3.1 The sets $\mathbf{Z}$, $\mathbf{Q}$, and $\mathbf{R}$ under the usual operations of addition and multiplication are commutative rings with unity.

Example 6.3.2 Let $R = \{x \in \mathbf{R} | x = a + b\sqrt{2}$ where $a, b \in \mathbf{Z}\}$. For example, $1, 1 + \sqrt{2}, 8 - 3\sqrt{2} \in R$ whereas $\sqrt{2}/2 \notin R$ (for if $\sqrt{2}/2 \in R$, then there exist $a, b \in \mathbf{Z}$ such that $\sqrt{2}/2 = a + b\sqrt{2}$, from which it follows that $\sqrt{2} = 2a + 2b\sqrt{2}$ and $\sqrt{2} = 2a/(1 - 2b) \in \mathbf{Q}$, which is a contradiction). R is a commutative ring with unity under the operations of addition and multiplication of real numbers. To verify this claim, we must first check that if $x, y \in R$, then $x + y \in R$ and $x \cdot y \in R$. In other words, we must check that addition and multiplication of real numbers are binary operations on R. For example, if $x = a + b\sqrt{2}$ and $y = c + d\sqrt{2}$ where $a, b, c, d \in \mathbf{Z}$, then $x + y = (a + c) + (b + d)\sqrt{2} \in R$ since $a + c \in \mathbf{Z}$ and $b + d \in \mathbf{Z}$, $x \cdot y = (a \cdot c + 2 \cdot b \cdot d) + (a \cdot d + b \cdot c)\sqrt{2} \in R$ since $a \cdot c + 2 \cdot b \cdot d \in \mathbf{Z}$ and $a \cdot d + b \cdot c \in \mathbf{Z}$. Properties 1, 4–6, and 8 hold in R with respect to $+$ and $\cdot$, since they hold in $\mathbf{R}$ with respect to $+$ and $\cdot$. Since $0 = 0 + 0 \cdot \sqrt{2}$ and $1 = 1 + 0 \cdot \sqrt{2}$, $0 \in R$ and $1 \in R$. Since for all $x \in R$, $x + 0 = 0 + x = x$ and $x \cdot 1 = 1 \cdot x = 1$, properties 2 and 6 hold for R. Finally, if $x = a + b\sqrt{2} \in R$ where $a, b \in \mathbf{Z}$, then $y = -a + (-b)\sqrt{2} \in R$ and $x + y = y + x = 0$, and hence property 3 holds for R. Therefore, R is a commutative ring with unity. The usual denotation for R is $\mathbf{Z}[\sqrt{2}]$.

Example 6.3.3 (i) Let $R = \mathscr{F}(\mathbf{R}) = \{f | f$ is a function from $\mathbf{R}$ to $\mathbf{R}\}$. Define the binary operations $+$ and $\cdot$ on R as follows: For $f, g \in R$,

$f + g$: $\mathbf{R} \to \mathbf{R}$ is the function defined by $(f + g)(x) = f(x) + g(x)$ for all $x \in \mathbf{R}$ and $f \cdot g$: $\mathbf{R} \to \mathbf{R}$ is the function defined by $(f \cdot g)(x) = f(x) \cdot g(x)$ for all $x \in \mathbf{R}$. We can check that R is a commutative ring with unity. The additive and multiplicative identities are, respectively, the constant functions f_0 and f_1 defined by $f_0(x) = 0$ for all $x \in \mathbf{R}$ and $f_1(x) = 1$ for all $x \in \mathbf{R}$.

(ii) Let $C(\mathbf{R}) = \{f \in F(\mathbf{R}) | f$ is continuous at x for all $x \in \mathbf{R}\}$. Then $C(\mathbf{R})$ is a commutative ring with unity under the operations $+$ and $\cdot$ defined in part (i). The significant point here is that if f and g are continuous at a point x, then $f + g$ and $f \cdot g$ are continuous at x. Thus $+$ and $\cdot$ are indeed binary operations on $C(\mathbf{R})$. One must also observe that the functions $f_0(x) = 0$ and $f_1(x) = 1$ are elements of $C(\mathbf{R})$. Since Axioms 1–8 hold in $F(\mathbf{R})$ and since $C(\mathbf{R}) \subseteq F(\mathbf{R})$, these axioms also hold in $C(\mathbf{R})$.

(iii) Let $D(\mathbf{R}) = \{f \in F(\mathbf{R}) | f$ is differentiable at x for all $x \in \mathbf{R}\}$. From elementary calculus, if $f, g \in D(\mathbf{R})$, then $f + g, f \cdot g \in D(\mathbf{R})$. Since $f_0, f_1 \in D(\mathbf{R})$, it follows as in (ii) that $D(\mathbf{R})$ is a commutative ring with unity. Moreover, $D(\mathbf{R}) \subset C(\mathbf{R}) \subset F(\mathbf{R})$.

Example 6.3.4 Let $\mathbf{R}[X]$ denote the set of all polynomials with real coefficients. Notice that $1, 1 + X, 1 + X + X^2 \in \mathbf{R}[X]$ while $1 + X^{1/2} \notin \mathbf{R}$. Then $\mathbf{R}[X]$ is a commutative ring with unity under the operations of usual polynomial addition and multiplication. Let us recall the definition of polynomial multiplication. If $f(X) \in \mathbf{R}[X]$, then $f(X)$ has the form $f(X) = a_0 + a_1 \cdot X + \cdots + a_n \cdot X^n = \sum_{i=0}^{n} a_i X^i$, where $a_i \in \mathbf{R}$ for $0 \le i \le n$. If $g(X) = b_0 + b_1 \cdot X + \cdots + b_m \cdot X^m$, then $f(X) \cdot g(X) = a_0 b_0 + (a_0 b_1 + a_1 b_0) \cdot X + (a_0 b_2 + a_i b_1 + a_2 b_1) X^2 + \cdots + a_n b_m \cdot X^{n+m}$. The proof that $\mathbf{R}[X]$ is a commutative ring with unity is straightforward and laborious. It is worth noting that the additive identity is the constant polynomial $z(X) = 0$ and the multiplicative identity is the constant $e(X) = 1$.

We can also consider the set, $\mathbf{Q}[X]$, of all polynomials with rational coefficients or the set, $\mathbf{Z}[X]$, of all polynomials with coefficients in $\mathbf{Z}$. Both $\mathbf{Q}[X]$ and $\mathbf{Z}[X]$ are commutative rings with unity. In general, if R is a commutative ring with unity, then the set $R[X]$ of polynomials with coefficients in R (defined analogously to $\mathbf{R}[X]$) is a commutative ring with unity.

Example 6.3.5 Let $R = \{a, b\}$ be any set with two elements. We define the operations of $+$ and $\cdot$ on R by the following tables:

$+$	a	b
a	a	b
b	b	a

$\cdot$	a	b
a	a	a
b	a	b

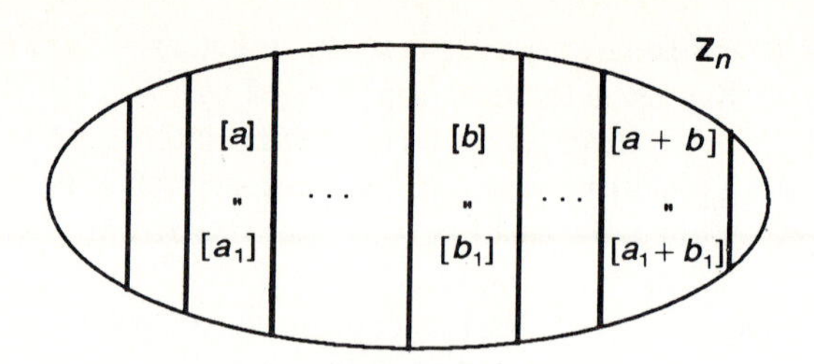

Figure 6.3.1

We can check that R is a commutative ring with unity. Notice that a is the additive identity and b is the multiplicative identity. Since a and b can be arbitrary objects, this example shows that a ring structure can be defined on any two-element set.

Example 6.3.6 Let $\mathbf{Z}_n$ denote the set of integers modulo n, $\mathbf{Z}_n = \{[0], [1], \ldots, [n-1]\}$ where $[r] = \{a \in \mathbf{Z} \mid a \equiv r \pmod{n}\}$. Let $x, y \in \mathbf{Z}_n$. We define $+$ and $\cdot$ in $\mathbf{Z}_n$ as follows: Write $x = [a]$ and $y = [b]$. Define $x + y = [a] + [b] = [a + b]$ and $x \cdot y = [a] \cdot [b] = [a \cdot b]$.

These definitions must be clarified. For example, if $x = [a] = [a_1]$ where $a_1 \in \mathbf{Z}$ and $y = [b] = [b_1]$ where $b_1 \in \mathbf{Z}$, then we must check that $[a + b] = [a_1 + b_1]$. Otherwise, $x + y$ is not a uniquely defined element of $\mathbf{Z}_n$. Similarly, we must show that $[a \cdot b] = [a_1 \cdot b_1]$.

We show that $[a + b] = [a_1 + b_1]$ and leave the proof that $[a \cdot b] = [a_1 \cdot b_1]$ as an exercise. (See Exercise 3.3.20.) Since $[a] = [a_1]$ and $[b] = [b_1]$, $a - a_1 = n \cdot k$ for some $k \in \mathbf{Z}$ and $b - b_1 = n \cdot m$ for some $m \in \mathbf{Z}$. Therefore, $(a - a_1) + (b - b_1) = a + b - (a_1 + b_1) = n \cdot (k + m)$, and hence $[a + b] = [a_1 + b_1]$.

The additive and multiplicative identities of $\mathbf{Z}_n$ are, respectively, [0] and [1]. Notice that in $\mathbf{Z}_6$, $[2] \cdot [3] = [2 \cdot 3] = [6] = [0]$; in other words, in $\mathbf{Z}_6$ the product of two nonzero elements can be zero.

Implicit in this example is an important idea—that of a *well-defined binary operation*. The set $\mathbf{Z}_n$ consists of equivalence classes of $\mathbf{Z}$ with respect to the equivalence relation $\equiv_n$. In order to define the operations $+$ and $\cdot$ on $\mathbf{Z}_n$, we naturally want to use the operations $+$ and $\cdot$ on $\mathbf{Z}$. Thus, for example, it is only reasonable to define the sum of two equivalence classes, $[a] + [b]$, to be $[a + b]$. In other words, the sum of two equivalence classes is defined in terms of representatives of the equivalence classes. Since a given equivalence class has infinitely many representatives, it is necessary to show that the equivalence class obtained when two equivalence classes are added is the same no matter what the representatives. Thus, if $[a] = [a']$ and $[b] = [b']$, then we must check that $[a + b] = [a' + b']$. Because this is indeed the case, we say that the operation $+$ is *well defined* on $\mathbf{Z}_n$. Similarly, $\cdot$ is a well-defined operation on $\mathbf{Z}_n$.

To illustrate how one can reason from the axioms for a ring, let us prove the following basic results.

Proposition 6.3.1 Let R be a commutative ring with unity. (i) R has a unique additive identity. (ii) Every element of R has a unique additive inverse.

Proof (i) Suppose 0 and $0'$ are additive identities in R. Then $0 = 0 + 0'$ since $0'$ is an additive identity and $0 + 0' = 0'$ since 0 is an additive identity. Therefore, $0 = 0'$.

(ii) Let $x \in R$ and suppose y and y' are additive inverses of x. Then, using associativity, we find that

$$y = y + 0 = y + (x + y') = (y + x) + y' = 0 + y' = y'. \qquad \blacksquare$$

Remark We usually denote the additive inverse of x by $-x$. Then we can define the difference $x - y$ of x and y to be $x - y = x + (-y)$. Loosely speaking, we can say that a ring is an algebraic system in which addition, subtraction, and multiplication are defined.

Factorization

As we stated earlier, our purpose in this chapter is to introduce several important algebraic structures. We intend to show how these structures arise naturally out of concrete mathematical considerations and how the study of a given type of mathematical structure (e.g., commutative rings with unity) can be guided by the results in a particular case (e.g., the ring of integers). We illustrate this process by describing one of the principal foci of the study of rings, the question of factorization of elements.

Henceforth, suppose R is a commutative ring with unity. Using as a guide the investigation of $\mathbf{Z}$ carried out in the previous section, we shall introduce several technical ideas that enable us to formulate the concepts of unique factorization in a commutative ring with unity.

Definition 6.3.2 *Divisibility in a Commutative Ring*

Let $x, y \in R$. (Recall that R is a commutative ring with unity.) We say that x *divides* y, written $x|y$, if there exists $z \in R$ such that $x \cdot z = y$. If $x|y$, then we also say that x is a *factor* of y, x is a *divisor* of y, or y is a *multiple* of x. If x does not divide y, then we write $x \nmid y$.

Example 6.3.7 (i) In $\mathbf{Z}$, $2 \nmid 3$, but in $\mathbf{Q}$, $2|3$ since $3 = 2 \cdot (3/2)$ and $3/2 \in \mathbf{Q}$.

(ii) Let $R = \mathbf{R}[X]$. Then $(X+1)|(X^2+4X+3)$ in R while $(X+1) \nmid (X^2+7X-4)$. Note that if $r \in \mathbf{R}^*$, then the polynomial $f_r(X)$ r has the property that for all $f(X) \in \mathbf{R}[X]$, $f_r(X)|f(X)$, for suppose that $f(X) = a_0 + a_1 \cdot X + \cdots + a_n \cdot X^n$. Then $f(X) = f_r(X) \cdot g(X)$ where $g(X) = (r^{-1} \cdot a_0) + (r^{-1} \cdot a_1) \cdot X + \cdots + (r^{-1} \cdot a_n) \cdot X^n \in \mathbf{R}[X]$.

(iii) In $\mathbf{Z}[\sqrt{2}]$, $(3+\sqrt{2})|7$ since $(3+\sqrt{2})(3-\sqrt{2}) = 9 - 2 = 7$ and $3 - \sqrt{2} \in \mathbf{Z}[\sqrt{2}]$.

(iv) In $\mathbf{Z}_6$, $[4]|[2]$ since $[4] \cdot [2] = [8] = [2]$.

In discussing divisibility or factoring in a general commutative ring with unity, we need an analog of the notion of prime number in $\mathbf{Z}$. Let us look closely at primes in $\mathbf{Z}$. If $p \in \mathbf{Z}$ is prime, then the only factorizations of p are: $p = p \cdot 1 = (-p) \cdot (-1)$. As we noted in the previous section, for any integer $n \neq 0$, $n = n \cdot 1 = (-n) \cdot (-1)$. Primes are significant because they possess no factorizations except these "trivial" ones. These factorizations are trivial because the elements 1 and -1 are factors of every integer. We now introduce analogs in a general commutative ring with unity of the elements 1 and -1 in $\mathbf{Z}$.

Definition 6.3.3 *Units in a Commutative Ring*

In a commutative ring with unity, an element u is a *unit* if u divides 1.

In other words, u is a unit in R if and only if there exists $v \in R$ such that $u \cdot v = 1$.

Let u be a unit in R and let $x \in R$. Then u is a factor of x since $x = 1 \cdot x = (u \cdot v) \cdot x = u \cdot (v \cdot x)$ for some $v \in R$. This observation means that any unit in R divides every element of R. This divisibility by a unit is a trivial matter in a commutative ring with unity since units divide all elements.

Example 6.3.8 (i) In $\mathbf{Z}$ the only units are ± 1.

(ii) In $\mathbf{Z}_8$ the elements [1], [3], [5], and [7] are units since $[1] \cdot [1] = [3] \cdot [3] = [5] \cdot [5] = [7] \cdot [7] = [1]$. No other elements of $\mathbf{Z}_8$ are units.

(iii) In $\mathbf{R}[X]$ any nonzero constant polynomial (e.g., $f_r(X) = r$ where $r \in \mathbf{R}^*$) is a unit. For $f_r(X) \cdot f_{1/r}(X) = r \cdot (1/r) = 1 \in \mathbf{R}[X]$. Conversely, we can show that any unit in $\mathbf{R}[X]$ must be a constant polynomial. (See Exercise 6.3.5.)

(iv) In $\mathbf{Z}[\sqrt{2}]$, $\sqrt{2} + 1$ is a unit since $(\sqrt{2}+1)(\sqrt{2}-1) = 2 - 1 = 1$. In fact, we can show that for all $n \in \mathbf{Z}$, $(\sqrt{2}+1)^n$ is a unit in $\mathbf{Z}[\sqrt{2}]$. (See Exercise 6.3.6(a).)

We next define the analog of prime integer in a general commutative ring with unity. We use the standard terminology of ring theory.

Definition 6.3.4 *Irreducible Element*

An element $r \in R$ is *irreducible* if (i) r is nonzero and not a unit, and (ii) whenever $r = a \cdot b$ where $a, b \in R$, then either a or b is a unit.

Thus an irreducible element is a nonzero nonunit that can be factored only trivially in the sense that one of the factors in any factorization must be a unit.

We have already seen that any prime integer is irreducible in **Z**. To give another example of an irreducible element, we look at the polynomial ring $\mathbf{R}[X]$.

Example 6.3.9 Let $f(X) \in \mathbf{R}[X]$ with $f(X) \neq 0$. Write $f(X) = a_0 + a_1 \cdot X + \cdots + a_n \cdot X^n$ where $a_n \neq 0$. We define the degree of $f(X)$, written $\deg(f)$, to be n. The degree of the polynomial $f(X) = 0$ is not defined. Thus the degree of $f(X)$ is the highest power of X appearing in f. For example, $\deg(X + 1) = 1$ and $\deg(1 + X^2 + X^3) = 3$. Notice that the polynomials of degree 0 are precisely the nonzero constant polynomials, i.e., the units of $\mathbf{R}[X]$.

From the definition of multiplication in $\mathbf{R}[X]$, it follows that for all $f(X), g(X) \in \mathbf{R}[X]$, $\deg(f \cdot g) = \deg(f) + \deg(g)$. We now use this important property of polynomials to show that $f(X) = X$ is irreducible in $\mathbf{R}[X]$, for if $f(X) = g(X) \cdot h(X)$ where $g(X), h(X) \in \mathbf{R}[X]$, then $\deg(g) + \deg(h) = \deg(f) = 1$. Since $\deg(g) \geq 0$ and $\deg(h) \geq 0$, either $\deg(g) = 0$ (and $\deg(h) = 1$) or $\deg(h) = 0$ (and $\deg(g) = 1$). Therefore, either $g(X)$ or $h(X)$ is a constant and hence a unit in $\mathbf{R}[X]$. Observe that this argument shows that any polynomial in $\mathbf{R}[X]$ of degree 1, i.e., any polynomial of the form $a \cdot X + b$ with $a \neq 0$, is irreducible. Another example of an irreducible polynomial in $\mathbf{R}[X]$ is $X^2 + 1$. Quick proof: If $X^2 + 1 = f(X) \cdot g(X)$ where neither $f(X)$ nor $g(X)$ is a unit, then, since $\deg(f) + \deg(g) = \deg(X^2 + 1) = 2$, $\deg(f) = \deg(g) = 1$. Hence $f(X) = a \cdot X + b$ where $a, b \in \mathbf{R}$ and $a \neq 0$. Then $f(r) = 0$ where $r = b/a$ and therefore $r^2 + 1 = f(r) \cdot g(r) = 0$, which contradicts the fact that $r^2 + 1 > 0$. Thus $X^2 + 1$ is irreducible in $\mathbf{R}[X]$.

Before introducing the concept of unique factorization, we impose an additional condition of the rings under consideration. This property (stated in Definition 6.3.6) is yet another of those that are possessed by the ring **Z**.

Definition 6.3.5 *Zero Divisor*

Let R be a commutative ring with unity. An element $a \neq 0$ in R is a *zero divisor* if there exists $b \in R$ such that $b \neq 0$ and $a \cdot b = 0$.

Definition 6.3.6 *Integral Domain*

An *integral domain* is a commutative ring with unity having no zero divisors.

Example 6.3.10 (i) The rings **Z**, **Q**, **R**, and **R**$[X]$ are all integral domains.

(ii) The ring $\mathbf{Z}_4$ is not an integral domain since $[2] \cdot [2] = [0]$ in $\mathbf{Z}_4$. The rings $\mathbf{Z}_2$ and $\mathbf{Z}_3$ are integral domains.

Let us now phrase the concept of unique factorization in the setting of an integral domain.

Definition 6.3.7 *Unique Factorization Domain*

An integral domain R is a *unique factorization domain* (UFD) if (i) for each nonzero nonunit $x \in R$, there exist irreducible elements $p_1, \ldots, p_r$ such that $x = p_1 \cdot \cdots \cdot p_r$, and (ii) if $x = p_1 \cdot \cdots \cdot p_r = q_1 \cdot \cdots \cdot q_s$ where the p's and q's are irreducible, then $r = s$ and (after renumbering the q's if necessary) $p_1 = q_1 \cdot u_1, \ldots, p_r = q_r \cdot u_r$ where for $1 \leq i \leq r$, u_i is a unit.

To summarize the definition in words, a UFD is an integral domain in which every element (except 0 and units) is either irreducible or expressible as a product of irreducibles and any two factorizations of a given element into irreducibles are identical except for the order of irreducible factors and for unit factors. For example, in **R**$[X]$,

$$X^2 + 3X + 3 = (X + 1) \cdot (X + 3) = (7 \cdot X + 7) \cdot (X/7 + 3/7)$$

are two factorizations of $X^2 + 3 \cdot X + 2$ that satisfy Definition 6.3.7(ii).

As we have seen, **Z** is a UFD. The next theorem provides another example.

Theorem 6.3.2 $\mathbf{R}[X]$ is a UFD.

Proof (Outline.) First, it must be checked that $\mathbf{R}[X]$ is an integral domain. We leave this step as an exercise.

Next it must be verified that if $f(X) \in \mathbf{R}[X]$ with $f(X) \neq 0$ and $f(X) \neq$ a unit, then $f(X) = p_1(X) \cdot \cdots \cdot p_r(X)$ where $p_1(X), \ldots, p_r(X)$ are irreducible polynomials. As in the case of $\mathbf{Z}$, we use the Principle of Strong Induction to prove this result.

For each $n \in \mathbf{N}^+$, define the statement $S(n)$: If $f(X) \in \mathbf{R}[X]$ and $\deg(f(X)) = n$, then there exist irreducible polynomials $p_1(X), \ldots, p_r(X)$ such that $f(X) = p_1(X) \cdot \cdots \cdot p_r(X)$.

Basis step If $\deg(f(X)) = 1$, then $f(X)$ is irreducible (see Example 6.3.9) and hence statement $S(1)$ is true.

Inductive step We suppose statements $S(1), \ldots, S(n)$ are true and we show that $S(n+1)$ is true. Let $f(x) \in \mathbf{R}[X]$ be a polynomial of degree $n+1$. If $f(X)$ is irreducible, then statement $S(n+1)$ holds for $f(X)$. If $f(X)$ is not reducible, then $f(X) = g(X) \cdot h(X)$ where $1 \leq \deg(g(X)) = a$, $b = \deg(h(X)) \leq n$. Since statements $S(a)$ and $S(b)$ are true, $g(X) = p_1(X) \cdot \cdots \cdot p_k(X)$ and $h(X) = p_{k+1}(X) \cdot \cdots \cdot p_r(X)$ where $p_1(X), \ldots, p_r(X)$ are irreducible. Therefore, $f(X) = p_1(X) \cdot \cdots \cdot p_r(X)$. Thus every element of $\mathbf{R}[X]$ of degree ≥ 1 either is irreducible or is a product of irreducibles.

Finally, we turn to the proof of the proof of the uniqueness of the decomposition into irreducibles of elements of $\mathbf{R}[X]$. Looking back at the proof of unique factorization in $\mathbf{Z}$, we see that the crucial ingredients are Lemma 6.2.2, from which the uniqueness of irreducible decomposition is deduced, and the Division Theorem, from which Lemma 6.2.2 is derived. We shall now state the analogs of these results for $\mathbf{R}[X]$ and indicate how the proofs for $\mathbf{Z}$ must be modified for $\mathbf{R}[X]$. Once the polynomial version of Lemma 6.2.2 is proved, the proof of Theorem 6.2.2 is almost identical in structure to the proof of Theorem 6.2.1. We leave the details as an exercise. ■

Theorem 6.3.3 *Division Theorem in* $\mathbf{R}[X]$. *Let* $f(X), g(X) \in \mathbf{R}[X]$ *with* $g(X) \neq 0$. *Then there exist* $q(X), r(X) \in \mathbf{R}[X]$ *with* $r(X) = 0$ *or* $\deg(r(X)) < \deg(g(X))$ *such that* $f(X) = g(X) \cdot q(X) + r(X)$.

Proof (Outline.) First write that if $f(X) = 0$, then the statement holds with $q(X) = r(X) = 0$. Henceforth, we assume $f(X) \neq 0$.

For $n \in \mathbf{N}$, define the statement $S(n)$: If $f(X), g(X) \in \mathbf{R}[X]$ with $\deg(f(X)) = n$ and $g(X) \neq 0$, then there exist $q(X), r(X) \in \mathbf{R}[X]$ with $r(X) = 0$ or $\deg(r(X)) < \deg(g(X))$ such that $f(X) = g(X) \cdot q(X) + r(X)$.

We argue by PSI. (Our proof parallels that of Theorem 1.3.6.)

Basis step If $\deg(f(X)) = 0$, then $f(X) = c \neq 0$ is constant. If $g(X) = d$ is constant, then $f(X) = c = d \cdot (c/d) = g(X) \cdot q(X) + r(X)$ where $q(X) = c/d$ and $r(X) = 0$. Otherwise, $f(X) = c = g(X) \cdot 0 + c = g(X) \cdot q(X) + r(X)$ where $q(X) = 0$ and $\deg(r(X)) = 0 < \deg(g(X))$.

Inductive step Suppose $S(0), \ldots, S(n)$ are true. We show $S(n+1)$ is true. Suppose $\deg(f(X)) = n + 1$. Write $f(X) = a \cdot X^{n+1}$ + lower order terms (abbreviated lots) and $g(X) = b \cdot X^d$ + lots. If $d = \deg(g(X)) > n + 1 = \deg(f(X))$, then we are done. (Just take $q(X) = 0$ and $r(X) = f(X)$.) Otherwise, $f(x) - g(X) \cdot (a/b) \cdot X^{n+1-d} = h(X)$ is a polynomial of degree $\leq n$. Apply the inductive hypothesis to $h(X)$ to find $q_1(X), r(X) \in \mathbf{R}[X]$ such that $h(X) = g(X) \cdot q_1(X) + r(X)$ where $r(X) = 0$ or $\deg(r(X)) < \deg(g(X))$. Therefore

$$f(X) = g(X) \cdot \underbrace{\left((a/b) \cdot X^{n+1-d} + q_1(X)\right)}_{q(X)} + r(X)$$

where $q(X), r(X) \in \mathbf{R}[X]$ and $r(X) = 0$ or $\deg(r(X)) < \deg(g(X))$.
By PSI the proof is complete. ∎

Lemma 6.3.4 *If $p(X) \in \mathbf{R}[X]$ is irreducible and $p(X) | f(X) \cdot g(X)$, then $p(X) | f(X)$ or $p(X) | g(X)$.*

Proof (Outline.) Once again, the result holds if $f(X) = 0$ (since $p(X) | f(X) = 0$), and hence we assume that $f(X) \neq 0$.
For $n \geq 0$, let $S(n)$ be the following statement: If $f(X), g(X) \in \mathbf{R}[X]$ with $\deg(f(X)) = a$, $\deg(g(X)) = b$ where $a + b = n$, then $p(X) | f(X)$ or $p(X) | g(X)$. We prove $S(n)$ holds for $n \geq 0$ by PSI.

Basis step Exercise.

Inductive step Suppose that $S(0), \ldots, S(n)$ are true and that $p(X) | f(X) \cdot g(X)$ where $\deg(f(X)) + \deg(g(X)) = a + b = n + 1$. Consider two cases:

Case 1. $\deg(p(X)) \leq \deg(f(X))$. Then use the Division Theorem to find $g(X), r(X) \in \mathbf{R}(X)$ such that $f(X) = p(X) \cdot q(X) + r(X)$ where $r(X) = 0$ or $\deg(r(X)) < \deg(p(X))$. Then $f(X) \cdot g(X) = p(X) \cdot q(X) \cdot g(X) + r(X) \cdot g(X)$ and hence $p(X) | r(X) \cdot g(X)$ with $\deg(r(X)) + \deg(g(X)) < \deg(p(X)) + \deg(g(X)) \leq \deg(f(X)) + \deg(g(X)) = n + 1$. Therefore, by inductive hypothesis $p(X) | r(X)$ or $p(X) | g(X)$. If $p(X) | r(X)$, then $r(X) = 0$ (since $\deg(r(X)) < \deg(p(X))$) and thus $p(X) | p(X) \cdot q(X) = f(X)$. Thus the conclusion of the theorem holds in Case 1.

Case 2. $\deg(p(X)) > \deg(f(X))$. Use the Division Theorem to write $p(X) = f(X) \cdot q(X) + r(X)$ where $r(X) = 0$ or $\deg(r(X)) < \deg(f(X))$.

Then $p(X)|r(X) \cdot g(X) = p(X) \cdot g(X) - f(X) \cdot g(X) \cdot q(X)$ and $\deg(r(X)) + \deg(g(X)) < \deg(f(X)) + \deg(g(X)) = n + 1$. Therefore, $p(X)|r(X)$ or $p(X)|g(X)$. If $r(X) = 0$, then $f(X)|p(X)$, which, since $p(X)$ is irreducible and $\deg(f(X)) < \deg(p(X))$, means that $f(X)$ is constant, and hence $p(X)|g(X)$. If $r(X) \neq 0$, then $\deg(r(X)) < \deg(f(X)) < \deg(p(X))$ and hence $p(X) \nmid r(X)$. Therefore, $p(X)|g(X)$.
Thus by PSI the proof is complete. ■

The question of factorization of elements of a commutative ring with unity became a central issue in algebra and number theory in the middle third of the nineteenth century. As noted in Section 6.2, Gauss proved the Unique Factorization Theorem for **Z** in 1801. In 1828 Gauss proved that the Unique Factorization Theorem holds for a certain ring contained in the set of complex numbers. Other mathematicians followed Gauss's lead by considering several other rings of complex numbers. In some cases, the unique factorization property holds, and in other cases this property fails to hold. In the work of Gauss and later mathematicians, the question of unique factorization was bound up with another of the principal concerns of algebra: the solving of polynomial equations. This topic is a primary focus of the next section.

Exercises §6.3

6.3.1. Let R be a commutative ring with unity. Prove:
(a) The multiplicative identity in R is unique.
(b) $a \cdot 0 = 0$ for all $a \in R$.
(c) $(a^2 - b^2) = (a - b) \cdot (a + b)$ for all $a, b \in R$.
(d) For each unit $x \in R$ the multiplicative inverse of x is unique.
(e) For all $a \in R$, $a|a$.
(f) For all $a, b, c \in R$, if $a|b$ and $b|c$, then $a|c$.
(g) For all $a, b, c \in R$, if $a|b$, then $a|b \cdot c$.

6.3.2. Find all units in (a) $\mathbf{Z}_4$, (b) $\mathbf{Z}_5$, (c) $\mathbf{Z}_6$, (d) $\mathbf{Z}_7$, (e) $\mathbf{Z}_{12}$.

6.3.3. Find all zero divisors in (a) $\mathbf{Z}_4$, (b) $\mathbf{Z}_5$, (c) $\mathbf{Z}_6$, (d) $\mathbf{Z}_7$, (e) $\mathbf{Z}_{12}$.

6.3.4. Does the ring $\mathscr{F}(\mathbf{R})$ contain zero divisors? Justify your answer.

6.3.5. (a) Prove that no nonconstant polynomial in $\mathbf{R}[X]$ is a unit in $\mathbf{R}[X]$.
(b) Is $f(X) = X^3 + 1$ irreducible in $\mathbf{R}[X]$?
(c) Find an irreducible polynomial $f(X) \in \mathbf{Q}[X]$ of degree 3.
(d) Is there an irreducible polynomial in $\mathbf{R}[X]$ having degree 3?

6.3.6. Let R be a commutative ring with unity and let u be a unit in R. (a) Prove that for all $n \in \mathbf{N}^+$, u^n is a unit.
(b) Prove that u is not a zero divisor.

6.3.7. Let R be an integral domain and let $a, b, c \in R$ with $a \neq 0$. Prove that if $a \cdot b = a \cdot c$, then $b = c$.

6.3.8. Prove that $\mathbf{Z}[\sqrt{2}] = \{x \in \mathbf{R} \mid x = a + b\sqrt{2} \text{ where } a, b \in \mathbf{Z}\}$ is an integral domain.

6.3.9. Let R be a commutative ring with unity. An element $a \in R$ is *idempotent* if $a^2 = a$:
(a) Find all idempotents in $\mathbf{Z}_6$.
(b) Prove that in an integral domain the only idempotents are 0 and 1.
(c) Exhibit an infinite set of rings each of which has an idempotent $\neq 0, 1$.

6.3.10. Investigate the following question: For which $n \in \mathbf{N}^+$ is $\mathbf{Z}_n$ an integral domain? Formulate a conjecture and prove what you can about your conjecture.

6.3.11. Prove that $\mathbf{R}[X]$ is an integral domain.

6.3.12. Complete the proof of Theorem 6.3.2.

6.3.13. Let R be a commutative ring with unity. For $a, b \in R$ we say that a is an *associate* of b if there exists a unit $u \in R$ such that $a = b \cdot u$.
(a) Prove that the relation $\sim$ on R defined by $a \sim b$ if a is an associate of b is an equivalence relation on R.
(b) Give an example of a ring R and an element $x \in R$ such that x has infinitely many associates.

6.3.14. Is it true that if R is a commutative ring with unity and $0 \neq x \in R$, then x is a zero divisor if and only if x divides 0? Explain.

§6.4 Fields

As we have seen, the sets **Z** and **Q** are rings under the usual operations of addition and multiplication. The rationals, however, enjoy an additional property: Every nonzero rational number has a multiplicative inverse: For all $x \in \mathbf{Q}^* = \mathbf{Q} - \{0\}$, there exists $x' \in \mathbf{Q}$ such that $x \cdot x' = 1$. This property causes the algebraic structure of **Q** to be quite different from that of **Z**. For example, if $x \in \mathbf{Q}^*$ and $y \in Q$, then x divides y since $y = 1 \cdot y = (x \cdot x') \cdot y = x \cdot (x'y)$. Because of this fact, the questions of factorization and factorization into primes in **Q** are meaningless. Nevertheless, many other interesting questions of an algebraic nature do arise with respect to **Q**. For the most part, these questions concern the solvability of polynomial equations with coefficients in **Q**. We shall consider these questions briefly after introducing the class of algebraic structures that possess the property discussed previously in this paragraph for the rational numbers. Such structures are called fields, and in this section we present several examples of fields and mention some of their important properties.

Definition 6.4.1 *Field*

A *field* is a commutative ring with unity in which every nonzero element is a unit.

We begin by recording an elementary fact about fields.

Proposition 6.4.1 A field is an integral domain.

Proof Let F be a field. Suppose $a, b \in F$ with $a \cdot b = 0$ and $a \neq 0$. We show $b = 0$. Since $a \neq 0$, a has a multiplicative inverse a'. Thus $0 = a' \cdot 0 = a' \cdot (a \cdot b) = (a' \cdot a) \cdot b = 1 \cdot b = b$. ■

We can interpret Proposition 6.4.1 as saying that multiplication is a binary operation on F^*.

We have already observed that **Q** and **R** are fields. Our next theorem provides infinitely many examples of fields, each of which is a finite set.

Theorem 6.4.2 If p is a positive prime integer, then $\mathbf{Z}_p$ is a field.

Proof We have noted that $\mathbf{Z}_p$ is a commutative ring with unity. Let $[0] \neq [a] \in \mathbf{Z}_p$. We show that there exists $[a'] \in \mathbf{Z}_p$ such that $[a] \cdot [a'] = [1]$, i.e., we find $a' \in \mathbf{Z}$ such that $a \cdot a' \equiv 1 \pmod{p}$ or equivalently such that $p|(aa' - 1)$.

Consider the first $p + 1$ powers of $[a]$: $[a], [a]^2 = [a^2], \ldots, [a]^{p+1} = [a^{p+1}]$. Since $\mathbf{Z}_p$ has p elements, by the Pigeonhole Principle there exist $i, j \in \mathbf{Z}$ with $1 \leq j < i \leq p + 1$ such that $[a^i] = [a^j]$. Therefore, $p|(a^i - a^j) = a^j(a^{i-j} - 1)$. Since p is prime, $p|a^j$ or $p|(a^{i-1} - 1)$. But if $p|a^j$, then $p|a$ and hence $[a] = [0]$. Therefore, $p|(a^{i-j} - 1)$ and $a \cdot a^{i-j-1} = a^{i-j} \equiv 1 \pmod{p}$. Thus $[a^{i-j-1}]$ is the multiplicative inverse of a. ■

What about the converse of Theorem 6.4.2? If $n \in \mathbf{Z}^+$ and $\mathbf{Z}_n$ is a field, then is n a prime integer? By considering examples such as $n = 4$ and $n = 6$, we might conjecture that this question must have an affirmative answer. We leave this problem as an exercise.

The Complex Numbers

We now construct what is probably the most important field in mathematics, the complex numbers. We take as given the field $\mathbf{R}$ of real numbers and define the complex numbers in terms of $\mathbf{R}$.

As a set we define the complex numbers $\mathbf{C}$ to be the Cartesian product $\mathbf{R} \times \mathbf{R}$. To show that $\mathbf{C}$ is a field, we have to define the two algebraic operations on $\mathbf{C}$ and then show that $\mathbf{C}$ is a commutative ring with unity in which every nonzero element is a unit.

Let $x = (a, b)$ and $y = (c, d)$ be in $\mathbf{C}$. Define addition and multiplication in $\mathbf{C}$ respectively by

$$(*) \qquad \begin{aligned} x + y &= (a + c, b + d) \\ x \cdot y &= (a \cdot c - b \cdot d, a \cdot d + b \cdot c). \end{aligned}$$

Recall that $a, b, c, d \in \mathbf{R}$ and that $a + c$, $a \cdot c$, etc., refer to the sum and product of real numbers. We are defining $+$ and $\cdot$ in $\mathbf{C}$ in terms of $+$ and $\cdot$ in $\mathbf{R}$. Again, we use the same symbols for addition and multiplication merely to minimize the number of symbols.

Before going on to the field properties, let us consider the motivation behind the definitions of $+$ and $\cdot$ in $\mathbf{C}$. The main idea is that we create $\mathbf{C}$ by adjoining to $\mathbf{R}$ an element i such that $i^2 = -1$. The existence of such an element means that equations like $X^2 + 1 = 0$, $X^2 + 5 = 0$, and in fact any quadratic equation $aX^2 + bX + c = 0$ where $a, b, c \in \mathbf{R}$ can be solved in $\mathbf{C}$. (See Exercise 6.4.1.) Of course, once we have i in $\mathbf{C}$, we also have $b \cdot i$ for all $b \in \mathbf{R}$ and $a + b \cdot i$ for all $a, b \in \mathbf{R}$. Of course, i is just a symbol for a complex number whose square is -1. To define the number i or any complex number satisfactorily, we must do so in terms of set theory. Thus the ordered pair (a, b) is a way of capturing formally the informal notion of a complex number as an expression of the form $a + b \cdot i$.

Now, given $a + b \cdot i, c + d \cdot i \in \mathbf{C}$, what should $(a + b \cdot i) + (c + d \cdot i)$ and $(a + b \cdot i) \cdot (c + d \cdot i)$ be? If $+$ and $\cdot$ are to be associative and commutative, and if $\cdot$ is to distribute over $+$, then

$$(a + b \cdot i) \cdot (c + d \cdot i) = (a + c) + (b + d) \cdot i$$

and

$$\begin{aligned}(a + b \cdot i) + (c + d \cdot i) &= a \cdot c + b \cdot c \cdot i + a \cdot d \cdot i + b \cdot d \cdot i^2 \\ &= (a \cdot c - b \cdot d) + (a \cdot d + b \cdot c) \cdot i.\end{aligned}$$

Thus the rationale behind definitions $(*)$. Notice that the element i is simply the ordered pair $(0, 1)$.

Theorem 6.4.3 *$\mathbf{C}$ is a field under the operations of $+$ and $\cdot$.*

Proof The proof that $\mathbf{C}$ is a commutative ring with unity is another of those tedious exercises that are left to the reader. (Note that the additive and multiplicative identities in $\mathbf{C}$ are, respectively, $(0, 0)$ and $(1, 0)$.) We show that if $z \in \mathbf{C}$, $z \neq 0$, then z is a unit.

Let $z = (a, b)$. Since $z \neq (0, 0)$, either $a \neq 0$ or $b \neq 0$. We must find $z' = (c, d)$ such that $z \cdot z' = (a \cdot c - b \cdot d, a \cdot d + b \cdot c) = (1, 0)$, i.e.,

$$\begin{aligned} a \cdot c - b \cdot d &= 1 \\ b \cdot c + a \cdot d &= 0. \end{aligned}$$

This 2×2 system of linear equations can be solved for c and d since $a^2 + b^2 \neq 0$. The solution is $c = a/(a^2 + b^2)$ and $d = -b/(a^2 + b^2)$. Thus z has multiplicative inverse $z' = (a/(a^2 + b^2), -b/(a^2 + b^2))$. ■

The set of complex numbers $\mathbf{C} = \mathbf{R} \times \mathbf{R}$ can, of course, be represented geometrically as the Cartesian plane. This representation happens to be compatible with the algebraic operations of addition and multiplication on $\mathbf{C}$.

First we consider addition. Let $\mathbf{0} = (0, 0)$. Let $x = (a, b) \neq \mathbf{0}$ and $y = (c, d) \neq \mathbf{0}$ be elements of $\mathbf{C}$. Suppose the points $\mathbf{0}$, x, and y are not collinear in $\mathbf{R}^2$. Then the points $\mathbf{0}$, x, y and $x + y = (a + c, b + d)$ are the vertices of a parallelogram in $\mathbf{R}^2$, as shown in Figure 6.4.1.

What about the product $x \cdot y$ where $x \neq \mathbf{0} \neq y$? We begin by defining the *norm* of x, written $|x|$, to be $|x| = \sqrt{x^2 + y^2}$. Since $x \neq \mathbf{0}$ and $y \neq \mathbf{0}$, $|x| \neq 0$ and $|y| \neq 0$. The real number $|x|$ represents the distance from 0 to x. (See Figure 6.4.2.)

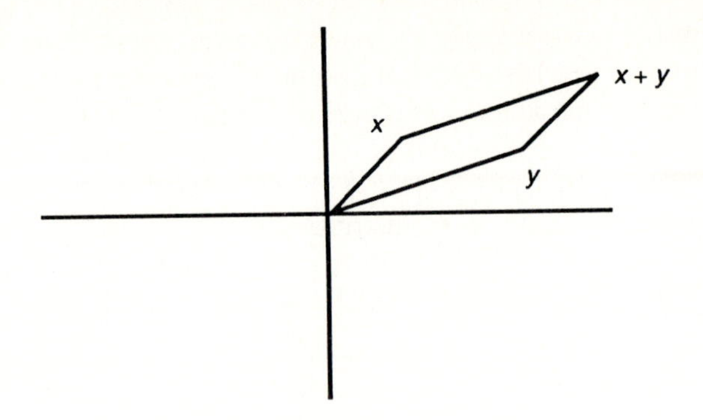

Figure 6.4.1

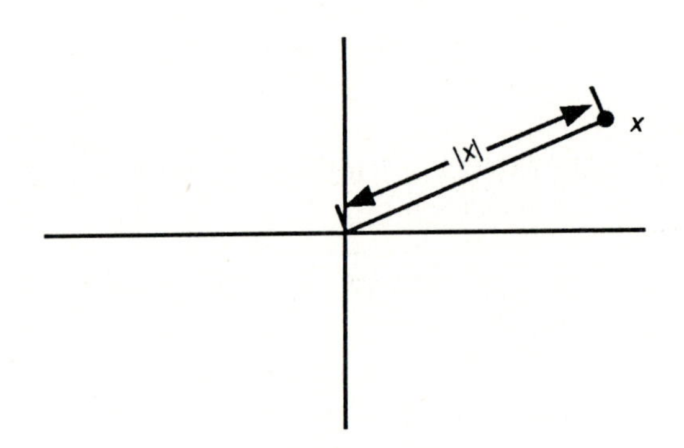

Figure 6.4.2

Now let us complete the norm of $x \cdot y$:

$$
\begin{aligned}
|x \cdot y| &= \sqrt{(a \cdot c - b \cdot d)^2 + (a \cdot d + b \cdot c)^2} \\
&= \sqrt{a^2 \cdot c^2 - 2a \cdot b \cdot c \cdot d + b^2 \cdot d^2 + a^2 \cdot d^2 + 2a \cdot b \cdot c \cdot d + b^2 \cdot c^2} \\
&= \sqrt{a^2 \cdot c^2 + a^2 \cdot d^2 + b^2 \cdot c^2 + b^2 \cdot d^2} \\
&= \sqrt{(a^2 + b^2) \cdot (c^2 + d^2)} \\
&= |x| \cdot |y|.
\end{aligned}
$$

Thus we have the miraculous equality $|x \cdot y| = |x| \cdot |y|$, which asserts that the distance from $x \cdot y$ to $\mathbf{0}$ is the product of the distance from x to $\mathbf{0}$ with the distance from y to $\mathbf{0}$. But even more can be said about multiplication.

Let $|x| = r_x$. Then the point $x = (a, b)$ has the polar coordinate representation $x = (r_x \cos(\theta_x), r_x \sin(\theta_x))$ where $0 \leq \theta_x < 2\pi$. (See Figure

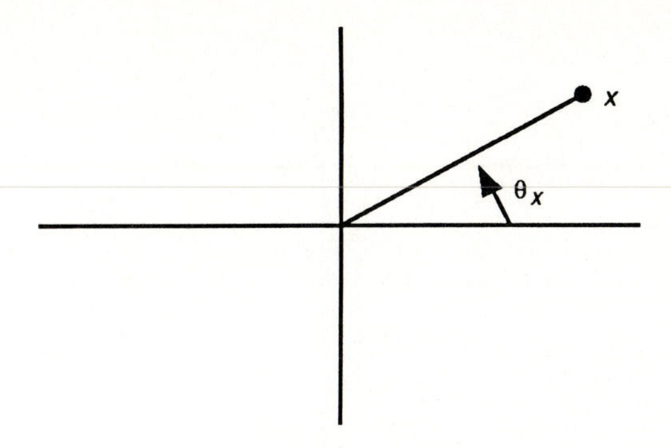

Figure 6.4.3

6.4.3.) The number θ_x is the radian measure of the angle between the positive branch of the x-axis and the segment joining $\mathbf{0}$ to x. We call θ_x the *polar angle* of x.

Now, if we multiply $x = (r_x \cos(\theta_x), r_x \sin(\theta_x))$ and $y = (r_y \cos(\theta_y), r_y \sin(\theta_y))$, then we find

$$\begin{aligned} x \cdot y &= \big(r_x \cdot r_y\big(\cos(\theta_x)\cos(\theta_y) - \sin(\theta_x)\sin(\theta_y)\big), \\ &\qquad r_x \cdot r_y\big(\cos(\theta_x)\sin(\theta_y) + \sin(\theta_x)\cos(\theta_y)\big)\big) \\ &= \big(r_x \cdot r_y \cos(\theta_x + \theta_y), r_x \cdot r_y \sin(\theta_x + \theta_y)\big), \end{aligned}$$

where the last equality comes from the trigonometric identities

$$\begin{aligned} \cos(\alpha + \beta) &= \cos(\alpha) \cdot \cos(\beta) - \sin(\alpha) \cdot \sin(\beta) \\ \sin(\alpha + \beta) &= \sin(\alpha) \cdot \cos(\beta) + \cos(\alpha) \cdot \sin(\beta). \end{aligned}$$

Thus $x \cdot y$ is the point in $\mathbf{R}^2$ whose distance to $\mathbf{0}$ is $r_{x \cdot y} = |x \cdot y| = r_x \cdot r_y$ and whose polar angle $\theta_{x \cdot y} = \theta_x + \theta_y$. (See Figure 6.4.4.)

Polynomial Rings over a Field

Until the middle of the nineteenth century, the principal concern of that portion of mathematics known as algebra was not the study of general algebraic structures but rather the investigation of solutions of polynomial equations. For example, the quadratic formula provides an explicit expression for those values of x for which

$$a \cdot x^2 + b \cdot x + c = 0$$

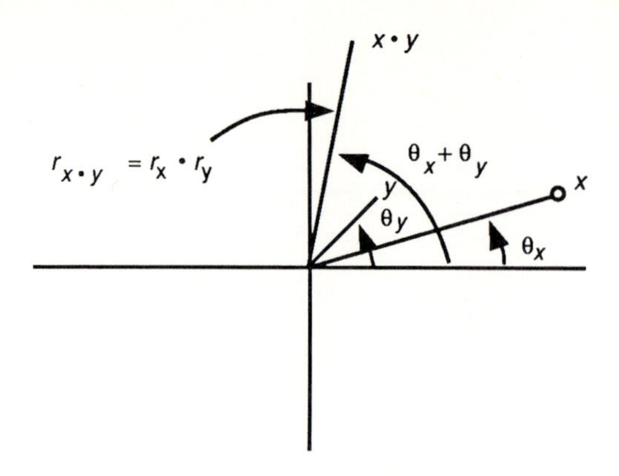

Figure 6.4.4

where $a, b, c, \in \mathbf{C}$ and $a \neq 0$:

$$x = \frac{-b \pm \sqrt{b^2 - 4a \cdot c}}{2 \cdot a}.$$

There are explicit formulas for complex solutions of the general cubic equation, $a \cdot x^3 + b \cdot x^2 + c \cdot x + d = 0$, and the general quartic equation, $a \cdot x^4 + b \cdot x^3 + c \cdot x^2 + d \cdot x + e = 0$. These formulas express the solutions in terms of cube roots and fourth roots, respectively, of expressions involving the coefficients a, b, c, d, and e. Perhaps the most important problem of classical algebra was the question of finding an explicit formula for the solutions of the general nth degree polynomial equation

$$a_n \cdot x^n + \cdots + a_1 \cdot x + a_0 = 0$$

where $a_i \in \mathbf{C}$ for $0 \leq i \leq n$ and $n \geq 5$. Specifically, the problem was to find an expression for solutions of this equation involving nth roots (or kth roots for $1 \leq k \leq n$) of expressions formed from the coefficients $a_0, a_1, \ldots, a_n$. As it happens, for any $n \geq 5$, no such formula exists! This remarkable fact was proved by E. Galois and N. Abel in the 1820s and 1830s. Their solutions depend on the concept of group (an algebraic structure to be studied in Section 6.5) and initiated the transition from classical algebra, the study of polynomial equations and their solutions, to modern algebra, the study of abstract algebraic structures.

We now embark upon the formal study of polynomials with coefficients in a field. Before doing so, let us look back to the ring $\mathbf{R}[X]$ of polynomials with real coefficients. A typical element of $\mathbf{R}[X]$ has the form

$a_0 + a_1 \cdot X + \cdots + a_n \cdot X^n$ where $a_i \in \mathbf{R}$ for $0 \le i \le n$. But we might legitimately wonder: What exactly is meant by this expression? What is X? What is X^2? How is $a_1 \cdot X$ defined? In all honesty, we must admit that we have not defined X, X^2, and $a_1 \cdot X$ and, in general, the notion of a polynomial on a set theoretic basis. Let us now consider this issue.

Our informal concept of a polynomial is that of an expression of the form $a_0 + a_1 \cdot X + \cdots + a_n \cdot X^n$. We can keep track of this polynomial simply by listing the coefficients $a_0, a_1, \ldots, a_n$ in order: Given the sequence $a_0, a_1, \ldots, a_n$, we can form the expression $a_0 + a_1 \cdot X + \cdots + a_n \cdot X^n$. Thus any polynomial in $\mathbf{R}[X]$ is determined by the sequence consisting of its coefficients. Hence it would seem that we can use the idea of a sequence to define formally the concept of polynomial. Of course, we usually think of a sequence as having infinitely many terms whereas a polynomial has only finitely many coefficients. This discrepancy can be remedied, however, if we consider those sequences that have only finitely many nonzero terms. Here, then, is the formal definition of a polynomial with coefficient in any field.

Definition 6.4.2 *Polynomials over a Field*

Let F be a field. A *polynomial with coefficients in* F is a function $s:\mathbf{N} \to F$ such that there exists $n \in \mathbf{N}$ such that $s(k) = 0$ for all $k > n$. Let $F[X]$ denote the set of all polynomials with coefficients in F.

A polynomial with coefficients in F is a sequence with values in F. For each $k \in \mathbf{N}$, let $a_k = s(k)$. By definition, $a_k = 0$ for all $k > n$. Henceforth, we represent each $s \in F[X]$, i.e., each polynomial with coefficients in F, as follows: If $s(k) = a_k = 0$ for all $k > n$, then we write $a_0 + a_1 \cdot X + \cdots + a_n \cdot X^n$ for the polynomial s. We emphasize that the expression $a_0 + a_1 \cdot X + \cdots + a_n \cdot X^n$ is just a symbol that represents the polynomial s, which is by definition a function $s:\mathbf{N} \to F$ such that $s(k) = 0$ for $k > n$ and $s(k) = a_k$ for $0 \le k \le n$. We follow the mathematical custom of denoting a polynomial $a_0 + a_1 \cdot X + \cdots + a_n X^n$ with a symbol such as $f(X)$.

We prefer the representation $a_0 + \cdots + a_n \cdot X^n$ for a polynomial to the sequential representation partly because of habit and partly because the former allows us to regard a polynomial as a function whose domain includes the field F. More on this later. For now, let us define the operations of addition and multiplication of polynomials. These definitions will be phrased in terms of sequences, but the motivation behind the definitions comes, of course, from our standard form for representing polynomials.

Definition 6.4.3 *Addition and Multiplication of Polynomials*

Let $s, t \in F[X]$.

(i) Define the *sum of s and t*, written $s + t$, to be the polynomial (i.e., the function) $s + t$: $\mathbf{N} \to F$ such that for all $k \in \mathbf{N}$, $(s + t)(k) = s(k) + t(k)$.

(ii) Define the *product of s and t*, written $s \cdot t$, to be the polynomial $s \cdot t$: $\mathbf{N} \to F$ such that for all $k \in \mathbf{N}$, $(s \cdot t)(k) = \sum_{j=0}^{k} s(j)t(k-j)$.

Let us consider these definitions in light of our standard representation of a polynomial. Suppose

$$f(X) = a_0 + a_1 \cdot X + \cdots + a_n \cdot X^n$$

and

$$g(X) = b_0 + b_1 \cdot X + \cdots + b_m \cdot X^m,$$

where for convenience we assume $n \leq m$. Then

$$f(X) + g(X) = (a_0 + b_0) + (a_1 + b_1) \cdot X + \cdots + (a_n + b_n) \cdot X^n + \cdots + b_m \cdot X^m,$$

and

$$f(X) \cdot g(X) = a_0 \cdot b_0 + (a_0 \cdot b_1 + a_1 \cdot b_0) \cdot X + (a_0 \cdot b_2 + a_1 \cdot b_1 + a_2 \cdot b_0) \cdot X^2 + \cdots + a_n \cdot b_m \cdot X^{n+m}.$$

Again we follow custom and denote the sum of $f(X)$ and $g(X)$ by $(f + g)(X)$ and the product of $f(X)$ and $g(X)$ by $(f \cdot g)(X)$.

Definition 6.4.4 *Degree of a Polynomial*

Let $f(X) = a_0 + \cdots + a_n \cdot X^n \in F[X]$ with $a_n \neq 0$. Then the degree of f, written $\deg(f)$, is the integer n. If $a_i = 0$ for all i, i.e., if f is the *zero polynomial*, then the degree of f is undefined.

Henceforth, we write $f(X) = 0$ if f is the zero polynomial. We can easily show that if $f(X) \neq 0$ and $g(X) \neq 0$, then $\deg(f \cdot g) = \deg(f) + \deg(g)$ and that $\deg(f + g) \leq \max(\deg(f), \deg(g))$, provided that $(f + g)(X) \neq 0$.

In Section 6.3 we studied the special case in which $F = \mathbf{R}$, the field of real numbers, and established several properties of $\mathbf{R}[X]$. All of the arguments presented in Section 6.3 concerning $\mathbf{R}[X]$ depend only on the fact that $\mathbf{R}$ is a field and not on the special nature of the real numbers. For example, the proof of the Division Theorem for $\mathbf{R}[X]$ carries over virtually word for word to the ring $F[X]$ where F is any field.

Theorem 6.4.4 *Division Theorem for $F[X]$. Let F be a field and let $f(X), g(X) \in F[X]$. There exist $q(X), r(X) \in F[X]$ with $r(X) = 0$ or $\deg(r(X)) < \deg(g(X))$ such that $f(X) = g(X) \cdot q(X) + r(X)$.*

Recall that the Division Theorem for $\mathbf{Z}$ asserts that if $a, b \in \mathbf{Z}$ with $b > 0$, then we can write $a = b \cdot q + r$ where $0 \leq r \leq b - 1$. Roughly speaking, this result says that we can divide b into a several times so as to obtain a nonnegative remainder that is smaller than b. In the case of $F[X]$, the degree function is used to measure the size of a polynomial. When viewed in this manner, the Division Theorem for $F[X]$ asserts that we can divide $g(X)$ into $f(X)$ so as to obtain a quotient $q(X)$ and a remainder $r(X)$ such that the remainder $r(X)$ either is 0 or is smaller in the sense of degree than $g(X)$.

Using the Division Theorem, we can prove the following theorem, which generalizes Theorem 6.3.2.

Theorem 6.4.5 *For any field F, $F[X]$ is a* UFD.

Thus far we have considered a polynomial $f(X) \in F[X]$ merely as an expression of the form: $f(X) = a_0 + a_1 \cdot X + \cdots + a_n \cdot X^n$ where $a_i \in F$ for $0 \leq i \leq n$. We can also, however, regard $f(X)$ as a function from F to F by noting that if $a \in F$, then $f(a) = a_0 + a_1 \cdot a + \cdots + a_n \cdot a^n \in F$. With this in mind, we introduce the concept of a root of a polynomial.

The next theorem associates to each root of a polynomial a linear factor of that polynomial.

Definition 6.4.5 *Root of a Polynomial*

Let $f(X) \in F[X]$ where F is a field. An element $a \in F$ is a *root* of f if $f(a) = 0$.

Theorem 6.4.6 *Let $f(X) \in F[X]$ where F is a field. If a is a root of f, then $(X - a) | f(X)$.*

Proof By the Division Theorem $f(X) = (X - a) \cdot q(X) + r(X)$ where $q(X), r(X) \in F[X]$ and either $r(X) = 0$ or $\deg(r(X)) < \deg(X - a) = 1$. Thus $r(X) = r$ is a constant polynomial. Therefore, $0 = f(a) = (a - a) \cdot q(a) + r = r$; hence $f(X) = (X - a) \cdot q(X)$. ■

Corollary 6.4.7 *Let F be a field and let $f(X) \in F[X]$. If $a_1, \ldots, a_n$ are distinct roots of $f(X)$ where $n \in \mathbf{Z}^+$, then $f(X) = (X - a_1) \cdot \cdots \cdot (X - a_n) \cdot g(X)$ for some polynomial $g(X) \in F[X]$.*

Proof We give a quick proof by induction on n. The basis step $n = 1$ is handled by Theorem 6.4.6. Suppose the assertion of Corollary 6.4.7 is true when $n = k \in \mathbf{Z}^+$ and suppose $f(X) \in F[X]$ is a polynomial with $k + 1$ distinct roots $a_1, \ldots, a_k, a_{k+1}$. By inductive hypothesis, since f has the k distinct roots $a_1, \ldots, a_k$, there exists $g(X) \in F[X]$ such that $f(X) = (X - a_1) \cdot \cdots \cdot (X - a_k) \cdot g(X)$. Since a_{k+1} is a root of f,

$$0 = f(a_{k+1}) = (a_{k+1} - a_1) \cdot \cdots \cdot (a_{k+1} - a_k) \cdot g(a_{k+1}).$$

Since $a_{k+1} \neq a_i$ for $1 \leq i \leq k$, $g(a_{k+1}) = 0$. From the basis case, we know that $g(X) = (X - a_{k+1}) \cdot h(X)$ where $h(X) \in F[X]$. Therefore, $f(X) = (X - a_1) \cdot \cdots \cdot (X - a_{k+1}) \cdot h(X)$. ■

Corollary 6.4.8 *If $f(X) \in F[X]$ has degree n, then f has at most n distinct roots.*

Example 6.4.1 (i) Since $5^2 \equiv -1 \pmod{13}$, $[5]$ is a root of $X^2 + [1] \in \mathbf{Z}_{13}[X]$. (Remember: $\mathbf{Z}_{13}$ is a field since 13 is prime.) Thus $X - [5]$ divides $X^2 + [1]$. In fact, since $8^2 \equiv 1 \pmod{13}$, $X^2 + [1] = (X - [5]) \cdot (X - [8])$.

(ii) Consider the polynomial $f(X) = X^4 - [1]_5 \in \mathbf{Z}_5[X]$. Since $1^4 \equiv 1 \pmod 5$, $2^4 \equiv 1 \pmod 5$, $3^4 \equiv 1 \pmod 5$, and $4^4 \equiv 1 \pmod 5$, $[1]_5$, $[2]_5$, $[3]_5$, and $[4]_5$ are all roots of $f(X)$. Thus

$$f(X) = x^4 - [1]_5 = (X - [1]_5) \cdot (X - [2]_5)$$
$$\cdot (X - [3]_5) \cdot (X - [4]_5) \cdot g(X)$$

for some $g(X) \in \mathbf{Z}_5[X]$. But clearly, $\deg(g(X)) = 0$, and thus $g(X) = c$ is a constant. Since

$$X^4 - [1]_5 = c \cdot x^4 + \cdots,$$

it follows that $c = [1]_5$. Therefore, in $\mathbf{Z}_5[X]$,

$$x^4 - [1]_5 = (X - [1]_5) \cdot (X - [2]_5) \cdot (X - [3]_5) \cdot (X - [4]_5).$$

Corollary 6.4.8 gives an upper bound on the number of distinct roots of a given polynomial. On the other hand, we can find fields F and polynomials $f(X) \in F[X]$ such that f has no roots in F. (The polynomial $X^2 + 1 \in \mathbf{R}[X]$ is an example.) Over the field $\mathbf{C}$, however, the situation is different. For example, any quadratic polynomial in $\mathbf{C}[X]$ has a root in $\mathbf{C}$. In fact, even more is true.

Theorem 6.4.9 *The Fundamental Theorem of Algebra. Every nonconstant polynomial in* $\mathbf{C}[X]$ *has a root in* $\mathbf{C}$.

The Fundamental Theorem of Algebra was known to, or at least believed by, Euler, Lagrange, and other mathematicians of the eighteenth century. Several incomplete proofs of the theorem were presented during that period. The first reasonably complete proof was given by Gauss in his doctoral dissertation written in 1799 when he was 22. Gauss eventually presented three proofs of the Fundamental Theorem of Algebra. (Important theorems have a way of being proved several times.) The following corollary provides a complete description of polynomials in $\mathbf{C}[X]$.

Corollary 6.4.10 *If* $f(X) \in \mathbf{C}[X]$ *is a nonzero polynomial, then there exist* $a, a_1, \ldots, a_n \in \mathbf{C}$ *such that* $f(X) = a \cdot (X - a_1) \cdot \cdots \cdot (X - a_n)$.

Exercises §6.4

6.4.1. Let $f(X) = a \cdot X^2 + b \cdot X + c \in \mathbf{C}[X]$ with $a \neq 0$. Prove that there exists $z \in \mathbf{C}$ such that $f(z) = 0$.

6.4.2. (a) Find two polynomials in $\mathbf{Z}_3[X]$ having no roots in $\mathbf{Z}_3$.
(b) Find two polynomials in $\mathbf{Z}_5[X]$ having no roots in $\mathbf{Z}_5$.

6.4.3. Let R be an integral domain. Prove that if R is a finite set, then R is a field. (Hint: Look at the proof of Theorem 6.4.2.)

6.4.4. Prove the Division Theorem for $F[X]$.

6.4.5. Prove that for any field F, $F[X]$ is a UFD.

6.4.6. Prove the converse of Theorem 6.4.6.

6.4.7. Prove Corollary 6.4.8.

6.4.8. Derive Corollary 6.4.10 from Theorem 6.4.9.

6.4.9. If $n \in \mathbf{Z}^+$ is not prime, then is $\mathbf{Z}_n$ a field? Explain.

6.4.10. (a) Factor the polynomial $X^2 + [1]$ in $\mathbf{Z}_{17}[X]$.
(b) Factor the polynomial $X^3 - [1]$ in $\mathbf{Z}_7[X]$.

6.4.11. Investigate the following question: For which primes p does the equation $X^2 + 1 = 0$ have a solution in $\mathbf{Z}_p$? (In this exercise we write a for $[a]_p$.) For example, in $\mathbf{Z}_5$, $2^2 + 1 = 4 + 1 = 0$ and thus $X^2 + 1 = 0$ has a solution in $\mathbf{Z}_5$. On the other hand, in $\mathbf{Z}_3$, $0^2 + 1 \neq 0$, $1^2 + 1 = 2 \neq 0$, and $2^2 + 1 = 2 \neq 0$ and thus $X^2 + 1$ has no solutions in $\mathbf{Z}_3$. Gather some experimental evidence and make a conjecture.

6.4.12. Use Corollary 6.4.10 to investigate the following statement: Every irreducible polynomial in $\mathbf{C}[X]$ has degree 1.

6.4.13. Prove that if F is a field, then there exist infinitely many irreducible polynomials in $F[X]$.

§6.5 Groups

The concepts of ring and field evolve naturally from the algebraic properties of the integers, the rationals, and the reals. Most of the problems that arise in the study of rings and fields are directly motivated by theorems and questions about the integers and the rationals. For example, the issue of unique factorization for multiplication in a general commutative ring with unity is a natural one to consider given that the unique factorization property holds for multiplication in $\mathbf{Z}$.

In this section, we concentrate on the notion of group. As algebraic structures go, groups are less complex than rings, fields, or Boolean algebras. (A group is an algebraic structure with just one binary operation rather than two.) In spite of the elementary nature of the group concept, group theory arose historically out of rather subtle mathematical considerations and the study of groups has led to some very deep mathematical results.

Definition 6.5.1 *Group*

A *group* is a set G together with a binary operation $*$ such that $*$ is associative, possesses an identity, and has inverses.

Recall that associativity means that for all $a, b, c \in G$, $a*(b*c) = (a*b)*c$; the existence of an identity means that there exists an element $e \in G$ such that for all $a \in G$, $a*e = e*a = a$; and the existence of inverses means that for each $a \in G$, there exists $b \in G$ such that $a*b = b*a = e$.

Example 6.5.1 The set $\mathbf{Z}$ is a group under $+$. In fact, if R is any ring, then R is a group under the operation of addition in R. Also notice that the subset $E = \{x \in \mathbf{Z} | x \text{ is even}\}$ of $\mathbf{Z}$ is a group under addition in $\mathbf{Z}$.

Example 6.5.2 (i) Let $\mathbf{R}^* = \{x \in \mathbf{R} | x \neq 0\}$ and $\mathbf{R}^+ = \{x \in \mathbf{R} | x > 0\}$. Then $\mathbf{R}^*$ and $\mathbf{R}^+$ are groups under multiplication in $\mathbf{R}$.

(ii) Let F be any field and let $F^* = \{x \in F | x \neq 0\}$. Then F^* is a group under multiplication in F. This follows from the fact that if $x, y \in F^*$, then $x \cdot y \in F^*$ by Proposition 6.4.1 and thus multiplication is a binary operation on F^*. From Definition 6.4.1, F^* is a group under multiplication.

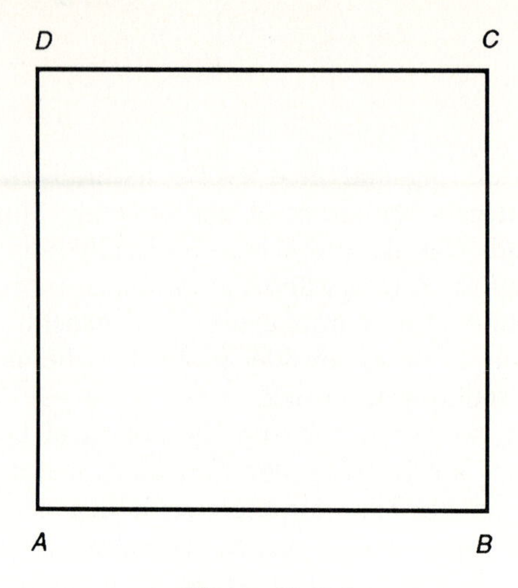

Figure 6.5.1

The previous examples illustrate how groups arise out of other types of algebraic structures. The next two examples show that groups appear naturally in geometric settings.

Example 6.5.3 Consider a square, S, in $\mathbf{R}^2$ with corners labeled as shown in Figure 6.5.1, and consider the set of all distance-preserving functions from $\mathbf{R}^2$ to $\mathbf{R}^2$ that map S to itself. This set is called the *symmetries of* S and is denoted by $\mathrm{Sym}(S)$. What are the elements of $\mathrm{Sym}(S)$?

First the square can be rotated by $\pi/2$ radians counterclockwise about its center. Call this mapping r_1. (See Figure 6.5.2.) The square can also be rotated counterclockwise by π radians and by $3\pi/2$ radians. Finally, the square can be rotated by 0 radians. This mapping is called the *identity mapping*. (See Figure 6.5.3.)

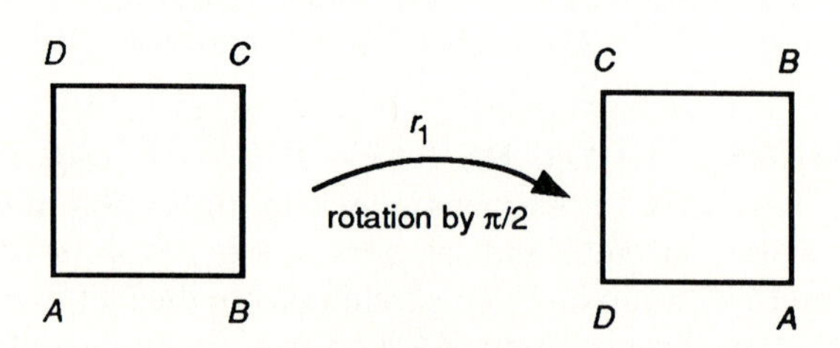

Figure 6.5.2

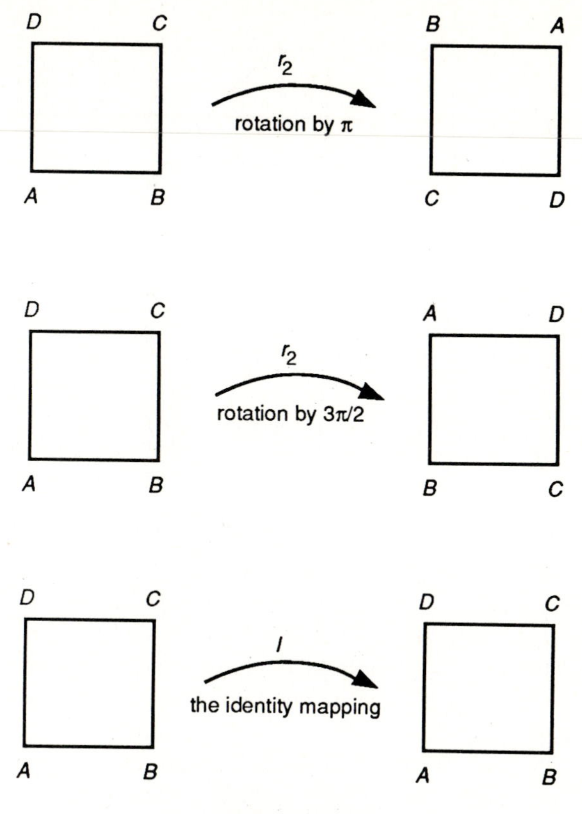

Figure 6.5.3

Thus far, we have identified four elements in the set Sym(S). But there are others. The square S can be reflected (or flipped) about its two diagonals and the two bisectors of opposite sides. Hence there are four reflections, labeled f_1, f_2, f_3, f_4 in Sym(S). (See Figure 6.5.4.)

We have now identified eight elements of Sym(S). Are there any others?

Let f be an element of Sym(S). We show that f is one of the eight elements described above. Hence Sym(F) = $\{I, r_1, r_2, r_3, f_1, f_2, f_3, f_4\}$. The mapping f sends A to either A, B, C, D. Let us assume that f sends A to C. Then B must go to a corner adjacent to C; hence B goes to either D or B. If B goes to D, then C must go to A under f and D goes to B. In this case, $f = r_2$. On the other hand, if B goes to B, then C goes to A and D goes to D under f. In this case, $f = f_2$. A similar argument shows that if f sends A to A, then $f = I$ or $f = f_1$; if f sends A to B, then $f = r_1$, or $f = f_4$; if f sends A to D, then $f = r_4$ or $f = f_3$.

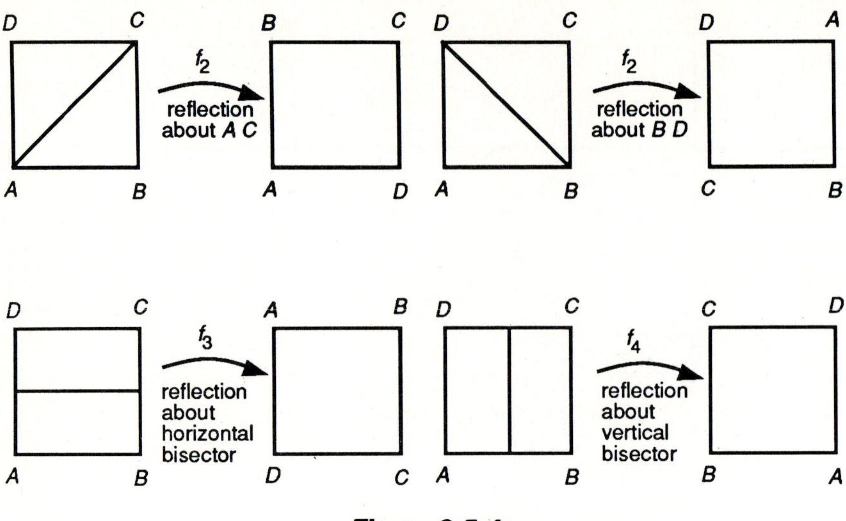

Figure 6.5.4

The set Sym(S) has the property that if f and g are functions in Sym(S), then the composition $g \circ f \in$ Sym(S). This fact can be seen either by checking all sixty-four possible compositions or by noting that the composition of two symmetries of the square must be a symmetry of the square and therefore must be in Sym(S). Thus composition of functions is a binary operation on Sym(S). Table 6.5.1 shows the operation table for Sym(S) for the operation of composition. It is clear from this table that $\circ$ has an identity and has inverses; since composition of functions is an associative operation, Sym(S) is a group under the operation $\circ$.

Example 6.5.4 We can generalize the previous example by considering the set of symmetries of a geometric figure in $\mathbf{R}^2$. Let F be any closed

	I	r_1	r_2	r_3	f_1	f_2	f_3	f_4
I	I	r_1	r_2	r_3	f_1	f_2	f_3	f_4
r_1	r_1	r_2	r_3	I	f_4	f_3	f_1	f_2
r_2	r_2	r_3	I	r_1	f_2	f_1	f_4	f_3
r_3	r_3	I	r_1	r_2	f_3	f_4	f_2	f_1
f_1	f_1	f_3	f_2	f_4	I	r_2	r_1	r_3
f_2	f_2	f_4	f_1	f_3	r_2	I	r_3	r_1
f_3	f_3	f_2	f_4	f_1	r_3	r_1	I	r_2
f_4	f_4	f_1	f_3	f_2	r_1	r_3	r_2	I

Table 6.5.1

plane curve; for example, F can be a circle, an ellipse, or a curve of arbitrary shape. (See Figure 6.5.5.) Let Sym(F) denote the set of functions from $\mathbf{R}^2$ to $\mathbf{R}^2$ that preserve distance and that send F onto itself. The operation of composition of functions is a binary operation on Sym(F); for if f and g preserve distance in $\mathbf{R}^2$ and send F to F, then $f \circ g$ preserves distance and sends F to F. Moreover, composition of functions is associative; the identity mapping on $\mathbf{R}^2$, $I(x) = x$, is in Sym(F); finally, for each $f \in$ Sym(F), f^{-1} is a distance-preserving function on $\mathbf{R}^2$ that sends F to F. Thus Sym(F) is a group under composition, called the *group of symmetries of F*.

To take an example, when F_1 is a circle in $\mathbf{R}^2$, Sym(F_1) contains the rotation about the center of F_1 through any angle α where $0 \leq \alpha < 2\pi$. Also Sym(F_1) contains all reflections of the plane about any line passing

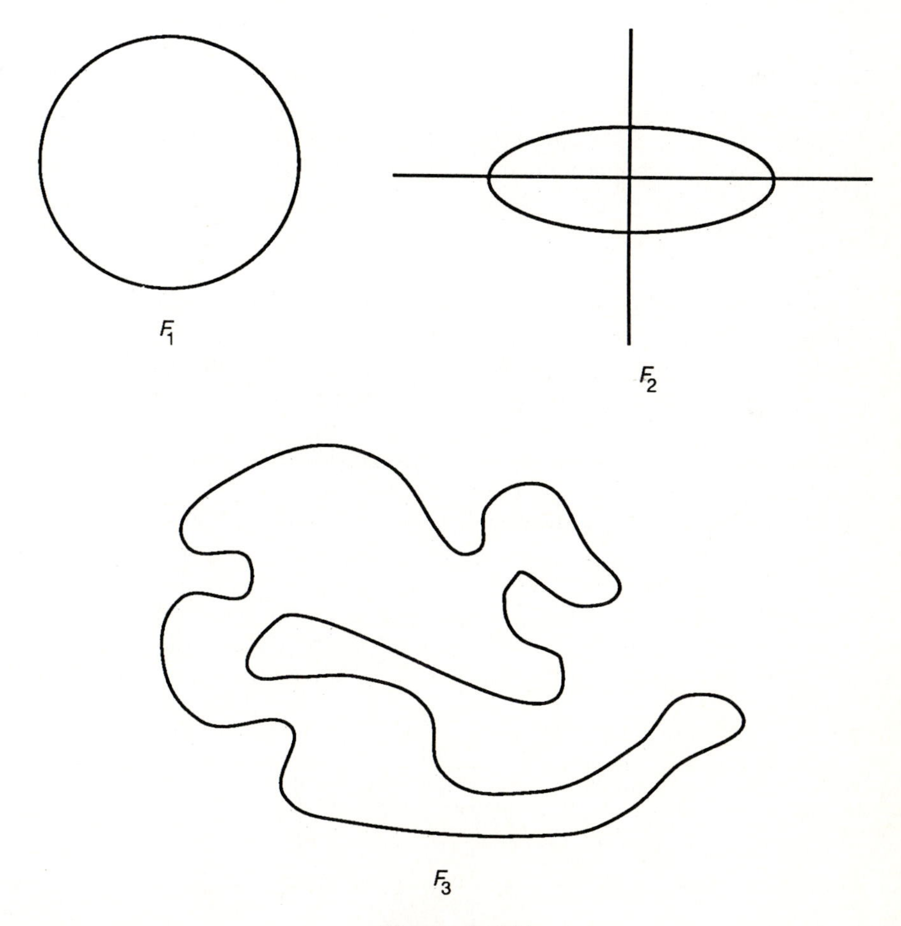

Figure 6.5.5

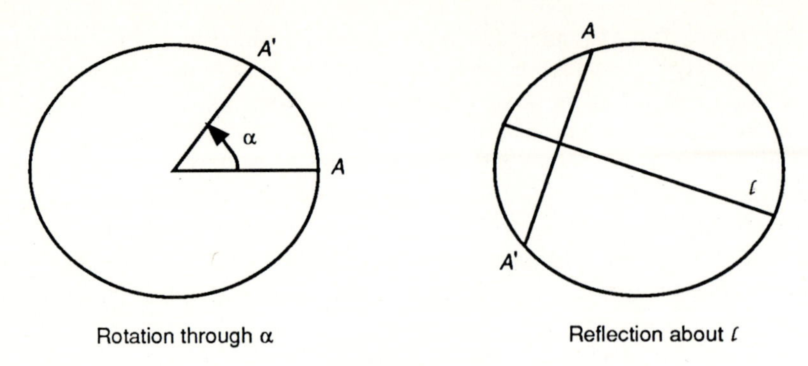

Rotation through α   Reflection about ℓ

Figure 6.5.6

through the center of F_1. Thus as a set, $\text{Sym}(F_1)$ is infinite. (See Figure 6.5.6.)

On the other hand, for the ellipse F_2, the group $\text{Sym}(F_2)$ contains exactly four elements: $I =$ the identity; $r_2 =$ rotation by π radians through the center of F_2; $f_2 =$ reflection about the major axis; $f_y =$ reflection about the minor axis. The operation table for $\text{Sym}(F_2)$ is shown in Table 6.5.2. The only symmetry of the curve F_3 is evidently the identity; thus $\text{Sym}(F_3) = \{I\}$.

Example 6.5.5 Let $n \in \mathbf{Z}^+$ and let $\mathbf{N}_n = \{1, \ldots, n\}$. Let $S_n = \{f: \mathbf{N}_n \to \mathbf{N}_n | f \text{ is a bijection}\}$. If $f, g \in S_n$, then by Theorem 3.4.1(iii), $f \circ g \in S_n$, and hence $\circ$ is a binary operation on S_n. It is easy to check that S_n is a group under $\circ$. We can regard a bijection of $\mathbf{N}_n$ to itself as a mapping that preserves the set theoretic structure of $\mathbf{N}_n$, and thus we can reasonably think of S_n as the group of symmetries of $\mathbf{N}_n$. Thus the group S_n is called the *symmetric group on n letters*.

Each element of S_n can be represented as shown in Table 6.5.3.

With this form of representing elements of S_n, we can compute the number of elements of S_n. The value of $f(1)$ can be any of the n numbers $1, 2, \ldots, n$. Once $f(1)$ is fixed, there are $n - 1$ choices for $f(2)$ since $f(2) \neq f(1)$. Thus there are $n \cdot (n - 1)$ ways of filling in the first two entries

	I	r_2	f_x	f_y
I	I	r_2	f_x	f_y
r_2	r_2	I	f_y	f_x
f_x	f_x	f_y	I	r_2
f_y	f_y	f_x	r_2	I

Table 6.5.2

x	1	2	$\cdots$	n
$f(x)$	$f(1)$	$f(2)$	$\cdots$	$f(n)$

Table 6.5.3

	f_1	f_2	f_3	f_4	f_5	f_6
f_1	f_1	f_2	f_3	f_4	f_5	f_6
f_2	f_2	f_1	f_5	f_6	f_3	f_4
f_3	f_3	f_4	f_1	f_2	f_6	f_5
f_4	f_4	f_3	f_6	f_5	f_1	f_2
f_5	f_5	f_6	f_2	f_1	f_4	f_3
f_6	f_6	f_5	f_4	f_3	f_2	f_1

Table 6.5.4

$f(1)$, $f(2)$ of the table. For each choice of $f(1)$ and $f(2)$, there are $n - 2$ choices for $f(3)$, and therefore $n \cdot (n - 1) \cdot (n - 2)$ ways of filling in the first three entries of the table. When carried out, this argument shows that Table 6.5.3 can be filled out in $n!$ ways. Thus S_n has $n!$ elements.

The elements of S_3 in tabular form and the operation table for S_3 are given below and in Table 6.5.4.

$$f_1 = I = \begin{pmatrix} 1 & 2 & 3 \\ 1 & 2 & 3 \end{pmatrix} \qquad f_2 = \begin{pmatrix} 1 & 2 & 3 \\ 1 & 3 & 2 \end{pmatrix} \qquad f_3 = \begin{pmatrix} 1 & 2 & 3 \\ 2 & 1 & 3 \end{pmatrix}$$

$$f_4 = \begin{pmatrix} 1 & 2 & 3 \\ 2 & 3 & 1 \end{pmatrix} \qquad f_5 = \begin{pmatrix} 1 & 2 & 3 \\ 3 & 1 & 2 \end{pmatrix} \qquad f_6 = \begin{pmatrix} 1 & 2 & 3 \\ 3 & 2 & 1 \end{pmatrix}$$

Various Types and Properties of Groups

The preceding examples suggest a number of definitions.

Definition 6.5.2 *Finite and Infinite Groups*

Let G be a group. If as a set G is finite, then G is called a *finite group*. In this case, the cardinality of G, $|G|$, is called the *order of* G. An *infinite group* is a group that is not finite.

Definition 6.5.3 *Abelian Group*

Let G be a group under a binary operation $*$. G is *commutative* or *abelian* if for all $x, y \in G$, $x * y = y * x$. G is called *noncommutative* or *nonabelian* if G is not abelian.

Remarks (1) For all $n \in \mathbf{N}^+$, S_n is finite. If $n = 1$ or 2, then S_n is abelian. For $n \geq 3$, S_n is nonabelian. For example, S_3 is nonabelian since (see Table 6.5.4) $f_2 \circ f_3 = f_5$ while $f_3 \circ f_2 = f_4$.

(2) The groups $\mathbf{R}$ under addition, $\mathbf{R}^*$ under multiplication, and $\text{Sym}(F_2)$ in Example 6.5.4 are abelian, while the groups $\text{Sym}(S)$ and $\text{Sym}(F_2)$ in Examples 6.5.3 and 6.5.4 are nonabelian.

(3) Let G be a finite group and let T be the operation table of G. Then G is abelian if and only if T is symmetric with respect to the diagonal running from upper left to lower right. In other words, G is abelian if and only if for all $a, b \in G$ the entry in row a, column b, of T equals the entry in row b, column a, of T.

(4) The operation table for a group is usually called the *Cayley table* after the mathematician who first introduced the concept. We shall use this terminology henceforth.

Example 6.5.6 Consider the integers modulo n, $\mathbf{Z}_n$. For $x, y \in \mathbf{Z}_n$, we define $x + y$ as follows (see Section 6.3): If $x = [a]$ and $y = [b]$, then $x + y = [a + b]$. As shown in Example 6.3.6, $x + y$ does not depend on the elements a and b used to represent the equivalence classes x and y. We leave it for the reader to check that $\mathbf{Z}_n$ is an abelian group under the operation $+$.

The addition table for $\mathbf{Z}_4$ is shown in Table 6.5.5.

Group theory breaks up into several subdisciplines. For example, there are the theory of abelian groups and the theory of nonabelian groups, and there are finite group theory and infinite group theory. (See Figure 6.5.7.) Finite abelian groups can be described explicitly. The so-called Fundamental Theorem for Finite Abelian Groups (in effect, a Unique Factorization Theorem for finite abelian groups) states that a finite abelian group must be

	[0]	[1]	[2]	[3]
[0]	[0]	[1]	[2]	[3]
[1]	[1]	[2]	[3]	[0]
[2]	[2]	[3]	[0]	[1]
[3]	[3]	[0]	[1]	[2]

Table 6.5.5

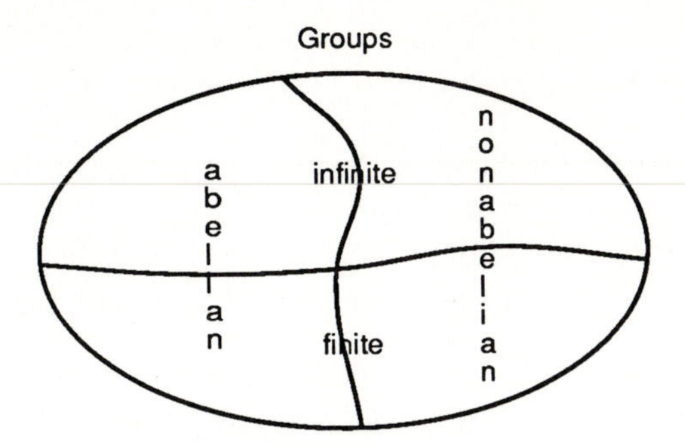

Figure 6.5.7

expressible in a certain specified form. (See Chapter 10 of the text *Contemporary Abstract Algebra* by J. Gallian for the statement of this theorem.) By contrast, infinite abelian groups are far from being classified. In recent years, research in infinite abelian groups has revealed surprising connections between mathematical logic and abelian group theory.

The theory of finite nonabelian groups has witnessed extensive activity in the past twenty-five years. A major step in this subject was completed in 1980 with the classification of all finite *simple* groups. Simple groups (which are defined in Exercise 6.5.14) are the most basic of all finite groups, since any finite group can in principle be built up from simple groups. (Again, the analogy with the Unique Factorization Theorem in **Z** is valuable: Simple groups are the prime numbers of group theory, and a decomposition theorem called the Jordan–Holder Theorem describes how any finite group can be expressed in terms of simple groups.) The classification of finite simple groups consists of a listing of all possible finite simple groups. The classification theorem asserts that any finite simple group must belong to one of 17 infinite families of simple groups or must be one of 26 exceptional simple groups. Even with the classification of simple groups, the theory of finite groups is far from complete, since, for example, the process by which an arbitrary finite group is built up from simple groups is not well understood.

Our first theorem records some elementary properties about groups, all of which follow directly from the axioms. Henceforth, we assume that G is a group with binary operation.

Theorem 6.5.1 *Let G be a group. Then*

(i) *there is a unique identity element, which we denote by e, in G;*
(ii) *each element of G has a unique inverse;*
(iii) *for each $x \in G$, $(x^{-1})^{-1} = x$ where x^{-1} denotes the inverse of x;*
(iv) *for all $g, x, y \in G$, if $g \cdot x = g \cdot y$, then $x = y$ and if $x \cdot g = y \cdot g$, then $x = y$;*

(v) *for each pair* $g, y \in G$, *there exists a unique* $x \in G$ *such that* $g \cdot x = y$ *and there exists a unique* $x' \in G$ *such that* $x' \cdot g = y$.

Proof We prove part of statement 4 and leave the proofs of the remaining statements as exercises.

Suppose $g, x, y \in G$ and $g \cdot x = g \cdot y$. Then $g^{-1} \cdot (g \cdot x) = g^{-1} \cdot (g \cdot y)$. By associativity, we have $(g^{-1} \cdot g) \cdot x = (g^{-1} \cdot g) \cdot y$; hence $e \cdot x = e \cdot y$ or $x = y$. ∎

Properties (iv) and (v) have the following important consequence for finite groups. Let T be the Cayley table for G. Then each row of T contains each element of G exactly once and each column of T contains each element of G exactly once, for suppose that in row g the elements in columns x and y are equal. Then $g \cdot x = g \cdot y$ and $x = y$ by property (iv) of Theorem 6.5.1. On the other hand, let g and y be arbitrary elements of G. There exists $x \in G$ such that $g \cdot x = y$ by property (v), which implies that in row g, column x of T, the element y appears.

Let us now use the axioms of group theory and the properties listed in Theorem 6.5.1 to determine groups of order 1, 2, and 4.

First, if $|G| = 1$, then $G = \{e\}$ and the Cayley table for G is simply

	e
e	e

Suppose $|G| = 2$. Then let e denote the identity of G and let g denote the nonidentity element of G. Then the Cayley table can be partially filled in:

	e	g
e	e	g
g	g	

To complete the table we need to determine $g \cdot g$. But this is easy: $g \cdot g$ must be e since the element e has to appear in row g of T. Thus T is

	e	g
e	e	g
g	g	e

In effect, T describes completely the binary operation in any two-element group: If $x, y \in G$ where $|G| = 2$, then:

(i) if $x = e$ (resp. $y = e$), then $x \cdot y = y$ (resp. $x \cdot y = x$);
(ii) if $x = y \neq e$, then $x \cdot y = e$.

We can easily show that the set $\{e, g\}$ under the binary operation described above is indeed a group. Thus we have exhibited an abelian group of order 2.

Let us now suppose that G is a group of order 4. Let $g \in G$ with $g \neq e$. Consider the elements $e, g, g^2 = g \cdot g, g^3 = g \cdot g^2, g^4 = g \cdot g^3$. Since $|G| = 4$, these elements cannot all be distinct. We now consider several possibilities.

Case 1. There exists $g \in G$ such that the elements $e, g, g^2, g^3 = g \cdot g^2$ are all distinct. Then since $|G| = 4$, $G = \{e, g, g^2, g^3\}$. We note that in this case $g^4 = e$, for otherwise, $g^4 = g$, $g^4 = g^2$, or $g^4 = g^3$. If $g^4 = g$, then by property (iv) of Theorem 6.5.1 it follows that $g^3 = e$. But by assumption, $g^3 \neq e$, and hence $g^4 \neq g$. Similar contradictions are reached from the assumptions that $g^4 = g^2$ or $g^4 = g^3$. Thus $g^4 = e$. We can think of the elements of G as consisting of the powers of the element g.

Case 2. Suppose that Case 1 does not apply but that there does exist $g \in G$ such that e, g, g^2 are all distinct. Since $|G| = 4$, there exists $h \in G$ such that $G = \{e, g, g^2, h\}$. As in Case 1, we can show that $g^3 = e$ from which it follows that $g^{-1} = g^2$. Consider the element $g \cdot h$. If $g \cdot h = e$, then $h = g^{-1} = g^2$, which is a contradiction since $h \neq g^2$. If $g \cdot h = g$, then $h = e$, again a contradiction. If $g \cdot h = g^2$, then $h = g$, which is also ruled out by the definition of h; if $g \cdot h = h$, then $g = e$, another absurd conclusion. Since we reach a contradiction in all cases, we must conclude that no such element $g \in G$ can exist.

Case 3. Suppose again that Case 1 does not hold. Then for all $g \in G$ with $g \neq e$, $g^2 = e$. Let g and h be two nonidentity elements of G. Then $g \cdot h \in G$ and $h \cdot g \in G$. It is easy to check that $g \cdot h \notin \{e, g, h\}$ and $h \cdot g \notin \{e, g, h\}$. Thus $g \cdot h = h \cdot g$. Thus $G = \{e, g, h, g \cdot h\}$ and G is an abelian group. (As an exercise, write out the Cayley table of G.)

Since these cases cover all the possibilities, we can conclude that if $|G| = 4$, then either there exists $g \in G$ such that $G = \{e, g, g^2, g^3\}$ or no such element exists but there exist $g, h \in G$ such that $G = \{e, g, h, g \cdot h\}$. Conversely, we can show that each of these sets is indeed a group under the given operation.

The group described in the first case is very similar to $\mathbf{Z}_4$ in structure. For example, the operation table of G will coincide with that of $\mathbf{Z}_4$ (see Table 6.5.5) if in the latter table we replace [0], [1], [2], and [3], respectively, by e, g, g^2, and g^3. In other words, G is isomorphic to $\mathbf{Z}_4$. Similarly, the group described in the second case is isomorphic to Sym(F_2) (see Example 6.5.4).

Subgroups

Let us look closely at the group of integers $\mathbf{Z}$ under addition. As noted in Example 6.5.1, the subset E of even integers is a group under $+$, for if $x, y \in E$, then $x + y \in E$, and thus $+$ is a binary operation on E. The associative property holds for addition in E since it holds for addition in $\mathbf{Z}$; $0 \in E$ and $x + 0 = 0 + x = x$ for all $x \in E$; finally, if $x \in E$, then $-x \in E$ and $x + (-x) = (-x) + x = 0$. There are other subsets of $\mathbf{Z}$ that are also groups under addition in $\mathbf{Z}$ (see Exercise 6.5.10). The next definition introduces the term that is commonly used to describe this situation.

Definition 6.5.4 *Subgroup*

Let G be a group with operation $\cdot$. A subset H of G is a *subgroup of* G if $H \neq \varnothing$ and H is a group under $\cdot$.

Many examples of subgroups are found within groups of numbers.

Example 6.5.7 (i) The sets E, $\mathbf{Z}$, and $\mathbf{Q}$ are all subgroups of $\mathbf{R}$ under addition. Notice that $\mathbf{R}^+ = \{x \in \mathbf{R} | x > 0\}$ is not a subgroup of $\mathbf{R}$ under addition since $\mathbf{R}^+$ has no identity element.

(ii) The group $\mathbf{Q}^* = \mathbf{Q} - \{0\}$ is a subgroup of $\mathbf{R}^* = \mathbf{R} - \{0\}$ under multiplication. The set $\mathbf{Z}^* = \mathbf{Z} - \{0\}$ is not, however, a subgroup of $\mathbf{R}^*$; for if so, then 1 is the identity of $\mathbf{Z}^*$ but then not every element of $\mathbf{Z}^*$ has an inverse in $\mathbf{Z}^*$.

Example 6.5.8 Let $G = Gl_2(\mathbf{R})$ be the set of 2×2 matrices with real entries and determinant $\neq 0$:

$$G = \left\{ \begin{pmatrix} a & b \\ c & d \end{pmatrix} \middle| D = ad - bc \neq 0 \right\}.$$

From elementary linear algebra, it follows that G is a group under matrix multiplication. For example, if $A, B \in G$ then $\det(A) \neq 0$, $\det(B) \neq 0$ and $\det(AB) = \det(A) \cdot \det(B) \neq 0$, which means that $A \cdot B \in G$. The identity element of G is $I = \begin{pmatrix} 1 & 0 \\ 0 & 1 \end{pmatrix}$ and if $A = \begin{pmatrix} a & b \\ c & d \end{pmatrix} \in G$, then

$$A^{-1} = \begin{pmatrix} d/D & -b/D \\ -c/D & a/D \end{pmatrix}$$

where $D = ad - bc$.

Let $H = \left\{ \begin{pmatrix} a & b \\ c & d \end{pmatrix} \in G | D = 1 \right\}$. Then H is a subgroup of G.

Proof (1) Let $A, B \in H$. Then $\det(A) = \det(B) = 1$ and therefore $\det(AB) = 1$. Thus H is closed under matrix multiplication.

(2) Since matrix multiplication is in general associative, matrix multiplication in H is associative.

(3) The matrix $I = \begin{pmatrix} 1 & 0 \\ 0 & 1 \end{pmatrix}$ is the identity of H.

(4) If $A = \begin{pmatrix} a & b \\ c & d \end{pmatrix} \in H$, then $B = \begin{pmatrix} d & -b \\ -c & a \end{pmatrix} \in H$ (since $\det(B) = 1$) and $AB = BA = I$. Thus each element of H has an inverse in H.

Notice that all the groups appearing in the last two examples are infinite. Let us look at subgroups of some finite groups.

Example 6.5.9 Consider the group of symmetries of a noncircular ellipse, F_2, described in Example 6.5.4: $\text{Sym}(F_2) = \{I, r_2, f_x, f_y\}$. It is easy to check that $\{I\}$ and $\text{Sym}(F_2)$ are subgroups of $\text{Sym}(F_2)$. Also the sets $\{I, r_2\}$, $\{I, f_x\}$, and $\{I, f_y\}$ are subgroups of $\text{Sym}(F_2)$. (See Exercise 6.5.8.) In fact, these five subgroups are the only subgroups of $\text{Sym}(F_2)$, for any subgroup H of G contains I. If H contains only one of the elements r_2, f_x, and f_y, then H is one of the three two-element subgroups listed above. If H contains at least two of the elements r_2, f_x, and f_y, call them g and h, then H contains $g \cdot h$, which is the remaining element and $H = \text{Sym}(F_2)$.

Observe that in the previous example, for each subgroup H of G, the order of H, $|H|$ divides the order of G, $|G|$. This phenomenon is a special case of a general result known as Lagrange's Theorem, one of the most important results in all of finite group theory.

Theorem 6.5.2 *Lagrange's Theorem. If G is a finite group and H is a subgroup of G, then $|H|$ divides $|G|$.*

We close this section by proving Lagrange's Theorem and deriving some corollaries of it. In order to prove the theorem, we shall require some general properties of subgroups. We list these properties in two lemmas, the second of which provides a way of characterizing subgroups. We leave the proofs of the lemmas as exercises.

Lemma 6.5.3 *Let H be a subgroup of a group G. Then the identity element of H is the identity element of G.*

Lemma 6.5.4 *Let G be a group and let H be a subset of G. Then H is a subgroup of G if and only if the following conditions hold:*

(i) *$e \in H$ (e is the identity of G);*

(ii) *if $a, b \in H$, then $a \cdot b \in H$ (i.e., H is closed under the operation of G);*

(iii) *if $a \in H$, then $a^{-1} \in H$ (i.e., H is closed under the taking of inverses in G).*

Proof of Theorem 6.5.2 We begin by outlining the key steps in the proof.

1. Using the subgroup H, we define an equivalence relation on G. As usual, this equivalence relation determines a partition Π_H of G.
2. Since G is a finite set, Π_H is a finite set; thus, $\Pi_H = \{P_1, \ldots, P_k\}$ where each P_i is a subset of G. We show that the sets $P_1, \ldots, P_k$ all have the same cardinality and that for $1 \le i \le k$, $|P_i| = |H|$.
3. We conclude that $|G| = k \cdot |H|$, and hence $|H|$ divides $|G|$.

We now carry out steps 1–3 in detail.

Step 1. We define a relation $\equiv_H$ on G as follows: For $a, b \in G$, $a \equiv_H b$ if $a \cdot b^{-1} \in H$. We show that $\equiv_H$ is an equivalence relation on G. (This claim is valid for any group G, finite or infinite, and any subgroup H of G.)

(a) $\equiv_H$ is reflexive. Let $a \in G$. Then $a \equiv_H a$ since $a \cdot a^{-1} = e \in H$ by Lemma 6.5.4.

(b) $\equiv_H$ is symmetric. Suppose $a, b \in G$ and $a \equiv_H b$. We show that $b \equiv_H a$. Since $a \equiv_H b$, $a \cdot b^{-1} \in H$. Because H is a subgroup of G, $(a \cdot b^{-1})^{-1} = b \cdot a^{-1} \in H$, and hence $b \equiv_H a$. (See Exercise 6.5.4(e) and (f).) Thus $\equiv_H$ is symmetric.

(c) $\equiv_H$ is transitive. Suppose $a \equiv_H b$ and $b \equiv_H c$. Then $a \cdot b^{-1} \in H$ and $b \cdot c^{-1} \in H$. Therefore, $(a \cdot b^{-1}) \cdot (b \cdot c^{-1}) = a \cdot b^{-1} \cdot b \cdot c^{-1} = a \cdot e \cdot c^{-1} = a \cdot c^{-1} \in H$ since H is a subgroup of G. But $a \cdot c^{-1} \in H$ implies $a \equiv_H c$. Thus $\equiv_H$ is transitive.

Step 2. Let Π_H be the partition determined by $\equiv_H$. Since G is a finite set, Π_H is also a finite set: $\Pi_H = \{P_1, \ldots, P_k\}$ where $P_i \subseteq G$ for $1 \le i \le k$. Recall that for $a, b \in G$, there exists i such that $a, b \in P_i$ if and only if $a \equiv_H b$, i.e., if and only if $a \cdot b^{-1} \in H$. Also recall that since Π_H is a partition of G, $\bigcup_{i=1}^{k} P_i = G$ and $P_i \cap P_j = \varnothing$ if $i \ne j$.

We show that for each i, $|P_i| = |H|$. We accomplish this by first showing that for each i there exist $a_i \in G$ such that $P_i = \{h \cdot a_i | h \in H\}$. In other words, P_i is obtained by multiplying the elements of H on the right by the fixed element a_i.

We choose a_i to be any element of P_i. Let $H \cdot a_i$ denote the set $\{h \cdot a_i | h \in H\}$. We show $P_i = H \cdot a_i$.

Let $b \in P_i$. Then since $b \equiv_H a_i$, $b \cdot a_i^{-1} = h \in H$. Therefore, $b = h \cdot a_i \in H \cdot a_i$. Thus $P_i \subseteq H \cdot a_i$.

Next let $x \in H \cdot a_i$. Then $x = h \cdot a_i$ for some $h \in H$. Thus $x \cdot a_i^{-1} = h \in H$ and $x \equiv_H a_i$. But since $a_i \in P_i$, $x \in P_i$ and $H \cdot a_i \subseteq P_i$; thus $P_i = H \cdot a_i$.

To complete step 2 we show that $|H| = |H \cdot a_i|$ for each i. This assertion follows from the fact that the function $r: H \to H \cdot a_i$, defined by $r(h) = h \cdot a_i$, is a bijection. (See Exercise 6.5.6.) Therefore, $|H| = |H \cdot a_i| = |P_i|$.

Step 3. Because Π_H is a partition of G,

$$|G| = \left| \bigcup_{i=1}^{k} P_i \right| = \sum_{i=1}^{k} |P_i| = \sum_{i=1}^{k} |H| = k \cdot |H|. \qquad \blacksquare$$

Remarks (1) Let $G = \mathbf{Z}_4$ and let $H = \{[0],[2]\}$. Then the partition $\Pi_H = \{H, \{[1],[3]\}\}$; observe that $\{[1],[3]\} = \{[0]+[1],[2]+[1]\} = H + [1]$, which is consistent with the proof of Lagrange's Theorem.

(2) Looking back at step 1, we can see that for any group G, any subgroup H, and all $a, b \in H$, $a \equiv_H b$. Conversely, if $a \in H$ and $b \in G$ and $a \equiv_H b$, then $b \in H$. Thus H is always one of the sets in the partition Π_H.

(3) We have already encountered the relation $\equiv_H$ in a special case. Let $G = \mathbf{Z}$ and $H = E$. Then for $a, b \in G$, $a \equiv_H b$ if and only if $a - b \in H = E$. (Note that the operation in G is ordinary addition and the inverse of $b \in \mathbf{Z}$ under addition is $-b$.) Thus $a \equiv_H b$ if and only if $a - b$ is even if and only if a and b are both odd or both even. Therefore, $\equiv_H$ is the equivalence relation $\equiv_2$.

We close this section by using Lagrange's Theorem to derive an important fact about the elements of a finite group.

Definition 6.5.5 *Order of an Element*

Let G be a finite group and let $g \in G$. The least positive integer m such that $g^m = e$ is called the *order of g*. The order of g is denoted by $o(g)$.

First observe that the order of each element of a finite group actually exists, for if $g \in G$ with $|G| = n$, then the set $\{e, g, \ldots, g^n\}$ has at most n elements. Hence there exist $i, j \in \mathbf{Z}$ with $0 \le i < j \le n$ and $g^j = g^i$. Therefore, $g^{j-i} = e$ where $j - i \in \mathbf{Z}^+$, and thus there exists a least positive integer $m \le j - i$ such that $g^m = e$.

Example 6.5.10 (i) In any finite group G, $o(g) = 1$ if and only if $g = e$.

(ii) In the group $\mathrm{Sym}(F_2)$ of Example 6.5.4, each element other than the identity has order 2.

(iii) In the group $\{e, g, g^2, g^3\}$ of order 4, $o(g) = o(g^3) = 4$ while $o(g^2) = 2$.

Proposition 6.5.5 Let G be a finite group and let $g \in G$. If $o(g) = m$, then $H = \{e, g, \ldots, g^{m-1}\}$ is a subgroup of G and $|H| = m$.

Proof First note that $e \in H$. Next, if $g^i, g^j \in H$, then $g^{i+j} \in H$ since either $i + j < m$ or $m \le i + j < 2m$, in which case $g^{i+j} = g^{i+j-m}$.

$g^m = g^{i+j-m} \in H$. Thus H is closed under the group operation in G. Finally, if $g^i \in H$, then $g^{m-i} \in H$ and $g^i \cdot g^{m-i} = g^m = e$. Thus H is closed under the taking of inverses. Therefore, by Lemma 6.5.4, H is a subgroup of G.

The proof that $|H| = m$ is left as an exercise. ■

Definition 6.5.6 *Finite Cyclic Groups and Subgroups*

Let G be a finite group and let $g \in G$ be an element of order m.

(i) The set $\{e, g, \ldots, g^{m-1}\}$ is called the *cyclic subgroup of* G *generated by* g and is denoted by $\langle g \rangle$.

(ii) The group G is *cyclic* if there exists $g \in G$ such that $G = \langle g \rangle$.

Notice that a finite group G of order n is cyclic if and only if G contains an element of order n. In general, $o(g) = |\langle g \rangle|$ by Proposition 6.5.5. The next theorem is a corollary of Lagrange's Theorem.

Theorem 6.5.6 *Let G be a finite group and let $g \in G$. Then $o(g)$ divides $|G|$.*

Proof As just noted, $o(g) = |\langle g \rangle|$. By Lagrange's Theorem $|\langle g \rangle|$ divides $|G|$. ■

Corollary 6.5.7 *Let G be a finite group and let $g \in G$. Then $g^{|G|} = e$.*

Proof By the definition of the order of g, $g^{o(g)} = e$. Since $o(g) \mid |G|$, there exists $k \in \mathbf{Z}$ such that $|G| = o(g) \cdot k$. Therefore, $g^{|G|} = g^{o(g) \cdot k} = (g^{o(g)})^{\cdot k} = e^k = e$.

Exercises §6.5

6.5.1. Verify Tables 6.5.1 and 6.5.2.

6.5.2. (a) Find Sym(R) where R is a nonsquare rectangle.
(b) Find Sym(T) where T is an equilateral triangle.
(c) Find Sym(T_1) where T_1 is a nonequilateral isosceles triangle.

6.5.3. Show that for $n \geq 3$, S_n is nonabelian.

6.5.4. Let G be a group.
(a) Show that G contains a unique identity element.
(b) Show that each element of G has a unique inverse.
(c) Show that the only idempotent element of G is the identity. ($g \in G$ is *idempotent* if $g^2 = g \cdot g = g$.)

(d) Show $e^{-1} = e$ where e is the identity of G.
(e) Show that for all $g \in G$, $(g^{-1})^{-1} = g$.
(f) Show that for all $g, h \in G$, $(g \cdot h)^{-1} = h^{-1} \cdot g^{-1}$.
(g) Show that for all $g, h \in G$, $(g \cdot h^{-1})^{-1} = h \cdot g^{-1}$.

6.5.5. (a) Prove Lemma 6.5.3.
(b) Prove Lemma 6.5.4.

6.5.6. Let G be a group. Suppose $S \subseteq G$ and $a \in G$. Let $r: S \to S \cdot a = \{g \cdot a \mid g \in S\}$ be the function defined by $r(g) = g \cdot a$ for all $g \in S$. Show that r is a bijection.

6.5.7. (a) Describe the structure of a group of order 3.
(b) Describe the structure of a group of order 5.

6.5.8. Let G be a group. Show that $\{e\}$ and G are subgroups of G.

6.5.9. Complete the proof of Proposition 6.5.5.

6.5.10. (a) Show that $\{\pm 1\}$ is a subgroup of $\mathbf{R}^*$ under multiplication.
(b) Show that the only finite subgroups of $\mathbf{R}^*$ are $\{1\}$ and $\{\pm 1\}$.

6.5.11. Show that $\{[0], [4]\}$ and $\{[0], [2], [4], [6]\}$ are subgroups of $\mathbf{Z}_8$ under addition. Are there any other subgroups of $\mathbf{Z}_8$? Explain.

6.5.12. (a) Let $3\mathbf{Z} = \{n \in \mathbf{Z} \mid n = 3 \cdot k \text{ for some } k \in \mathbf{Z}\}$. Show: $3\mathbf{Z}$ is a subgroup of $\mathbf{Z}$.
(b) Let $4\mathbf{Z} = \{n \in \mathbf{Z} \mid n = 4 \cdot k \text{ for some } k \in \mathbf{Z}\}$. Show: $4\mathbf{Z}$ is a subgroup of $\mathbf{Z}$. Find the partition determined by the equivalence relation $\equiv_{4\mathbf{Z}}$.
(c) Generalize the results in (a) and (b).
(d) Determine all subgroups of $\mathbf{Z}$.

6.5.13. (a) Find three subgroups of order 2 in S_3.
(b) For each of these subgroups H, find the partition determined by the equivalence relation $\equiv_H$.

6.5.14. Let G and H be groups with operations $*_1$ and $*_2$, respectively. Let $K = G \times H$ be the Cartesian product of G and H. Let $(g_1, h_1), (g_2, h_2) \in K$. Define $(g_1, h_1) \cdot (g_2, h_2) = (g_1 *_1 g_2, h_1 *_2 h_2)$.
(a) Show that K is a group under the operation $\cdot$.
(b) Write out the Cayley table for the group $\mathbf{Z}_2 \times \mathbf{Z}_2$.

6.5.15. Let G be a group and let H be a subgroup of G. H is a *normal subgroup* of G if for all $g \in G$ and all $h \in H$, $g \cdot h \cdot g^{-1} \in H$.
(a) Show that every subgroup of an abelian group is normal.
(b) Show that $\{e\}$ and G are normal subgroups of G.
(c) Show that none of the three subgroups of order 2 in S_3 is normal.

6.5.16. A group G is called *simple* if the only normal subgroups of G are $\{e\}$ and G. Show that any group of prime order is simple.

6.5.17. Prove, or disprove and salvage: For each $n \in \mathbf{Z}^+$, the group $\mathbf{Z}_n$ is a cyclic group under $+$.

§6.6 Homomorphisms and Isomorphisms

In this section we concentrate on structure-preserving functions between algebraic structures of a given type. Such functions, called homomorphisms, were introduced in Section 6.1. Our goal is to discuss some basic properties of homomorphisms. In order to prevent our presentation from becoming excessively general and broad, we focus primarily on the concepts of homomorphism and isomorphism of groups. (For readers familiar with linear algebra, we present the notion of a homomorphism between vector spaces, namely, that of a linear transformation.) We shall leave as an exercise the formulation of the definition of a homomorphism for commutative rings with unity. Many of the basic results for homomorphisms of groups have analogs for homomorphisms of other types of algebraic structures.

Definition 6.6.1 *Homomorphism and Isomorphism of Groups*

Let $(G_1, *_1)$ and $(G_2, *_2)$ be groups. Let $f: G_1 \to G_2$ be a function.

(i) f is a *homomorphism* if for all $x, y \in G_1$, $f(x *_1 y) = f(x) *_2 f(y)$.
(ii) f is an *isomorphism* if f is a bijective homomorphism.

As illustrated in Figure 6.6.1, if we combine x and y in G_1 to obtain $x *_1 y$ and then carry $x *_1 y$ to G_2 via f, then we obtain the same result as if we carry x and y to G_2 via f and then combine the resulting elements in G_2.

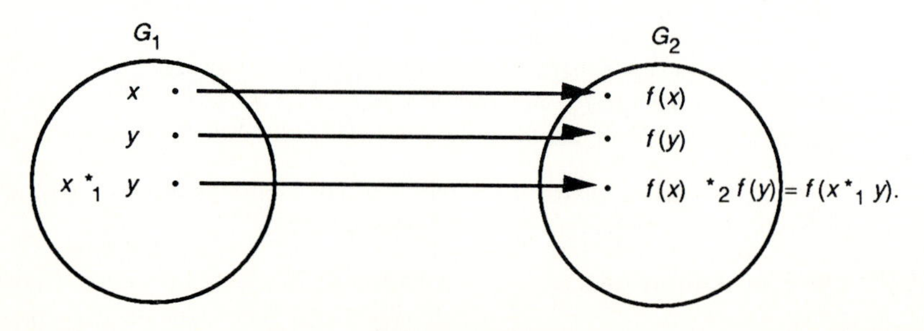

Figure 6.6.1

Example 6.6.1 The sets $\mathbf{R}^*$ and $\{\pm 1\}$ are both groups under multiplication. Define $f:\mathbf{R}^* \to \{\pm 1\}$ by $f(x) = \begin{cases} 1 & \text{if} \quad x > 0 \\ -1 & \text{if} \quad x < 0 \end{cases}$. (In other words, $f(x) = x/|x|$.) Then f is a homomorphism of groups.

Other examples of group homomorphisms are given in Examples 6.1.4 and 6.1.5.

For readers who have been exposed to linear algebra, we include a definition of homomorphism between vector spaces. This definition is given only to illustrate the form of the homomorphism concept in an algebraic structure different from a group.

> Definition 6.6.2 *Homomorphism of Vector Spaces*
>
> Let V_1 and V_2 be vector spaces (in which the scalars are real numbers). A function $f:V_1 \to V_2$ is a *homomorphism of vector spaces* or a *linear transformation* if (i) for all $x, y \in V_1$, $f(x + y) = f(x) + f(y)$, and (ii) for all $x \in V$ and all $r \in \mathbf{R}$, $f(r \cdot v) = r \cdot f(v)$.

Thus a linear transformation is a function between vector spaces that preserves both addition and scalar multiplication.

Properties of Group Homomorphisms

The remainder of this section is devoted to some basic properties of group homomorphisms.

Proposition 6.6.1 Let G_1 and G_2 be groups with identity elements e_1 and e_2, respectively. Let $f:G_1 \to G_2$ be a homomorphism of groups. Then

(i) $f(e_1) = e_2$,
(ii) for all $x \in G_1$, $f(x^{-1}) = f(x)^{-1}$, and
(iii) for all $x \in G_1$, and all $n \in \mathbf{Z}^+$, $f(x^n) = f(x)^n$.

Proof (i) Since e_1 is the identity of G_1 and since f is a homomorphism, $f(e_1) = f(e_1 \cdot e_1) = f(e_1) \cdot f(e_1)$. Applying $f(e_1)^{-1}$, we conclude that $e_2 = f(e_1)$.

(ii) Let $x \in G$. By (i) $e_2 = f(e_1) = f(x \cdot x^{-1})$. Thus (recalling that $f(x) \in G_2$), we have

$$\begin{aligned} f(x)^{-1} &= f(x)^{-1} \cdot e_2 = f(x)^{-1} \cdot \left(f(x) \cdot f(x^{-1})\right) \\ &= \left(f(x)^{-1} \cdot f(x)\right) \cdot f(x^{-1}) \\ &= e_2 \cdot f(x^{-1}) = f(x^{-1}). \end{aligned}$$

The proof of (iii) can be accomplished by induction and is left as an exercise. ■

Definition 6.6.3 *Kernel of a Homomorphism*

Let $f:G \to H$ be a homomorphism of groups. The *kernel of f*, Ker(f), is the set of elements of G sent by f to the identity of H:

$$\text{Ker}(f) = \{x \in G | f(x) = e\}.$$

Definition 6.6.4 *Range of a Homomorphism*

Let $f:G \to H$ be a homomorphism of groups. The *range of f*, or *image of f*, written Ran(f), is the set Ran(f) = $\{y \in H |$ There exists $x \in G$ such that $f(x) = y\}$.

Example 6.6.2 Let $f:\mathbf{Z}_8 \to \mathbf{Z}_8$ be the homomorphism $f[x] = [2x]$. (It is easy to check that f is actually a homomorphism.) The following table gives the values of f:

$[x]$	[0]	[1]	[2]	[3]	[4]	[5]	[6]	[7]
$f([x])$	[0]	[2]	[4]	[6]	[0]	[2]	[4]	[6]

Thus Ker(f) = {[0], [4]} and Ran(f) = {[0], [2], [4], [6]}. Notice that both these sets are subgroups of $\mathbf{Z}_8$. This fact is a special case of the next theorem.

Theorem 6.6.2 *Let $f:G \to H$ be a homomorphism of groups. Then*

(i) Ran(f) *is a subgroup of H;*
(ii) Ker(f) *is a subgroup of G;*
(iii) *for all $x \in G$ and all $y \in$* Ker(f), $xyx^{-1} \in$ Ker(f).

Proof We leave the proof of (i) as an exercise.

(ii) Let e and e' be the identities of G and H respectively. By Lemma 6.5.4, we must show that (i) $e \in$ Ker(f); (ii) if $x, y \in$ Ker(f), then $x \cdot y \in$ Ker(f); and (iii) if $x \in$ Ker(f), then $x^{-1} \in$ Ker(f).

First, by Proposition 6.6.1(i), $e \in$ Ker(f). Next let $x, y \in$ Ker(f); then $f(x \cdot y) = f(x) \cdot f(y) = e' \cdot e' = e'$ and $x \cdot y \in$ Ker(f). Finally, let $x \in$ Ker(f). Then, by Proposition 6.6.1(ii), $f(x^{-1}) = f(x)^{-1} =$ $(e')^{-1} =$

(iii) Let $x \in G$ and $y \in \text{Ker}(f)$. Then

$$\begin{aligned} f(x \cdot y \cdot x^{-1}) &= f(x) \cdot f(y) \cdot f(x^{-1}) \\ &= f(x) \cdot e' \cdot f(x^{-1}) \\ &= f(x) \cdot f(x^{-1}) \\ &= f(x) \cdot f(x)^{-1} = e'. \end{aligned}$$

Therefore, $x \cdot y \cdot x^{-1} \in \text{Ker}(f)$. ■

Remarks (1) Suppose G and H are finite groups. Then, by Lagrange's Theorem, $|\text{Ker}(f)|$ divides $|G|$ and $|\text{Ran}(f)|$ divides $|H|$. In fact, the numbers $|\text{Ker}(f)|$, $|\text{Ran}(f)|$, and $|G|$ are related by the equality $|\text{Ker}(f)| \cdot |\text{Ran}(f)| = |G|$. (See Theorem 6.6.3 below.) This pretty fact is analogous to a theorem of linear algebra: If $L: V_1 \to V_2$ is a linear transformation between finite dimensional vector spaces, then $\dim(\text{Ker}(L)) + \dim(\text{Ran}(L)) = \dim(V_1)$.

(2) A subgroup N of a group G with the property that $xyx^{-1} \in N$ for all $x \in G$ and all $y \in N$ is called a *normal subgroup of* G. Thus the kernel of any homomorphism is a normal subgroup of the domain of the homomorphism. We can show conversely that if N is a normal subgroup of a group G, there exists a homomorphism f having domain G such that $\text{Ker}(f) = N$.

Theorem 6.6.3 *Let G and H be finite groups and let $f: G \to H$ be a homomorphism. Then $|G| = |\text{Ker}(f)| \cdot |\text{Ran}(f)|$.*

Remark Lagrange's Theorem asserts that the order of a subgroup of a finite group divides the order of the group. Theorem 6.6.3 is a refinement of Lagrange's Theorem in case the subgroup is the kernel of a homomorphism, for Theorem 6.6.3 not only asserts that $|\text{Ker}(f)|$ divides $|G|$ but also identifies the ratio $|G|/|\text{Ker}(f)|$ as being equal to $|\text{Ran}(f)|$. The proof of Theorem 6.6.3 is itself a refinement of the proof of Lagrange's Theorem.

Proof Let $K = \text{Ker}(f)$. For $y \in \text{Ran}(f)$, define $f^{-1}(y) =$ the inverse image of $y = \{x \in G | f(x) = y\}$. We begin by showing:

(1) The sets $\{f^{-1}(y) | y \in \text{Ran}(f)\}$ partition G. First, if $x \in G$, then $x \in f^{-1}(y)$ where $y = f(x)$, and thus $\bigcup_{y \in \text{Ran}(f)} f^{-1}(y) = G$. Second, if $x \in f^{-1}(y_1) \cap f^{-1}(y_2)$, then $y_1 = f(x) = y_2$ (see Figure 6.6.2). Next we show:

(2) If $x \in f^{-1}(y)$ is fixed, then $f^{-1}(y) = K \cdot x = \{k \cdot x | k \in K\}$.

For let $z \in f^{-1}(y)$. Then $f(z) = y = f(x)$. Therefore, $f(z \cdot x^{-1}) = f(z \cdot x^{-1}) = f(z) \cdot f(x^{-1}) = f(z) \cdot f(x)^{-1} = y \cdot y^{-1} = e$ (the identity of H), which implies that $z \cdot x^{-1} = k \in K$, and hence $z = k \cdot x \in K \cdot x$. Thus $f^{-1}(y) \subseteq K \cdot x$. On the other hand, if $k \cdot x \in K \cdot x$, then $f(k \cdot x) = f(k) \cdot f(x) = e \cdot y = y$ and $k \cdot x \in f^{-1}(y)$. Thus $K \cdot x \subseteq f^{-1}(y)$.

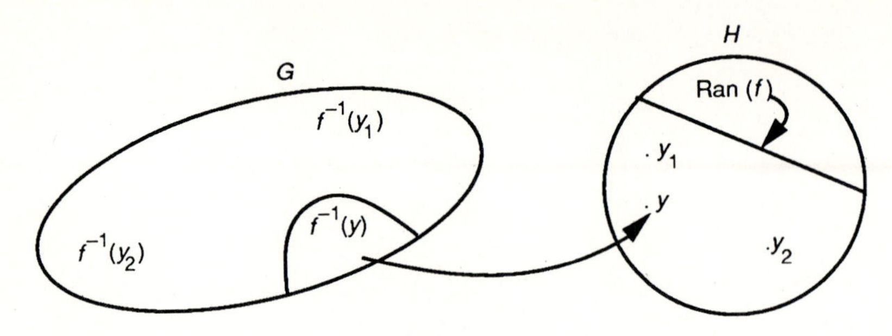

Figure 6.6.2

Now we can complete the proof. By (2), for each $y \in \text{Ran}(f)$, $|f^{-1}(y)| = |K \cdot x| = |K|$. Therefore,

$$|G| = \sum_{y \in \text{Ran}(f)} |f^{-1}(y)| = \sum_{\substack{f(x)=y \\ y \in \text{Ran}(f)}} |K \cdot x|$$

$$= \sum_{y \in \text{Ran}(f)} |K| = |K| \cdot |\text{Ran}(f)|. \qquad \blacksquare$$

We close this section by considering an important class of groups, cyclic groups. The definition that we now propose generalizes Definition 6.5.6.

> **Definition 6.6.5** *Cyclic Group*
>
> A group G is *cyclic* if there exist $g \in G$ such that for each $x \in G$ there is $n \in \mathbf{Z}$ such that $x = g^n$. The element g is called a *generator* of G.

Thus a group G is cyclic if every element can be represented as a positive power of some fixed element or as the inverse of a positive power of that fixed element. Note that when the operation of G is written additively (e.g., as if $g = \mathbf{R}$), then for $n \in \mathbf{Z}^+$, $g^n = g + \cdots + g$, which is usually denoted by $n \cdot g$. In other words, the nth "power" of g is g added to itself n times.

Example 6.6.3 $\mathbf{Z}$ is cyclic since if $x \in \mathbf{Z}^+$, $\underbrace{x = 1 + \cdots + 1}_{x \text{ times}}$ while if $x < 0$, then $x = \underbrace{-(1 + \cdots + 1)}_{-x \text{ times}} =$ inverse of 1 added to itself $-x$ times.

Example 6.6.4 For each $n \in \mathbf{Z}^+$, $\mathbf{Z}_n$ is cyclic since if $[x] \in \mathbf{Z}_n$ where $1 \le x \le n$, then $[x] = \underbrace{[1] + \cdots + [1]}_{x \text{ times}}$.

Our last theorem asserts that these examples are in effect the only cyclic groups.

Theorem 6.6.4 Any cyclic group is isomorphic either to $\mathbf{Z}$ or to $\mathbf{Z}_n$ for some $n \in \mathbf{Z}^+$.

Proof We first suppose that G is a finite cyclic group and show that G is isomorphic to $\mathbf{Z}_n$ where $n = |G|$. We leave as an exercise the proof that any infinite cyclic group is isomorphic to $\mathbf{Z}$.

Suppose $|G| = n$. Let g be a generator of g. Then the elements $e, g, g^2, \ldots, g^{n-1}$ are all distinct, for if $g^i = g^j$ where $0 \le i < j \le n - 1$, then $g^{j-i} = e$; it then follows that for each $k \in \mathbf{Z}$, $g^k = g^r$ where $k = (j - i)q + r$ and $0 \le r \le j - i$ and hence $G \subseteq \{e, g, \ldots, g^{j-i-1}\}$. Therefore $|G| \le j - i < n$. This contradiction implies that $|\{e, g, \ldots, g^{n-1}\}| = n$. Since $|G| = n$ and $G \supseteq \{e, g, \ldots, g^{n-1}\}$, $G = \{e, g, \ldots, g^{n-1}\}$.

We now claim that the function $f{:}\mathbf{Z}_n \to G$ defined by $f([i]) = g^i$ for $0 \le i \le n - 1$ is an isomorphism. Clearly, f is a bijection. Also for $[i], [j] \in \mathbf{Z}_n$, $f([i] + [j]) = f([i + j]) = g^{i+j} = g^i \cdot g^j = f([i] \cdot f[j])$; hence f is a homomorphism. ■

Exercises §6.6

6.6.1. Prove Proposition 6.6.1(iii).

6.6.2. Prove: If $f, g{:}G \to G$ are homomorphisms of the group G to itself, then $g \circ f$ is a homomorphism of G to itself.

6.6.3. Show: The function $f{:}\mathbf{R} \to \mathbf{R}$ defined by $f(x) = x^2$ is not a homomorphism of the additive group of reals to itself.

6.6.4. Show: $f{:}\mathbf{Z}_6 \to \mathbf{Z}_6$ defined by $f([x]_6) = [3x]_6$ is a homomorphism.

6.6.5. Give several examples of homomorphisms $f{:}\ \mathbf{Z}_{10} \to \mathbf{Z}_{10}$.

6.6.6. Let $f{:}\mathbf{Z}_{12} \to \mathbf{Z}_{12}$ be defined by $f([x]) = [4x]$. Show f is a homomorphism. Compute $\text{Ker}(f)$ and $\text{Ran}(f)$.

6.6.7. Let G be an abelian group and let $n \in \mathbf{Z}^+$ be fixed. Define $f_n{:}G \to G$ by $f_n(x) = x^n$.
(a) Show: f_n is a group homomorphism.
(b) Give an example of a group G and integer n for which f_n is an isomorphism.
(c) Give an example of a group G and integer n for which f_n is not an isomorphism.

6.6.8. Prove Theorem 6.6.2(i).

6.6.9. Let $f: G \to H$ be a homomorphism of groups. Let K be a subgroup of H and let A be a subgroup of G.
(a) Show: $f^{-1}(K) = \{x \in G | f(x) \in K\}$ is a subgroup of G.
(b) Show: $f(A) = \{y \in H | y = f(x)$ for some $x \in G\}$ is a subgroup of H.
(c) Show: If G is abelian, then $f(G)$ is abelian.
(d) Show: If G is cyclic, then $f(G)$ is cyclic.

6.6.10. Prove that any infinite cyclic group is isomorphic to $\mathbf{Z}$.

6.6.11. An isomorphism $f: G \to G$ of a group G with itself is called an *automorphism* of G. Show: For fixed $a \in G$, the function $f_a: G \to G$ defined by $f_a(x) = axa^{-1}$ is an automorphism of G. If G is abelian, what is f_a?

Notes

The algebraic structures discussed in this chapter have interesting and complex histories. For example, the theory of groups arose in part from the study of solutions of polynomial equations with coefficients in the rational numbers. One of the great problems of classical algebra concerned the solvability of a polynomial equation $P(X) = a_n \cdot X^n + \cdots + a_1 \cdot X + a_0$ where $a_i \in \mathbf{Q}$ via *radicals* or *roots*. By the Fundamental Theorem of Algebra, this equation has n roots, $r_1, \ldots, r_n$, in the field of complex numbers. The problem was to express these roots in terms of the numbers $a_0, \ldots, a_n$ and roots (square roots, cube roots, up to nth roots) of combinations of these numbers. An example is the *quadratic formula*: $a_2 \cdot X^2 + a_1 \cdot X + a_0 = 0$ with $a_2 \neq 0$ if and only if $X = (-a_1 \pm \sqrt{a_1^2 - 4 \cdot a_0 \cdot a_2})/2 \cdot a_2$. The roots of cubic and quartic polynomials (i.e., polynomials of degrees 3 and 4, respectively) can also be expressed in terms of radicals. Until the 1820s, the principal problem in algebra concerned the solvability of 5th degree polynomial equations by radicals. In 1826 a young Norwegian mathematician, Niels Henrik Abel (in honor of whom abelian groups are named), proved that not all equations of degree 5 or greater can be solved by radicals. Abel was one of the shining stars of mathematics when he died in poverty at age 27 in 1829. In 1832 a brilliant young French mathematician, Evariste Galois, extended Abel's work on the solvability of equations. Galois showed how to assign to each polynomial $P(X)$ a group that we shall call $G(P)$. Galois proved that the polynomial equation $P(X) = 0$ is solvable in radicals if and only if the group $G(P)$ possesses a certain property that is called, reasonably enough, *solvability*. It turns out that not all groups $G(P)$ have the solvability property and, therefore, not all polynomial equations $P(X) = 0$ are solvable by radicals. Galois outlined his proof on the evening of May 30, 1832, the night before he participated in a duel that he unfortunately lost. Galois died at age 21 on the evening of May 31, 1832, after lying wounded on the dueling field all day. A brief, colorful account of Galois's life can be found in E. T. Bell's *Men of Mathematics* [1].

Another source of group theory was the theory of numbers. Special cases of groups (e.g., $\mathbf{Z}_n$) appeared in the work of Gauss and later in the works of number theorists such as E. E. Kummer and R. Dedekind. Credit for the definition of an abstract group is given to the English mathematician Arthur Cayley. Cayley's definition, appearing in an 1854 paper, culminates a long developmental period. Until Cayley, special cases of groups (abelian groups, permutation groups) were studied and theorems such as Lagrange's Theorem were proved in each case. After Cayley, such general theorems were deduced for all groups from the defining axioms of a group.

The other algebraic structures discussed in this chapter underwent a similar development. Examples of rings and fields were encountered and studied throughout the nineteenth century. For instance, Gauss proved the unique factorization property in $\mathbf{Z}$ (in 1801) and in the *ring of Gaussian integers* (apparently not so named by the master himself) $\mathbf{Z}[i] = \{a + b \cdot i \in \mathbf{C} | a, b \in \mathbf{Z}\}$ (in 1828). Thereafter, the question of unique factorization in a ring became an important issue in algebra. The theory of fields was pushed along by several problems including the solvability of polynomial equations described above and some construction problems of Greek geometry. Among these problems were: (1) For each integer $n > 1$, using a straight edge and compass, construct a polygon having n sides of equal length (a so-called *regular n-gon*). (2) Given an arbitrary circle, using a straight edge and compass, construct a square having the same area as the circle. The first problem was solved by Gauss in his *Disquisitiones*; the second was answered by Lindemann in 1882. Among the mathematicians who laid the foundation for the abstract study of rings and fields were Kummer, Dedekind, and Kronecker. It is unclear who first defined the concepts of ring and field as abstract algebraic structures. Special credit, however, for the development of modern algebra must be given to Dedekind. In part because of his emphasis on axiomatics and his role in the development of algebra, Dedekind is one of the central figures in the history of mathematics.

Bibliography

1. E. T. Bell, *Men of Mathematics*, Simon & Schuster, 1937.
2. J. Gallian, *Contemporary Abstract Algebra*, D. C. Heath, 1986.
3. C. F. Gauss, *Disquisitiones Arithmeticae*, Yale University Press, 1965.
4. M. Kline, *Mathematical Thought from Ancient to Modern Times*, Oxford University Press, 1972.
5. L. Novy, *Origins of Modern Algebra*, Noordhoff, 1973.

Miscellaneous Exercises

1. Let G be a finite group of prime order. Show that G is cyclic and that any nonidentity element generates G.

2. Let G be a group and let X be a set. We say that G acts on X if there exists a function from $G \times X$ to X sending (g, x) to $g \cdot x$ satisfying (i) $g \cdot (h \cdot x) = (g \cdot h) \cdot x$ for all $g, h \in G$ and all $x \in X$, and (ii) $e \cdot x = x$ for all $x \in X$ where e is the identity of G. Examples: (1) For each $n \in \mathbf{Z}$, the group $n\mathbf{Z} = \{k \in \mathbf{Z} | n | k\}$ acts on the set $\mathbf{Z}$: $(nz, y) \to nz + y$ for $z, y \in \mathbf{Z}$. (2) The

group $\{\pm 1\}$ acts on the sphere $S^2 = \{x, y, z) \in \mathbf{R}^3 | x^2 + y^2 + z^2 = 1\}$: $1 \cdot (x, y, z) = (x, y, z)$ and $-1 \cdot (x, y, z) = (-x, -y, -z)$.

(a) Give two other examples of groups acting on sets.

(b) Suppose a group G acts on a set X. If $x, y \in X$, then we define $x \sim y$ to mean there exists $g \in G$ such that $y = g \cdot x$. Show that $\sim$ is an equivalence relation on X.

(c) Describe the set of equivalence classes when $n\mathbf{Z}$ acts on $\mathbf{Z}$.

3. (a) Prove $\log_4(10) \notin \mathbf{Q}$.
   (b) Prove $\log_7(14) \notin \mathbf{Q}$.
   (c) Prove $\log_{10}(55) \notin \mathbf{Q}$.
   (d) Prove $\log_4(8) \in \mathbf{Q}$.

4. Fill in the blank and prove the statement: $\log_a(b) \in \mathbf{Q}$ if and only if ____________________.

5. Let R be a commutative ring. An *ideal in R* is a subset I of R satisfying the following conditions:
   (i) I is a subgroup of R under $+$.
   (ii) For all $r \in R$ and all $x \in I$, $r \cdot x \in I$.
   (a) Find some examples of ideals.
   (b) Is $\mathbf{Z}$ an ideal in $\mathbf{Q}$?
   (c) Is $\mathbf{Q}$ an ideal in $\mathbf{R}$?
   (d) Let R be an arbitrary commutative ring with unity. Let $r_0 \in R$ be fixed. Show that $\{x \in R | r_0 | x\}$ is an ideal of R.

6. Let R and S be commutative rings. Propose a definition of the concept of homomorphism from R to S. Give at least two examples illustrating your definition.

7. Let $f: R \to S$ be a homomorphism of commutative rings. What can you say about the kernel of f:$\text{Ker}(f) = \{x \in R | f(x) = 0\}$?

8. Find all ring isomorphisms $f: \mathbf{Z} \to \mathbf{Z}$.

9. Let $f: \mathbf{Z}[\sqrt{2}] \to \mathbf{Z}[\sqrt{2}]$ be defined by $f(a + b\sqrt{2}) = a - b\sqrt{2}$. Show that f is a ring isomorphism.

10. Let $(G_1, *_1)$ and $(G_2, *_2)$.
    (a) If G_1 and G_2 are abelian, then is $G_1 \times G_2$ abelian?
    (b) If G_1 and G_2 are cyclic, then is $G_1 \times G_2$ cyclic?

11. Let R_1 and R_2 be commutative rings with unity.
    (a) Show how to define a ring structure on the set $R_1 \times R_2$.
    (b) If R_1 and R_2 are fields, then is $R_1 \times R_2$ a field under the ring structure that you defined in (a)?

12. Let R be a commutative ring with unity. Let $U(R)$ denote the set of units of R. Prove that $U(R)$ is a group with respect to multiplication in R.

13. Let p be a positive prime. Let $n = p^a$ where $a \in \mathbf{Z}^+$. Find all zero divisors in $\mathbf{Z}_n$.

14. Let p be a positive prime. Find two nonisomorphic groups of order p^2.

15. Let $a \in \mathbf{Z}_p = \{1, \ldots, p-1\}$ where p is a positive prime integer. (Again we denote the element $[a]_p$ of $\mathbf{Z}_p$ by a.)
    (a) Show that $a^{p-1} = 1$ for all $a \in \mathbf{Z}_p^*$.
    (b) Show that the polynomial equation $X^{p-1} - 1 = 0$ has $p - 1$ solutions in $\mathbf{Z}_p$.
    (c) Show that $X^{p-1} - 1 = (X-1) \cdot (X-2) \cdot \cdots \cdot (X-(p-1))$ in $\mathbf{Z}_p[X]$.
    (d) Show that $(p-1)! \equiv -1 \pmod{p}$.
    (e) Show that if $p \equiv 1 \pmod{4}$ then $[((p-1)/2)!]^2 \equiv -1 \pmod{p}$.

16. Let $n, m \in \mathbf{Z}^+$. Suppose $n | m$. Define $f: \mathbf{Z}_m \to \mathbf{Z}_n$ by the rule $f([a]_m) = [a]_n$.
    (a) Show that f is a well-defined function; i.e., show that if $[a]_m = [b]_m$, then $f([a]_m) = f([b]_m)$.
    (b) Prove that f is a group homomorphism.
    (c) Calculate $\operatorname{Ker}(f)$.

CHAPTER SEVEN

The Real Number System

During the nineteenth century great advances were made in the theory and applications of functions of a real variable. Significant developments in the theory of Fourier series, the study of differential equations, and approximation theory rapidly expanded the frontiers of mathematical research. As progress in these areas was occurring, mathematicians recognized the need for the establishment of a firm foundation on which to rest the theory of functions. A function of a real variable, of course, assigns to each real number x in its domain a unique real number y. Properties of functions such as differentiability and continuity depend on properties of real numbers. Up to 1850, however, a clear definition of a real number had not been given. Thus a *rigorous* development of the real number system was deemed to be of utmost importance.

In the period from 1865 to 1900 this rigorous construction of the real numbers was accomplished. Major contributors to this program were R. Dedekind, G. Peano, C. Weierstrass, B. Russell, and A. N. Whitehead. These mathematicians in turn based their work on that of A. Cauchy and B. Bolzano, mathematicians who had, in the period 1815–1850, published many important papers and books on the real number system and on functions of a real variable. By 1902 a construction of the real number system that was acceptable to mathematicians of the day had been provided.

In this chapter we present a development of the real number system. We begin by stating the axioms for the natural number system essentially as presented by G. Peano. We do not actually construct a set that satisfies the Peano axioms (i.e., we do not construct a model of the Peano axioms).

Instead, we assume the existence of such a set and call it the set **N** of natural numbers. We do, however, derive properties of **N** from the Peano axioms. In particular, we define the arithmetic operations of addition and multiplication on **N** and the order relation $<$ on **N**.

Next we use the set **N** to construct the set of integers, **Z**, and we extend the arithmetic operations from **N** to **Z**. From **Z** we construct the set of rational numbers **Q**. Finally, we use **Q** as a basis on which to define the set of real numbers **R** and to show that **R** satisfies Axioms 1–14 of Section 1.2. We note that, back in Chapter 6, we constructed the field of complex numbers, **C**, from the real numbers. Thus, by the end of this chapter, we shall have traveled from **N** to **Z** to **Q** to **R** to **C**.

We might note that historically, the development of the real numbers occurred in exactly the reverse order. Mathematicians assumed the existence of the rational number system, **Q**, and defined **R** in terms of **Q**. At that point, the exact nature of the rational number system was questioned, leading mathematicians to define **Q** in terms of **Z** and **Z** in terms of **N**. Finally, with the work of Peano in the 1890s and Russell and Whitehead in the early 1900s, the set of natural numbers was rigorously axiomatized and constructed on the basis of set theory.

§7.1 The Natural Numbers

We begin by defining the natural numbers to be a set satisfying five properties that are known as the Peano axioms. Our goal is to deduce further properties of the set of natural numbers from the Peano axioms.

Definition 7.1.1 *The Natural Numbers*

The *set of natural numbers* is a set $\mathbf{N}$ that satisfies the following five axioms:

(P1) There exists an element $0 \in \mathbf{N}$.

(P2) For each $n \in \mathbf{N}$, there exists an element $s(n) \in \mathbf{N}$ called *the successor of* n.

(P3) For each $n \in \mathbf{N}$, $s(n) \neq 0$.

(P4) If $m, n \in \mathbf{N}$ and $s(m) = s(n)$, then $m = n$.

(P5) If A is a subset of $\mathbf{N}$ such that (i) $0 \in A$, and (ii) $s(n) \in A$ whenever $n \in A$, then $A = \mathbf{N}$.

This definition raises at least two questions. First, does there exist a set that actually satisfies Axioms P1–P5? In other words, does the system of Peano axioms have a model? It is at least conceivable that no set can satisfy simultaneously the five conditions stated in Definition 7.1.1. To answer this question affirmatively, we would have to use the Zermelo–Fraenkel axioms of set theory (see Chapter 2) or some other axiom system for set theory to define a set $\mathbf{N}$ and to show that the set $\mathbf{N}$ possesses P1–P5. This task is beyond the scope of this text. We refer the interested reader to [3] or [4]. For our part, we assume the existence of a set $\mathbf{N}$ satisfying P1–P5. We write the elements of $\mathbf{N}$ in the usual fashion: Given the element 0 from Axiom P1, we define $1 = s(0), 2 = s(1), 3 = s(2)$, etc.

A second question also arises: Do there exist two distinct sets that satisfy the Peano axioms? More specifically, if $\mathbf{N}_1$ and $\mathbf{N}_2$ are sets satisfying the Peano axioms, then are $\mathbf{N}_1$ and $\mathbf{N}_2$ isomorphic? In the context of the Peano axioms, the concept of isomorphism takes the following form: Let 0_1 and 0_2 be the elements of $\mathbf{N}_1$ and $\mathbf{N}_2$, respectively, each of whose existence is guaranteed by Axiom P1. For each $n_1 \in \mathbf{N}_1$, let $s_1(n_1)$ denote the successor of n_1 in $\mathbf{N}_1$ and for each $n_2 \in \mathbf{N}_2$, let $s_2(n_2)$ denote the successor of n_2 in $\mathbf{N}_2$. Then $\mathbf{N}_1$ is *isomorphic* to $\mathbf{N}_2$ if there exists a one-to-one correspondence $f: \mathbf{N}_1 \to \mathbf{N}_2$ such that $f(0_1) = 0_2$ and for all $n_1 \in \mathbf{N}_1$, $f(s_1(n_1)) = s_2(f(n_1))$. The following theorem, which we shall not prove, answers the question raised at the start of this paragraph.

Theorem 7.1.1 *Any two models of the Peano axioms are isomorphic.*

Theorem 7.1.1 asserts that if $\mathbf{N}_1$ and $\mathbf{N}_2$ are models of the Peano axioms, then $\mathbf{N}_1$ and $\mathbf{N}_2$ have the same structure as sets (since f is a bijection) and behave in exactly the same way with regard to the Peano axioms: The one-to-one correspondence f from $\mathbf{N}_1$ to $\mathbf{N}_2$ sends 0_1 to 0_2 and sends the successor of each $n_1 \in \mathbf{N}_1$ to the successor of $f(n_1)$ in $\mathbf{N}_2$. Because of this fact, we are justified in thinking of $\mathbf{N}$ in our customary fashion: $\mathbf{N} = \{0, 1, 2, \ldots, n, \ldots\}$.

(By way of contrast, notice that the situations with the group axioms (Section 6.5) and field axioms (Section 1.2 or Section 6.4) are quite different. There exist models of the group axioms, such as $\mathbf{Z}_2$ and $\mathbf{Z}_3$ under addition, which are not isomorphic as groups; and there exist models of the field axioms, such as $\mathbf{Q}$ and $\mathbf{R}$, which are not isomorphic as fields.)

Several other comments on the Peano axioms are in order. The first axiom implies that the set $\mathbf{N}$ is nonempty. The element 0 given by P1 is called the *zero element*. The second axiom asserts the existence of a function $s\colon \mathbf{N} \to \mathbf{N}$. As we shall see in the next section, for each $n \in \mathbf{N}$, we can think of $s(n)$ as $n + 1$. Axioms P4 and P3 imply, respectively, that the successor function is injective but not surjective (since 0 is not the successor of any element). Finally, axiom P5 is simply a formalization of the Principle of Mathematical Induction. For example, to prove that the formula

$$1 + 2 + \cdots + n = n(n + 1)/2$$

holds for all positive natural numbers, define the set

$$A = \{n \in \mathbf{N} \mid n = 0 \text{ or } n \neq 0 \text{ and } 1 + \cdots + n = n(n + 1)/2\}.$$

Then we can show that $0 \in A$, and, in general, if $n \in A$, then $s(n) = n + 1 \in A$. Thus, by P5, $A = \mathbf{N}$, and hence the given formula holds for all $n \in \mathbf{N}$ such that $n \neq 0$.

Let us now illustrate how we can use the Peano axioms in a formal proof. Although the proof is straightforward, the following result will be very useful in the sequel.

Theorem 7.1.2 *For each $k \in \mathbf{N}$, either $k = 0$ or there exists $n \in \mathbf{N}$ such that $s(n) = k$.*

Proof Let $A = \{k \in \mathbf{N} \mid k = 0$ or there exists $n \in \mathbf{N}$ such that $s(n) = k\}$. We show that (i) $0 \in A$ and (ii) if $n \in A$, then $s(n) \in A$. By P5, it will follow that $A = \mathbf{N}$.

By the definition of A, $0 \in A$. Suppose next that $n \in A$. We show that $s(n) \in A$. Since $s(n) = s(n)$, there exists $m \in \mathbf{N}$ (namely, $m = n$) such that $s(m) = s(n)$. Thus $s(n) \in A$.

By P5, we have that $A = \mathbf{N}$. ■

Theorem 7.1.2 has the following interpretation in the language of functions: The range of the successor function, Ran(s), is the set $\mathbf{N} - \{0\} = \{1, 2, \ldots\}$.

As we have observed throughout this text, the Principle of Mathematical Induction (Axiom P5) is one of the most important proof techniques in mathematics. We can also use the concept of induction in definitions. For example, let $a \in \mathbf{R}$. Suppose we wish to define the successive powers $a^1, a^2, a^3, \ldots$. We do so by defining $a^1 = a$, $a^2 = a \cdot a$, and for $n \geq 2$, $a^{n+1} = a^n \cdot a$. Although it might seem obvious that this procedure defines a^n for all $n \in \mathbf{N}^+$, a rigorous justification of this fact requires the next theorem, which is itself a consequence of P5.

Theorem 7.1.3 *Recursion Theorem. Let A be a set and let f: $A \to A$ be a function. Let $a_0 \in A$. There exists a unique function F: $\mathbf{N} \to A$ such that $F(0) = a_0$ and for all $n \in \mathbf{N}$, $F(s(n)) = f(F(n))$.*

Perhaps the following diagram will help to visualize the Recursion Theorem:

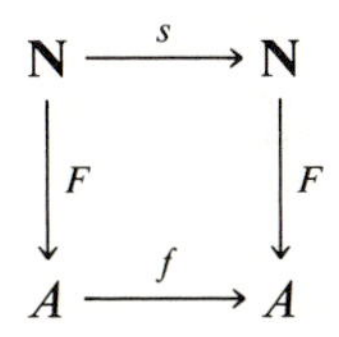

The theorem asserts that if we start in the upper left corner of the diagram and choose any $n \in \mathbf{N}$, and if we follow that diagram to the lower right corner, then we end up with the same element of A whether we travel the "high road" (s first, then F) or the "low road" (F first, then f): For all $n \in \mathbf{N}$, $F(s(n)) = f(F(n))$.

Let us apply the Recursion Theorem to show how to define a^n for $n \geq 1$. Let $a \in \mathbf{R}$ be fixed. Define f: $\mathbf{R} \to \mathbf{R}$ by $f(x) = a \cdot x$. (Thus we are applying the Recursion Theorem when the set $A = \mathbf{R}$.) The Recursion Theorem guarantees the existence of a function F: $\mathbf{N} \to \mathbf{R}$ such that $F(0) = 1$ and for all $n \in \mathbf{N}$, $F(s(n)) = F(n + 1) = f(F(n)) = a \cdot F(n)$. Thus $F(1) = a \cdot F(0) = a$, $F(2) = a \cdot F(1) = a \cdot a$, and $F(n + 1) = F(s(n)) = f(F(n)) = a \cdot F(n) = a \cdot a^n$. A definition such as the one just presented is called an *inductive* or *recursive definition*.

A proof of the Recursion Theorem will not be given here. A proof can be found in the books of van Dalen et al. and Hamilton listed at the end of this chapter. In Exercise 7.1.3 a proof of Theorem 7.1.1, using the Recursion Theorem and mathematical induction, is outlined.

Exercises §7.1

7.1.1. Show that for all $n \in \mathbf{N}$, $s(n) \neq n$.

7.1.2. Let $a \in \mathbf{N}$.
- (a) Give an informal recursive definition (i.e., in the spirit of our first definition of a^n) of $n \cdot a$ where $n \in \mathbf{N}$.
- (b) Use the Recursion Theorem to give a formal recursive definition of $n \cdot a$ where $n \in \mathbf{N}$.

7.1.3. The purpose of this exercise is that we show, using the Recursion Theorem, that any two models of the Peano axioms are isomorphic. Let $\mathbf{N}_1$ and $\mathbf{N}_2$ be models of the Peano axioms with successor functions s_1 and s_2, respectively, and zero elements 0_1 and 0_2, respectively.
- (a) Use the Recursion Theorem (taking $\mathbf{N} = \mathbf{N}_1$ and $A = \mathbf{N}_2$) to find a function F: $\mathbf{N}_1 \to \mathbf{N}_2$ such that $F(0_1) = 0_2$ and $F(s_1(n_1)) = s_2(F(n_1))$ for all $n_1 \in \mathbf{N}_1$.
- (b) Use the Recursion Theorem to find a function G: $\mathbf{N}_2 \to \mathbf{N}_1$ such that $G(0_2) = 0_1$ and $G(s_2(n_2)) = s_1(G(n_2))$ for all $n_2 \in \mathbf{N}_2$.
- (c) Use P5 to show that $G(F(n_1)) = n_1$ for all $n_1 \in \mathbf{N}_1$ and $F(G(n_2)) = n_2$ for all $n_2 \in \mathbf{N}_2$.
- (d) Conclude that $\mathbf{N}_1$ and $\mathbf{N}_2$ are isomorphic.

7.1.4. Let $\mathbf{N} = \{0, 1, 2, \dots\}$. Show that the set $E = \{0, 2, 4, 6, \dots\}$ is also a model of the Peano axioms. Find an isomorphism f: $\mathbf{N} \to E$.

§7.2 Arithmetic and Order Properties of the Natural Numbers

At this point we have a set **N** that satisfies the Peano axioms. We are writing **N** as $\mathbf{N} = \{0, 1, 2, \dots\}$, using the familiar base 10 numerals. We are not assuming, however, any knowledge of how to add or multiply elements of **N**. In fact, our next task is to use the Peano axioms (and the Recursion Theorem) to define addition and multiplication on **N**. We shall subsequently show that these operations possess the commutative, associative, and distributive properties.

Let us set the ground rules. The set **N**, together with its successor function s, satisfies Axioms P1–P5. Let s be the successor function on **N**. For the present, it is convenient to write $s(n) = n^+$ for all $n \in \mathbf{N}$. With this notation, Axiom P5 reads: If A is a subset of **N** such that (i) $0 \in A$ and (ii) $n^+ \in A$ whenever $n \in A$, then $A = \mathbf{N}$.

The following definitions of addition and multiplication rely on Theorems 7.1.2 and 7.1.3.

Definition 7.2.1 *Addition in N*

Let $n, m \in \mathbf{N}$. We define the *sum of n and m*, written $n + m$, as follows:

(i) If $m = 0$, then $n + m = n$.

(ii) If $m \neq 0$, then $m = k^+$ for some $k \in \mathbf{N}$ and $n + m = (n + k)^+$.

For example, $n + 1 = n + 0^+ = (n + 0)^+ = n^+$ and $n + 2 = n + 1^+ = (n + 1)^+ = (n^+)^+$. The operation $+$, as given in Definition 7.2.1, is clearly a binary operation on **N** and is called *addition* in **N**. We now use the Peano axioms to prove that addition is associative and commutative.

Theorem 7.2.1. *Addition in* **N** *is associative.*

Proof We prove that for all $n, m, k \in \mathbf{N}$, $(n + m) + k = n + (m + k)$. We must prove this fact using only the Peano axioms, the most powerful of which is P5, the Principle of Mathematical Induction. Since three variables are involved in the statement of the associative law, it is not altogether clear how to use P5 to prove associativity. In this case, the key is to use induction on one of the variables and to quantify matters properly.

Define $A = \{k \in \mathbf{N} | \text{For all } n, m \in \mathbf{N}, (n + m) + k = n + (m + k)\}$. We use P5 to show that $A = \mathbf{N}$, thereby proving that the associative property holds in $\mathbf{N}$.

Basis step We show that $0 \in A$. Let $n, m \in \mathbf{N}$. By Definition 7.2.1, $(n + m) + 0 = n + m$ while $n + (m + 0) = n + m$. Thus, $0 \in A$.

Inductive step Suppose $k \in A$. We show that $k^+ \in A$. By the definition of A, $k^+ \in A$ if and only if $(n + m) + k^+ = n + (m + k^+)$. Now, by Definition 7.2.1(ii) (applied three times) and by inductive hypothesis,

$$\begin{aligned}(n + m) + k^+ &= ((n + m) + k)^+ \\ &= (n + (m + k))^+ \\ &= n + (m + k)^+ \\ &= n + (m + k^+)\end{aligned}$$

■

We next turn to the commutativity of addition in $\mathbf{N}$.

Theorem 7.2.2 *Addition in* $\mathbf{N}$ *is commutative.*

Proof We must show that for all $m, n \in \mathbf{N}$, $n + m = m + n$. A reasonable way of attacking this proof is to mimic the previous proof. To do so, define

$$B = \{m \in \mathbf{N} | \text{For all } n \in \mathbf{N}, n + m = m + n\}.$$

Our goal is to show that $B = \mathbf{N}$ by verifying that (1) $0 \in B$, and (2) if $k \in B$, then $k^+ \in B$.

To prove that $0 \in B$, we must check that $n + 0 = 0 + n$ for all $n \in \mathbf{N}$. By Definition 7.2.1(i), $n + 0 = n$; hence we must show that $0 + n = n$ for all $n \in \mathbf{N}$. This fact is established in the next lemma, whose proof is (not surprisingly) carried out by induction.

Lemma 7.2.3 *For all* $n \in \mathbf{N}$, $0 + n = n$.

Proof Let $A' = \{n \in \mathbf{N} | 0 + n = n\}$. We show that A' satisfies P5, which proves that $A' = \mathbf{N}$.

Basis step Since $0 + 0 = 0$, $0 \in A'$.

Inductive step Suppose $n \in A'$. We show that $n^+ \in A'$. By Definition 7.2.1(ii) and by inductive hypothesis

$$0 + n^+ = (0 + n)^+ = n^+.$$

Therefore, $n^+ \in A'$.

By P5, $A' = \mathbf{N}$. ■

As it happens, to complete the inductive step of the proof of the theorem, we shall need the special case $m = 1$.

Lemma 7.2.4 *For all $n \in \mathbf{N}$, $1 + n = n + 1$.*

Proof Exercise.

We are ready to complete the proof of Theorem 7.2.2 by showing that if $m \in B$, then $m^+ \in B$. We must prove that for all $n \in \mathbf{N}$, $n + m^+ = m^+ + n$. Now

$$\begin{aligned} n + m^+ &= (n + m)^+ \\ &= (m + n)^+ \\ &= m + n^+ \\ &= m + (n + 1) \\ &= m + (1 + n) \\ &= (m + 1) + n \\ &= m^+ + n. \end{aligned}$$ ■

Notice that the proof of the commutative property of addition uses the associativity of addition.

To summarize our work thus far, we have defined a binary operation, $+$, on $\mathbf{N}$ and have proved that this operation is associative, is commutative, and possesses an identity. We complete our discussion of the arithmetic operations on $\mathbf{N}$ by defining multiplication and proving the associative and commutative properties for multiplication and the distributivity of multiplication over addition.

Definition 7.2.2 *Multiplication in* $\mathbf{N}$

Let $n, m \in \mathbf{N}$. We define the *product of n and m*, written $n \cdot m$, as follows:

(i) If $m = 0$, then $n \cdot m = 0$.

(ii) If $m \neq 0$, then $m = k^+$ for some $k \in \mathbf{N}$, and $n \cdot m = n \cdot k^+ = n \cdot k + n$.

Notice that from Definition 7.2.2 it follows that for all $n \in \mathbf{N}$, $n \cdot 1 = n \cdot 0^+ = n \cdot 0 + n = n$. We begin with the distributive property that is used in the proof of commutativity. Actually, we prove that two distributive laws hold in $\mathbf{N}$.

Theorem 7.2.5 *For all* $n, m, k \in \mathbf{N}$,
(i) $n \cdot (m + k) = n \cdot m + n \cdot k$;
(ii) $(n + m) \cdot k = n \cdot k + m \cdot k$.

Proof (i) Let $A = \{k \in \mathbf{N} \mid \text{For all } n, m \in \mathbf{N},\ n \cdot (m + k) = n \cdot m + n \cdot k\}$. We verify that A satisfies the hypotheses of P5.

Basis step $0 \in A$ since for all $n, m \in \mathbf{N}$, $n \cdot (m + 0) = n \cdot m = n \cdot m + 0 = n \cdot m + n \cdot 0$.

Inductive step Suppose $k \in A$. We show $k^+ \in A$:

$$\begin{aligned} n \cdot (m + k^+) &= n \cdot (m + k)^+ \\ &= n \cdot (m + k) + n \\ &= (n \cdot m + n \cdot k) + n \\ &= n \cdot m + (n \cdot k + n) \\ &= n \cdot m + n \cdot k^+. \end{aligned}$$

Thus $k^+ \in A$.

By P5, $A = \mathbf{N}$.

The proof of (ii) is left as an exercise. ■

We now state the associative law for multiplication. The proof of the associative property is similar to the proofs that have already been presented in this section and is left as an exercise.

Theorem 7.2.6 *For all* $n, m, k \in \mathbf{N}$, $n \cdot (m \cdot k) = (n \cdot m) \cdot k$.

The final arithmetic property that we establish in **N** is commutativity of multiplication.

Theorem 7.2.7 *For all* $n, m \in \mathbf{N}$, $n \cdot m = m \cdot n$.

In order to prove Theorem 7.2.7, we shall require two special cases of the theorem. We state these results in the following lemma, whose proof is left as an exercise.

Lemma 7.2.8 (i) *For all* $n \in \mathbf{N}$, $n \cdot 0 = 0 \cdot n = 0$. (ii) *For all* $n \in \mathbf{N}$, $n \cdot 1 = 1 \cdot n = n$.

Proof of Theorem 7.2.7. Let $A = \{m \in \mathbf{N} \mid \text{For all } n \in \mathbf{N},\ n \cdot m = m \cdot n\}$. By Lemma 7.2.8(i), $0 \in A$; thus to show that $A = \mathbf{N}$, we must prove that if $m \in A$, then $m^+ \in A$.

Suppose that for all $n \in \mathbf{N}$, $n \cdot m = m \cdot n$. Then

$$\begin{aligned} n \cdot m^{+} &= n \cdot m + n && \text{by the definition of mutliplication,} \\ &= m \cdot n + n && \text{by the inductive hypothesis,} \\ &= m \cdot n + 1 \cdot n && \text{by Lemma 7.2.8(ii),} \\ &= (m + 1) \cdot n && \text{by Theorem 7.2.5(ii),} \\ &= m^{+} \cdot n. \end{aligned}$$

■

Order in N

Using the operation of addition in **N**, we can define several order relations on **N**.

Definition 7.2.3 *Order in N*

Let $n, m \in \mathbf{N}$.

(i) n is *less than* m, written $n < m$, if there exists $k \in \mathbf{N}$, $k \neq 0$ such that $n + k = m$.

(ii) n is *less than or equal to* m, written $n \leq m$, if $n < m$ or $n = m$.

(iii) n is *greater than* (*or equal to*) m, written $n > m$ ($n \geq m$) if $m < n$ ($m \leq n$).

From the definition it follows immediately that for all $n \in \mathbf{N}$, $0 \leq n$. Thus 0 is the *smallest* element of **N**. It also follows easily that for all $n \in \mathbf{N}$, $n < n^{+}$. Our principal goal is to prove the trichotomy property for $<$: For all $n, m \in \mathbf{N}$, exactly one of the following holds, $n < m$, $n = m$, or $n > m$. Note that the trichotomy property holds if and only if the relation $\leq$ is a total ordering on **N**. We shall derive the trichotomy property from the following important property of the natural number system.

Theorem 7.2.9 *The Well-Ordering Principle. If A is a nonempty subset of* **N**, *then there exists $n_0 \in A$ such that for all $n \in A$, $n_0 \leq n$.*

In words, the Well-Ordering Principle (WOP for short) asserts that *every nonempty subset of* **N** *contains a least element*. Not surprisingly, the proof of WOP depends on Axiom P5.

Proof We argue by contraposition. Suppose A is a subset of **N** that does not contain a least element. Consider the set $B = \{m \in \mathbf{N} \mid m \leq n$ for all $n \in A\}$. We use P5 to show that $B = \mathbf{N}$.

Basis step $0 \in B$ since for all $n \in \mathbf{N}$, $0 \leq n$.

Inductive step Suppose $m \in B$. Then $m \leq n$ for all $n \in A$. If $m \in A$, then m would be the least element of A, which is impossible by our assumption. Thus $m \notin A$ and $m < n$ for all $n \in A$. It follows that $m^+ \leq n$ for all $n \in A$ (see Exercise 7.2.7), and hence $m^+ \in B$.

Therefore, $B = \mathbf{N}$, which implies that for all $m \in \mathbf{N}$, $m \leq n$ for all $n \in A$. This condition holds only if $A = \varnothing$. (Proof is left as an exercise.) This contradiction implies that A must have a least element. ∎

We next illustrate a typical application of WOP by proving the trichotomy property in **N**. As with most, if not all, proofs using WOP, we can show that all natural numbers have a certain property using an indirect argument. We suppose that there exists at least one natural number that does not have the desired property. By WOP there exists a least such natural number. We then use this number to derive a contradiction. For example, we can handle Exercises 7.2.10 and 7.2.11 nicely by WOP.

Theorem 7.2.10 *The Trichotomy Law in* **N**. *Let* $n, m \in \mathbf{N}$. *Then exactly one of the following holds*: $n < m$, $n = m$, *or* $n > m$.

Proof We begin by proving that at least for each pair of natural numbers, at least one of the three conditions given in the theorem holds. We argue by contradiction: We assume that the condition fails to hold and use WOP to derive a contradiction.

Let $A = \{n \in \mathbf{N} \mid \text{There exists } m \in \mathbf{N} \text{ such that } n \not< m,\ n \neq m, \text{ and } n \not> m\}$.

Suppose $A \neq \varnothing$. By WOP, A has a least element n_0. From the definition of A, there exists $m_0 \in \mathbf{N}$ such that $n_0 \not< m_0$, $n_0 \neq m_0$, and $n_0 \not> m_0$. Note that $n_0 \neq 0$ since for all $m \in \mathbf{N}$ either $0 = m$ or $0 < m$. Thus there exists $k \in \mathbf{N}$ such that $n_0 = k^+$. Since $k < n_0$, $k \notin A$, and hence either $k < m_0$, $k = m_0$, or $k > m_0$. We consider the cases separately.

Case 1. $k < m_0$. Then $k + a = m_0$ for some $a \in \mathbf{N}$ with $a \neq 0$. If $a = 1$, then $n_0 = k^+ = m_0$, which is impossible. If $a \neq 1$, then $a > 1$ and $a = b^+$ for some $b \in \mathbf{N}$ with $b \neq 0$. Thus $m_0 = k + a = k + b^+ = k^+ + b = n_0 + b$, which means that $n_0 < m_0$. This conclusion contradicts our assumption on n_0.

Case 2. $k = m_0$. Then $n_0 = k^+ = m_0^+$, which means that $m_0 < n_0$, yet another contradiction.

Case 3. $k > m_0$. Then there exists $a \in \mathbf{N}$, $a \neq 0$ such that $k = m_0 + a$. Thus $n_0 = k^+ = (m_0 + a)^+ = m_0 + a^+$, implying that $m_0 < n_0$.

Since each of Cases 1–3 leads to a contradiction, our original assumption that $A \neq \varnothing$ must be invalid. Thus $A = \varnothing$. It follows that for all $n, m \in \mathbf{N}$, either $n < m$, $n = m$, or $n > m$.

We now argue that at most one of these conditions can hold. For example, suppose $n < m$ and $m < n$. Then $m = n + k$ for some $k \in \mathbf{N}$, $k \neq 0$, and $n = m + j$ for some $j \in \mathbf{N}$, $j \neq 0$. Thus $n + 0 = n = m + j = (n + k) + j = n + (k + j)$, which implies that $k + j = 0$. (See Exercise 7.2.13.) But by Exercise 7.2.3, if $k > 0$ and $j > 0$, then $k + j \neq 0$. Thus the inequalities $n < m$ and $m < n$ cannot hold simultaneously. A similar argument shows that neither $n < m$ and $n = m$, nor $m < n$ and $m = n$ can hold simultaneously. ∎

We have succeeded in our goal of using the Peano axioms to define the binary operations of addition and multiplication on **N** and the order relations $<$ and $\leq$ on **N**. We have also used the Peano axioms to establish some fundamental properties of these operations and relations. Our next goal is to use the set **N** of natural numbers to define the set **Z** of integers. Once the set **Z** is defined, we shall define binary operations on **Z** based on the binary operations defined on **N** in this section. Thus we are constructing **Z** from **N**. Later, we shall construct the set of rational numbers **Q** from **Z** and then build the set of real numbers **R** from **Q**. As we saw in Chapter 6, the set of complex numbers, **C**, can be defined in terms of **R**. In effect, we have a bootstrap construction (see Chapter 1, Section 3) of the complex number system beginning with the natural numbers.

Exercises §7.2

7.2.1. Prove Lemma 7.2.4.

7.2.2. (a) Prove by induction that for all $n, m \in \mathbf{N}$, $n + m^+ = n^+ + m$.
(b) Prove, using associativity and commutativity of addition, that for all $n, m \in \mathbf{N}$, $n + m^+ = n^+ + m$.

7.2.3. Show: If $n, m \in \mathbf{N}$ with $n \neq 0$, then $n + m \neq 0$.

7.2.4. (a) Prove: For all $n \in \mathbf{N}$, $0 \cdot n = n \cdot 0 = 0$.
(b) Prove: For all $n \in \mathbf{N}$, $1 \cdot n = n \cdot 1 = n$.

7.2.5. Prove Theorem 7.2.5(ii) by induction.

7.2.6. Prove Theorem 7.2.6.

7.2.7. (a) Prove: If $n < m$, then $n^+ \leq m$.
(b) Prove: If $n \in \mathbf{N}$ and $n \neq 0, 1$, then $n > 1$.

7.2.8. Use WOP to prove the Principle of Strong Induction: If $A \subseteq \mathbf{N}$ such that (i) $0 \in A$, and (ii) $n^+ \in A$ whenever $0, 1, \ldots, n \in A$, then $A = \mathbf{N}$.

7.2.9. (a) Prove: If $n, m \in \mathbf{N}$ with $n \neq 0$ and $m \neq 0$, then $n \cdot m \neq 0$.
(b) Prove: If $n, m, k \in \mathbf{N}$ with $k \neq 0$ and $n \cdot k \leq m \cdot k$, then $n \leq m$.

7.2.10. Use WOP to prove that every integer $n \geq 2$ is divisible by a prime.

7.2.11. Use WOP to prove that every integer $n \geq 2$ either is a prime or is expressible as a product of primes.

7.2.12. Prove that the product of two nonzero natural numbers is nonzero.

7.2.13. Prove that if $n, m, k \in \mathbf{N}$ and $n + k = m + k$, then $n = m$.

7.2.14. Prove that if $n, m, k \in \mathbf{N}$ and $n < m$ and $m < k$, then $n < k$.

7.2.15. Prove that if $n, m \in \mathbf{N}$ with $n \neq 0$, then there exists $k \in \mathbf{N}$ such that $n \cdot k > m$.

§7.3 The Integers

To summarize our work thus far, we have described a list of axioms, the Peano axioms, for the set of natural numbers. We have assumed the existence of a set **N** satisfying the Peano axioms. On the basis of these axioms, we introduced the binary operations of addition and multiplication and the order relation $<$ and established some basic arithmetic and order properties of **N**. In particular, we proved that addition is an associative, commutative binary relation for which an identity element, namely 0, exists.

If you have studied Section 6.5, you will recall that a group is an algebraic system $(G; *)$ consisting of a set G and a binary operation $*$ satisfying:

1. $*$ is associative: For all $x, y, z \in G$, $(x * y) * z = x * (y * z)$.
2. $*$ has an identity: There exists $e \in G$ such that for all $x \in g$, $x * e = e * x = x$.
3. $*$ has inverses: For each $x \in G$, there exists $y \in G$ such that $x * y = y * x = e$.

Notice that the algebraic system $(\mathbf{N}; +)$ is not a group since statement 3 does not hold for $(\mathbf{N}; +)$. For example, for no $y \in \mathbf{N}$ is it true that $1 + y = y + 1 = y^{+} = 0$. (Note that 0 is an identity element for $+$ in **N**.) Therefore, statement 3 fails to hold for $x = 1$. Since statements 1 and 2 do hold for $(\mathbf{N}; +)$, $(\mathbf{N}; +)$ fails to be a group because inverses do not exist for each element of **N**. It is natural to ask if the set **N** can be "enlarged" into a set that is a group under an operation with the property that this new operation coincides with addition of natural numbers when restricted to natural numbers. From our experience with real numbers, we might surmise that this "enlargement" of **N** is the set of integers, **Z**.

In this section we address the problem of defining the set of integers, **Z**, defining the algebraic operations and order relations in **Z**, and establishing the many properties of these operations and relations.

In order to obtain a rigorous definition of **Z**, let us look back at our informal way of describing **Z**: $\mathbf{Z} = \{0, \pm 1, \pm 2, \pm 3, \ldots\}$. The number -3 can be obtained in several ways: $-3 = 1 + (-4) = 1 - 4 = 2 - 5 = 3 - 6$. Thus -3 can be determined from several pairs of natural numbers $(1, 4), (2, 5), (3, 6), \ldots$. Of course, the notion of "subtraction" has not been rigorously defined, but the key idea is that an integer such as -3 can be represented by a pair of natural numbers and several pairs of natural numbers can represent -3.

With these remarks in mind, we consider $A = \mathbf{N} \times \mathbf{N}$, the set of all ordered pairs of natural numbers. On A we wish to define a relation that expresses the fact that the pairs (a, b) and (c, d) are equivalent of $a - b =$

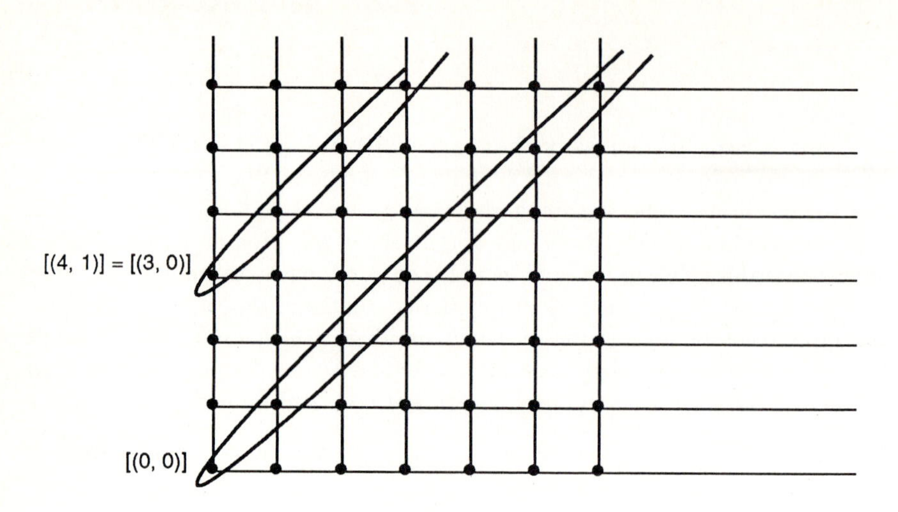

Figure 7.3.1

$c - d$. As noted above, subtraction has not been defined in **N**; we can, however, express the notion of equivalence in terms of addition, an operation that has been defined. Thus, on a, we *define* a relation $\sim$ as follows: $(a, b) \sim (c, d)$ if $a + d = b + c$. For example, $(1, 4) \sim (2, 5)$, $(1, 4) \sim (3, 6)$ and $(2, 5) \sim (3, 6)$. The next theorem describes the most significant property of $\sim$.

Theorem 7.3.1 *The relation* $\sim$ *is an equivalence relation on* $\mathbf{N} \times \mathbf{N}$. *Moreover, the set of equivalence classes,* $\mathbf{N} \times \mathbf{N}/\sim$, *is the set*: $\mathbf{N} \times \mathbf{N}/\sim \; = \{[(n, 0)] | n \in \mathbf{N}\} \cup \{[(0, n)] | n \in \mathbf{N},\ n \neq 0\}$.

Proof We leave the proof that $\sim$ is an equivalence relation as an exercise (see Exercise 3.3.8). Proving the second assertion amounts to showing that $B = \{(n, 0) | n \in \mathbf{N}\} \cup \{(0, n) | n \in \mathbf{N} - \{0\}\}$ is a complete set of representatives for $\mathbf{N} \times \mathbf{N}/\sim$, i.e., that each element of $\mathbf{N} \times \mathbf{N}$ is equivalent to exactly one element of B. (See Figure 7.3.1.)

Let $(a, b) \in \mathbf{N} \times \mathbf{N}$. We consider three cases:

Case 1. $a = b$. Then $(a, b) = (a, a) \sim (0, 0) \in B$.

Case 2. $a > b$. Then there exists $k \in \mathbf{N}$ such that $a = b + k$. Thus $(a, b) \sim (k, 0) \in B$.

Case 3. $a < b$. Then there exists $k \in \mathbf{N}$ such that $b = a + k$ and thus $(a, b) \sim (0, k) \in B$.

Hence each element of $\mathbf{N} \times \mathbf{N}$ is equivalent to at least one element of B. Since no two elements of B are equivalent (as can easily be checked by considering cases), each element of $\mathbf{N} \times \mathbf{N}$ is equivalent to exactly one element of B and B is a complete set of representatives of $\mathbf{N} \times \mathbf{N}/\sim$. ■

Definition 7.3.1 *The Integers*

The set $\mathbf{N} \times \mathbf{N}/\sim$ is the *set of integers* and is denoted by $\mathbf{Z}$.

By Theorem 7.3.1, $\mathbf{Z} = \{[(n,0)] | n \in \mathbf{N}\} \cup \{[(0,n)] | n \in \mathbf{N} - \{0\}\}$. Notice that the set $\mathbf{Z}$ is indeed built from $\mathbf{N}$ using the axioms or constructions, if you will, of set theory. In practice, the notation $[(n,0)]$ is quite cumbersome. We simplify matters by identifying $[(n,0)]$ with the natural number n, and then writing $[(n,0)] = n$. We also introduce the symbol $-n$ and write $[(0,n)] = -n$. (Note that $-0 = 0$.) With this notation, we have $\mathbf{Z} = \{n | n \in \mathbf{N}\} \cup \{-n | n \in \mathbf{N} - \{0\}\}$.

Arithmetic Operations in Z

We have defined $\mathbf{Z}$ to be the set of equivalence classes of $\mathbf{N} \times \mathbf{N}$ with respect to the equivalence relation $\sim$. Let us now turn to the arithmetic operations in $\mathbf{Z}$.

We want to define the sum and product of two elements of $\mathbf{Z}$. Let $x, y \in \mathbf{Z}$. Then from the definition of $\mathbf{Z}$ (to which we must appeal if we are to be precise), $x = [(a,b)]$ and $y = [(c,d)]$ for some $a, b, c, d \in \mathbf{N}$. How shall we define $x + y$ and $x \cdot y$? For guidance, let us revert to our informal picture of $[(a,b)]$ as $a - b$. Then $x + y = (a - b) + (c - d)$, which should equal $(a + c) - (b + d)$. Also $x \cdot y = (a - b) \cdot (c - d)$, which should equal $a \cdot c + b \cdot d - a \cdot d - b \cdot c = (a \cdot c + b \cdot d) - (a \cdot d + b \cdot c)$. These comments motivate the following definition.

Definition 7.3.2 *Addition and Multiplication in* $\mathbf{Z}$

Let $x = [(a,b)]$ and $y = [(c,d)]$ be elements of $\mathbf{Z}$. Then the elements $x + y$ and $x \cdot y$ of $\mathbf{Z}$, called, respectively, the *sum* and *product* of x and y, are defined to be

$$x + y = [(a + c, b + d)]$$

$$x \cdot y = [(a \cdot c + b \cdot d, a \cdot d + b \cdot c)].$$

Let us compute a few examples. For instance, let us find $7 + (-2)$. Since $7 = [(7,0)]$ and $-2 = [(0,2)]$, $7 + (-2) = [(7 + 0, 0 + 2)] = [(7,2)] = [(5,0)] = 5$. Also, $7 \cdot (-2) = [(7,0)] \cdot [(0,2)] = [(7 \cdot 0 + 0 \cdot 2, 7 \cdot 2 + 0 \cdot 0)] = [(0,14)] = -14$. So far, so good.

While the definitions of addition and multiplication seem clear enough, they do suffer from a potential defect. Consider again the sum $7 + (-2)$. Suppose that we had represented 7 as $[(10,3)]$ and -2 as $[(18,20)]$. Then is it true that $[(10,3)] + [(18,20)] = [(5,0)] = 5$? Let us check:

$$[(10,3)] + [(18,20)] = [(10 + 18, 3 + 20)] = [(28,23)] = [(5,0)].$$

This is fine, but a problem arises. We know that the element 7 in $\mathbf{Z}$ can be represented in infinitely many ways: $7 = [(7,0)] = [(8,1)] = \dots$; similarly -2 can be represented in infinitely many ways. Do we know that no matter how the integers 7 and -2 are represented, then sum $7 + (-2)$, given by Definition 7.3.2, will always yield the integer $5 = [(5,0)]$? The following proposition answers this question.

Proposition 7.3.2 If (a, b), (a', b'), (c, d), $(c', d') \in \mathbf{N} \times \mathbf{N}$ and $(a, b) \sim (a', b')$ and $(c, d) \sim (c', d')$, then
(i) $(a + c, b + d) \sim (a' + c', b' + d')$, and
(ii) $(a \cdot c + b \cdot d, a \cdot d + b \cdot c) \sim (a' \cdot c' + b' \cdot d', a' \cdot d' + b' \cdot c')$.

Proof (i) Since $(a, b) \sim (a', b')$ and $(c, d) \sim (c', d')$, $a + b' = b + a'$ and $c + d' = d + c'$. Therefore

$$(a + b') + (c + d') = (b + a') + (d + c').$$

By the associative and commutative laws of addition in $\mathbf{N}$,

$$(a + c) + (b' + d') = (b + d) + (a' + c'),$$

which implies that $(a + c, b + d) \sim (a' + c', b' + d')$.

(ii) We must show that if $(a, b) \sim (a', b')$ and $(c, d) \sim (c', d')$, then $(a \cdot c + b \cdot d, a \cdot d + b \cdot c) = (a' \cdot c' + b' \cdot d', a' \cdot d' + b' \cdot c')$. In other words, we must prove that

$$(a \cdot c + b \cdot d) + (a' \cdot d' + b' \cdot c') = (a \cdot d + b \cdot c) + (a' \cdot c' + b' \cdot d'),$$

given that $a + b' = b + a'$ and $c + d' = d + c'$.

From $a + b' = b + a'$ and the distributive property, we have

$$a \cdot c + b' \cdot c = b \cdot c + a' \cdot c$$

and

$$a \cdot d + b' \cdot d = b \cdot d + a' \cdot d.$$

From this set of equations it follows (from the associativity and commutativity of addition) that

$$(1) \quad a \cdot c + b \cdot d + b' \cdot c + a' \cdot d = a \cdot d + b \cdot c + a' \cdot c + b' \cdot d.$$

Similarly, from $c + d' = d + c'$, we find

$$a' \cdot c + a' \cdot d' = a' \cdot d + a' \cdot c'$$

and

$$b' \cdot c + b' \cdot d' = b' \cdot d + b' \cdot c'.$$

Therefore,

$$(2) \quad a' \cdot d' + b' \cdot c' + a' \cdot c + b' \cdot d = a' \cdot c' + b' \cdot d' + a' \cdot d + b' \cdot c.$$

Adding the left sides of (1) and (2) and the right sides of (1) and (2), we find that

$$\begin{aligned} a \cdot c + b \cdot d + a' \cdot d' + b' \cdot c' + b' \cdot c + a' \cdot d + a' \cdot c + b' \cdot d \\ = a \cdot d + b \cdot c + a' \cdot c' + b' \cdot d' + a' \cdot c + b' \cdot d + a' \cdot d + b' \cdot c. \end{aligned}$$

From the cancellation property of addition in **N** (Exercise 7.2.13), it follows that

$$(a \cdot c + b \cdot d) + (a' \cdot d' + b' \cdot c') = (a \cdot d + b \cdot c) + (a' \cdot c' + b' \cdot d').$$

The proof is now complete. ∎

Let us reemphasize the significance of Proposition 7.3.2. The set **Z** is by definition a set of equivalence classes. Specifically, **Z** is the set of equivalence classes of $\mathbf{N} \times \mathbf{N}$ with respect to the equivalence relation $\sim$. To define addition and multiplication on **Z**, we must show how to assign to any pair of elements x, y of **Z**, the elements $x + y$ and $x \cdot y$ in **Z**. The elements x and y, however, are equivalence classes of $\mathbf{N} \times \mathbf{N}$ relative to $\sim$ and the natural way to define the sum (and product) of x and y is in terms of the representatives of x and y, i.e., ordered pairs of natural numbers, using the operations of addition and multiplication in **N**. Nevertheless, even though addition and multiplication in **Z** must be defined in terms of representatives of the equivalence classes, we have to prove that the results of these operations depend only on the equivalence classes and not on the representatives of the equivalence classes. Proposition 7.3.2 establishes this statement. In this case, we say that $+$ and $\cdot$ are *well-defined* operations on **Z**.

The next theorem lists the basic properties of addition and multiplication in **Z**. In the language of Chapter 6, properties 1–4 assert that $(\mathbf{Z}; +)$ is an abelian group and properties 1–9 show that $(\mathbf{Z}; +, \cdot)$ is an integral domain. The proof of Theorem 7.3.3 is left as an exercise.

Theorem 7.3.3 *Let* $x, y, z \in \mathbf{Z}$.

1. $(x + y) + z = x + (y + z)$.
2. $x + 0 = 0 + x = x$ *where* $0 = [(0,0)]$.
3. *For all* $x \in \mathbf{Z}$, *there exists a unique* $x' \in \mathbf{Z}$ *such that* $x + x' = x' + x = 0$. *Henceforth, we denote* x' *by* $-x$.
4. $x + y = y + z$.
5. $(x \cdot y) \cdot z = x \cdot (y \cdot z)$.
6. $x \cdot 1 = 1 \cdot x = x$ *where* $1 = [(1,0)]$.
7. $x \cdot y = y \cdot x$.
8. $x \cdot (y + z) = x \cdot y + x \cdot z$.
9. *If* $x \neq 0$ *and* $y \neq 0$, *then* $x \cdot y \neq 0$.

Order in Z

We conclude this section with a discussion of order in $\mathbf{Z}$. First, we single out the set of positive integers, $\mathbf{Z}^+$, and then we use $\mathbf{Z}^+$ to define a total ordering in $\mathbf{Z}$. It is not surprising that the idea of order in $\mathbf{Z}$ depends on the idea of order in $\mathbf{N}$.

Definition 7.3.3 *Order in* $\mathbf{Z}$

(i) Define $\mathbf{Z}^+ = \{[(n,0)] | n \in \mathbf{N} - \{0\}\}$. $\mathbf{Z}^+$ is called the set of positive integers.

(ii) If $x, y \in \mathbf{Z}$, then x *is less than* y, written $x < y$, if $y - x = y + (-x) \in \mathbf{Z}^+$.

(iii) If $x, y \in \mathbf{Z}$, then x *is less than or equal to* y, written $x \leq y$, if $x < y$ or $x = y$.

(iv) If $x, y \in \mathbf{Z}$, then x *is greater than* (*or equal to*) y, written $x > y$ $(x \geq y)$, if $y < x$ $(y \leq x)$.

Theorem 7.3.4

(i) *Closure under* $+$ *and* $\cdot$. *If* $x, y \in \mathbf{Z}^+$, *then* $x + y$, $x \cdot y \in \mathbf{Z}^+$.

(ii) *If* $x \in \mathbf{Z}$, *then exactly one of the following holds*: $x \in \mathbf{Z}^+$, $x = 0$, *or* $-x \in \mathbf{Z}^+$.

(iii) *Trichotomy Law*. *If* $x, y \in \mathbf{Z}$, *then exactly one of the following holds*: $x < y$, $x = y$, *or* $x > y$.

Proof (i) Let $x, y \in \mathbf{Z}^+$. Then $x = [(n,0)]$ and $y = [(m,0)]$ where $n, m \in \mathbf{N} - \{0\}$.

Since $m \in \mathbf{N} - \{0\}$, $m = k^+$ for some $k \in \mathbf{N}$ by Theorem 7.1.1. Thus $n + m = n + k^+$ for some $k \in \mathbf{N}$ by Theorem 7.1.1. Hence in $\mathbf{N}$, $n + m = n + k^+ = (n + k)^+ \neq 0$ by Axiom P3. Therefore, $x + y = [(n + m, 0)] \in \mathbf{Z}^+$. Also, in $\mathbf{N}$, $n \cdot m = n \cdot k^+ = n \cdot k + n \neq 0$ since $n \neq 0$. Therefore, $x \cdot y = [(n \cdot m, 0)] \in \mathbf{Z}^+$.

(ii) If $x \in \mathbf{Z}$, then, by Theorem 7.3.1, either $x = [(n, 0)]$ where $n \in \mathbf{N} - \{0\}$, $x = [(0, 0)] = 0$, or $x = [(0, n)]$ where $n \in \mathbf{N} - \{0\}$. In the first case, $x \in \mathbf{Z}^+$; in the second case, $x = 0$; and in the third case $-x = [(n, 0)] \in \mathbf{Z}^+$.

(iii) Statement (iii) follows immediately from (ii). ∎

One final comment on the relationship between the sets $\mathbf{N}$ and $\mathbf{Z}$. Observe that the set $\mathbf{Z}_0 = \{[(n, 0)] | n \in \mathbf{N}\} = \{[(0, 0)]\} \cup \mathbf{Z}^+$ is quite similar in its arithmetic and order structure to $\mathbf{N}$. In fact, it is easy to check that the function $f\colon \mathbf{N} \to \mathbf{Z}_0$ defined by $f(n) = [(n, 0)]$ is a bijection that preserves both addition and multiplication: For all $n, m \in \mathbf{N}$, $f(n + m) = f(n) + f(m)$ and $f(n \cdot m) = f(n) \cdot f(m)$. In the language of Chapter 6, f is an isomorphism. Because of this bijection, we can regard $\mathbf{N}$ and $\mathbf{Z}_0$ as being identical, in effect regarding $\mathbf{N}$ as a subset of $\mathbf{Z}$. This fact justifies our writing $[(n, 0)] = n$. Since $[(0, n)]$ is the additive inverse of $[(n, 0)]$, we write $[(0, n)] = -n$. Another way of phrasing these remarks is to say that $\mathbf{Z}_0$ is a model for the Peano axioms. As a result any property that is provable from the Peano axioms holds for $\mathbf{Z}_0$. In particular, as proved in Chapter 6, the unique factorization property for multiplication holds in $\mathbf{Z}_0$ and the ring $\mathbf{Z}$ is a unique factorization domain. Finally, we can rephrase these remarks in yet another way. We have formed the set $\mathbf{Z}$ by adding to the set $\mathbf{N}$ the additive inverses of all the nonzero elements of $\mathbf{N}$: $\mathbf{Z} = \mathbf{N} \cup \{x | \text{There exists } n \in \mathbf{N}, n \neq 0, \text{ such that } x \text{ is the additive inverse of } n\}$. Thus we have resolved the issue raised at the start of this section by showing how to "enlarge" $\mathbf{N}$ to form the set $\mathbf{Z}$ in which the operation of addition possesses inverses.

In the next section we use the set $\mathbf{Z}$ to construct the set of rational numbers $\mathbf{Q}$. We then use the concepts of addition, multiplication, and order in $\mathbf{Z}$ to define analogous concepts in $\mathbf{Q}$.

Exercises §7.3

7.3.1. Prove: For all $x \in \mathbf{Z}$, $x \cdot 0 = 0$.

7.3.2. Prove Theorem 7.3.3.

7.3.3. Prove: For $x, y, z \in \mathbf{Z}$, $x < y$ if and only if $x + z < y + z$.

7.3.4. (a) Prove: $(-1) \cdot (-1) = 1$.
(b) Prove: For all $x \in \mathbf{Z}$, $-x = (-1) \cdot x$.
(c) Prove: For all $x, y \in \mathbf{Z}$, $x \cdot y = (-x) \cdot (-y)$.
(d) Prove: For all $x, y \in \mathbf{Z}$, $-x \cdot y = (-x) \cdot y$.

7.3.5. Prove: If $x \in \mathbf{Z}$, then $x \cdot x = x^2 \geq 0$.

7.3.6. Prove: If $x, y \in \mathbf{Z}$ and $z \in \mathbf{Z}^+$, then $x < y$ if and only if $x \cdot z < y \cdot z$.

7.3.7. Prove: If $x = [(n, 0)]$ for $n \in \mathbf{N}$, then $-x = [(0, n)]$.

7.3.8. Define $\mathbf{Z}^- = \{x \in \mathbf{Z} \mid -x \in \mathbf{Z}^+\}$.

(a) Prove: If $x, y \in \mathbf{Z}^-$, then $x + y \in \mathbf{Z}^-$.

(b) Prove: If $x, y \in \mathbf{Z}^-$, then $x \cdot y \in \mathbf{Z}^+$.

(c) Prove: If $x \in \mathbf{Z}^+$ and $y \in \mathbf{Z}^-$, then $x \cdot y \in \mathbf{Z}^-$.

§7.4 The Rational Numbers

Our development of the real number system began with the set of natural numbers, **N**. This set, which can be constructed using the Zermelo–Fraenkel axioms of set theory, satisfies the Peano axioms. Using these axioms, we defined addition and multiplication in **N**. We noted that the operation of addition is associative, is commutative, and has 0 as an identity element. In general, however, additive inverses of elements of **N** do not exist in **N**. (In fact, only 0 has an additive inverse in **N**.)

To remedy this situation, we constructed the set of integers **Z** with the property that under addition **Z** is associative, is commutative, has identity elements, and has inverses. Thus **Z** is an abelian group under addition. Moreover, **Z** contains a subset, $\mathbf{Z}_0$, that also satisfies the Peano axioms and is isomorphic to **N**. Finally, $\mathbf{Z} = \mathbf{Z}_0 \cup \{-n | n \in \mathbf{Z}_0 - \{0\}\}$. In effect, we enlarged **N** by adjoining the additive inverses of all nonzero elements of **N**. As it happens, the set **Z** has many other pleasant properties, the most notable being that **Z** is a unique factorization domain—an integral domain with the property that every integer $\neq 0, \pm 1$ can be factored uniquely into a product of primes. (See Theorem 6.2.1.)

Let us consider the operation of multiplication in **Z** more closely. As we know, multiplication is a binary operation on **Z** which is associative, is commutative, and has 1 as an identity element. As for multiplicative inverses, it is easy to show that only 1 and -1 have multiplicative inverses in **Z**. (The only units in **Z** are ± 1.) Thus we ask: Can the set **Z** be enlarged to obtain a set in which multiplicative inverses of elements of **Z** do exist? More specifically, we want a set F such that

1. two binary operations, $\oplus$ and $\odot$, are defined on F,
2. F contains a subset F_0 that is isomorphic to **Z** in the sense that there exists a bijective function $f\colon \mathbf{Z} \to F_0$ such that $f(x + y) = f(x) \oplus f(y)$ and $f(x \cdot y) = f(x) \odot f(y)$, and
3. each nonzero element of F has a multiplicative inverse.

Of course, we know or at least believe that such a set exists; it is the set of rational numbers, **Q**. Our task is to define this set formally and to establish its properties on the basis of this definition. Our plan is to use the set **Z** along with the apparatus of set theory to construct the set **Q**.

From our past experience, we think of a rational number, x, as being determined by a pair of integers: $x = a/b$ where $a, b \in \mathbf{Z}$ and $b \neq 0$. Moreover, two distinct pairs of integers might determine the same rational number; for example, $1/2 = 2/4$. In general, the fractions a/b and a'/b' are regarded as equal if $a \cdot b' = b \cdot a'$.

To formalize these concepts, we first define

$$\mathscr{F} = \mathbf{Z} \times (\mathbf{Z} - \{0\}) = \{(a, b) | a \in \mathbf{Z}, b \in \mathbf{Z} - \{0\}\}.$$

On $\mathscr{F}$, we define a relation $\equiv$ by the rule:

$$(a, b) \equiv (a', b') \text{ if } a \cdot b' = b \cdot a'.$$

It is easy to check that $\equiv$ is an equivalence relation on $\mathscr{F}$. The resulting set of equivalence classes is, by definition, the set of rational numbers.

Proposition 7.4.1 $\equiv$ is an equivalence relation on $\mathscr{F}$.

Proof Exercise or see Section 3.3.

Definition 7.4.1 *The Rational Numbers*

The set $\mathscr{F}/\equiv$ is the set of *rational numbers* and is denoted by $\mathbf{Q}$.

Arithmetic Operations in Q

We define the binary operations of addition and multiplication in $\mathbf{Q}$. Each of these operations must assign an element of $\mathbf{Q}$ to each pair of elements of $\mathbf{Q}$. Since the elements of $\mathbf{Q}$ are equivalence classes of pairs of integers, the operations of addition and multiplication on $\mathbf{Q}$ will be defined using representatives of the equivalence classes and in terms of addition and multiplication in $\mathbf{Z}$. It will, therefore, be necessary to prove that addition and multiplication in $\mathbf{Q}$ are well defined.

Definition 7.4.2 *Addition and Multiplication in* $\mathbf{Q}$

Let $x = [(a, b)]$ and $y = [(c, d)] \in \mathbf{Q}$. The *sum of x and y*, written $x + y$, is defined to be

$$x + y = [(a \cdot d + b \cdot c, b \cdot d)].$$

The *product of x and y*, written $x \cdot y$, is defined to be

$$x \cdot y = \{(a \cdot c, b \cdot d)\}.$$

We should note that since $b \neq 0$ and $d \neq 0$, $b \cdot d \neq 0$ by Theorem 7.3.3(8). Thus $(a \cdot d + b \cdot c, b \cdot d) \in \mathbf{Z} \times (\mathbf{Z} - \{0\})$ and $(a \cdot c, b \cdot d) \in$

$\mathbf{Z} \times (\mathbf{Z} - \{0\})$, and it makes sense to take the equivalence classes of $(a \cdot d + b \cdot c, b \cdot d)$ and $(a \cdot c, b \cdot d)$. Again, it is worth emphasizing that addition and multiplication in $\mathbf{Q}$ are defined in terms of addition and multiplication in $\mathbf{Z}$. The next result shows that the sum and product of x and y are well defined for all $x, y \in \mathscr{F}/\equiv$.

Proposition 7.4.1 Suppose $(a, b), (c, d), (a', b'), (c', d') \in \mathscr{F}$ with $(a, b) \equiv (a', b')$ and $(c, d) \equiv (c', d')$. Then

$$(a \cdot d + b \cdot c, b \cdot d) \equiv (a' \cdot d' + b' \cdot c', b' \cdot d')$$

and

$$(a \cdot c, b \cdot d) \equiv (a' \cdot c', b' \cdot d').$$

The proposition asserts that if $x = [(a, b)] = [(a', b')]$ and $y = [(c, d)] = [(c', d')]$, then $[(a \cdot d + b \cdot c, b \cdot d)] = [(a' \cdot d' + b' \cdot c', b' \cdot d')]$ and $[(a \cdot c, b \cdot d)] = [(a' \cdot c', b' \cdot d')]$, and hence the equivalence classes $x + y$ and $x \cdot y$ do not depend on the representatives of the equivalence classes x and y. Thus the operations $+$ and $\cdot$ are well defined in $\mathbf{Q}$.

Proof Since $(a, b) \equiv (a', b')$ and $(c, d) \equiv (c', d')$, $a \cdot b' = b \cdot a'$ and $c \cdot d' = d \cdot c'$. To prove that $[(a \cdot d + b \cdot c, b \cdot d)] = [(a' \cdot d' + b' \cdot c', b' \cdot d')]$, we must show that $(a \cdot d + b \cdot c)(b' \cdot d') = (b \cdot d)(a' \cdot d' + b' \cdot c')$. From $a \cdot b' = b \cdot a'$ and $c \cdot d' = d \cdot c'$, we find that

$$a \cdot b' \cdot d \cdot d' = b \cdot a' \cdot d \cdot d' \quad \text{and} \quad c \cdot d' \cdot b \cdot b' = d \cdot c' \cdot b \cdot b'.$$

Hence

$$a \cdot b' \cdot d \cdot d' + c \cdot d' \cdot b \cdot b' = b \cdot a' \cdot d \cdot d' + d \cdot c' \cdot b \cdot b'.$$

From the associative, commutative laws for multiplication in $\mathbf{Z}$ and from the distributivity of multiplication over addition, we conclude that

$$(a \cdot d + b \cdot c) \cdot (b' \cdot d') = (b \cdot d) \cdot (a' \cdot d' + b' \cdot c'),$$

which is our desired conclusion.

We leave the proof that $[(a \cdot c, b \cdot d)] = [(a' \cdot c', b' \cdot d')]$ as an exercise. ∎

Our next theorem lists the basic algebraic properties of $\mathbf{Q}$. In the language of Section 6.4, it asserts that $\mathbf{Q}$ is a field under addition and multiplication.

Theorem 7.4.2 *For all $x, y, z \in \mathbf{Q}$,*

1. $(x + y) + z = x + (y + z)$;
2. $x + 0 = 0 + x = x$ *where* $0 = [(0, 1)]$;
3. *there exists* $x' \in \mathbf{Q}$ *such that* $x + x' = x' + x = 0$;

4. $x + y = y + x$;
5. $(x \cdot y) \cdot z = x \cdot (y \cdot z)$;
6. $x \cdot 1 = 1 \cdot x = x$ *where* $1 = [(1, 1)]$;
7. *if* $x \neq 0$, *then there exists* $\bar{x} \in \mathbf{Q}$ *such that* $x \cdot \bar{x} = \bar{x} \cdot x = 1$;
8. $x \cdot y = y \cdot x$;
9. $x \cdot (y + z) = x \cdot y + x \cdot z$.

The proof of Theorem 7.4.2 relies on the properties of addition and multiplication in **Z** that are listed in Theorem 7.3.3. We leave the proof as an exercise except to note that if $x = [(a, b)] \in \mathbf{Q}$, then the additive inverse of x, x', whose existence is asserted in property 3, is $x' = [(-a, b)]$; and if $x = [(a, b)] \neq 0$, then the multiplicative inverse of x, $\bar{x}$, is the element $\bar{x} = [(b, a)]$. Henceforth, we denote the multiplicative inverse of x by x^{-1}.

We now follow the usual practice of denoting the equivalence class $[(a,b)]$ by a/b. Thus, with this notation, $1/2 = 2/4$, etc., and addition and multiplication become $a/b + c/d = (a \cdot d + b \cdot c)/(b \cdot d)$ and $(a/b) \cdot (c/d) = (a \cdot c)/(b \cdot d)$.

Since **Q** is a set of equivalence classes of certain pairs of integers, the sets **Z** and **Q** are disjoint. Nevertheless, there exists an obvious function $g\colon \mathbf{Z} \to \mathbf{Q}$ defined by $g(n) = n/1$ for all $n \in \mathbf{Z}$. We can easily check that g is a one-to-one function such that for all $n, m \in \mathbf{Z}$,

$$g(n + m) = g(n) + g(m),$$

and

$$g(n \cdot m) = g(n) \cdot g(m).$$

Thus g carries **Z** to a subset of **Q** in such a way as to preserve the set theoretic and algebraic structure of **Z**. Because of the existence of the function g, we regard **Z** as a subset of **Q**. Hence we have $\mathbf{N} \subset \mathbf{Z} \subset \mathbf{Q}$.

Order in Q

Next we consider the concepts of positivity and order in **Q**. Our approach follows closely the development of these concepts in **Z**. We begin by defining the concept of positivity in **Q**. Then we use positivity to define the usual order relations on **Q**. Our first task, therefore, is to propose a definition of positivity in **Q**. Intuitively, we think of a rational number a/b as being positive if the integers a and b are either both positive or both negative. This condition can be succinctly stated by saying that $a \cdot b > 0$. (See Exercise 7.3.8.) Thus we have the following definition.

Definition 7.4.3 Positivity and Order

(i) An element $a/b \in \mathbf{Q}$ is positive if $a \cdot b$ is a positive integer. Let $\mathbf{Q}^+$ denote the set of positive rationals: $\mathbf{Q}^+ = \{a/b \in \mathbf{Q} | a \cdot b \in \mathbf{Z}^+\}$.

(ii) Let $x, y \in \mathbf{Q}$. Then x *is less than* y, written $x < y$, if $y - x \in \mathbf{Q}^+$.

(iii) Let $x, y \in \mathbf{Q}$. Then x *is less than or equal to* y, written $x \leq y$, if $x < y$ or $x = y$.

(iv) If $x, y \in \mathbf{Q}$, *then* x *is greater than* (*or equal to*) y, written $x > y$ $(x \geq y)$, if $y > x$ $(y \leq x)$.

Recall that the elements of $\mathbf{Q}$ are equivalence classes. In Definition 7.4.1, positivity is defined in terms of representatives of equivalence classes. Thus it is necessary to show that positivity in $\mathbf{Q}$ is a well-defined concept. In other words, we must check that if $a/b = c/d$ and a/b is positive, then c/d is positive. We leave this task as an exercise. (See Exercise 7.4.1(c).)

Theorem 7.4.3

(i) *Closure under* $+$ *and* $\cdot$. *If* $x, y \in \mathbf{Q}^+$, *then* $x + y$, $x \cdot y \in \mathbf{Q}^+$.

(ii) *If* $x \in \mathbf{Q}$, *then exactly one of the following holds*: $x \in \mathbf{Q}^+$, $x = 0$, *or* $-x \in \mathbf{Q}^+$.

(iii) *Trichotomy Law. If* $x, y \in \mathbf{Q}$, *then exactly one of the following holds*: $x < y$, $x = y$, *or* $x > y$.

The proof of Theorem 7.4.3 is left as an exercise. Not surprisingly, its proof depends on the proof of the corresponding result for $\mathbf{Z}$ (Theorem 7.3.4).

Theorems 7.4.2 and 7.4.3 prove that $\mathbf{Q}$ satisfies Axioms 1–13 of Section 1.2, thereby establishing that $\mathbf{Q}$ is an ordered field. There are many other connections between the algebraic operations and the order relations in $\mathbf{Q}$. Two of the most important are listed in the next theorem, whose straightforward proof is left as an exercise.

Theorem 7.4.4 *Let* $x, y, z \in \mathbf{Q}$.

(i) $x < y$ *if and only if* $x + z < y + z$.

(ii) *If* $z \in \mathbf{Q}^+$, *then* $x < y$ *if and only if* $x \cdot z < y \cdot z$.

Our next theorem asserts that between any two rational numbers, another rational can be found. This property is often called the *denseness property* of $\mathbf{Q}$, since it shows that from the geometric viewpoint, the rationals are spread quite thickly along the real number line. Observe that this result does not hold in **Z**: If **Q** is replaced by **Z** in the statement of Theorem 7.4.5, then the resulting statement is false.

Theorem 7.4.5 *If $x, y \in \mathbf{Q}$ with $x < y$, then there exists $z \in \mathbf{Q}$ such that $x < z < y$.*

Proof Let $z = (x + y)/2$. (Thus z is the average of x and y.) We show that $x < z$ and $z < y$.

By definition, $x < z$ if and only if $z - x \in \mathbf{Q}^+$. Now $z - x = (x + y)/2 - x = y/2 - x/2 = (y - x)/2$, and $(y - x)/2 \in \mathbf{Q}^+$ since $y - x \in \mathbf{Q}^+$ and $1/2 = 2^{-1} \in \mathbf{Q}^+$. (See Exercise 7.4.6(d).)

Also, $z < y$ since $y - z = (y - x)/2 \in \mathbf{Q}^+$.

Therefore, $x < z < y$. ■

The final theorem of this section asserts the so-called Archimedean property of **Q**: Given any two positive rationals, the smaller can be added to itself some finite number of times to obtain a rational greater than the larger of the original two numbers. We use this theorem, which relies on the Division Theorem in **Z**, to prove the Archimedean property in **R**, which in turn has several important consequences.

Theorem 7.4.6 *If $x, y \in \mathbf{Q}^+$, then there exists $n \in \mathbf{N}$ such that $n \cdot x > y$.*

Proof Let $x = a/b$ and $y = c/d$ where $a, b, c, d \in \mathbf{Z}$. Since $x, y \in \mathbf{Q}^+$, we can assume that $a, b, c, d \in \mathbf{Z}^+$. (See Exercise 7.4.1(b).)

Let us analyze the condition $n \cdot x > y$ in detail. For any integer n, $n \cdot x > y$ if and only if $n \cdot x - y \in \mathbf{Q}^+$ if and only if $n(a/b) - (c/d) = (n \cdot a \cdot d - b \cdot c)/b \cdot d \in \mathbf{Q}^+$ if and only if $n \cdot a \cdot d - b \cdot c \in \mathbf{Z}^+$ (since $b \cdot d \in \mathbf{Z}^+$).

Now, by the Division Theorem in **N** (Theorem 1.3.6), there exist $q, r \in \mathbf{N}$ such that $b \cdot c = (a \cdot d) \cdot q + r$ where $0 \leq r < a \cdot d$. Therefore, $(a \cdot d) \cdot q - b \cdot c = -r$ and $(a \cdot d) \cdot (q + 1) - b \cdot c = a \cdot d - r > 0$. Thus if we let $n = q + 1$, then $n(a \cdot d) - b \cdot c \in \mathbf{Z}^+$ and hence $n \cdot x > y$. ■

We should note the similarities between the sets **Z** and **Q**. First, by its existence, each set remedies a certain deficiency: **Z** provides additive inverses for elements of **N**; **Q** provides multiplicative inverses for nonzero elements of **Z**. Second, each set is a set of equivalence classes: **Z** from the set $\mathbf{N} \times \mathbf{N}$, **Q** from the set $\mathbf{Z} \times (\mathbf{Z} - \{0\})$. Finally, the algebraic and order properties of **Z** and **Q** are established, respectively, using the algebraic and order properties of **N** and **Z**. Thus, each of **Z** and **Q** is build upon an existing mathematical object as it enhances that object. The pattern described here recurs in the next four sections as we construct the real numbers from the rational numbers.

Exercises §7.4

7.4.1. (a) Prove: For all $a, b, c \in \mathbf{Z}$ with $b \neq 0$ and $c \neq 0$, $a/b = (a \cdot c)/(b \cdot c)$.
(b) Prove: Every rational number can be represented in the form a/b where $b > 0$.
(c) Prove: The concept of positivity in $\mathbf{Q}$ is well defined.

7.4.2. Prove that if $(a, b) \equiv (a', b')$ and $(c, d) \equiv (c', d')$, then $(a \cdot c, b \cdot d) \equiv (a' \cdot c', b' \cdot d')$.

7.4.3. Prove Theorem 7.4.2.

7.4.4. Prove Theorem 7.4.3.

7.4.5. Prove Theorem 7.4.4.

7.4.6. Prove: If $x, y \in \mathbf{Q}$ and $x \cdot y = 0$, then $x = 0$ or $y = 0$.

7.4.7. Let $\mathbf{Q}^- = \{x \in \mathbf{Q} \mid -x \in \mathbf{Q}^+\}$.
(a) Prove: If $x, y \in \mathbf{Q}^-$, then $x + y \in \mathbf{Q}^-$ and $x \cdot y \in \mathbf{Q}^+$.
(b) Prove: If $x \in \mathbf{Q}^+$ and $y \in \mathbf{Q}^-$, then $x \cdot y \in \mathbf{Q}^-$.
(c) Prove: If $x \in \mathbf{Q}^+$, then $x^{-1} \in \mathbf{Q}^+$ and if $x \in \mathbf{Q}^-$, then $x^{-1} \in \mathbf{Q}^-$.
(d) Prove: If $x, y \in \mathbf{Q}^+$, then $x < y$ if and only if $x^{-1} > y^{-1}$.

7.4.8. Let $x = a/b$ and $y = c/d$ with $b > 0$ and $d > 0$. Prove that $x > y$ in $\mathbf{Q}$ if and only if $a \cdot d - b \cdot c > 0$ in $\mathbf{Z}$.

7.4.9. (a) Prove: If $x \in \mathbf{Q}$, then there exists $n \in \mathbf{Z}$ such that $x < n$.
(b) Prove: If $x \in \mathbf{Q}$, then there exists $n_0 \in \mathbf{Z}$ such that $2^{n_0} > x$.

7.4.10. Absolute value in $\mathbf{Q}$. For $x \in \mathbf{Q}$, define

$$|x| = \begin{cases} x & \text{if} \quad x \geq 0 \\ -x & \text{if} \quad x < 0 \end{cases}.$$

$|x|$ is called the *absolute value of* x. The following properties of absolute value will be used in the next four sections.
(a) Prove $|x \cdot y| = |x| \cdot |y|$ for all $x, y \in \mathbf{Q}$.
(b) Prove: $|x + y| \leq |x| + |y|$ for all $x, y \in \mathbf{Q}$. (This fact is known as the Triangle Inequality.)
(c) Prove: $|x| - |y| \leq |x - y|$ for all $x, y \in \mathbf{Q}$.

§7.5 The Real Numbers

We continue toward our stated goal of defining the set of real numbers and establishing its fundamental properties. Our next task, which will be carried out over the following four sections, is to define the set of real numbers, **R**. In addition to prescribing the set **R**, we aim to define the arithmetic operations on **R** and show that under these operations, **R** is a field. Then we shall define the notion of positivity on **R** and the related concepts of $<$ and $\leq$ on **R**, and show that **R** is an ordered field with respect to $<$. Finally, we shall prove that **R** satisfies the completeness property. (See Section 1.2, Axiom 14.) Thus we shall verify that **R** satisfies Axioms 1–14 of Section 1.2. Therefore, any property that can be derived from these axioms holds for the field **R** that we are about to construct. For example, by Theorem 1.2.3, it follows that there exists $x_0 \in \mathbf{R}$ such that $x_0^2 = 2$.

Before embarking upon our construction of **R**, let us consider some aspects of the rational number system that suggest the need for the introduction of a larger collection of numbers. First, there are algebraic reasons. Recall that the equation $X + 1 = 0$ has no solution in the set of natural numbers. In fact, for all $n \in \mathbf{N} - \{0\}$, $X + n = 0$ has no solution in **N**. Once **N** is enlarged, however, to the set of integers **Z**, then $X + n = 0$ has a unique solution for all $n \in \mathbf{N}$. Similarly, the equation $2 \cdot X = 1$ (or in general $a \cdot X = b$ for $a, b \in \mathbf{Z}$ where $a \neq 0$ and a does not divide b) has no solution in **Z**, but does have a unique solution in the set of rational numbers **Q**. Consider next the equation $X^2 - 2 = 0$. As shown in Section 1.3, there exists no $x \in \mathbf{Q}$ such that $x^2 = 2$; i.e., $X^2 - 2 = 0$ has no solutions in **Q**.

Nevertheless, there are rational numbers that are close to being solutions to the equation $X^2 - 2 = 0$. For instance, $(1.4)^2 = 1.96$, $(1.41)^2 = 1.988$, $(1.414)^2 = 1.9993$, $(1.4141)^2 = 1.999996$. In a sense, each term in the sequence of rationals, $1.4, 1.41, 1.414, 1.4141$ is an approximate solution of the equation $X^2 - 2 = 0$. Moreover, it would be convenient, at the very least, if a number system were to exist in which this sequence of rationals would converge to an actual solution of the given equation. In fact, as we shall see in this section, each real number will be defined by a certain type of sequence of rationals.

Second, there are geometric reasons for considering numbers that are not rational. To this end, recall the geometric representation of rational numbers as points on the number line. Now the Pythagorean Theorem states that for any right triangle with sides a, b, c with $c > a, b$, $a^2 + b^2 = c^2$. Thus, if T is the right triangle with $a = b = 1$, then $c^2 = 1^2 + 1^2 = 2$, i.e., c is the positive square root of 2. (See Figure 7.5.1.) As we have already noted, c represents a distance that does not correspond to a rational

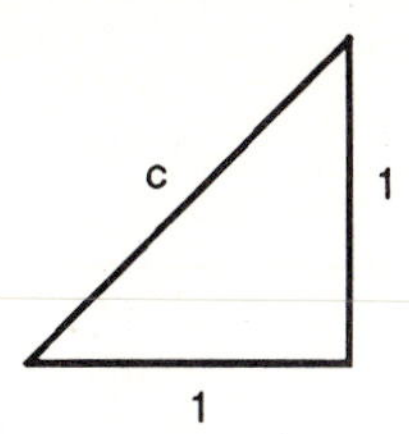

Figure 7.5.1

number. Also, if C is a circle of radius 1, then the diameter of C is $2 \cdot \pi$ where $\pi = 3.14159\ldots$. It can be proved that π also represents a distance that does not correspond to a rational number. Now it seems reasonable to insist that each point on the number line correspond to a "number." Since some points on the number line do not correspond to rational numbers, we wish to enlarge the rational number system to obtain a set of numbers whose elements are in one-to-one correspondence with the points in the number line.

The purpose of this section is merely to define the set of real numbers, **R**. Our method of defining **R** is the so-called sequential method. We begin, however, with a brief discussion of some alternative ways of defining the set **R**.

Of the several approaches to the task of defining the real numbers, one of the first was proposed by the eminent mathematician Richard Dedekind. Dedekind's definition is based on the geometric representation of numbers via the number line. Suppose that the set of rational numbers **Q** has been defined and that each rational number is represented by a point on the number line as in Section 1.2. Guided by the belief that each point on the number line should correspond to a real number and vice versa, Dedekind observes that each point, x, on the number line divides the rational numbers (or rational points) into two disjoint sets: (i) those rationals $\leq x$ and (ii) those rationals $> x$. Dedekind then uses this observation to define the set of real numbers. A *cut* of **Q** consists of an ordered pair of nonempty sets (A, B) such that (i) $A \subset \mathbf{Q}$, $B \subset \mathbf{Q}$, (ii) $A \cap B = \phi$, (iii) $A \cup B = \mathbf{Q}$, (iv) for all $x \in A$ and for all $y \in B$, $x < y$, and (v) there exists no element $y_0 \in B$ such that for all $y \in B$, $y_0 \leq y$.

In words, a cut of **Q** is a partition of **Q** into two nonempty subsets such that each element in one of the sets is less than every element in the other set and the set of larger numbers does not have a least element. (The last condition is included for technical reasons.)

Here are two examples of cuts of **Q**: (i) (A_0, B_0), where $A_0 = \{x \in \mathbf{Q} | x \leq 0\}$ and $B_0 = \{x \in \mathbf{Q} | x > 0\}$; and (ii) (A_1, B_1), where $A_1 = \{x \in \mathbf{Q} | x \leq 0 \text{ or } x^2 < 2\}$ and $B_1 = \mathbf{Q} - A_1$. The pair (A_2, B_2) where $A_2 = \{x \in \mathbf{Q} | x^2 < 2\}$ and $B_2 = \mathbf{Q} - A_2$ does not define a cut of **Q** since $-2 \in B_2$ and $-2 < 0$ where $0 \in A_2$. (See Figure 7.5.2.)

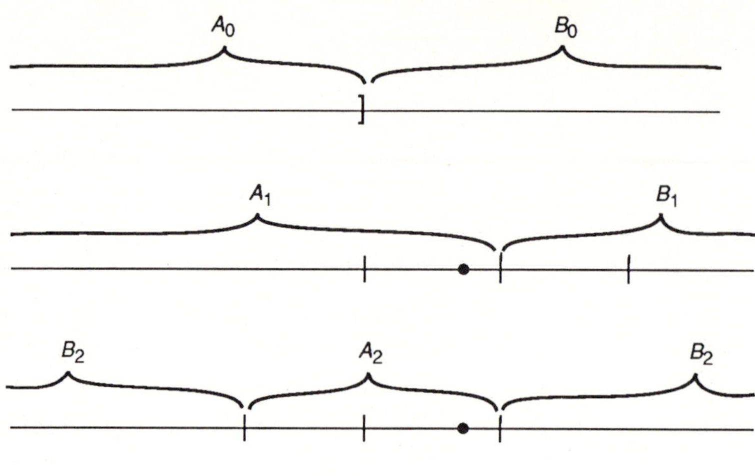

Figure 7.5.2

Dedekind *defines* the set of real numbers to be the set, **R**, of all cuts of **Q**. As it happens, the cut (A_0, B_0) defined above gives the real number 0 and the cut (A_1, B_1) gives the real number $\sqrt{2}$. To show that the set **R** satisfies Axioms 1–14 of Section 1.2, we must define the operations of addition and multiplication and the relation $<$ on **R**. These definitions, especially the one for multiplication, are somewhat subtle. On the basis of these definitions, each of the axioms can be checked. In some cases, the proofs are straightforward. In other cases, for example, for the Completeness Property (Axiom 14), the proofs are rather difficult. Details can be found in Parzinski and Zipse [6].

A second method of defining the set of real numbers entails the formalization of the familiar decimal representation of real numbers. This approach is not widely used, and hence we refer readers interested in this method to the book by Burrill [2], where full details are given.

The method that we present for defining the set of real numbers is based on the idea that any real numbers can be approximated by a sequence of rational numbers. For instance, the real number $\sqrt{2} = 1.4142\ldots$ can be approximated by the sequence of rationals: $1, 14/10, 141/100, \ldots$. Intuitively, we think of the sequence $1, 14/10, 141/100, \ldots$ as "converging" to $\sqrt{2}$. Similarly, we think of the sequence $1, 1/10, 1/100, \ldots, 1/10^n, \ldots$ as a sequence that "converges" to or determines 0. Thus we picture a real number as an object being specified by a sequence of rational numbers. In order to obtain a precise definition of a real number, we have to study the sequences of rational numbers and especially those sequences of rationals that will in fact yield real numbers. To begin, we recall the definition of a sequence.

Definition 7.5.1 *Sequence of Rationals*

A *sequence of rational numbers* is a function $s: \mathbf{N} \to \mathbf{Q}$.

We follow the usual custom of writing for all $n \in \mathbf{N}$, $s(n) = a_n$ and denote the sequence s by $s = \{a_n | n \in \mathbf{N}\}$. The numbers a_n are called the *terms* of the sequence s.

The sequences of rationals that interest us are those such as $\{1/10^n | n \in \mathbf{N}\}$, $\{1, 14/10, 141/100, \dots\}$, or $\{1 + 1/n | n \in \mathbf{N}\}$, whose terms get arbitrarily close together as we move out through the sequence. For example, if we choose a very small positive number, say $1/10^{100}$, then we can find a point in the sequence beyond which all the terms of the sequence are within $1/10^{100}$ of each other. Moreover, this assertion should be true no matter which positive rational we choose in place of $1/10^{100}$. Here is the formal definition of the type of sequence that we shall study. Note: For $x \in \mathbf{Q}$, $|x|$ denotes the *absolute value* of x:

$$|x| = \begin{cases} x \text{ if } x \geq 0 \\ -x \text{ if } x < 0 \end{cases}.$$

Definition 7.5.2 *Cauchy Sequence*

A sequence $\{a_n | n \in \mathbf{N}\}$ of rational numbers is called a *Cauchy sequence* if for each $\epsilon \in \mathbf{Q}^+$, there exists $N \in \mathbf{N}$ such that if $n, m > N$, then $|a_n - a_m| < \epsilon$. Let $C(\mathbf{Q})$ denote the set of all Cauchy sequences of rational numbers.

Suppose $\{a_n | n \in \mathbf{N}\}$ is a Cauchy sequence of rationals. If ϵ is a given positive rational number (usually thought of as being a very small rational number), then the natural number N can be found with the property that any pair of terms in the sequence beyond the term a_N are within ϵ of each other. A somewhat cruder way of putting matters is to say that $|a_n - a_m|$ is small for n and m sufficiently large.

By the way, we use the absolute value of $a_n - a_m$, since we are interested only in the distance between a_n and a_m. (We do not care if $a_n \leq a_m$ or $a_n > a_m$.) Also note that N depends on ϵ in that once ϵ is given, N can be determined. The following examples illustrate the definitions and the process of finding N for a given ϵ.

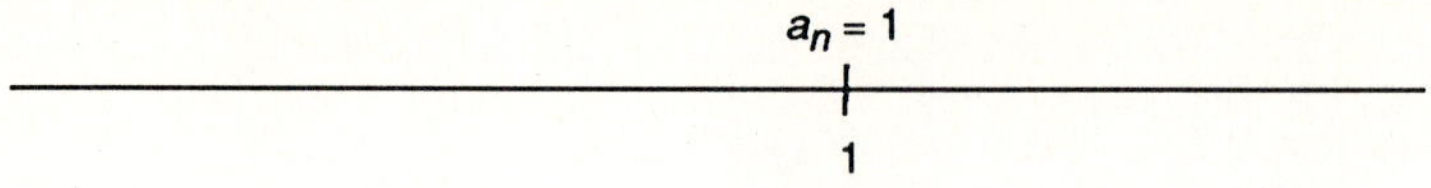

$a_n = 1$

1

Figure 7.5.3

Example 7.5.1 Our first example is trivial but helpful. Let $a_n = 1$ for all $n \in \mathbf{N}$. (See Figure 7.5.3.) Then $\{a_n | n \in \mathbf{N}\}$ is a Cauchy sequence: For each $\epsilon \in \mathbf{Q}^+$, take $N = 0$. If $n, m > N = 0$, then $|a_n - a_m| = |1 - 1| = 0 < \epsilon$. Therefore, $\{a_n | n \in \mathbf{N}\}$ is Cauchy. (Note: We could take $N = 50$ or $N =$ any fixed element of $\mathbf{N}$.) In general, if r is any rational number, then $\{a_n | a_n = r \text{ for all } n\}$ is a Cauchy sequence of rationals.

Example 7.5.2 Let $a_n = 1/2^n$ for each $n \in \mathbf{N}$. We show that $\{a_n | n \in \mathbf{N}\}$ is Cauchy. (See Figure 7.5.4.) Let ϵ be a positive rational. Then $1/\epsilon \in \mathbf{Q}^+$ and by Exercise 7.4.7(b), there exists $n_0 \in \mathbf{N}$ such that $2^{n_0} > 1/\epsilon$. Therefore, $\epsilon > 1/2^{n_0}$. Let $N = n_0 + 1$. Then for all $n, m \in \mathbf{N}$ such that $n, m > N$,

$$\begin{aligned} |a_n - a_m| &= |1/2^n - 1/2^m| \leq 1/2^n + 1/2^m < 1/2^{n_0+1} + 1/2^{n_0+1} \\ &= 1/2^{n_0} < \epsilon. \end{aligned}$$

Example 7.5.3 Let

$$a_n = (-1)^n = \begin{cases} 1 & \text{if} \quad n \text{ is even} \\ -1 & \text{if} \quad n \text{ is odd.} \end{cases}$$

We claim that $\{a_n | n \in \mathbf{N}\}$ is not Cauchy. What must be done to show that $\{a_n | n \in \mathbf{N}\}$ is not Cauchy? We must show that it is not the case that for every $\epsilon \in \mathbf{Q}^+$, there exists $N \in \mathbf{N}$ such that if $n, m \in \mathbf{N}$ with $n > N$ and $m > N$, then $|a_n - a_m| < \epsilon$. Therefore, we must show that there exists $\epsilon \in \mathbf{Q}^+$ such that for all $N \in \mathbf{N}$, there exist $n, m \in \mathbf{N}$ with $n > N$ and $m > N$ and $|a_n - a_m| \geq \epsilon$. In fact, we show that if $\epsilon = 1/2$, then for all $N \in \mathbf{N}$, there exist $n, m \in \mathbf{N}$ with $n > N$ and $m > N$ and $|a_n - a_m| \geq \epsilon = 1/2$. Let N be any element of $\mathbf{N}$. Then choose $n, m \in \mathbf{N}$ such that

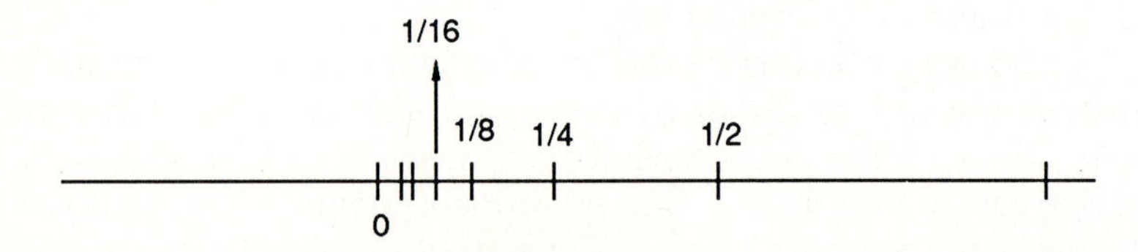

Figure 7.5.4

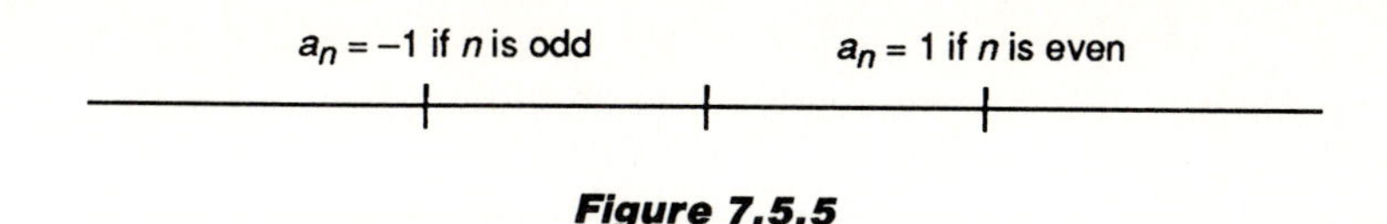

Figure 7.5.5

$n > N$, $m > N$, n is even, and m is odd. Then $a_n = 1$ while $a_m = -1$. Therefore, $|a_n - a_m| = |1 - (-1)| = 2 > 1/2 = \epsilon$. Hence $\{a_n | n \in \mathbf{N}\}$ is not Cauchy.

We think of a Cauchy sequence of rationals as a sequence whose terms eventually become close to each other. Recall that our strategy is to represent each real number as a sequence of rationals and in fact as a Cauchy sequence of rationals. Intuitively, our guiding idea is that each real number, including each rational number, is a *limit of a Cauchy sequence of rationals*.

To develop a sense of how an arbitrary real number might be represented by a Cauchy sequence of rationals, let us represent the rational number 0 by a sequence of rational numbers. Perhaps the simplest way is that via the sequence $s_1 = \{a_n | n \in \mathbf{N}\}$ where $a_n = 0$ for all $n \in \mathbf{N}$. Another way is by the sequence $s_2 = \{b_n | n \in \mathbf{N}\}$ where $b_n = 1/2^n$. We think of the second sequence as representing 0 because the elements of the sequence $b_n = 1/2^n$ approach 0 as n grows without bound. (Formally, we would say that for each $\epsilon \in \mathbf{Q}^+$ there exists $N \in \mathbf{N}$ such that $|b_n - 0| < \epsilon$ for all $n > N$.) Although the sequences s_1 and s_2 are distinct, we want to say that both sequences represent 0. Similarly, we would assert that the sequences $s_3 = \{c_n | n \in \mathbf{N}\}$ where $c_n = 1$ for all $n \in \mathbf{N}$ and $s_4 = \{d_n | n \in \mathbf{N}\}$ where $d_n = 1 + 1/2^n$ both represent 1. Therefore, distinct sequences can represent the same rational number. Emerging from this discussion is the concept of equivalence of Cauchy sequences, which we now formalize.

Definition 7.5.3 *Null Sequence*

A sequence $s = \{a_n | n \in \mathbf{N}\}$ of rational numbers is a *null sequence* if for each $\epsilon \in \mathbf{Q}^+$, there exists $N \in \mathbf{N}$ such that $|a_n| < \epsilon$ for all $n > N$.

The sequences $\{1/2^n | n \in \mathbf{N}\}$ and $\{0 | n \in \mathbf{N}\}$ are null sequences while $\{1 | n \in \mathbf{N}\}$ is not a null sequence. Null sequences are those whose terms become arbitrarily small in absolute value and hence can be regarded as sequences that approach 0. We can show that any null sequence is a Cauchy sequence. (Quick proof. Let $\{a_n | n \in \mathbf{N}\}$ be a null sequence and let $\epsilon \in \mathbf{Q}^+$ be given. We must find $N \in \mathbf{N}$ such that for all $n, m > N$, $|a_n - a_m| < \epsilon$.

Since $\{a_n | n \in \mathbf{N}\}$ is a null sequence, there exists $N \in \mathbf{N}$ such that $|a_k| < \epsilon/2$ for all $k > N$. Thus, by properties of absolute values, for $n, m > N$

$$|a_n - a_m| < |a_n| + |a_n| < \epsilon/2 + \epsilon/2 = \epsilon.$$

Therefore, $\{a_n | n \in \mathbf{N}\}$ is Cauchy.)

> **Definition 7.5.4** *Equivalent Cauchy Sequences*
>
> Let $s_1 = \{a_n | n \in \mathbf{N}\}$ and $s_2 = \{b_n | n \in \mathbf{N}\}$ be Cauchy sequences. We say that s_1 is *equivalent to* s_2, written $s_1 \simeq s_2$, if the sequence $\{a_n - b_n | n \in \mathbf{N}\}$ is a null sequence.

Theorem 7.5.1 *The relation* $\simeq$ *is an equivalence relation on* $C(\mathbf{Q})$.

Proof (1) $\simeq$ is reflexive. Let $s = \{a_n | n \in \mathbf{N}\} \in C(\mathbf{Q})$. Then $s \simeq s$ since $\{a_n - a_n = 0 | n \in \mathbf{N}\}$ is a null sequence.

(2) $\simeq$ is symmetric. Let $s_1 = \{a_n | n \in \mathbf{N}\}$ and $s_2 = \{b_n | n \in \mathbf{N}\}$ be Cauchy sequences such that $\{a_n - b_n | n \in \mathbf{N}\}$ is a null sequence. Then $\{b_n - a_n | n \in \mathbf{N}\}$ is a null sequence: Let $\epsilon \in \mathbf{Q}^+$ be given. Since $\{a_n - b_n | n \in \mathbf{N}\}$ is a null sequence, there exists $N \in \mathbf{N}$ such that for $n, m > N$, $|a_n - b_n| < \epsilon$ for all $n, m > N$. But $|b_n - a_n| = |a_n - b_n| < \epsilon$ for all $n, m > N$. Thus $\{b_n - a_n | n \in \mathbf{N}\}$ is a null sequence.

(3) $\simeq$ is transitive. Suppose $s_1 \simeq s_2$ and $s_2 \simeq s_3$ where $s_1 = \{a_n | n \in \mathbf{N}\}$ and $s_2 = \{b_n | n \in \mathbf{N}\}$ and $s_3 = \{c_n | n \in \mathbf{N}\}$. We know that $\{a_n - b_n | n \in \mathbf{N}\}$ and $\{b_n - c_n | n \in \mathbf{N}\}$ are null sequences. We must show that $\{a_n - c_n | n \in \mathbf{N}\}$ is a null sequence.

Let $\epsilon \in \mathbf{Q}^+$ be given. We must show that there exists $N \in \mathbf{N}$ such that for all $n > N$, $|a_n - c_n| < \epsilon$. Since $\{a_n - b_n | n \in \mathbf{N}\}$ and $\{b_n - c_n | n \in \mathbf{N}\}$ are null sequences, there exist $N_1, N_2 \in \mathbf{N}$ such that $|a_n - b_n| < \epsilon$ for all $n > N_1$ and $|b_n - c_n| < \epsilon$ for all $n > N_2$. Our present task is to relate $|a_n - c_n|$ to the numbers $|a_n - b_n|$ and $|b_n - c_n|$.

We can do so by adding 0 to $a_n - c_n$ in the form $b_n - b_n$: $|a_n - c_n| = |a_n - b_n + b_n - c_n|$. Now, by the Triangle Inequality (see Exercise 7.4.9(b)),

$$|a_n - c_n| = |a_n - b_n + b_n - c_n| \leq |a_n - b_n| + |b_n - c_n| < \epsilon + \epsilon = 2\epsilon,$$

as long as $n > N_1$ and $n > N_2$. To summarize, we have shown that $|a_n - c_n| < 2\epsilon$ for $n > N_1$ and $n > N_2$. We want to show, however, that $|a_n - c_n| < \epsilon$ for n sufficiently large. To achieve this conclusion, we can adjust that preceding argument slightly.

Since $\{a_n - b_n | n \in \mathbf{N}\}$ and $\{b_n - c_n | n \in \mathbf{N}\}$ are null sequences, there exist $N_1', N_2' \in \mathbf{N}$ such that $|a_n - b_n| < \epsilon/2$ for all $n > N_1'$ and $|b_n - c_n|$

$< \epsilon/2$ for all $n > N_2'$. Let $N = \max\{N_1', N_2'\}$. Then, for all $n > N$ (hence for $n > N_1'$ and $n > N_2'$),

$$|a_n - c_n| = |a_n - b_n + b_n - c_n| \le |a_n - b_n| + |b_n - c_n|$$

$$< \epsilon/2 + \epsilon/2 = \epsilon.$$

Therefore, $\{a_n - c_n | n \in \mathbf{N}\}$ is a null sequence.

We have proved that $\simeq$ is an equivalence relation on $C(\mathbf{Q})$. ∎

Remark Two observations are worth noting. First, in the course of establishing the symmetry of $\simeq$, we proved in effect that if $\{c_n | n \in \mathbf{N}\}$ is a null sequence, then $\{-c_n | n \in \mathbf{N}\}$ is a null sequence. Secondly, we can easily modify the proof of the transitivity of $\simeq$ to prove that if $\{c_n | n \in \mathbf{N}\}$ and $\{d_n | n \in \mathbf{N}\}$ are null sequences, then $\{c_n + d_n | n \in \mathbf{N}\}$ is a null sequence.

Given the equivalence relation $\simeq$ on $C(\mathbf{Q})$, we can form the set of equivalence classes of $C(\mathbf{Q})$ modulo $\simeq$. Two Cauchy sequences determine the same equivalence class if the sequence formed by taking the difference between corresponding terms of the sequences is a null sequence. (Intuitively, we think of the two sequences as having the same limit.) We have now arrived at our goal of defining the set of real numbers.

Definition 7.5.5 *The Real Numbers*

The set of *real numbers*, $\mathbf{R}$, is the set of equivalence classes, $\mathbf{R} = C(\mathbf{Q})/\simeq$.

In the next three sections, we establish the algebraic, order, and completeness properties for $\mathbf{R}$, thereby showing that $\mathbf{R}$ is a model for the axiom scheme presented in Section 1.2.

Exercises §7.5

7.5.1. Prove: $\{1/3^n | n \in \mathbf{N}\}$ is Cauchy.

7.5.2. Prove: $\{(-1)^n/2^n | n \in \mathbf{N}\}$ is Cauchy.

7.5.3. Prove: $\{1/n | n \in \mathbf{N}^+\}$ is Cauchy.

7.5.4. Prove: $\{n/(n+1) | n \in \mathbf{N}\}$ is Cauchy.

7.5.5. Prove: $\{(n-1)/(n+1) | n \in \mathbf{N}\}$ is Cauchy.

7.5.6. Which of the sequences in Exercises 7.5.1–7.5.5 are null sequences?

7.5.7. (a) Let $a_n = 0$ for $n \leq 100$ and $a_n = 1$ for $n > 100$. Show $\{a_n | n \in \mathbf{N}\} \simeq \{b_n | b_n = 1 \text{ for all } n\}$.
(b) Let $a_n = 2 + 1/2^n$ for n odd and $a_n = 2$ for n even. Show $\{a_n | n \in \mathbf{N}\} \simeq \{b_n | b_n = 2 \text{ for all } n\}$.

7.5.8. (a) Prove: If $\{a_n | n \in \mathbf{N}\}$ is a null sequence and $r \in \mathbf{Q}$, then $\{r \cdot a_n | n \in \mathbf{N}\}$ is a null sequence.
(b) Prove: If $\{a_n | n \in \mathbf{N}\}$ and $\{b_n | n \in \mathbf{N}\}$ are null sequences, then $\{a_n + b_n | n \in \mathbf{N}\}$ is a null sequence.

7.5.9. Prove, or disprove and salvage: If $\{a_n | n \in \mathbf{N}\}$ is a Cauchy sequence and $r \in \mathbf{Q}$, then $\{r \cdot a_n | n \in \mathbf{N}\}$ is a Cauchy sequence.

7.5.10. Prove, or disprove and salvage: If $\{a_n | n \in \mathbf{N}\}$ is a Cauchy sequence, then $\{(-1)^n \cdot a_n | n \in \mathbf{N}\}$ is a Cauchy sequence.

7.5.11. (a) Let $\{a_n | n \in \mathbf{N}\}$ be a sequence of rationals and let $r \in \mathbf{Q}$. We say then $\{a_n | n \in \mathbf{N}\}$ *converges to* r if, for each $\epsilon \in \mathbf{Q}^+$, there exists $N \in \mathbf{N}$ such that $|a_n - r| < \epsilon$ for all $n > N$. Prove: If $\{a_n | n \in \mathbf{N}\}$ converges to r, then $\{a_n | n \in \mathbf{N}\}$ is Cauchy.
(b) Find three distinct Cauchy sequences that converge to $3/2$.

7.5.12. Let $r_1, r_2 \in \mathbf{Q}$ and let $s_1 = \{a_n | a_n = r_1 \text{ for all } n\}$ and $s_2 = \{b_n | b_n = r_2 \text{ for all } n\}$. Show $s_1 \simeq s_2$ if and only if $r_1 = r_2$.

§7.6 Algebraic Properties of the Real Numbers

In this section we consider the algebraic aspects of **R**. Our aim is to define two binary operations, addition and multiplication, on **R** and to show that under these operations **R** satisfies Axioms 1–11 of Section 1.2; i.e., that **R** is a field under these operations.

By definition, $\mathbf{R} = C(\mathbf{Q})/\simeq$ is a set of equivalence classes of Cauchy sequences. In order to define addition and multiplication on **R**, we shall first define these operations on the set $C(\mathbf{Q})$ of Cauchy sequences of rationals. We shall then show that these operations are compatible with the equivalence relation $\simeq$, thus showing that these operations can be defined on the set of equivalence classes of $C(\mathbf{Q})$ modulo $\simeq$.

Definition 7.6.1 *Addition and Multiplication of Sequences*

Let s_1 and s_2 be sequences of rationals. Define the *sum of* s_1 and s_2, written $s_1 + s_2$, to be the sequence

$$(s_1 + s_2)(n) = s_1(n) + s_2(n),$$

and the *product of* s_1 *and* s_2, written $s_1 \cdot s_2$, to be the sequence

$$(s_1 \cdot s_2)(n) = s_1(n) \cdot s_2(n).$$

Thus, if $s_1 = \{a_n | n \in \mathbf{N}\}$ and $s_2 = \{b_n | n \in \mathbf{N}\}$, then $s_1 + s_2 = \{a_n + b_n | n \in \mathbf{N}\}$ and $s_1 \cdot s_2 = \{a_n \cdot b_n | n \in \mathbf{N}\}$.

We emphasize that the sum and product are defined for any pair of sequences. A natural question arises when we restrict our attention to Cauchy sequences. Is the set of Cauchy sequences closed under the operations of addition and multiplication of sequences? To conjecture an answer to this question, let us suppose that $\{a_n | n \in \mathbf{N}\}$ and $\{b_n | n \in \mathbf{N}\}$ are Cauchy sequences. Then we ask if $\{a_n + b_n | n \in \mathbf{N}\}$ and $\{a_n \cdot b_n | n \in \mathbf{N}\}$ are Cauchy sequences. Let us consider a special case by supposing in addition that $\{a_n | n \in \mathbf{N}\}$ and $\{b_n | n \in \mathbf{N}\}$ converge, respectively, to rational numbers 2 and 3. Informally, this means that as n becomes arbitrarily large, a_n moves arbitrarily close to 2 and b_n becomes arbitrarily close to 3. (Formal definition: For each $\epsilon \in \mathbf{Q}^+$, there exists $N \in \mathbf{N}$ such that $|a_n - 2| < \epsilon$ for $n > N$.) It thus seems clear that as n becomes arbitrarily large, $a_n + b_n$ becomes arbitrarily close to $2 + 3 = 5$ (Figure 7.6.1) and $a_n \cdot b_n$ becomes arbitrarily close to $2 \cdot 3 = 6$. It follows, there-

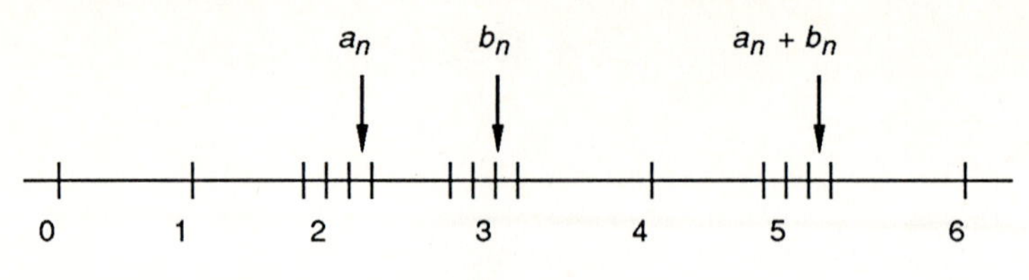

Figure 7.6.1

fore, that the terms of the sequence $\{a_n + b_n | n \in \mathbf{N}\}$ (and $\{a_n \cdot b_n | n \in \mathbf{N}\}$) become arbitrarily close to each other (if the terms $c_n = a_n + b_n$ get close to 5 for n large, then surely the terms of $\{c_n | n \in \mathbf{N}\}$ get close to each other) and hence each is a Cauchy sequence. (See Exercise 7.5.10.) The next theorem handles the general situation.

Theorem 7.6.1 *Let s_1 and s_2 be Cauchy sequences. Then $s_1 + s_2$ and $s_1 \cdot s_2$ are Cauchy sequences.*

Proof We first prove that $s_1 + s_2$ is Cauchy given that s_1 and s_2 are Cauchy. Let $s_1 = \{a_n | n \in \mathbf{N}\}$ and $s_2 = \{b_n | n \in \mathbf{N}\}$, and let $\epsilon \in \mathbf{Q}^+$ be given. We must show that $|(a_n + b_n) - (a_m + b_m)| < \epsilon$ if n and m are sufficiently large. Specifically, our task is to find $N \in \mathbf{N}$ such that if $n, m \in \mathbf{N}$ with $n > N$ and $m > N$, then

$$|(a_n + b_n) - (a_m + b_m)| < \epsilon.$$

How do we find such an N? Let us explore this question by considering the number $|(a_n + b_n) - (a_m + b_m)|$ with an eye toward using the fact that $s_1 = \{a_n | n \in \mathbf{N}\}$ and $s_2 = \{b_n | n \in \mathbf{N}\}$ are Cauchy sequences.

Since s_1 and s_2 are Cauchy, $|a_n - a_m| < \epsilon$ if n and m are sufficiently large and $|b_n - b_m| < \epsilon$ if n and m are sufficiently large. Thus it is natural to group the a's and b's together:

$$|(a_n + b_n) - (a_m + b_m)| = |(a_n - a_m) + (b_n - b_m)|.$$

Now we can use the triangle inequality for absolute values: $|x + y| \leq |x| + |y|$ for all $x, y \in \mathbf{Q}$. We conclude that

$$\begin{aligned} |(a_n + b_n) - (a_m + b_m)| &= |(a_n - a_m) + (b_n - b_m)| \\ &\leq |(a_n - a_m)| + |(b_n - b_m)|. \end{aligned}$$

Remember our goal is to find $N \in \mathbf{N}$ such that if $n, m \in \mathbf{N}$ with $n > N$ and $m > N$, then $|(a_n + b_n) - (a_m + b_m)| < \epsilon$. Since s_1 and s_2 are Cauchy, there exists $N_1 \in \mathbf{N}$ such that for all $n, m \in \mathbf{N}$ with $n > N_1$ and

$m > N_1$, $|a_n - a_m| < \epsilon$, and there exists $N_2 \in \mathbf{N}$ such that for all $n, m \in \mathbf{N}$ with $n > N_2$ and $m > N_2$, $|b_n - b_m| < \epsilon$. Thus, if we let N be the largest of N_1 and N_2, then for $n > N$ and $m > N$, $|a_n - a_m| < \epsilon$ and $|b_n - b_m| < \epsilon$. Therefore, $|(a_n + b_n) - (a_m + b_m)| \leq |(a_n - a_m) + (b_n - b_m)| < \epsilon + \epsilon = 2\epsilon$ for $n, m \in \mathbf{N}$ with $n > N$ and $m > N$.

Now we want $|(a_n + b_n) - (a_m + b_m)|$ to be less than ϵ rather than $2 \cdot \epsilon$. But, as in the proof of Theorem 7.5.1, we can modify the preceding argument to achieve this goal: Since s_1 and s_2 are Cauchy, there exist $N_1, N_2 \in \mathbf{N}$ such that if $n, m \in \mathbf{N}$ with $n > N_1$ and $m > N_1$, then $|a_n - a_m| < \epsilon/2$, and if $n, m \in \mathbf{N}$ with $n > N_2$ and $m > N_2$, then $|b_n - b_m| < \epsilon/2$. Therefore, if $N = \max\{N_1, N_2\}$ and $n, m \in \mathbf{N}$ with $n > N$ and $m > N$, then

$$|(a_n + b_n) - (a_m + b_m)| \leq |(a_n - a_m)| + |(b_n - b_m)| < \epsilon/2 + \epsilon/2 = \epsilon.$$

Thus $s_1 + s_2$ is a Cauchy sequence. ■

The proof that $s_1 \cdot s_2$ is Cauchy relies on the following lemma.

Lemma 7.6.2 *If $s = \{a_n | n \in \mathbf{N}\}$ is a Cauchy sequence, then there exists $M \in \mathbf{Q}^+$ such that $|a_n| \leq M$ for all $n \in \mathbf{N}$.*

Remark In words, Lemma 7.6.2 asserts that any Cauchy sequence is *bounded*.

Proof Since $\{a_n | n \in \mathbf{N}\}$ is Cauchy, there exists $N \in \mathbf{N}$ such that for all $n, m > N$, $|a_n - a_m| < 1$. (We are taking $\epsilon = 1$ in the definition of Cauchy sequence.) In particular, for all $n > N$, $|a_n - a_{N+1}| < 1$; hence $-1 < a_n - a_{N+1} < 1$ or $a_{N+1} - 1 < a_n < a_{N+1} + 1$. (See Figure 7.6.2.) Therefore, for all $n > N$, $|a_n| < \max(|a_{N+1} - 1|, |a_{N+1} + 1|)$. Let $M = \max\{|a_0|, |a_1|, \ldots, |a_N|, |a_{N+1} - 1|, |a_{N+1} + 1|\}$. Then, for all $n \in \mathbf{N}$, $|a_n| \leq M$. ■

Now back to the proof of Theorem 7.6.1. Since $s_1, s_2 \in C(\mathbf{Q})$, s_1 and s_2 are bounded. Thus there exist $M_1, M_2 \in \mathbf{Q}$ such that $|a_n| \leq M_1$ for all $n \in \mathbf{N}$ and $|b_n| \leq M_2$ for all $n \in \mathbf{N}$.

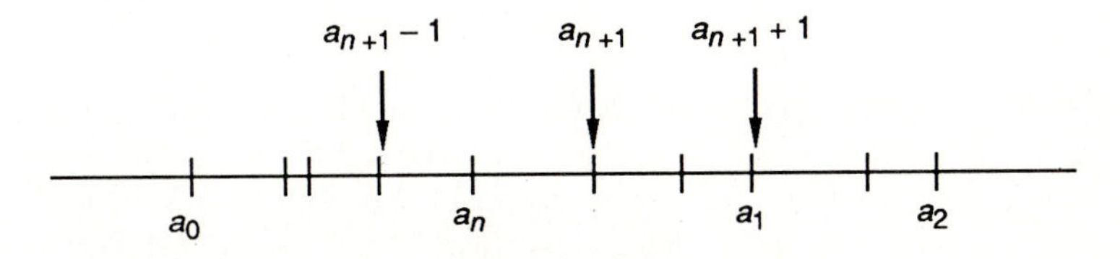

Figure 7.6.2

Let $\epsilon \in \mathbf{Q}^+$ be given. We want to find $N \in \mathbf{N}$ such that for all $n, m > N$, $|a_n \cdot b_n - a_m \cdot b_m| < \epsilon$. We can manipulate $|a_n \cdot b_n - a_m \cdot b_m|$ as follows:

$$\begin{aligned} |a_n \cdot b_n - a_m \cdot b_m| &= |a_n \cdot b_n - a_m \cdot b_n + a_m \cdot b_n - a_m \cdot b_m| \\ &= |(a_n - a_m) \cdot b_n + a_m \cdot (b_n - b_m)| \\ &\leq |(a_n - a_m) \cdot b_n| + |a_m \cdot (b_n - b_m)| \\ &= |a_n - a_m| \cdot |b_n| + |a_m| \cdot |b_m - b_m|. \end{aligned}$$

Now $|b_n| \leq M_2$ and $|a_m| \leq M_1$, and hence

$$|a_n \cdot b_n - a_m \cdot b_m| \leq |a_n - a_m| \cdot M_2 + M_1 \cdot |b_n - b_m|.$$

Since $\{a_n | n \in \mathbf{N}\}$ and $\{b_n | n \in \mathbf{N}\}$ are Cauchy, we can find $N_1, N_2 \in \mathbf{N}$ such that for $n, m > N_1$, $|a_n - a_m| < \epsilon/(2 \cdot M_2)$ and for $n, m > N_2$, $|b_n - b_m| < \epsilon/(2 \cdot M_1)$. Therefore

$$\begin{aligned} |a_n \cdot b_n - a_m \cdot b_m| &\leq |a_n - a_m| \cdot M_2 + M_1 \cdot |b_n - b_m| \\ &< (\epsilon/(2 \cdot M_2)) \cdot M_2 + M_1 \cdot (\epsilon/(2 \cdot M_1)) = \epsilon. \end{aligned}$$ ■

We have now proved that on the set $C(\mathbf{Q})$ two binary operations, addition and multiplication, are defined. What properties do these operations possess? Since $C(\mathbf{Q})$ consists of sequences of rationals, many of the algebraic properties of $\mathbf{Q}$ are inherited by $C(\mathbf{Q})$.

Theorem 7.6.3 *$C(\mathbf{Q})$ is a commutative ring with unity with respect to addition and multiplication. That is, for all $r, s, t \in C(\mathbf{Q})$, the following properties hold:*

1. $(r + s) + t = r + (s + t)$.
2. $r + 0 = 0 + r = r$ *where* $0 = \{a_n | a_n = 0$ *for all* $n \in \mathbf{N}\}$.
3. *There exists* $r' \in C(\mathbf{Q})$ *such that* $r + r' = r' + r = 0$.
4. $r + s = s + r$.
5. $(r \cdot s) \cdot t = r \cdot (s \cdot t)$.
6. $r \cdot 1 = 1 \cdot r = 1$ *where* $1 = \{a_n | a_n = 1$ *for all* $n \in \mathbf{N}\}$.
7. $r \cdot s = s \cdot r$.
8. $r \cdot (s + t) = r \cdot s + r \cdot t$.

Proof The proofs of most parts of Theorem 7.6.3 follow from the corresponding properties in $\mathbf{Q}$ and are quite straightforward. We leave the proofs of properties 1, 2, and 4–8 as exercises. For property 3 we note that if $r = \{a_n | n \in \mathbf{N}\} \in C(\mathbf{Q})$, then $r' = \{-a_n | n \in \mathbf{N}\} \in C(\mathbf{Q})$ and $r + r' = r' + r = \{b_n | b_n = 0$ for all $n \in \mathbf{N}\} = 0$. ■

Now that we have defined the binary operations of addition and multiplication on $C(\mathbf{Q})$, we can ask if these operations can be carried over to the equivalence classes of $C(\mathbf{Q})$ modulo $\simeq$. If so, then we shall have the operations of addition and multiplication defined on $\mathbf{R}$.

Theorem 7.6.4 *Suppose $r, r', s, s' \in C(\mathbf{Q})$ with $r \simeq r'$ and $s \simeq s'$. Then $r + s \simeq r' + s'$ and $r \cdot s \simeq r' \cdot s'$.*

Proof Let $r = \{a_n | n \in \mathbf{N}\}$, $r' = \{a'_n | n \in \mathbf{N}\}$, $s = \{b_n | n \in \mathbf{N}\}$, and $s' = \{b'_n | n \in \mathbf{N}\}$.

We first show that $r + s \simeq r' + s'$. Since $r \simeq r'$ and $s \simeq s'$, $r - r' = \{a_n - a'_n | n \in \mathbf{N}\}$ and $s - s' = \{b_n - b'_n | n \in \mathbf{N}\}$ are null sequences. Therefore,

$$\{(a_n - a'_n) + (b_n - b'_n) | n \in \mathbf{N}\} = \{(a_n + b_n) - (a'_n + b'_n) | n \in \mathbf{N}\}$$

is a null sequence. Hence $r + s \simeq r' + s'$. (See the proof of Theorem 7.5.1 (3) or Exercise 7.5.8(b).)

We show that $r \cdot s \simeq r' \cdot s'$ via an argument similar to the proof that the product of two Cauchy sequences is a Cauchy sequence (Theorem 7.6.1). We aim to prove that $r \cdot s - r' \cdot s' = \{a_n \cdot b_n - a'_n \cdot b'_n | n \in \mathbf{N}\}$ is a null sequence.

Let $\epsilon \in \mathbf{Q}^+$ be given. We must find $N \in \mathbf{N}$ such that for all $n > N$, $|a_n \cdot b_n - a'_n \cdot b'_n| < \epsilon$. Again we manipulate the number $|a_n \cdot b_n - a'_n \cdot b'_n|$, hoping to use the fact that $a_n - a'_n$ and $b_n - b'_n$ are null sequences:

$$\begin{aligned} |a_n \cdot b_n - a'_n \cdot b'_n| &= |a_n \cdot b_n - a_n \cdot b'_n + a_n \cdot b'_n - a'_n \cdot b'_n| \\ &= |a_n(b_n - b'_n) + (a_n - a'_n) \cdot b'_n| \\ &\le |a_n| \cdot |b_n - b'_n| + |a_n - a'_n| \cdot |b'_n|. \end{aligned}$$

By Lemma 7.6.2, there exist $M_1, M_2 \in \mathbf{Q}^+$ such that for all $n \in \mathbf{N}$, $|a_n| < M_1$ and $|b'_n| < M_2$. Thus $|a_n \cdot b_n - a'_n \cdot b'_n| < M_1 \cdot |b_n - b'_n| + M_2 \cdot |a_n - a'_n|$. Now, since $\{a_n - a'_n | n \in \mathbf{N}\}$ and $\{b_n - b'_n | n \in \mathbf{N}\}$ are null sequences, there exist $N_1, N_2 \in \mathbf{N}$ such that for all $n > N_1$, $|a_n - a'_n| < \epsilon/(2 \cdot M_2)$ and for all $n > N_2$, $|b_n - b'_n| < \epsilon/(2 \cdot M_1)$.

Let $N = \max(N_1, N_2)$. Then, for all $n > N$,

$$\begin{aligned} |a_n \cdot b_n - a'_n \cdot b'_n| &< M_1 \cdot |b_n - b'_n| + M_2 \cdot |a_n - a'_n| \\ &< M_1 \cdot (\epsilon/(2 \cdot M_1)) + M_2 \cdot (\epsilon/(2 \cdot M_2)) = \epsilon. \end{aligned}$$

Therefore $r \cdot s - r' \cdot s'$ is a null sequence and $r \cdot s \simeq r' \cdot s'$. ■

Definition 7.6.2 *Addition and Multiplication in* $\mathbf{R}$

Let $x, y \in \mathbf{R} = C(\mathbf{Q})/\simeq$. If $x = [r]$ and $y = [s]$ where $r, s \in C(\mathbf{Q})$, then define the *sum of* x *and* y, written $x + y$, to be $x + y = [r + s]$ and the *product of* x *and* y, written $x \cdot y$, to be $x \cdot y = [r \cdot s]$.

The gist of Theorem 7.6.4 is, of course, that these operations are well defined: If $x = [r] = [r']$ and $y = [s] = [s']$, then $[r + s] = [r' + s']$ and $[r \cdot s] = [r' \cdot s']$, and therefore $x + y$ and $x \cdot y$ do not depend on the choice of representatives of x and y.

We close this section by noting that $\mathbf{R}$ is a field under the operations of addition and multiplication. Most of the field properties (Axioms 1–12 of Section 1.2) will directly follow from Theorem 7.6.3. Let $0 = [\{a_n | a_n = 0$ for all $n\}]$ and $1 = [\{a_n | a_n = 1$ for all $n\}]$.

Theorem 7.6.5 *For all* $x, y, z \in \mathbf{R}$,

1. $(x + y) + z = x + (y + z)$.
2. $x + 0 = 0 + x = x$.
3. *There exists* $x' \in \mathbf{R}$ *such that* $x + x' = x' + x = 0$.
4. $x + y = y + x$.
5. $(x \cdot y) \cdot z = x \cdot (y \cdot z)$.
6. $x \cdot 1 = 1 \cdot x = x$.
7. *If* $x \neq 0$, *then there exists* $\bar{x} \in \mathbf{R}$ *such that* $x \cdot \bar{x} = \bar{x} \cdot x = 1$.
8. $x \cdot y = y \cdot x$.
9. $x \cdot (y + z) = x \cdot y + x \cdot z$.

As mentioned above, most of the field properties for $\mathbf{R}$ follow from Theorem 7.6.3. The only subtle step is property 7, the existence of multiplicative inverses for nonzero real numbers. Recall that the real number 0 consists of the equivalence class $0 = [\{a_n | a_n = 0$ for all $n\}]$. Thus, if $s = \{b_n | n \in \mathbf{N}\}$ and $s \simeq \{a_n | a_n = 0$ for all $n\}$, then s is a null sequence. Conversely, if s is a null sequence, then $s \simeq \{a_n | a_n = 0$ for all $n\}$. Therefore, if $x \in \mathbf{R}$ and $x \neq 0$, then $x = [s]$ where s is not a null sequence. Then, intuitively, s is a sequence whose terms do not approach 0. Actually, even more can be said about the terms of s.

Lemma 7.6.6 *Let* $s = \{a_n | n \in \mathbf{N}\}$ *be a nonnull Cauchy sequence. Then there exist* $c \in \mathbf{Q}^+$ *and* $N \in \mathbf{N}$ *such that for all* $n > N$, $|a_n| \geq c$.

In other words, any nonnull Cauchy sequence is *bounded away from* 0: There exists a positive rational c such that beyond some point, all the terms in the sequence are, in absolute value, of size c or greater.

Proof Since s is not a null sequence, there exists $\epsilon \in \mathbf{Q}^+$ such that for each $N \in \mathbf{N}$, there exists $m > N$ for which $|a_m| \geq \epsilon$. (This conclusion is obtained by negating the statement that defines a null sequence: For each $\epsilon \in \mathbf{Q}^+$, there exists $N \in \mathbf{N}$ such that for all $m > N$, $|a_m| < \epsilon$.)

Since s is a Cauchy sequence, there exists $N_1 \in \mathbf{N}$ such that for all $n, m > N_1$, $|a_n - a_m| < \epsilon/2$. Now, by the previous observation, there exists $m \in \mathbf{N}$ such that $m > N_1$ and $|a_m| \geq \epsilon$. Therefore, for all $n > N_1$, $|a_n| = |a_m + a_n - a_m| \geq |a_m| - |a_n - a_m| \geq \epsilon - \epsilon/2 = \epsilon/2$. (Here we are using another fact about absolute values: For all $x, y \in \mathbf{Q}$, $|x - y| \geq |x| - |y|$.) By taking $c = \epsilon/2$ and $N = N_1$, we have verified the statement of the lemma. ∎

Proof of Theorem 7.6.5. We prove property 7 and leave the remaining statements as exercises. Let $x = [s] \neq 0$. If $s = \{a_n | n \in \mathbf{N}\}$, then by Lemma 7.6.6, there exist $c \in \mathbf{Q}^+$ and $N \in \mathbf{N}$ such that $|a_n| \geq c$ for all $n > N$. Define

$$b_n = \begin{cases} 1 & \text{if} \quad n \leq N \\ 1/a_n & \text{if} \quad n > N \end{cases}$$

We show that $t = \{b_n | n \in \mathbf{N}\}$ is a Cauchy sequence. Let $\epsilon \in \mathbf{Q}^+$ be given. We must find $N_0 \in \mathbf{N}$ such that for all $n, m > N_0$, $|b_n - b_m| < \epsilon$. Now, for $n, m > N$,

$$|b_n - b_m| = |1/a_n - 1/a_m| = \frac{|a_n - a_m|}{|a_n \cdot a_m|} \leq \frac{|a_n - a_m|}{c^2}.$$

Since $\{a_n | n \in \mathbf{N}\}$ is a Cauchy sequence, there exists $N' \in \mathbf{N}$ such that for $n, m > N'$, $|a_n - a_m| < c^2 \cdot \epsilon$. Therefore, for $n, m > N_0 = \max(N, N')$,

$$|b_n - b_m| \leq \frac{|a_n - a_m|}{c^2} = \frac{c^2 \cdot \epsilon}{c^2} = \epsilon.$$

Hence $t = \{b_n | n \in \mathbf{N}\}$ is a Cauchy sequence.

It follows that if $\bar{x} = [t] \in \mathbf{R}$, then

$$\begin{aligned} x \cdot \bar{x} = [s \cdot t] &= [\{c_n | c_n = a_n \text{ for } n \leq N \text{ and } c_n = 1 \text{ for } n > N\}] \\ &= [\{d_n | d_n = 1 \text{ for all } n\}] = 1. \end{aligned}$$
∎

Let us summarize our work in this section. We have constructed the real numbers, $\mathbf{R}$, from the rational numbers, $\mathbf{Q}$, and have proved that $\mathbf{R}$ is a field under the operations of addition and multiplication. At this point it is a simple matter to "embed" $\mathbf{Q}$ in $\mathbf{R}$. We define a function $\phi: \mathbf{Q} \to \mathbf{R}$ by $\phi(r) = [\{a_n | a_n = r \text{ for all } n\}]$ for all $r \in \mathbf{Q}$. In other words, the rational number r is sent to the equivalence class of the Cauchy sequence, each of whose terms equals r. It is easy to verify that ϕ is a one-to-one function such that for all $r, r' \in \mathbf{Q}$, $\phi(r + r') = \phi(r) + \phi(r')$ and $\phi(r \cdot r') = \phi(r) \cdot \phi(r')$. Therefore, in the language of Chapter 6, ϕ is an injective homomorphism of the field **Q** into the field **R**. Thus we say that **Q** is embedded in **R**, and henceforth we regard **Q** as a subset of **R**. The exact way in which the elements of **Q** lie inside **R** will be one of the topics of the next section.

Exercises §7.6

7.6.1. Give an example of a sequence that is bounded but is not Cauchy.

7.6.2. Prove: If s_1 is Cauchy and s_2 is null, then $s_1 \cdot s_2$ is null.

7.6.3. Prove Theorem 7.6.3.

7.6.4. Investigate the following questions:
(a) If s_1 is bounded and s_2 is null, then $s_1 \cdot s_2$ is null.
(b) If s_1 is bounded and s_2 is Cauchy, then $s_1 \cdot s_2$ is Cauchy.

7.6.5. Complete the proof of Theorem 7.6.5.

7.6.6. Is $C(\mathbf{Q})$ an integral domain under the operations of addition and multiplication of sequences?

§7.7 Order Properties of the Real Numbers

In this section we discuss order properties of the real numbers. We follow the pattern that was established in our discussion of order in **Z** and **Q**. We first define the concept of positivity for real numbers and then use this concept to define the order relations $<$ and $\leq$.

To define positivity in **R**, we must, of course, work with equivalence classes of Cauchy sequences. In carrying out this task, we follow what has become our standard procedure: We begin by defining positivity for Cauchy sequences and then show that this concept extends to equivalence classes of Cauchy sequences and hence to **R**.

How can we define the concept of positivity for Cauchy sequences? Perhaps the first idea that comes to mind is the following: Define $\{a_n | n \in \mathbf{N}\} \in C(\mathbf{Q})$ to be positive if $a_n > 0$ for all $n \in \mathbf{N}$. This definition, however, unreasonably excludes certain Cauchy sequences from being positive. For example, $s = \{a_n | a_0 = -1,\ a_n = 1 \text{ for } n \geq 1\}$ would not be positive even though $[s] = 1$ in **R**. An immediate way to remedy this deficiency is to define $\{a_n | n \in \mathbf{N}\}$ to be positive if there exists $N \in \mathbf{N}$ such that $a_n > 0$ for all $n > N$. In this sense, roughly speaking, a sequence is positive if its terms are "eventually" all positive rational numbers. Nevertheless, even with this modification, a difficulty arises. For instance, according to this definition, $t = \{1/n | n \in \mathbf{N}\}$ is positive, yet $[t] = 0$ in **R**; therefore, it is probably unwise to call t a positive Cauchy sequence.

From these comments and examples, we can perhaps see that although all the terms of a Cauchy sequence that we wish to call positive need not themselves be positive rational numbers, beyond some point all the terms should be positive; moreover, beyond some point all the terms should be bounded away from 0. With these thoughts in mind, we have the following formal definition of positivity of Cauchy sequences.

Definition 7.7.1 *Positive Cauchy Sequences*

A Cauchy sequence of rationals $s = \{a_n | n \in \mathbf{N}\}$ is *positive* if there exist $\delta \in \mathbf{Q}^+$ and $N \in \mathbf{N}$ such that for all $n > N$, $a_n > \delta$.

Figure 7.7.1 provides a picture of a generic positive Cauchy sequence.

Example 7.7.1 The sequence $\{1/10 - 1/n | n \geq 1\}$ is positive since for $n > 20$, $1/10 - 1/n > 1/20$. The sequences $\{1/n | n \geq 1\}$ and $\{1/2^n | n \in \mathbf{N}\}$ are not positive. Thus we repeat: A Cauchy sequence all of whose

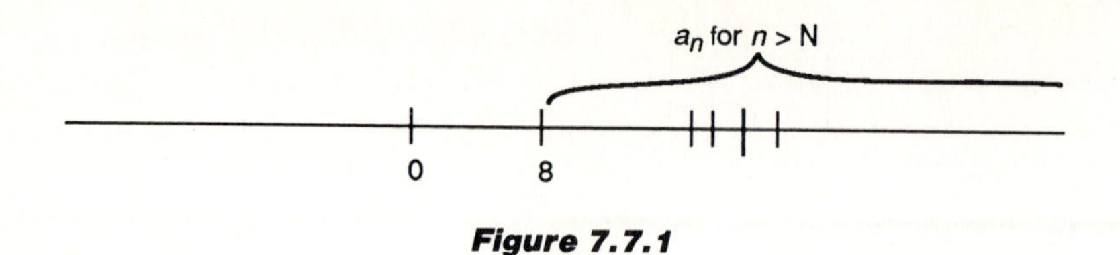

Figure 7.7.1

terms are positive is not necessarily a positive Cauchy sequence. Note that for $r \in \mathbf{Q}$, $\{a_n | a_n = r \text{ for } n \in \mathbf{N}\}$ is positive if and only if $r \in \mathbf{Q}^+$.

Next we wish to extend the definition of positivity to real numbers. The next lemma allows us to do so, for it asserts that any Cauchy sequence equivalent to a positive sequence is itself positive. Thus, in a given equivalence class of Cauchy sequences, either all the sequences are positive or none is positive.

Lemma 7.7.1 *If s is a positive Cauchy sequence and s' is a Cauchy sequence such that $s' \simeq s$, then s' is a positive Cauchy sequence.*

Proof The idea of the proof is fairly straightforward. Since s is positive beyond some point, the terms of s are positive and bounded away from 0. Since $s' \simeq s$, the terms of s' are eventually very close to those of s. Thus, beyond some point, the terms of s' are positive and bounded away from 0. Now the details.

Let $s = \{a_n | n \in \mathbf{N}\}$ and $s' = \{a'_n | n \in \mathbf{N}\}$. To show that s' is positive, we must find $\delta' \in \mathbf{Q}^+$ and $N' \in \mathbf{N}$ such that for all $n > N'$, $a'_n > \delta'$.

Since s is positive, there exist $\delta \in \mathbf{Q}^+$ and $N \in \mathbf{N}$ such that $a_n > \delta$ for all $n > N$. Let $\epsilon = \delta/2$. Then, since $s' \simeq s$, there exists $N_1 \in \mathbf{N}$ such that $|a'_n - a_n| < \epsilon = \delta/2$ for all $n > N_1$. Therefore, $-\delta/2 < a'_n - a_n < \delta/2$ for all $n > N_1$. Let $N' = \max(N_1, N)$. Then for $n > N'$

$$a'_n = a_n + (a'_n - a_n) > \delta - \delta/2 = \delta/2.$$

Thus, for $\delta' = \delta/2$ and for $N' = \max(N_1, N)$, $a'_n > \delta'$ for all $n > N'$. ∎

Definition 7.7.2 *Positive Real Numbers*

A real number $x = [s]$ is *positive* if s is a positive Cauchy sequence. Let $\mathbf{R}^+ = \{x \in \mathbf{R} | x \text{ is positive}\}$.

By Lemma 7.7.1, if $x = [s] = [s']$ where s is positive, then s' is also positive. Thus the concept of positivity for a real number x is independent of the choice of representative for x. Or, equivalently, the concept of positivity for real numbers as given in Definition 7.7.2 is well defined.

We now verify that the order axioms (Axioms 12 and 13) of Section 1.2 hold for $\mathbf{R}^+$.

Theorem 7.7.2 *If $x, y \in \mathbf{R}^+$, then $x + y, x \cdot y \in \mathbf{R}^+$.*

Proof Let $x, y \in \mathbf{R}^+$. We prove that $x + y \in \mathbf{R}^+$ and leave the proof that $x \cdot y \in \mathbf{R}^+$ as an exercise.

Let $x = [\{a_n | n \in \mathbf{N}\}]$ and $y = [\{b_n | n \in \mathbf{N}\}]$. Then $x + y = [\{a_n + b_n | n \in \mathbf{N}\}]$. To show that $x + y \in \mathbf{R}^+$, we must find $\delta \in \mathbf{Q}^+$ and $N \in \mathbf{N}$ such that $a_n + b_n > \delta$ for all $n > N$.

Since $x, y \in \mathbf{R}^+$, there exist $N_1 \in \mathbf{N}$, $\delta_1 \in \mathbf{Q}^+$ such that $a_n > \delta_1$ for all $n > N_1$, and $N_2 \in \mathbf{N}$, $\delta_2 \in \mathbf{Q}^+$ such that $b_n > \delta_2$ for all $n > N_2$. Let $N = \max(N_1, N_2)$. Then, for $n > N$, $a_n + b_n > \delta_1 + \delta_2$. Therefore, with $\delta = \delta_1 + \delta_2 \in \mathbf{Q}^+$ and $N = \max(N_1, N_2)$, $a_n + b_n > \delta$ for all $n > N$. Hence $\{a_n + b_n | n \in \mathbf{N}\}$ is a positive Cauchy sequence and $x + y \in \mathbf{R}^+$. ∎

Theorem 7.7.2 takes care of Axiom 12. We now establish Axiom 13.

Theorem 7.7.3 *Let $x \in \mathbf{R}$. Then exactly one of the following holds: $x \in \mathbf{R}^+$, $x = 0$, or $-x \in \mathbf{R}^+$.*

Proof Let $x \in \mathbf{R}$ and suppose that $x \notin \mathbf{R}^+$ and $x \neq 0$. We show that $-x \in \mathbf{R}^+$. In other words, we show that if x is neither zero nor positive, then the additive inverse of x is positive. We begin by writing out the meaning of our assumptions about x. Let $x = [s]$ where $s = \{a_n | n \in \mathbf{N}\} \in C(\mathbf{Q})$.

1. Since $x \neq 0$, by Lemma 7.6.6 there exist $c \in \mathbf{Q}^+$ and $N_1 \in \mathbf{N}$ such that $|a_n| > c$ for all $n > N_1$. (Actually, Lemma 7.6.6 asserts that there exist $c' \in \mathbf{Q}^+$ and $N_1 \in \mathbf{N}$ such that $|a_n| \geq c'$ for all $n > N$. Taking $c = c'/2$, we have $|a_n| > c$ for all $n > N_1$.)
2. Since $x \notin \mathbf{R}^+$, for each $\delta > 0$ and for each $N \in \mathbf{N}$, there exists $n_0 > N$ such that $a_{n_0} \leq \delta$.
3. Since $\{a_n | n \in \mathbf{N}\}$ is a Cauchy sequence, there exists $N_2 \in \mathbf{N}$ such that for all $n, m > N_2$, $|a_n - a_m| < \epsilon = c/2$.

Thus the terms of the sequence s are close together (statement 3); they are bounded away from 0 in absolute value (statement 1); yet beyond no point are all terms greater than any fixed positive number (statement 2).

Let $N' = \max(N_1, N_2)$. Taking $\delta = c$ and $N = N'$ in statement 2, we find that there exists $n_0 > N'$ such that $a_{n_0} < \delta = c$. By statement 1, however, $|a_n| > c$ for all $n > N'$. Therefore, $|a_{n_0}| > c$ and $a_{n_0} < c$; it follows that $a_{n_0} < -c$. Finally, by statement 3, $|a_n - a_m| < \epsilon = c/2$ for all $n, m > N'$.

Thus, for $n > N'$,

$$a_n = a_n - a_{n_0} + a_{n_0} < (a_n - a_{n_0}) - c < \frac{c}{2} - c = -\frac{c}{2}.$$

Therefore, for all $n > N'$, $-a_n > c/2$; i.e., $-x = [\{-a_n | n \in \mathbf{N}\}] \in \mathbf{R}^+$. ∎

Corollary 7.7.4 $\mathbf{R}$ *is an ordered field.*

Once the concept of positivity is established in $\mathbf{R}$, the usual order relations can be introduced.

Definition 7.7.3 *Order in* $\mathbf{R}$

Let $x, y \in \mathbf{R}$.

(i) *x is less than y*, written $x < y$, if $y - x \in \mathbf{R}^+$.

(ii) *x is less than or equal to y*, written $x \leq y$, if $x < y$ or $x = y$.

(iii) *x is greater than (or equal to) y*, written $x > y$ $(x \geq y)$ if $y < x$ $(y \leq x)$.

The following three theorems can be deduced easily from the definition of $<$ and Theorems 7.7.2 and 7.7.3.

Theorem 7.7.5 *Trichotomy Law. For all $x, y \in \mathbf{R}$, exactly one of the following holds: $x < y$, $x = y$, or $x > y$.*

Theorem 7.7.6 *Let $x, y, z \in \mathbf{R}$.*

(i) *$x < y$ if and only if $x + z < y + z$.*

(ii) *If $z \in \mathbf{R}^+$, then $x < y$ if and only if $x \cdot z < y \cdot z$.*

Theorem 7.7.7 *Denseness Property of $\mathbf{R}$. Let $x, y \in \mathbf{R}$ with $x < y$. Then there exists $z \in \mathbf{R}$ such that $x < z < y$.*

The next theorem, which is a refinement of the preceding result, establishes the denseness of $\mathbf{Q}$ in $\mathbf{R}$. It asserts that the rationals are spread out densely along the real number line in the sense that between any two real numbers is found a rational number.

Theorem 7.7.8 *Denseness of $\mathbf{Q}$ in $\mathbf{R}$. If $x, y \in \mathbf{R}$ with $x < y$, then there exists $z \in \mathbf{Q}$ such that $x < z < y$.*

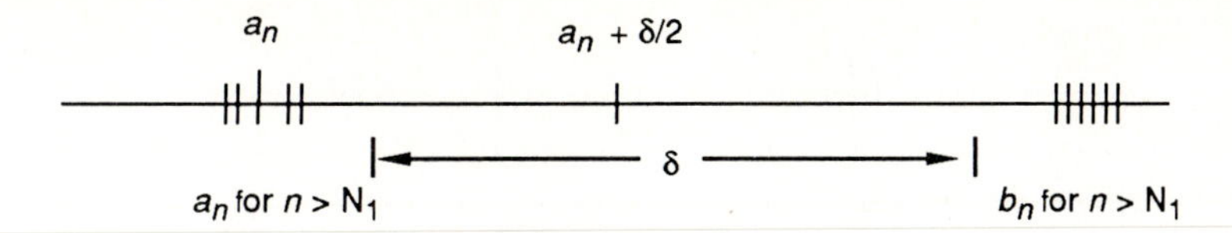

Figure 7.7.2

Proof Let $x = [\{a_n | n \in \mathbf{N}\}]$ and $y = [\{b_n | n \in \mathbf{N}\}]$. Since $x < y$, the Cauchy sequence $\{b_n - a_n | n \in \mathbf{N}\}$ is positive; therefore, there exist $\delta \in \mathbf{Q}^+$ and $N_1 \in \mathbf{N}$ such that $b_n - a_n > \delta$ for all $n > N_1$. Since $\{a_n | n \in \mathbf{N}\} \in C(\mathbf{Q})$, there exists $N_2 \in \mathbf{N}$ such that for all $n, m > N_2$, $|a_n - a_m| < \delta/4$.

Let $N = \max(N_1, N_2)$. Then, for all $n > N$, $|a_n - a_{N+1}| < \delta/4$, and hence $-\delta/4 < a_n - a_{N+1} < \delta/4$. It follows that $(a_{N+1} + \delta/2) - a_n > \delta/4$ for all $n > N$. Therefore, if we let $r = \{c_k | c_k = a_{N+1} + \delta/2\}$, then r is a (constant) Cauchy sequence determined by the rational number $a_{N+1} + \delta/2$. (See Figure 7.7.2.)

Let $z = [r]$. Then z is a rational number and we shall show that $x < z < y$.

The Cauchy sequence $r - \{a_n | n \in \mathbf{N}\}$ is positive, since $c_n - a_n = a_{N+1} + \delta/2 - a_n > \delta/4$ for $n > N$; thus the rational number $z = [r]$ has the property that $z - x \in \mathbf{R}^+$. Thus $x < z$.

Next we show that $z < y$. To do so, we consider $b_n - c_n = b_n - (a_{N+1} + \delta/2)$ for $n > N$:

$$\begin{aligned} b_n - (a_{N+1} + \delta/2) &= (b_n - a_n) - (a_{N+1} + \delta/2 - a_n) \\ &> \delta - \delta/4 = 3 \cdot \delta/4. \end{aligned}$$

Therefore, $y - z \in \mathbf{R}^+$ and $z < y$. ■

The final result of this section is the Archimedean property of $\mathbf{R}$, a result that, as we saw in Theorem 7.4.6, combines the algebraic operations with the order relations in $\mathbf{R}$.

Theorem 7.7.9 *Archimedean Property of* $\mathbf{R}$. *If* $x, y \in \mathbf{R}^+$, *then there exists* $n_0 \in \mathbf{N}$ *such that* $n_0 \cdot x > y$.

Proof Let $x = [\{a_n | n \in \mathbf{N}\}]$ and $y = [\{b_n | n \in \mathbf{N}\}]$. Since $x \in \mathbf{R}^+$, there exist $\delta \in \mathbf{Q}^+$ and $N \in \mathbf{N}$ such that $a_n > \delta$ for all $n > N$. Since $\{b_n | n \in \mathbf{N}\} \in C(\mathbf{Q})$, there exists, by Lemma 7.6.2, $M \in \mathbf{Q}^+$ such that $|b_n| < M$ for all $n \in \mathbf{N}$. (See Figure 7.7.3.)

From the Archimedean property in $\mathbf{Q}$ (Theorem 7.4.6), there exists $n_0 \in \mathbf{N}$ such that $n_0 \cdot \delta > M + 1$. Now, for all $n > N$,

$$n_0 \cdot a_n - b_n > n_0 \cdot \delta - M > 1,$$

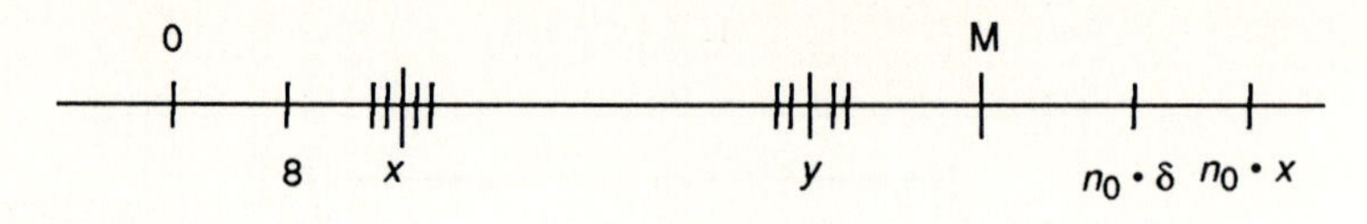

Figure 7.7.3

which means that the Cauchy sequence $\{n_0 \cdot a_n - b_n | n \in \mathbf{N}\}$ is positive. But $[\{n_0 \cdot a_n - b_n | n \in \mathbf{N}\}] = n_0 \cdot x - y$. Therefore, $n_0 \cdot x > y$. ∎

Exercises §7.7

7.7.1. (a) Prove: If $r \in \mathbf{Q}$, then $\{a_n | a_n = r \text{ for all } n\}$ is positive if and only if $r \in \mathbf{Q}^+$.

(b) Prove: If $r, s \in \mathbf{Q}$, then $r < s$ in $\mathbf{Q}$ if and only if $[\{a_n | a_n = r \text{ for all } n\}] < [\{b_n | b_n = s \text{ for all } n\}]$ in $\mathbf{R}$. Thus the function $\phi: \mathbf{Q} \to \mathbf{R}$ defined at the end of Section 7.6 is an *order-preserving* injective homomorphism of the field $\mathbf{Q}$ into the field $\mathbf{R}$.

7.7.2. Prove: If $x, y \in \mathbf{R}^+$, then $x \cdot y \in \mathbf{R}^+$.

7.7.3. Prove Theorem 7.7.5.

7.7.4. Prove Theorem 7.7.6.

7.7.5. Prove Theorem 7.7.7.

7.7.6. Without using Theorem 7.7.8, prove that if $x \in \mathbf{R}^+$, then there exists $r \in \mathbf{Q}^+$ such that $r < x$.

7.7.7. Prove: If $x \in \mathbf{R}$, then $x^2 \geq 0$.

7.7.8. Use Theorem 7.7.8 to show that if $x, y \in \mathbf{R}$ with $x < y$, then there exists $z \in \mathbf{R} - \mathbf{Q}$ such that $x < z < y$. (Hint: Recall that $\sqrt{2} \in \mathbf{R} - \mathbf{Q}$.)

§7.8 The Completeness Property of the Real Numbers

In this section we complete our journey through the real number system. Our goal is to establish the completeness property for **R**. Let us recall the relevant definitions.

> **Definition 7.8.1** *Boundedness*
>
> Let $A \subseteq \mathbf{R}$. A is *bounded from above* if there exists $x \in \mathbf{R}$ such that for all $a \in A$, $x \geq a$. Such an element x is called an *upper bound* for A.

> **Definition 7.8.2** *Least Upper Bound*
>
> Let A be a subset of **R** having an upper bound. An element $x_0 \in \mathbf{R}$ is called a *least upper bound* of A if (i) x_0 is an upper bound for A, and (ii) for any upper bound x of A, $x_0 \leq x$.

Some subsets of **R**, such as **R** itself, have no upper bound. If a subset $A \subseteq \mathbf{R}$ has an upper bound, then A has infinitely many upper bounds in **R**, for if x is an upper bound for A, then $x + y$ is an upper bound for A for all $y > 0$. Also, any subset of **R** has at most one least upper bound in **R**. See Exercise 7.8.1. Finally, if A has a least upper bound x, then x can be, but is not necessarily, an element of A.

Theorem 7.8.1 *Any nonempty subset of* **R** *having an upper bound has a least upper bound.*

Proof Let $A \subseteq \mathbf{R}$ be a nonempty subset having an upper bound x. The idea of the proof is to find two Cauchy sequences of rational numbers $\{a_n | n \in \mathbf{N}\}$ and $\{b_n | n \in \mathbf{N}\}$ such that

(1) for each n, a_n is an upper bound for A,
(2) for each n, b_n is not an upper bound for A,
(3) for each n, $a_n - b_n = 1/n$.

By (1) and (2), the least upper bound of A, if it exists, is trapped between a_n and b_n for each n. By (3), $[\{a_n | n \in \mathbf{N}\}] = [\{b_n | n \in \mathbf{N}\}]$. It will follow that the real number $x_0 = [\{a_n | n \in \mathbf{N}\}]$ is the least upper bound for A.

To begin, let $a, b \in \mathbf{Q}$ be chosen so that $a < x < b$. (By Theorem 7.7.8, such elements a and b exist. Also remember that x is an upper bound for A.) For each $n \in \mathbf{N}^+$, we appeal to Theorem 7.7.9 to find a positive integer k such that $k \cdot (1/n) = k/n > b - a$. Since $a + k/n > b > x$, $a + k/n$ is an upper bound for A. For each $n \in \mathbf{N}^+$, let

$$A_n = \{m \in \mathbf{N} | a + m/n \text{ is an upper bound for } A\}.$$

Since $A_n \neq \varnothing$, A_n has a least member by WOP (Theorem 7.2.9). Call this least element k_n and let $a_n = a + k_n/n$ and $b_n = a + (k_n - 1)/n$. By the defining property of k_n, for each $n \in \mathbf{N}$, a_n is an upper bound for A and b_n is not an upper bound for A. Therefore, for all $n, m \in \mathbf{N}^+$, $b_m < a_n$.

We show that $\{a_n | n \in \mathbf{N}^+\} \in C(\mathbf{Q})$ and $\{b_n | n \in \mathbf{N}^+\} \in C(\mathbf{Q})$ and that $\{a_n | n \in \mathbf{N}^+\} \simeq \{b_n | n \in \mathbf{N}^+\}$.

For $n, m \in \mathbf{N}^+$, $a_m - a_n < a_m - b_m = 1/m$ since $b_m < a_n$. Also, $a_n - a_m < a_n - b_n = 1/n$. Therefore, $-1/n < a_m - a_n < 1/m$ and $|a_m - a_n| < 1/k$ where $k < m, n$. Thus, for any given $\epsilon \in \mathbf{Q}^+$, we find $k \in \mathbf{N}$ such that $1/k < \epsilon$. Then, for all $m, n > k$, $|a_m - a_n| < 1/k < \epsilon$. Therefore, $\{a_n | n \in \mathbf{N}^+\} \in C(\mathbf{Q})$. A similar argument shows that $\{b_n | n \in \mathbf{N}^+\} \in C(\mathbf{Q})$. Finally, notice that for each n, $a_n - b_n = 1/n$; thus, for $\epsilon \in \mathbf{Q}^+$, find $k \in \mathbf{N}$ such that $1/k < \epsilon$. Then, for $n > k$, $|a_n - b_n| = 1/n < 1/k < \epsilon$. Therefore, $\{a_n | n \in \mathbf{N}^+\} \simeq \{b_n | n \in \mathbf{N}^+\}$.

Let $x_0 = [\{a_n | n \in \mathbf{N}^+\}] = [\{b_n | n \in \mathbf{N}^+\}]$. We show that x_0 is the least upper bound for A. To show that x_0 is an upper bound for A, we argue by contradiction. If x_0 is not an upper bound for A, then there exists $c = [\{c_n | n \in \mathbf{N}\}] \in A$ such that $x_0 < c$. It follows that the sequence $\{c_n - a_n | n \in \mathbf{N}^+\}$ is positive, and thus there exist $\delta \in \mathbf{Q}^+$ and $N_1 \in \mathbf{N}$ such that $c_n - a_n > \delta$ for all $n > N_1$. In addition, $\{a_n | n \in \mathbf{N}^+\}$ is Cauchy, and so there exists $N_2 \in \mathbf{N}$ such that $|a_n - a_m| < \delta/2$ for all $n, m > N_2$. Let $N = \max(N_1, N_2)$. Then, for $n > N$, $|a_n - a_{N+1}| < \delta/2$ and $c_n - a_n > \delta$. Therefore, $c_n > a_n + \delta > a_{N+1} + \delta/2$. Thus $c_n - a_{N+1} > \delta/2$ for all $n > N$. Therefore, if we identify $a_{N+1} \in \mathbf{Q}$ with the real number $[\{a_{N+1} | n \in \mathbf{N}\}]$ defined by the constant sequence $\{a_{N+1} | n \in \mathbf{N}\}$, then $c - a_{N+1} > 0$. But a_{N+1} is, by its definition, an upper bound for A and $c \in A$. Thus we have reached a contradiction; hence we conclude that x_0 is an upper bound for A.

Next we show that x_0 is the least upper bound for A. Suppose to the contrary that there exists $x_1 \in \mathbf{R}$ such that $x_1 < x_0$ and x_1 is an upper bound for A. Let $x_1 = [\{d_n | n \in \mathbf{N}\}]$. Since $x_1 < x_0 = [\{b_n | n \in \mathbf{N}\}]$, there exist $\delta \in \mathbf{Q}^+$ and $N_1 \in \mathbf{N}$ such that $b_n - d_n > \delta$ for all $n > N_1$. Also there exists $N_2 \in \mathbf{N}$ such that $|b_n - b_m| < \delta/2$ for all $n, m > N_2$. Let $N = \max(N_1, N_2)$. Then it follows that $b_{N+1} - d_n > \delta/2$ for all $n > N$. Hence $b_{N+1} = [\{b_{N+1} | n \in \mathbf{N}\}] > x_1$. Since x_1 is an upper bound for A and

$x_1 < b_{N+1}$, b_{N+1} is an upper bound for A. But this conclusion contradicts the fact that for all $n \in \mathbf{N}$, b_n is not an upper bound for A. Therefore, x_0 is the least upper bound of A. ■

We have proved that $\mathbf{R}$ is an ordered field satisfying the completeness property, thereby completing our construction of the real number system. Among the consequences of the completeness property is the fact that there exists $x_0 \in \mathbf{R}$ such that $x_0^2 = 2$. (See Section 1.2.) Another consequence of completeness is the fact that any bounded sequence of reals, $\{x_n \in \mathbf{R} | n \in \mathbf{N}\}$, such that $x_n < x_{n+1}$ for all $n \in \mathbf{N}$ has a limit in $\mathbf{R}$: There exists $L \in \mathbf{R}$ such that for each $\epsilon \in \mathbf{R}^+$, there exists $N \in \mathbf{N}$ such that $|x_n - L| < \epsilon$ for all $n > N$. In general, any Cauchy sequence of reals has a limit in $\mathbf{R}$. We can say, therefore, that the real number system has no holes in the sense that all Cauchy sequences of elements of $\mathbf{R}$ converge to an element of $\mathbf{R}$. (The analogous statement does not hold for $\mathbf{Q}$ since there exist sequences of rationals that do not converge to a rational number.)

Nevertheless, in an algebraic sense, $\mathbf{R}$ is deficient. For example, the equation $x^2 + 1 = 0$ has no roots in $\mathbf{R}$. In order to have a number system in which this equation has a root, the field $\mathbf{R}$ can be enlarged or extended by defining the field of complex numbers $\mathbf{C}$. This task was carried out in Section 6.4. As noted in that section, $\mathbf{C}$ has the property that any polynomial equation, $P(x) = 0$, where the coefficients of P are in $\mathbf{C}$ has all its roots in $\mathbf{C}$. Thus $\mathbf{C}$ is an *algebraically closed field*. Moreover, one can show that any Cauchy sequence in $\mathbf{C}$ has a limit in $\mathbf{C}$. Thus $\mathbf{C}$ is "complete" in both an algebraic sense and a limit sense.

Back in Section 1.2, we asserted that the axiom system given by Axioms 1–14 of that section has a unique model, or, to put it another way, any two models of that axiom system are isomorphic. If, for example, we had defined $\mathbf{R}'$ to be the set of all Dedekind cuts, had defined the arithmetic operations and order relations in $\mathbf{R}'$, and had verified that Axioms 1–14 hold in $\mathbf{R}'$, then we could be sure that $\mathbf{R}'$ is isomorphic to the set $\mathbf{R}$ that we developed via the method of Cauchy sequences. We close this section with a formal statement of the uniqueness property of the real numbers.

Theorem 7.8.2 *Let $\mathbf{R}_1$ and $\mathbf{R}_2$ be sets satisfying Axioms 1–14 of Section 1.2. Let $+, \cdot$ and $\oplus, \odot$ denote the operations of addition and multiplication in $\mathbf{R}_1$ and $\mathbf{R}_2$, respectively. Then there exists a bijection $\phi: \mathbf{R}_1 \to \mathbf{R}_2$ such that for all $x, y \in \mathbf{R}_1$, $\phi(x + y) = \phi(x) \oplus \phi(y)$ and $\phi(x \cdot y) = \phi(x) \odot \phi(y)$. Moreover, if $x < y$ in $\mathbf{R}_1$, then $\phi(x) < \phi(y)$ in $\mathbf{R}_2$.*

In words, Theorem 7.8.2 asserts that any two models of the axiom system for the real numbers are *order isomorphic*. The proof of Theorem 7.8.2 is lengthy but fairly straightforward. Here is brief outline of the argument.

The function ϕ is defined in stages. We begin by defining subsets $\mathbf{N}_1 \subseteq \mathbf{R}_1$ and $\mathbf{N}_2 \subseteq \mathbf{R}_2$, each of which is a model for the Peano axioms: Let 0_1 and 1_1 be the additive and multiplicative identities for $\mathbf{R}_1$, and let 0_2 and 1_2 denote the additive and multiplicative identities of $\mathbf{R}_2$. Then let $\mathbf{N}_1 = \{x \in \mathbf{R}_1 | x = 0_1,\ x = 1_1$, or there exists $n \in \mathbf{N}$ such that $n > 1$ and $x = \underbrace{1_1 + \cdots + 1_1}_{n \text{ times}}$. The set $\mathbf{N}_2 \subseteq \mathbf{R}_2$ is defined analogously. Now define $\phi\colon \mathbf{N}_1 \to \mathbf{N}_2$ by $\phi(0_1) = 0_2$, $\phi(1_1) = 1_2$, and, in general,

$$\phi\Big(\underbrace{1_1 + \cdots + 1_1}_{n \text{ times}}\Big) = \underbrace{1_2 + \cdots + 1_2}_{n \text{ times}}.$$

Next define subsets $\mathbf{Z}_1$ and $\mathbf{Q}_1$ of $\mathbf{R}_1$ and $\mathbf{Z}_2$ and $\mathbf{Q}_2$ of $\mathbf{R}_2$ as follows:

$$\mathbf{Z}_1 = \{\pm x | x \in \mathbf{N}_1\},$$

$$\mathbf{Q}_1 = \{m/n | m, n \in \mathbf{Z}_1,\ n \neq 0\},$$

$$\mathbf{Z}_2 = \{\pm y | y \in \mathbf{N}_2\},$$

$$\mathbf{Q}_2 = \{m'/n' | m', n' \in \mathbf{Z}_2,\ n' \neq 0\}.$$

Extend ϕ to a function from $\mathbf{Z}_1$ to $\mathbf{Z}_2$ by $\phi(-x) = -\phi(x)$ for $x \in \mathbf{N}_1$; then extend ϕ to a function from $\mathbf{Q}_1$ to $\mathbf{Q}_2$ by:

$$\phi(m/n) = \phi(m)/\phi(n).$$

We can easily check that $\phi\colon \mathbf{Q}_1 \to \mathbf{Q}_2$ is an order-preserving isomorphism. Finally, to extend ϕ to a function from $\mathbf{R}_1$ to $\mathbf{R}_2$, we can proceed as follows: For $x \in \mathbf{R}_1$ let $A = \{y \in \mathbf{Q}_1 | y \leq x\}$. Let $A' = \{\phi(y) | y \in A\}$. Then $A' \subseteq \mathbf{R}_2$, $A' \neq \varnothing$, and A' has an upper bound. (For example, there exists $x_0 \in \mathbf{Q}_1$ such that $x_0 > y$ for all $y \in A$, and it can be shown that the element $z_0 = \phi(x_0) \in \mathbf{Q}_2$ is an upper bound for A'.) Let z be the least upper bound for A' in $\mathbf{R}_2$. (Since $\mathbf{R}_2$ satisfies Axiom 14, the element z exists.) Define $\phi(x) = z$. It remains to be shown that ϕ is a bijection that preserves the algebraic operations and the order relations. We leave the details of the proof as an exercise. (A good reference is Volume 1, pp. 144–146 of Behnke et al. [1].)

Our treatment of the real numbers is now complete. The set, $\mathbf{R}$, has been constructed and has been shown to be a complete ordered field. Several consequences of this fact, such as the existence of $\sqrt{2}$, have been derived. Finally, the uniqueness of $\mathbf{R}$ as a complete ordered field has been established. Thus the many properties and features of the real number system, familiar to us intuitively from elementary algebra and calculus, have been finally established on the basis of set theory.

Exercises §7.8

7.8.1. Prove that if x is an upper bound for A and $y > x$, then y is an upper bound for A.

7.8.2. (a) Prove: Any subset of **R** has at most one least upper bound.
(b) Prove: If x is the least upper bound for a nonempty set $A \subseteq \mathbf{R}$ and if $y < x$, then there exists $a \in A$ such that $y < a \leq x$.

7.8.3. (a) Propose definitions of the concepts of lower bound and greatest lower bound. (See Section 1.2 for help.)
(b) Using Theorem 7.8.1, show that any nonempty subset of **R** having a lower bound has a greatest lower bound.

7.8.4. Let $n \in \mathbf{N}^+$ and $a \in \mathbf{R}$. Prove, or disprove and salvage: There exists $x_0 \in \mathbf{R}$ such that $x_0^n = a$.

7.8.5. For each $n \in \mathbf{N}^+$, let $a_i \in \mathbf{N}$ with $0 \leq a_i \leq 9$. Let $s_n = a_1/10 + \cdots + a_n/10^n$. Show that $\{s_n | n \in \mathbf{N}\} \in C(\mathbf{Q})$.

7.8.6. The purpose of this exercise is to show that each positive real x has a representation in the form

$$x = m + a_1/10 + a_2/10 + a_3/10 + \cdots + a_n/10 + \cdots$$

where $m \in \mathbf{N}$, for all n, $a_n \in \mathbf{N}$ and $0 \leq a_n \leq 9$. The assertion that $x = m + a_1/10 + \cdots a_n/10^n + \cdots$ means that $x = [\{s_k | k \in \mathbf{N}\}]$ where s is the Cauchy sequence whose terms are $s_0 = m$, $s_1 = m + a_1/10, \ldots, s_n = m + \sum_{i=1}^{n} a_1/10^i, \ldots$. (See Exercise 7.8.5.)
(a) Show: There exist $m \in \mathbf{N}$ and $x_1 \in \mathbf{R}$ such that $0 \leq x_1 < 1$ and $x = m + x_1$.
(b) Show: There exist $a_1 \in \mathbf{N}$ with $0 \leq a_1 \leq 9$ and $x_2 \in \mathbf{R}$ such that $0 \leq x_2 < 1/10$ and $x = m + a_1/10 + x_2$. (Hint: Consider $10 \cdot x_1$ and use (a).)
(c) Show: For each $n \in \mathbf{N}$ there exist $a_1, \ldots, a_n \in \mathbf{N}$ such that

$$x = m + \sum_{1}^{n} \left(\frac{a_i}{10^i} \right) + x_{n+1} = s_n + x_{n+1}$$

where $x_{n+1} \in \mathbf{R}$ and $0 \leq x_{n+1} < 1/10^n$.
(d) Conclude that $x = [\{s_n | n \in \mathbf{N}\}]$. The number x is usually represented in the so-called *decimal form* $x = m.a_1 a_2 \cdots a_n \cdots$.

7.8.7. Find the decimal expansion of each of the following real numbers:
(a) 2/5.
(b) 2/7.
(c) 2/9.
(d) 12/99.

7.8.8. Prove that if the decimal expansion of x eventually repeats, then x is rational; that is, if $x = .a_1 \cdots a_k a_{k+1} \cdots a_n a_{k+1} \cdots a_n a_{k+1} \cdots a_n \cdots$, then $x \in \mathbf{Q}$.

7.8.9. Write each of the following real numbers in the form m/n where $m, n \in \mathbf{Z}$:
(a) $.23717171 \cdots$.
(b) $8.4351351351 \cdots$.

Notes

Peano presented his axioms for the natural numbers in 1889. In Peano's original development, the first natural number is 1. Thus, to obtain Peano's original axioms from ours, we must replace 0 by 1 in Definition 7.1.1. As noted earlier, Peano's axiomatic treatment of the natural numbers completed the rigorous development of the real and complex number systems. Back in the 1830s, William Hamilton (of Hamilton cycles fame) had defined a complex number to be an ordered pair of real numbers. Thus complex numbers are defined once real numbers are defined. In the 1870s, both Cantor and Dedekind defined irrational numbers in terms of rational numbers—Cantor via the sequential approach presented in this chapter and Dedekind through the concept of cut mentioned in Section 7.5. Therefore, the definition of an irrational number requires the existence of rational numbers. In the 1860s, Karl Weierstrass defined rational numbers as pairs of integers. Thus, once integers are defined, rational numbers are also defined. Some mathematicians, including Weierstrass and Kronecker, did not see a need to define integers rigorously, since integers seem to be the most basic of mathematical objects. Kronecker's view is evident in his famous remark: "God made the integers, all else is the work of man."

Undeterred by such a lofty pronouncement, Dedekind presented in 1888 a complicated set theoretic development of the integers. Dedekind's effort, however, attracted little notice, especially given the appearance of Peano's elegant work shortly thereafter.

Bibliography

1. H. Behnke et al., *Fundamentals of Mathematics*, vol. 1, MIT Press, 1974.
2. C. Burrill, *Foundations of Real Numbers*, McGraw-Hill, 1967.
3. D. van Dalen, H. C. Doets, H. DeSwart, *Sets: Naive, Axiomatic, and Applied*, Pergamon, 1978.
4. A. G. Hamilton, *Numbers, Sets and Axioms*, Cambridge University Press, 1982.
5. N. Hamilton and J. Dandin, *Set Theory and the Structure of Arithmetic*, Allyn & Bacon, 1961.
6. W. R. Parzynski and P. W. Zipse, *Introduction to Mathematical Analysis*, McGraw-Hill, 1982.
7. R. L. Wilder, *Introduction to the Foundations of Mathematics*, J. Wiley, 1952.

Miscellaneous Exercises

Definitions. A sequence of real numbers $\{x_n | n \in \mathbf{N}\}$ is *increasing* if $x_n \leq x_{n+1}$ for all $n \in \mathbf{N}$ and is *bounded* if there exists $M \in \mathbf{R}$ such that $|x_n| < M$ for all

$n \in \mathbf{N}$. A sequence of real numbers $\{x_n | n \in \mathbf{N}\}$ has a *limit* L if for each $\epsilon \in \mathbf{R}^+$, there exists $N \in \mathbf{N}$ such that $|x_n - L| < \epsilon$ for all $n > N$.

1. Prove that a bounded increasing sequence of real numbers has a limit in $\mathbf{R}$.

2. (a) Prove that a sequence of real numbers $\{x_n | n \in \mathbf{N}\}$ has at most one limit in $\mathbf{R}$.
   (b) Does every bounded increasing sequence of rational numbers have a limit in $\mathbf{Q}$?

3. (a) Propose a definition for the concept of a *decreasing* sequence of real numbers.
   (b) Prove that a bounded decreasing sequence of real numbers has a limit in $\mathbf{R}$.

4. Assume that Axioms P1–P4 and WOP hold in $\mathbf{N}$. Deduce Axiom P5, the Principle of Mathematical Induction.

5. (For readers who have studied Sections 6.3 and 6.4.) Let R be an integral domain with operations $+$ and $\cdot$. Let $A = R \times (R - \{0\})$. (0 is the additive identity of R.) Define a relation $\sim$ on A as follows: $(a, b) \sim (c, d)$ if $a \cdot d = b \cdot c$ in R.
   (a) Prove that $\sim$ is an equivalence relation on R.
   (b) Let $F = A/\sim$ be the set of equivalence classes of A modulo $\sim$. Define operations $\oplus$ and $\odot$ on F as follows: $[(a, b)] \oplus [(c, d)] = [(a \cdot d + b \cdot c, b \cdot d)]$ and $[(a, b)] \odot [(c, d)] = [(a \cdot c, b \cdot d)]$. Prove that $\oplus$ and $\odot$ are well-defined operations on F.
   (c) Prove that F is a field under the operations $\oplus$ and $\odot$. The field F is called the *quotient field* of the integral domain R. The construction of F from R generalizes the construction of $\mathbf{Q}$ from $\mathbf{Z}$. For a final challenge:
   (d) Prove that if R is itself a field, then the quotient field of R is isomorphic to R.

Hints and Solutions to Selected Exercises

Chapter 1

Section 1.1

1.1.1. (b)

P	Q	$P \Rightarrow Q$	$P \wedge$ not-Q	$(P \Rightarrow Q) \Rightarrow (P \wedge$ not-$Q)$
T	T	T	F	F
T	F	F	T	T
F	T	T	F	F
F	F	T	F	F

(f)

P	Q	$P \Rightarrow Q$	not-$P \Rightarrow$ not-Q	$(P \Rightarrow Q) \Rightarrow ($not-$P \Rightarrow$ not-$Q)$
T	T	T	T	T
T	F	F	T	T
F	T	T	F	F
F	F	T	T	T

1.1.3. (a) From the following partial truth table,

P	Q	$P \Rightarrow Q$
T	T	T
T	F	F

it follows that Q must be true.

(b) The following truth table

P	Q	$P \wedge Q$	$P \vee Q$
T	T	T	T
T	F	F	T
F	T	F	T
F	F	F	F

shows that $P \wedge Q$ is false and $P \vee Q$ is true precisely when exactly one of P and Q is true.

(c) P must be false.

(d) If $P \Leftrightarrow Q$ is true, then P and Q are either both true or both false. Thus exactly one of the statements P and not-Q is true; hence $P \Leftrightarrow$ (not-Q) is false.

1.1.4. If P is false and Q is true, then $P \Rightarrow Q$ is true while (not-P) $\Rightarrow$ (not-Q) is false. Thus the statements are not logically equivalent.

1.1.10. (a) Contrapositive: For all real numbers a and b, if $a \cdot b \not> 0$, then $a \not> 0$ or $b \not> 0$.

(b) Converse: For all real numbers a and b, if $a \cdot b > 0$, then $a > 0$ and $b > 0$. The converse is false since $(-1) \cdot (-1) > 0$ yet $-1 \not> 0$.

1.1.22. (a) There exists x such that x is an integer and x is a perfect square.

(b) Negation: For all integers x, x is not a perfect square.

1.1.26. (a) For all real numbers x, if x is irrational, then $\sqrt{x}$ is irrational.

(b) There exists a number x such that x is irrational and $\sqrt{x}$ is rational.

1.1.31. (a) $(\exists x)$ (not-$P(x) \wedge$ not-$Q(x)$).

(d) $(\forall x)$ (not-$P(x) \vee$ not-$Q(x)$).

Section 1.2

1.2.1. By Axiom 9, for x in **R**, there exists at least one element y in **R** such that $x \cdot y = y \cdot x = 1$. Suppose there exists another element y' in **R** such that $x \cdot y' = y' \cdot x = 1$. Since $x \cdot y = 1$, $y' \cdot (x \cdot y) = y' \cdot 1 = y'$. But by associativity, $y' = y' \cdot (x \cdot y) = (y' \cdot x) \cdot y = 1 \cdot y = y$. Thus there exists exactly one element of **R** having the desired property.

1.2.2. (a) Let x be in **R**. Then $x \cdot 0 = x \cdot (0 + 0) = x \cdot 0 + x \cdot 0$ by Axioms 3 and 11. By Axiom 4, there exists y in **R** such that $x \cdot 0 + y = 0$. Therefore, $0 = x \cdot 0 + y = (x \cdot 0 + x \cdot 0) + y = x \cdot 0 + (x \cdot 0 + y) = x \cdot 0 + 0 = x \cdot 0$ by Axioms 2 and 3.

(c) Let x and y be in **R**. Then $0 = x \cdot 0 = x \cdot (y + (-y)) = x \cdot y + x \cdot (-y)$. Therefore, $x \cdot (-y)$ is the additive inverse of $x \cdot y$; i.e., $x \cdot (-y) = -(x \cdot y)$. A similar argument shows that $(-x) \cdot y = -(x \cdot y)$.

1.2.6. (a) If x and y are in **R** and $x < y$, then $y - x$ is in $\mathbf{R}^+$. We show that $(-x) - (-y)$ is in $\mathbf{R}^+$:

$$\begin{aligned} y - x = y + (-x) &= (-x) + y && \text{by Axiom 5} \\ &= (-x) - (-1) \cdot y && \text{by 1.2.3 (b)} \\ &= (-x) + (-1) \cdot (-y) && \text{by 1.2.2 (c)} \\ &= (-x) - (-y) && \text{by 1.2.2 (c)} \end{aligned}$$

Therefore, $(-x) - (-y)$ is in $\mathbf{R}^+$ and hence $(-x) > (-y)$.

1.2.7. (a) By Axiom 13, either 1 is in $\mathbf{R}^+$ or -1 is in $\mathbf{R}^+$; however, it is not the case that both 1 and -1 are in $\mathbf{R}^+$. Suppose 1 is not in $\mathbf{R}^+$. Then (-1) must be in $\mathbf{R}^+$. But by Axiom 12, it follows that $(-1) \cdot (-1) = 1$ is in $\mathbf{R}^+$, which is impossible since 1 and -1 cannot both be in $\mathbf{R}^+$.

1.2.12. Axioms 1–8 and 10–13 hold when **R** is replaced by **Z**.

1.2.15. Hint: To show that $x_0 = x_1$, first show that $x_0 \le x_1$ and $x_1 \le x_0$. It then follows that $x_0 = x_1$.

1.2.19. (a) Let P be any person. Then by Axiom (i) there exists a committee C of which P is a member. By Axiom (iii) there exists a committee C' that is disjoint from C. By the definition of committee, there exists a person Q who is a member of C'. Note that $P \neq Q$ since C and C' are disjoint committees. From Axiom (ii) it follows that there exists a committee C'' that has P and Q as members. Since Q is a member of C'' and not a member of C, the committees C and C'' are not equal. However, C and C'' both contain P as a member. Thus any person is a member of at least two committees.

Section 1.3

1.3.1. (a) Let x and y be even integers. Then there exist integers m and n such that $x = 2 \cdot m$ and $y = 2 \cdot n$. Therefore, by the distributive law, $x + y = 2 \cdot m + 2 \cdot n = 2 \cdot (m + n)$ is an even integer. Also, by the associative law of multiplication, $x \cdot y = (2 \cdot m) \cdot (2 \cdot n) = 2 \cdot (m \cdot 2 \cdot n)$ is an even integer.

1.3.4. (b) Let x and y be negative real numbers. We must show that $x \cdot y$ is positive. Then $-x$ and $-y$ are positive real numbers. By Axiom 12, $(-x) \cdot (-y) = x \cdot y$ is also positive.

1.3.7. Suppose x and y are positive reals such that $x^2 = y^2$. Then $0 = x^2 - y^2$. By the distributive law and the commutative law for multiplication, $x^2 - y^2 = (x - y) \cdot (x + y)$. Therefore, $(x - y) \cdot (x + y) = 0$. By Corollary 1.3.2, either $x - y = 0$ or $x + y = 0$. But x and y are positive and hence $x + y$ is positive by Axiom 12. Therefore, $x - y = 0$ which implies that $x = y$.

1.3.10. Proof by contradiction. Suppose that a is positive and that $a \cdot (-1) \not< 0$. Then either $a \cdot (-1) = 0$ or $a \cdot (-1) > 0$. First consider the case that $a \cdot (-1) = 0$. If $a \cdot (-1) = 0$, then by Corollary 1.3.2, either $a = 0$ or $-1 = 0$. But $-1 \neq 0$ and by assumption, $a \neq 0$. Therefore, $a \cdot (-1) \neq 0$, which means that $a \cdot (-1) > 0$. Since $a > 0$, $a + a \cdot (-1) > 0$. But $a + a \cdot (-1) = a \cdot 1 + a \cdot (-1) = a \cdot (1 + (-1)) = a \cdot 0 = 0$, implying that $0 > 0$, which is a violation of Axiom 13. Since the assumption that $a \cdot (-1) \not< 0$ leads to a contradiction, it follows that $a \cdot (-1) < 0$.

1.3.12. Suppose $\sqrt{8}$ is rational. Then there exist integers a and b such that $b \neq 0$ and $\sqrt{8} = a/b$. By Exercise 1.3.8, $\sqrt{8} = \sqrt{4} \cdot \sqrt{2} = 2\sqrt{2}$, and hence $\sqrt{2} = a/(2 \cdot b)$ is a rational number. This conclusion, however, contradicts Example 1.3.6. Therefore, $\sqrt{8}$ is irrational.

1.3.14. (a) Suppose $\log_2(3)$ is rational. Then there exist integers a and b with $b \neq 0$ such that $\log_2(3) = a/b$. Since $\log_2(3)$ is positive, we can assume that a and b are both positive integers. From the definition of the base 2 logarithm, $2^{a/b} = 3$. By properties of exponents, $2^a = (2^{a/b})^b = 3^b$. Thus $2^a = 3^b$ is an integer that is both even and odd. Since this conclusion is impossible, the assumption that $\log_2(3)$ is rational is not valid. Therefore, $\log_2(3)$ is irrational.

1.3.22. For an integer $x \geq 1$, let $S(x)$ be the statement: $(1/5)x^5 + (1/3)x^3 + (7/15)x$ is an integer.

Basis step $S(1)$ is the statement: $(1/5) \cdot 1^5 + (1/3) \cdot 1^3 + (7/15) \cdot 1$ is an integer. Since $(1/5) \cdot 1^5 + (1/3) \cdot 1^3 + (7/15) \cdot 1 = 1$, statement $S(1)$ is valid.

Inductive step Suppose statement $S(x)$ holds. We show statement $S(x + 1)$ holds. To do so, we must prove that $(1/5)(x + 1)^5 + (1/3)(x + 1)^3 + (7/15)(x + 1)$ is an integer. Now

$$\begin{aligned}
&(1/5)(x+1)^5 + (1/3)(x+1)^3 + (7/15)(x+1) \\
&\quad = (1/5)(x^5 + 5x^4 + 10x^3 + 10x^2 + 5x + 1) \\
&\qquad + (1/3)(x^3 + 3x^2 + 3x + 1) + (7/15)(x+1) \\
&\quad = (1/5)x^5 + (1/3)x^3 + (7/15)x + (1/5 + 1/3 + 7/15) \\
&\qquad + (x^4 + 2x^3 + 2x^2 + x) + (x^2 + x).
\end{aligned}$$

From the closure of **Z** under addition and multiplication, $x^4 + 2x^3 + 2x^2 + x$ and $x^2 + x$ are integers. Also $1/5 + 1/3 + 7/15 = 1$. Finally, by hypothesis, $S(x)$ holds, and thus $(1/5)x^5 + (1/3)x^3 + (7/15)x$ is an integer. It follows that $(1/5)(x + 1)^5 + (1/3)(x + 1)^3 + (7/15)(x + 1)$ is an integer.

1.3.26. For $n \geq 1$, $s_n = (2n + 1)^2$.

1.3.28. For $n \geq 1$, $b_n = n/(2n + 1)$.

1.3.30. For $n \geq 1$, $e_n = (n + 1)! - 1$.

1.3.34. (a) For each positive integer n, let $S(n)$ be the statement: 3 divides $n^3 - n$.

Basis step $S(1)$ holds since 3 divides $1^3 - 1 = 0$.

Inductive step Suppose $S(n)$ holds. To show that $S(n+1)$ holds, we must prove that 3 divides $(n+1)^3 - (n+1)$. But $(n+1)^3 - (n+1) = n^3 + 3n^2 + 3n + 1 - n - 1 = (n^3 - n) + 3(n^2 + n)$. Since $S(n)$ holds, 3 divides $n^3 - n$. Clearly, 3 divides $3(n^2 + n)$. Therefore, by Lemma 1.3.3, 3 divides $(n+1)^3 - (n+1)$.

1.3.38. $|x+1| < |x-1|$ if and only if $x < 0$.

1.3.39. $|x+1| < |x^2 - 1|$ if and only if $x < -1$ or $-1 < x < 0$ or $x > 2$.

1.3.45. $\sqrt{2}$ is irrational, yet $\sqrt{2} \cdot \sqrt{2}$ is rational.

1.3.57. Use PMI.

1.3.58. Use direct proof.

1.3.59. Use direct proof.

1.3.60. Use proof by contraposition.

1.3.61. Use direct proof.

1.3.62. Use proof by contradiction.

1.3.63. Use proof by contraposition.

1.3.72. First find a formula that expresses the statement precisely and then establish this formula via a direct calculation.

Section 1.4

1.4.1. The unknown is a pair of numbers a, b that $a > 0$, $b > 0$, $a + b = 10$, and $a \cdot b$ is as large as possible. Another description: Since $a + b = 10$, $b = 10 - a$. Thus the unknown is a number a such that $0 < a < 10$ and $a \cdot (10 - a)$ is as large as possible.

1.4.4. (a) The unknown consists of the collection of all vectors $\mathbf{v}$ such that $\mathbf{v}$ is perpendicular to $(1, 2, 3)$. In algebraic terms, the unknown is the set of all vectors $\mathbf{v} = (x, y, z)$ such that

$$\mathbf{v} \cdot (1,2,3) = x + 2y + 3z = 0.$$

1.4.11. Represent each person by a point in the plane. Join two points by a solid line if the corresponding people are friends and join two points by a dotted line if the corresponding people are strangers.

1.4.19. Related problem: Find a nonzero vector $\mathbf{v}$ such that $\mathbf{v}$ makes an angle of $\pi/6$ with $(2, 3)$. Another: Find a unit vector $\mathbf{u}$ that makes an angle of $\pi/2$ with $(2, 3)$.

1.4.20. Related problem: Prove that the infinite series $\Sigma_{n=1}^{\infty} 1/n^2$ converges.

1.4.21. Related problem: The area of the circle $x^2/a^2 + y^2/a^2 = 1$ is πa^2.

1.4.26. Generalization: Among all n-gons (polygons with n sides) with a given perimeter, the regular n-gon (the one whose sides are of equal length) has the largest area.

1.4.27. Generalization: Suppose n is a positive integer and that $3 \cdot n$ points in the plane are given such that no three of the points are collinear. Prove that there exist n disjoint triangles having the given points as vertices.

1.4.30. The general pattern can be described in words: The product of four consecutive integers is one less than a perfect square. In quantitative terms: For each positive integer n, $n \cdot (n + 1) \cdot (n + 2) \cdot (n + 3) = (n^2 + 3n + 1)^2 - 1$.

Miscellaneous Exercises

2. To prove the assertion, argue by PMI. The basis step is straightforward. For the inductive step, let B be a $2^{n+1} \times 2^{n+1}$ defective chessboard. Then B can be regarded as four $2^n \times 2^n$ chessboards joined together at the central point P of B. (In effect, B can be divided into four quadrants, each consisting of a $2^n \times 2^n$ chessboard.) The square that has been removed lies in one of these four chessboards, call it B'. Then B' is a defective $2^n \times 2^n$ chessboard that by inductive hypothesis can be tiled by triominos. Now a single triomino can be placed so as to cover exactly one square in each of the remaining three chessboards at the central point P. Since each of the remaining $2^n \times 2^n$ chessboards has one square covered, each of these boards can, by inductive hypothesis, be tiled with triominos. Thus any defective $2^{n+1} \times 2^{n+1}$ chessboard can be tiled by triominos. The inductive proof is now complete.

3. Data: $1 + 2 = 3$, $2 + 3 = 5$, $1 + 2 + 3 = 6$, $3 + 4 = 7$, $4 + 5 = 9$, $1 + 2 + 3 + 4 = 10$, $5 + 6 = 11$, $3 + 4 + 5 = 12$, $6 + 7 = 13$, $2 + 3 + 4 + 5 = 14$, $7 + 8 = 15$, $8 + 9 = 17$, $3 + 4 + 5 + 6 = 18$, $9 + 10 = 19$, $2 + 3 + 4 + 5 + 6 = 20$. Thus if $n \leq 20$, then n is representable as a sum of consecutive positive integers if and only if $n \neq 1, 2, 4, 8, 16$.

5. Hint: Try analogous special cases. For example, consider the same problem with two or three couples.

6. Hint: Try a proof by contradiction.

Chapter 2

Section 2.1

2.1.2. (a) $\{1,2,5\} = \{5,2,1\}$ since the two sets have precisely the same elements.

(b) $\{\{0\}, \{0,1\}\} \neq \{\{0\}, \{1\}\}$ since $\{1\} \in \{\{0\}, \{1\}\}$ yet $\{1\} \notin \{\{0\}, \{0,1\}\}$.

(c) $\{a, a\} = \{a\}$ since, if $x \in \{a, a\}$, then $x = a$ and $x \in \{a\}$, and if $x \in \{a\}$, then $x = a$ and $x \in \{a, a\}$.

(d) $\{\{a\}, \{\{a\}\}\} \neq \{\{a\}\}$ since $\{\{a\}\} \in \{\{a\}, \{\{a\}\}\}$, but $\{\{a\}\} \notin \{\{a\}\}$.

2.1.4. (b) Proof by contradiction. Suppose $A \subseteq B$, $A \not\subseteq C$, and $B \subseteq C$. By Theorem 2.1.1(iii), $A \subseteq B$ and $B \subseteq C$ imply that $A \subseteq C$. But by assumption,

$A \not\subseteq C$. Thus we have a contradiction. Therefore, if $A \subseteq B$ and $A \not\subseteq C$, then $B \not\subseteq C$.

2.1.6. (a) 00000. (b) 11111. (c) 11100. (d) 10101. (e) 01110.

2.1.8. (a) To show that $\{\varnothing\}$ is transitive, we must show that if $x \in A$, then $x \subseteq A$. If $x \in \{\varnothing\}$, then $x = \varnothing$; by Theorem 2.1.1(i), $\varnothing \subseteq \{0\}$. Therefore, $\{\varnothing\}$ is transitive.

2.1.10. (a) The number of elements of A is the number of ones in the sequence $a_1 a_2 \cdots a_n$. Since each a_i is either 0 or 1, this number equals $a_1 + a_2 + \cdots + a_n$.
(b) 00------0.
(c) 11------1.

Section 2.2

2.2.1. (a) $\{1,2,3,4\}$. (b) $\{2,3\}$. (c) $\{4,5,6\}$. (d) $\{1,5,6\}$. (e) $\{4\}$. (f) $\{2,3\}$. (g) $\{2,3\}$. (h) $\{1,2,3\}$. (i) $\{1,2,3\}$. (j) $\{\varnothing,\{1\},\{2\},\{1,2\}\}$.

Note: In Exercises 2.2.3–2.2.5, Venn diagrams are helpful.

2.2.3. (a) First, suppose $A \subseteq B$ and show $A \cup B = B$. To do so, we show (i) $A \cup B \subseteq B$ and (ii) $B \subseteq A \cup B$. (i) Let $x \in A \cup B$. We show that $x \in B$. Then $x \in A$ or $x \in B$; if $x \in A$, then, since $A \subseteq B$, $x \in B$; on the other hand, if $x \in B$, then $x \in B$. Thus, if $x \in A \cup B$, then $x \in B$. (ii) Let $x \in B$. Then certainly, $x \in A$ or $x \in B$. Thus $B \subseteq A \cup B$. Therefore, if $A \subseteq B$, then $A \cup B = B$.

For the converse, suppose $A \cup B = B$ and show $A \subseteq B$. We must show that if $x \in A$, then $x \in B$. If $x \in A$, then $x \in A$ or $x \in B$, i.e., $x \in A \cup B$. But $A \cup B = B$ and hence $x \in B$. Therefore, if $A \cup B = B$, then $A \subseteq B$.

2.2.3. (c) Suppose $A \subseteq B$ and $A \subseteq C$. Show that $A \subseteq B \cap C$. We show that if $x \in A$, then $x \in B \cap C$. By hypothesis, if $x \in A$, then $x \in B$ and $x \in C$, and thus $x \in B \cap C$.

2.2.4. (c) $(A - B)^c = (A \cap B^c)^c$ by Exercise 2.2.4(a).
$= A^c \cup (B^c)^c$ by Theorem 2.2.3(iv).
$= A^c \cup B$ by Theorem 2.2.3(ii).

2.2.8. $u_1(A) = A$; $u_2(A) = -A$; and $u_3(A) = A^2$. All define unary operations on $M_n(A)$.

2.2.11. $\mathscr{P}(\{1\}) = \{\varnothing,\{1\}\}$, $\mathscr{P}(\mathscr{P}(\{1\})) = \{\varnothing,\{\varnothing\},\{\{1\}\},\{\varnothing,\{1\}\}\}$.

2.2.12. We show that if $A \cap B = \varnothing$, then $\mathscr{P}(A) \cap \mathscr{P}(B) = \{\varnothing\}$. Since $\varnothing \subseteq A$ and $\varnothing \subseteq B$, $\varnothing \in \mathscr{P}(A)$ and $\varnothing \in \mathscr{P}(B)$, and hence $\varnothing \in \mathscr{P}(A) \cap \mathscr{P}(B)$. Therefore, $\{\varnothing\} \subseteq \mathscr{P}(A) \cap \mathscr{P}(B)$. (Note that this fact holds for all sets A and B.) We now show that $\mathscr{P}(A) \cap \mathscr{P}(B) \subseteq \{\varnothing\}$. (In demonstrating this fact, we shall evidently need to use the assumption that $A \cap B = \varnothing$.) Suppose $x \in \mathscr{P}(A) \cap \mathscr{P}(B)$. Then $x \in \mathscr{P}(A)$ and $x \in \mathscr{P}(B)$ and thus $x \subseteq A$ and $x \subseteq B$. By Exercise 2.2.3(c), $x \subseteq A \cap B = \varnothing$. Therefore, $x = \varnothing$ and $\mathscr{P}(A) \cap \mathscr{P}(B) \subseteq \{\varnothing\}$.

The converse is also true. In fact, the converse is a special case of Exercise 2.2.13.

2.2.14. (a) n subsets of A have exactly one element.
(b) n subsets of A have exactly $n - 1$ elements.

2.2.19. (a) Hint: Reason by contradiction and use the Axiom of Regularity.

Section 2.3

2.3.1. (a) $\{(1,3),(1,4),(1,5),(2,3),(2,4),(2,5),(3,3),(3,4),(3,5)\}$.
(c) $\{((a, b), a), ((a, b), b)\}$.

2.3.4. (a) By definition, $(a, a) = \{\{a\},\{a, a\}\}$. But for any x, $\{x, x\} = \{x\}$. Thus $(a, b) = \{\{a\},\{a\}\} = \{\{a\}\}$.
(b) $\{a\} \times \{a\} = \{(a, a)\} = \{\{\{a\}\}\}$ by part (a).

2.3.5. False, since $(a, b) \cap (b, a) = \{\{a\},\{a, b\}\} \cap \{\{b\},\{b, a\}\} = \{\{a, b\}\}$.

2.3.11. $A^2 - B^2 = ((A - B) \times A) \cup ((A \cap B) \times (A - B))$.

2.3.15. (a) Generalizing the definition of (a, b), we can define $(a, b, c) = \{\{a\},\{a, b\},\{a, b, c\}\}$. Another possibility is to define $(a, b, c) = ((a, b), c)$. In terms of sets, the second definition reads $(a, b, c) = \{\{\{\{a\},\{a, b\}\}\},\{\{\{a\},\{a, b\}\}, c\}\}$.

Miscellaneous Exercises

2. Proof by contradiction. Suppose there exists $a \in A$ such that $D = R_a$. We derive a contradiction by showing that if $a \in D$, then $a \notin D$ and if $a \notin D$, then $a \in D$. First, suppose $a \in D$. Then $a \in R_a$ and aRa, i.e., $(a, a) \in R$. But if $(a, a) \in R$, then $a \notin D$ by the definition of D. Thus if $a \in D$, then $a \notin D$. Next suppose $a \notin D$. Then $a \notin R_a$ and hence $(a, a) \notin R$, which implies that $a \in D$. Thus if $a \notin D$, then $a \in D$. Thus $a \notin D \cup D^c = A$, which is a contradiction.

7. $\bigcap_{n=1}^{\infty} A_n = \{0\}$ and $\bigcup_{n=1}^{\infty} A_n = \{x \in \mathbf{R} | 0 < x < 1\}$.

8. Suppose $\{x \in A | x \in x\} \in B$. Then there exists $a \in A$ such that $\{x \in A | x \in x\} = \{a\}$. Then the set $\{a\} \cap a \neq \varnothing$ since $a \in \{a\}$ and $a \in a$. Thus $\{a\}$ is a set that is not disjoint from any of its elements, which contradicts the Axiom of Regularity.

Chapter 3

Section 3.1

3.1.1. $\mathbf{R}_{<} \subset \mathbf{R}_{\leq}, \mathbf{R}_{>} \subset \mathbf{R}_{\geq}, \mathbf{R}_{\leq} \cap \mathbf{R}_{\geq} = \{(x, x) | x \in \mathbf{R}\}, \mathbf{R}_{<} \cap \mathbf{R}_{\geq} = \varnothing$, $\mathbf{R}_{\leq} \cap \mathbf{R}_{>} = \varnothing$.

3.1.3. (a) There are four reflective relations on $\{1,2\}$.
(b) There are eight symmetric relations on $\{1,2\}$.

(c) There are two reflexive and symmetric relations on $\{1,2\}$.
(d) There are six relations on $\{1,2\}$ that are neither reflexive nor symmetric.

3.1.6. (a) Symmetric.
(b) Symmetric.
(c) Reflexive, symmetric, transitive.
(d) None.
(e) Reflexive, symmetric, transitive.

3.1.8. (a) $\varnothing$ is symmetric since the statement "if $(a, b) \in \varnothing$, then $(b, a) \in \varnothing$" is true, because its hypothesis is false. $\varnothing$ is transitive for the same reason.
(c) If $a, b \in A$ with $a \neq b$, then $(a, b) \in A \times A$ and $(b, a) \in A \times A$. Since $a \neq b$, the relation $A \times A$ is not antisymmetric and therefore is not a partial ordering.

3.1.10. (a) $(R^{-1})^{-1} = \{(x, y) \in A \times A \mid (y, x) \in R^{-1}\}$
$= \{(x, y) \in A \times A \mid (x, y) \in R\} = R.$
(b) $(x, y) \in (R \cap S)^{-1}$ if and only if $(y, x) \in R \cap S$ if and only if $(y, x) \in R$ and $(y, x) \in S$ if and only if $(x, y) \in R^{-1}$ and $(x, y) \in S^{-1}$ if and only if $(x, y) \in R^{-1} \cap S^{-1}$.

3.1.11. To obtain the sketch of R^{-1}, simply reverse the direction of each directed segment in the sketch of R.

3.1.13. (a) Hint: Try some examples. For instance, on $\mathbf{R}$ consider the relation $R = \leq$ and $S = \geq$.
(b) Hint: Again, try some examples.

3.1.14. (a) $\leq^{-1} = \geq$.
(b) $<^{-1} = >$.
(c) $f^{-1} = \{(y, x) \mid y = x^2 + 1\}$
$= \{(y, x) \mid y \geq 1 \text{ and } x = \pm\sqrt{y - 1}\}.$
(d) $f^{-1} = \{(y, x) \mid y = e^{e^x}\}$
$= \{(y, x) \mid y \geq e \text{ and } x = \ln(\ln(y))\}.$

3.1.16. (a) $R^{-1} = R$ and $S^{-1} = S$.
(b) $R \circ R = R$ and $S \circ S = S$.
(c) $R \circ S = S$.

3.1.23. To prove that $S \subseteq (S \circ R) \cap (R \circ S)$ we show that (i) $S \subseteq S \circ R$ and (ii) $S \subseteq R \circ S$. (i) Let $u \in S$. Then $u = (x, y)$ for some $x, y \in A$. Since R is reflexive, $(x, x) \in R$. Thus $(x, x) \in R$ and $(x, y) \in S$, which implies that $(x, y) \in S \circ R$. Therefore, $S \subseteq S \circ R$. (ii) is proved by a similar argument.

Section 3.2

3.2.2. (a) The relation is not transitive ((1, 2) and (2, 3) are in the relation, but (1, 3) is not) and hence is not a partial ordering.

3.2.4. (a) $n = 12$ or $n = 32$.
(b) $n = 2^{k-1}$.

3.2.6. (a) Reflexivity. For all $(a, b) \in A \times B$, $(a, b) \leq (a, b)$ since $a \leq_1 a$ and $b \leq_2 b$.

Antisymmetry. Suppose $(a, b) \leq (c, d)$ and $(c, d) \leq (a, b)$. Then by the definition of $\leq$, $a \leq_1 c$ and $c \leq_1 a$, and $b \leq_2 d$ and $d \leq_2 b$. Since $\leq_1$ and $\leq_2$ are antisymmetric, $a = c$ and $b = d$. Thus if $(a, b) \leq (c, d)$ and $(c, d) \leq (a, b)$ for $(a, b), (c, d) \in A \times B$, then $(a, b) = (c, d)$.

Transitivity. Suppose $(a, b) \leq (c, d)$ and $(c, d) \leq (e, f)$ for $(a, b), (c, d), (e, f) \in A \times B$. Then $a \leq_1 c$ and $c \leq_1 e$, which by the transitivity of $\leq_1$, implies that $a \leq_1 e$. Similarly, it follows that $b \leq_2 d$ and $d \leq_2 f$ and hence $b \leq_2 f$. Therefore, $(a, b) \leq (e, f)$ and $\leq$ is transitive. We conclude that $\leq$ is a partial ordering.

(b) Let $\leq_1$ and $\leq_2$ both be the usual ordering $\leq$ on $\mathbf{R}$. Then the ordering $\leq$ defined in part (a) on $\mathbf{R} \times \mathbf{R}$ is not a total ordering since $(1, 2) \nleq (2, 1)$ and $(2, 1) \nleq (1, 2)$.

3.2.8. A straightforward argument shows that $\leq_1 \cap \leq_2$ is a partial ordering. Compare Exercise 3.1.14.

3.2.15. Suppose A has n elements. The diagram of I_A consists of n points in the plane, no two of which are connected by a line segment.

Section 3.3

3.3.1. (a) There are five equivalence relations on $\{a, b, c\}$.
(b) Let $A = \{a, b, c, d\}$. The following are equivalence relations on A: I_A, $A \times A$, $I_A \cup \{(a, b), (b, a)\}$, $I_A \cup \{(a, c), (c, a)\}$.

3.3.2. $\perp$ is not a reflexive relation and thus is not an equivalence relation.

3.3.5. The relations defined in (a) and (c) are equivalence relations. The others are not.

3.3.7. $\sim$ is an equivalence relation on $C[a, b]$.

3.3.9. $\sim$ is an equivalence relation on $M_n(\mathbf{R})$.

3.3.11. Compare Exercise 3.1.14.

3.3.13. (b) Compare Exercise 3.2.1.

3.3.14. (a) For $a, b \in \mathbf{R}$, define $a \sim b$ to mean $a \geq 0$ and $b \geq 0$, or $a < 0$ and $b < 0$.

3.3.17. (b) One equivalence class is the origin. Each of the remaining equivalence classes consists of a square with vertices on the coordinate axes and center at the origin.

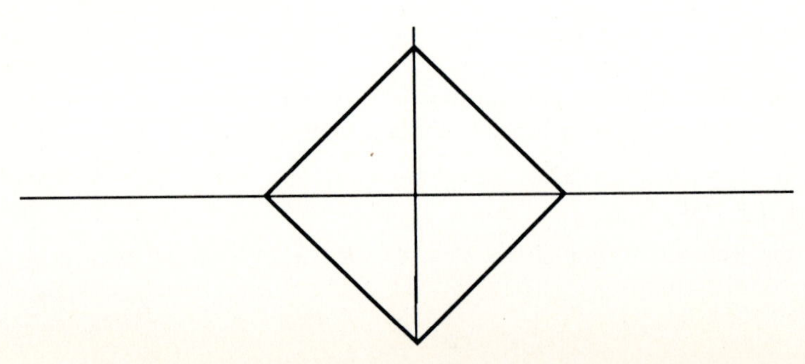

(c) The set $A = \{(a,0) | a \geq 0\}$ is a complete set of representatives for $\mathbf{R}^2/R$, for, if $(x, y) \in \mathbf{R}$ and $a = |x| + |y|$, then $(x, y) R (a,0)$; and, if $a \geq 0$ and $b \geq 0$ with $a \neq b$, then $(a,0) \not R (b,0)$. Thus every element of $\mathbf{R}^2$ is equivalent with respect to R to exactly one element of A. It follows that A is a complete set of representatives for $\mathbf{R}^2/R$.

3.3.22. (b) Let $A = \{a, b, c\}$ and let $R = \{(a, b), (b, c), (a, c)\}$. Then R is circular but R is not an equivalence relation on A.

3.3.24. $\{(1,1), (1,2), (2,1), (2,2), (3,3), (4,4), (4,5), (5,4), (5,5)\}$.

3.3.26. $k_1^2 + \cdots + k_n^2$.

Section 3.4

3.4.1. (c) $f: A \to B$ defined by $f(x) = b$ for all $x \in A$ is the only function from A to B.
(e) $f_1(1) = 1$, $f_2(1) = 2$, $f_3(1) = 3$ define the only functions from A to B.

3.4.2. (a) f is bijective.
(b) f is bijective.
(c)–(e) f is neither injective nor surjective.

3.4.3. (f) $g^{-1}: \mathbf{N} \to \mathbf{Z}$ is defined by

$$g^{-1}(x) = \begin{cases} x/2 & \text{if } x \text{ is even} \\ -(x+1)/2 & \text{if } x \text{ is odd.} \end{cases}$$

3.4.7. (a) $S_{n+1} = S_n + n + 1$.

3.4.8. (a) There are two bijections from A to A.

3.4.11. (a) $(g \circ f)^{-1} = f^{-1} \circ g^{-1}$.
(d) We show that f is (i) injective and (ii) surjective. (i) Suppose $x_1, x_2 \in A$ with $f(x_1) = f(x_2)$. Then $(g \circ f)(x_1) = g(f(x_1)) = g(f(x_2)) = (g \circ f)(x_2)$. But $g \circ f$ is bijective and hence is injective. Thus $x_1 = x_2$, and f is injective. (ii) Let $y \in B$. Then we must find $x \in A$ such that $f(x) = y$. Let $z = g(y)$. Then $z \in C$; since $g \circ f: A \to C$ is bijective, there exists $x \in A$ such that $(g \circ f)(x) = z$. Since $g(y) = z = (g \circ f)(x) = g(f(x))$ and since g is injective, we conclude that $y = f(x)$.

3.4.16. $$f(x) = \begin{cases} -(1/4) \cdot x & \text{for } x \leq 0 \\ -4 \cdot x & \text{for } x > 0. \end{cases}$$

3.4.17. (a) $f^{-1}(x) = f(x) = 1/x$
(b) $g^{-1}(x) = 1/\sqrt{x}$.

Miscellaneous Exercises

1. If f is injective, then each equivalence class has exactly one element. The identity relation on any set, $\equiv_n$ on $\mathbf{Z}$, and the relations in Example 3.3.4 and

Exercises 3.3.7, 3.3.8, 3.3.16, and 3.3.17 are among the equivalence relations that arise in this way.

3. We show that f is injective and surjective. Suppose $f(a, b) = f(a_1, b_1)$. Then $2^a \cdot (2b + 1) = 2^{a_1} \cdot (2b_1 + 1)$. Suppose $a \neq a_1$. Then we can assume $a < a_1$. By cancellation, $(2b + 1) = 2^{a_1 - a} \cdot (2b_1 + 1)$. Since $a_2 - a > 0$, $2^{a_1 - a} \cdot (2b_1 + 1)$ is even; but $2b + 1$ is odd. This contradiction shows that $a = a_1$. Therefore, $2b + 1 = 2b_1 + 1$ that implies that $b = b_1$. Therefore, if $f(a, b) = f(a_1, b_1)$, then $(a, b) = (a_1, b_1)$. To show that f is surjective, let $n \in \mathbf{Z}^+$. Then n can be written in the form $n = 2^a \cdot m$ where m is odd. Writing m in the form $m = 2b + 1$, we see that $n = f(a, b)$. Thus f is both injective and surjective.

6. (b) $f(x) = x \cdot \sin(x)$ defines one function with the stated property.

7. $f(a, b) = (b, a)$ for all $(a, b) \in A \times B$ defines a one-to-one correspondence from $A \times B$ to $B \times A$.

11. Hints: (a) Try a direct proof.
    (b) and (c) Consider examples on the set $\{a, b, c\}$.

15. First, note that $R \cap R^{-1}$ is reflexive and transitive. It remains to show that $R \cap R^{-1}$ is symmetric. Let $(a, b) \in R \cap R^{-1}$. We show that $(b, a) \in R \cap R^{-1}$. Since $(a, b) \in R$, $(b, a) \in R^{-1}$. Since $(a, b) \in R^{-1}$, $(b, a) \in (R^{-1})^{-1} = R$. Therefore, $(b, a) \in R \cap R^{-1}$ and $R \cap R^{-1}$ is symmetric.

Chapter 4

Section 4.1

4.1.1. (a) The function $f\colon \mathbf{N} \to C$ defined by $f(k) = k^3$ is a bijection from $\mathbf{N}$ to C. Thus $\mathbf{N} \approx C$. By Example 4.1.2(iv), $\mathbf{Z} \approx \mathbf{N}$, and hence by Theorem 4.1.1(iii), $\mathbf{Z} \approx C$.
(b) $\mathbf{Z}^+ \approx \mathbf{Z}^-$ since the function $f\colon \mathbf{Z}^+ \to \mathbf{Z}^-$ defined by the rule $f(k) = -k$ for $k \in \mathbf{Z}^+$ is a bijection.

4.1.2. (b) Define $f\colon (0,1) \to (3,6)$ by the rule $f(x) = 3x + 3$ for any $x \in (0,1)$. First note that if $0 < x < 1$, then $3 < 3x + 3 < 6$ and thus f is a function from $(0,1)$ to $(3,6)$. Then f is one-to-one since $f(x_1) = f(x_2)$ for $x_1, x_2 \in (0,1)$, then $3x_1 + 3 = 3x_2 + 3$ and therefore $3x_1 = 3x_2$, which implies that $x_1 = x_2$. Also, f is onto, since if $y \in (3,6)$, then $x = (y - 3)/3 \in (0,1)$ and $f(x) = 3x + 3 = 3 \cdot (y - 3)/3 + 3 = y - 3 + 3 = y$. Since f is a one-to-one correspondence, $(0,1) \approx (3,6)$.
(c) The function $g\colon (0,1) \to (a, a + 1)$ defined by $g(x) = x + a$ for $x \in (0,1)$ is a one-to-one correspondence.

4.1.3. (a) $\mathbf{R}^+ \approx (1, \infty)$ since the function $f\colon \mathbf{R}^+ \to (1, \infty)$ defined by $f(x) = x + 1$ is a bijection.
(b) Define a function f by the rule $f(x) = 1/x$ for $x \in (0,1)$. Observe that if $0 < x < 1$, then $1/x > 1$; therefore, f is indeed a function from $(0,1)$

to $(1, \infty)$. To show that f is a bijection, we show that f is (i) injective and (ii) surjective. (i) Suppose $f(x_1) = f(x_2)$ for $x_1, x_2 \in (0,1)$; then $1/x_1 = 1/x_2$. Taking the reciprocals, we see that $x_1 = x_2$; thus f is injective. (ii) Let $y \in (1, \infty)$. Let $x = 1/y$. Then $0 < x < 1$ and $f(x) = 1/x = 1/(1/y) = y$. Therefore, f is surjective. We have proved that f is a bijection.

4.1.7. Since $A \approx B$, there exists a one-to-one correspondence $f\colon A \to B$. To show that $\mathscr{P}(A) \approx \mathscr{P}(B)$, we must define a bijective function $F\colon \mathscr{P}(A) \to \mathscr{P}(B)$. Let $A_1 \in \mathscr{P}(A)$. Then $A_1 \subseteq A$. Let $B_1 = \{ y \in B |$ There exists $x \in A_1$ such that $y = f(x)\}$. In words, B_1 is the image of A_1 under the function f. Define $F\colon \mathscr{P}(A) \to \mathscr{P}(B)$ by the rule $F(A_1) = B_1$ for all $A_1 \in \mathscr{P}(A)$.

We show that F is one-to-one. Suppose $A_1, A_1' \in \mathscr{P}(A)$ with $F(A_1) = F(A_1')$. We show that $A_1 = A_1'$. To do so, we show that $A_1 \subseteq A_1'$ and $A_1' \subseteq A$. First $A_1 \subseteq A_1'$: Let $x \in A_1$, then $f(x) \in F(A_1) = F(A_1')$. Thus there exists $x' \in A_1'$ such that $f(x) = f(x')$. Since f is a one-to-one function, $x = x'$ and $x \in A_1'$. Thus $A_1 \subseteq A_1'$. An identical argument shows that $A_1' \subseteq A_1$. Thus F is one-to-one. We leave the proof that F is onto for the reader.

4.1.11. (a) $f(x) = 1/2 + (x - 1/2)/2(x - 1)$.

4.1.13. (a) Let $A = E \cup (\mathbf{Q} - \mathbf{Z})$. Define $f\colon A \to \mathbf{Q}$ by

$$f(x) = \begin{cases} x/2 & \text{if } x \in E \\ x & \text{if } x \in \mathbf{Q} - \mathbf{Z}. \end{cases}$$

Note that A consists of the union of the set of even integers with the set of nonintegral rationals. The function f does not disturb $\mathbf{Q} - \mathbf{Z}$ and maps E onto $\mathbf{Z}$. We can show that f is a bijection. Alternatively, we can use Theorem 4.1.2 to show that $A \approx \mathbf{Q}$: $A = E \cup (\mathbf{Q} - \mathbf{Z})$ and $\mathbf{Q} = \mathbf{Z} \cup (\mathbf{Q} - \mathbf{Z})$. Since $E \approx \mathbf{Z}$, $\mathbf{Q} - \mathbf{Z} \approx \mathbf{Q} - \mathbf{Z}$ and $E \cap (\mathbf{Q} - \mathbf{Z}) = \mathbf{Z} \cap (\mathbf{Q} - \mathbf{Z}) = \varnothing$, $A \approx \mathbf{Q}$.

Section 4.2

4.2.2. Proof by contradiction. Suppose that A is finite, B is infinite, $A \subseteq B$, and $B - A$ is not infinite. Then $B - A$ is finite and by Theorem 4.2.3, $B = A \cup (B - A)$ is finite. Thus B is both finite and infinite, a contradiction. Therefore, $B - A$ is infinite.

4.2.3. Define a function $\phi\colon A \to \mathscr{P}(A)$ by $\phi(a) = \{a\}$ for all $a \in A$. Since $a \in A$, $\{a\} \subseteq A$ and hence $\{a\} \in \mathscr{P}(A)$. Also, if $a, b \in A$ and $a \neq b$, then $\{a\} \neq \{b\}$. Therefore, ϕ is an injective function. Hence $A \approx A' = \{\{a\} | a \in A\}$. Since $\mathscr{P}(a)$ is finite and $A' \subseteq \mathscr{P}(A)$, A' is finite by Theorem 4.2.2. Since $A \approx A'$, A is finite.

4.2.5. (a) Proof by induction. For $n \geq 2$, let $S(n)$ be the statement: If $A_1, \ldots, A_n$ are pairwise disjoint finite sets, then $|A_1 \cup \cdots \cup A_n| = |A_1| + \cdots + |A_n|$.

Basis step The case $n = 2$ is covered by Theorem 4.2.3.

Inductive step Suppose that $S(n)$ holds. We show that $S(n+1)$ holds. To do so, assume that $A_1, \ldots, A_{n+1}$ are pairwise disjoint finite sets. Let $A = A_1 \cup \cdots \cup A_{n+1}$. Then $A = (A_1 \cup \cdots \cup A_n) \cup A_{n+1}$ and $(A_1 \cup \cdots \cup A_n) \cap A_{n+1} = (A_1 \cap A_{n+1}) \cup \cdots \cup (A_n \cap A_{n+1}) = \varnothing$ since A_{n+1} is disjoint from each of $A_1, \ldots, A_n$. Therefore, by Theorem 4.2.3 (i.e., by $S(2)$),

$$|A| = |A_1 \cup \cdots \cup A_n| + |A_{n+1}|.$$

By hypothesis, $S(n)$ holds, and hence

$$|A_1 \cup \cdots \cup A_n| = |A_1| + \cdots + |A_n|.$$

Therefore,

$$|A| = |A_1 \cup \cdots \cup A_{n+1}| = |A_1| + \cdots + |A_{n+1}|.$$

4.2.8. Use PMI with Theorem 4.2.4 as the basis step. The proof of the inductive step is patterned after the solution to Exercise 4.2.5(a).

4.2.9. Proof by contradiction (outline). Note that $A = \cup_{b \in B} f^{-1}(b)$ (see Proposition 3.4.5). Thus if $f^{-1}(b)$ is finite for all $b \in B$, then $A = \cup_{b \in B} f^{-1}(b)$ is a union of a finite number of finite sets, and by Exercise 4.2.5(b), is also a finite set. This contradicts the assumption that A is infinite. Thus $f^{-1}(b)$ is infinite for at least one $b \in B$.

4.2.14. (a) Let three integers be given. Then either two of these integers are even or two of these integers are odd. Call these integers x and y. Then $x - y$ is even since it is either a difference of two even integers or a difference of two odd integers.

(b) Let a_1, a_2, a_3, a_4 be integers. By the Division Theorem, for $1 \le i \le 4$, there exist integers q_i and r_i, $1 \le i \le 4$ such that

$$a_i = 3 \cdot q_i + r_i \qquad \text{where } 0 \le r_i \le 2 \quad \text{for } 1 \le i \le 4.$$

Thus the four integers r_1, r_2, r_3, r_4 are members of the set $\{0, 1, 2\}$. Hence $r_i = r_j$ for some pair i, j such that $i \ne j$ and $1 \le i, j \le 4$. Let $x = a_i$ and $y = a_j$. Then $x - y = 3 \cdot q_i + r_i - (3 \cdot q_j + r_j) = 3(q_i - q_j)$ is divisible by 3.

Section 4.3

4.3.1. (a) Since $\mathbf{Z} \approx \mathbf{N}$, $\mathbf{Z} \times \mathbf{N} \approx \mathbf{N} \times \mathbf{N}$ by Theorem 4.1.4. By Theorem 4.3.2, $\mathbf{N} \times \mathbf{N} \approx \mathbf{N}$. Therefore, $\mathbf{Z} \times \mathbf{N} \approx \mathbf{N}$ and $\mathbf{N}$ is denumerable.

4.3.3. (a) Define $f\colon \mathbf{N} \times \mathbf{N} \times \mathbf{N} \to \mathbf{N}$ by the rule $f(a, b, c) = 2^a \cdot 3^b \cdot 5^c$ for $(a, b, c) \in \mathbf{N} \times \mathbf{N} \times \mathbf{N}$. By an argument similar to the first proof of Theorem 4.3.2, it follows that f is one-to-one. Thus $\mathbf{N} \times \mathbf{N} \times \mathbf{N} \approx A$

where $A = \{n \in \mathbf{N} \mid \text{There exist } a, b, c \in \mathbf{N} \text{ such that } n = 2^a \cdot 3^b \cdot 5^c\}$.

(b) Note that A is an infinite set since for all $k \in \mathbf{N}$, $2^k \in A$. Since A is infinite and $A \subseteq \mathbf{N}$, A is denumerable by Theorem 4.3.1. Therefore, $\mathbf{N} \times \mathbf{N} \times \mathbf{N} \approx A$ is also denumerable.

4.3.6. (a) $A \sim A$: $(A - A) \cup (A - A) = \varnothing$ is finite and hence countable. Therefore $A \sim A$.

(b) If $A \sim B$, then $B \sim A$. Since $A \sim B$, $(A - B) \cup (B - A)$ is countable. But $(A - B) \cup (B - A) = (B - A) \cup (A - B)$; thus $(B - A) \cup (A - B)$ is countable and $B \sim A$.

(c) If $A \sim B$ and $B \sim C$, then $A \sim C$. We must show that $(A - C) \cup (C - A)$ is countable. But $A - C \subseteq (A - B) \cup (B - C)$ and $C - A \subseteq (C - B) \cup (B - A)$ (proof: exercise), which implies that $(A - C) \cup (C - A) \subseteq [(A - B) \cup (B - A)] \cup [(B - C) \cup (C - B)]$. Since $A \sim B$ and $B \sim C$, the set on the right is the union of two countable sets and is itself countable by Theorem 4.3.5(ii). Thus $(A - C) \cup (C - A)$ is countable and $A \sim C$.

4.3.7. Proof by contradiction. If $B - A$ is countable and A is denumerable, then $B = (B - A) \cup (B \cap A)$ is the union of two countable sets and is countable by Theorem 4.3.5(ii). But by assumption, B is uncountable. This contradiction means that the assumption that $B - A$ is countable is incorrect. Therefore, $B - A$ is uncountable.

4.3.8. We show that B is countable by finding a subset A' of A such that $B \approx A'$. Since A is denumerable, the set A' is either finite or denumerable and hence is countable.

Since f is surjective, for each $b \in B$ there exists $a_b \in A$ such that $f(a_b) = b$. Let $A' = \{a_b \mid b \in B\}$. Clearly, the function ϕ: $B \to A'$ defined by $\phi(b) = a_b$ is a bijection. Thus $B \approx A'$.

Section 4.4

4.4.1. If B is countable and $A \subseteq B$, then by Theorem 4.3.1(ii), A is countable. By assumption, A is uncountable, and, therefore, the assumption that B is countable is incorrect. Thus B is uncountable.

4.4.2. First, we show that if A is uncountable, then $A \times A$ is uncountable. As is often the case in proofs of uncountability, an indirect argument is appropriate. Suppose $A \times A$ is countable. Let $a_0 \in A$. Then the set $A_0 = \{(a_0, a) \mid a \in A\}$ is a subset of $A \times A$, which is equinumerous with A. (The function f: $A_0 \to A$ defined by $f(a_0, a) = a$ is a bijection.) But since $A \times A$ is countable and $A_0 \subseteq A \times A$, A_0 is countable. Since $A \approx A_0$, A is countable. This conclusion contradicts the hypothesis that A is uncountable. Therefore, $A \times A$ is uncountable.

Conversely, we must show that if $A \times A$ is uncountable, then A is uncountable. We argue by contraposition. Suppose that A is countable. Then by Theorem 4.2.4 and Corollary 4.3.3, $A \times A$ is countable. This completes the proof.

4.4.3. (a) Proof by contradiction. If $(0,1]$ is countable, then its subset $(0,1)$ is also countable by Theorem 4.3.1(ii). This conclusion contradicts Theorem 4.4.1. Thus $(0,1]$ is uncountable.

(c) $(a,b) \approx (0,1)$ by Exercise 4.1.2(f). Thus by Theorem 4.4.1, (a,b) is uncountable.

4.4.10. (a) Reflexivity. For all $x \in \mathbf{R}$, $x \sim x$ since $x - x = 0 \in \mathbf{Q}$.

Symmetry. For $x, y \in \mathbf{R}$ if $x \sim y$, then $x - y \in \mathbf{Q}$ and hence $-(x-y) = y - x \in \mathbf{Q}$. Therefore, $y \sim x$.

Transitivity. If $x \sim y$ and $y \sim z$, then $x - y, y - z \in \mathbf{Q}$. It follows that $(x-y) + (y-z) = x - z \in \mathbf{Q}$ and $x \sim z$.

Therefore, $\sim$ is an equivalence relation on $\mathbf{R}$.

(b) For all $x \in \mathbf{R}$, $[x]_\sim = \{y \mid y \sim x\} = \{y \mid y - x \in \mathbf{Q}\} = \{y \mid y = x + r$ where $r \in \mathbf{Q}\}$. Now $[x]_\sim \approx \mathbf{Q}$ since the function $f\colon [x]_\sim \to \mathbf{Q}$ defined by $f(y) = r$ if $y = x + r$ where $r \in \mathbf{Q}$ is a bijection.

(c) The set of equivalence classes is uncountable. If not, then $\mathbf{R}$, being the union of the set of equivalence classes by Theorem 3.3.2, is a union of a countable collection of sets. By part (b), each of these sets is countable. Therefore, if the set of equivalence classes is countable, then $\mathbf{R}$ is a union of a countable number of countable sets and is countable. Therefore, the set of equivalence classes is uncountable.

Section 4.5

4.5.4. (a) To show that $(0,1) \approx [0,1)$ we use Cantor–Bernstein. Since $(0,1) \subseteq [0,1)$, we need only show that $[0,1) \approx A$ where $A \subseteq (0,1)$. Let $A = [\frac{1}{4}, \frac{3}{4})$. If $f\colon [0,1) \to A$ is defined by $f(x) = (\frac{1}{2})x + (\frac{1}{4})$, then f is a bijection and $[0,1) \approx A$. Therefore, by Cantor–Bernstein $(0,1) \approx [0,1)$.

4.5.5. To show $(0,1) \times (0,1) \approx [0,1] \times [0,1]$, we appeal to Cantor–Bernstein. Define $F\colon [0,1] \times [0,1] \to (0,1) \times (0,1)$ by $F(x,y) = ((\frac{1}{2})x + (\frac{1}{4}), (\frac{1}{2})y + (\frac{1}{4}))$. It is easy to check that F is a one-to-one correspondence between $[0,1] \times [0,1]$ and $[\frac{1}{4}, \frac{3}{4}] \times [\frac{1}{4}, \frac{3}{4}] \subseteq (0,1) \times (0,1)$. Since $(0,1) \times (0,1) \subseteq [0,1] \times [0,1]$, each of $(0,1) \times (0,1)$ and $[0,1] \times [0,1]$ is equinumerous with a subset of the other. Thus by the Cantor–Bernstein Theorem, $(0,1) \times (0,1) \approx [0,1] \times [0,1]$.

Chapter 5

Section 5.1

5.1.2. There are six ways of placing three nonattacking indistinguishable rooks on a 3×3 chessboard.

5.1.6.

n	1	2	3	4	5
p_n	2	4	7	11	16

5.1.8. By considering examples in which $n \leq 6$, we might conjecture that if A has exactly n elements, then A contains n subsets having exactly one element, n subsets having exactly $n - 1$ elements, and $n(n - 1)/2$ subsets having exactly two elements.

Section 5.2

5.2.1. (a) $(2n!)/n! = [1 \cdot 3 \cdot 5 \cdot \cdots \cdot (2n - 1)][2 \cdot (2 \cdot 2)(2 \cdot 3) \cdots (2 \cdot n)]/n! = 2^n[1 \cdot 3 \cdot \cdots \cdot (2n - 1)]$, and hence $x = 2^n$.
(b) $(n + 1) \cdot (n + 2) \cdot \cdots \cdot (2n) = (2n)!/n! = 2^n \cdot (1 \cdot 3 \cdot 5 \cdot \cdots \cdot (2n - 1)) = 2 \cdot 6 \cdot 10 \cdot \cdots \cdot 2(2n - 1)$. Thus $y = 2 \cdot (2n - 1)$.

5.2.2. (a) $|A| = 19$.

5.2.3. (a) $|A_1 \cup A_2| = |A_1| + |A_2| - |A_1 \cap A_2| = 8 + 7 - 3 = 12$.
(b) $|A_1 \cap A_2| = |A_1| + |A_2| - |A_1 \cup A_2| = 5 + 4 - 7 = 2$.

5.2.4. For $n = 4$, the Inclusion–Exclusion Principle reads: $|A_1 \cup A_2 \cup A_3 \cup A_4| = |A_1| + |A_2| + |A_3| + |A_4| - |A_1 \cap A_2| - |A_1 \cap A_3| - |A_1 \cap A_4| - |A_2 \cap A_3| - |A_2 \cap A_4| - |A_3 \cap A_4| + |A_1 \cap A_2 \cap A_3| + |A_1 \cap A_2 \cap A_4| + |A_1 \cap A_3 \cap A_4| + |A_2 \cap A_3 \cap A_4| - |A_1 \cap A_2 \cap A_3 \cap A_4|$.

5.2.5. There are 20 choices for the president. Once the president is chosen, the vice-president can be picked in 19 ways. After these officers are chosen, the secretary can be selected in 18 ways. Thus the three officers can be picked in $20 \cdot 19 \cdot 18$ ways.

5.2.7. (a) Every four-digit palindrome has the form *abba* where $a, b \in \mathbf{Z}$, $1 \leq a \leq 9$, and $0 \leq b \leq 9$. Thus there are $9 \cdot 10 = 90$ four-digit palindromes.

5.2.11. The number is $14 \cdot 13 \cdot 12$.

5.2.12. (a) The first rook can be placed in any of 64 squares. The second rook can be placed on any of the 63 remaining squares. After the first seven rooks have been placed, the eighth rook can be put down on any of the 57 remaining squares. Thus there are $64 \cdot 63 \cdot 62 \cdot 61 \cdot 60 \cdot 59 \cdot 58 \cdot 57$ ways of placing eight indistinguishable rooks on an 8×8 chessboard.
(b) The eight rooks must be placed in different columns. The rook that goes in the first column has eight possible placements, one for each row. Then the rook that goes in the second column can be placed in any of the seven remaining rows. In general, the rook in column k can be placed in any of the $(8 - k + 1)$ rows in which a rook has not been placed. Thus there are $8 \cdot 7 \cdot 6 \cdot 5 \cdot 4 \cdot 3 \cdot 2 \cdot 1 = 8!$ ways of placing eight nonattacking rooks on an 8×8 chessboard.

5.2.15. Proof by contradiction. Let $B = \text{Ran}(f) = \{y \in A \mid y = f(x) \text{ for some } x \in A\}$. If f is not surjective, then $B \neq A$ and $|B| < |A|$. Since f is a function from A to B, by the Pigeonhole Principle f is not injective, which contradicts our assumption. Thus f is surjective.

Section 5.3

5.3.2. (a) The letters in *land* can be reordered in 4! ways.

(b) If the letters of *mara* were distinct, then the letters could be reordered in $4! = 24$ different ways. For the moment, regard the a's in *mara* as distinct. For example, let us write the word as ma_1ra_2. For each rearrangement of these letters, such as a_1mra_2, we can obtain a different arrangement by interchanging a_1 and a_2: a_2mra_1. Now if we identify the letters a_1 and a_2 with the letter a, then we obtain the same rearrangement of *mara*: *amra*. Since each rearrangement of *mara* arises from two permutations of ma_1ra_2, there are $4!/2 = 12$ ways of reordering the letters of *mara*.

5.3.4. The women can be placed in positions 1, 3, 5, 7, 9, 11, and 13 in 7! ways. For each such placement, the men can be put into positions 2, 4, 6, 8, 10, 12, and 14 in 7! ways. Thus by the multiplication principle, there are $(7!) \cdot (7!)$ ways of arranging the people as stated.

5.3.5. (a) When the eight rooks are placed on the board, exactly one row and one column contain no rook. There are $9 \cdot 9$ ways of choosing this row and column. Once the row and column that contain no rook are chosen, the eight rooks can be placed on the remaining nine rows and columns in 8! ways. Thus there are $9^2 \cdot 8!$ ways of placing the rooks on the board.

5.3.6. (a) If the people were seated in a row, then there would be 4! ways of seating them. However, four distinct linear arrangements determine the same circular arrangement: 1234, 2341, 3412, and 4123 all determine the same seating pattern around the table. Thus, there are $4!/4 = 6$ ways of seating four people around a table.

5.3.9. $C(n,2) = n!/2!(n-2)! = n(n-1)/2$ by Corollary 5.3.4.

5.3.10. By Corollary 5.3.4,

$$\begin{aligned} C(n,k) = n!/k!(n-k)! &= n!/((n-k)! \cdot k!) \\ &= n!/((n-k)! \cdot (n-(n-k))! \\ &= C(n, n-k) \end{aligned}$$

5.3.11. Suppose $|A| = n$. We know from Chapter 2 that $|\mathscr{P}(A)| = 2^n$. On the other hand, we can determine $|\mathscr{P}(A)|$ by counting the number of subsets of A having k elements as k ranges from 0 to n. This number is $C(n,0) + C(n,1) + \cdots + C(n,n) = 2^n$.

Section 5.4

5.4.3. Using the recurrence relation, we can evaluate the left side:

$$\binom{n}{k} + 2\binom{n}{k-1} + \binom{n}{k-2} = \binom{n}{k} + \binom{n}{k-1} + \binom{n}{k-1} + \binom{n}{k-2}$$

$$= \binom{n+1}{k} + \binom{n+1}{k-1} = \binom{n+2}{k}.$$

Hence $y = n + 2$.

5.4.5. Setting $x = 2$ in the Binomial Theorem, we find that

$$3^n = (1 + 2)^n = \sum_{k=0}^{n} \binom{n}{k} 2^k.$$

5.4.6. (a) A generalization is: For $n \geq 2$,

$$\binom{2}{2} + \binom{3}{2} + \binom{4}{2} + \cdots + \binom{n}{2} = \binom{n+1}{3}.$$

(b) Combinatorially, the result asserts that the number of 3-element subsets of the set $\{1, 2, \ldots, n + 1\}$ equals the sum of the number of 2-element subsets of $\{1, \ldots, k\}$ as k varies from 2 to n.

5.4.10. (a) $(2x + y)^3 = 8x^3 + 12x^2y + 6xy^2 + y^3$.
(b) $(2x - y)^4 = 16x^4 - 32x^3y + 48x^2y^2 - 8xy + y^4$.

5.4.11. Hint: Use the Binomial Theorem.

5.4.12. (Outline.) For $n \geq 0$, let $S(n)$ be the statement $S(n)$: For $0 \leq k \leq n$, $a_{n,k} = \binom{n}{k}$. Show by PMI that $S(n)$ holds for $n \geq 0$. The basis step is immediate. The inductive step follows from the recurrence relation for $\binom{n}{k}$ and the conditions satisfied by $a_{n,k}$.

5.4.13. Let $A = \{1, \ldots, n\}$. Let $X = \{(x, B) | x \in A,\ B \subseteq A \text{ and } x \in B\}$. The element 1 appears in any set of the form $B = \{1\} \cup B'$ where $B' \subseteq \{2, \ldots, n\}$. Thus 1 appears in 2^{n-1} subsets and is therefore counted 2^{n-1} times. The same result holds for any of the numbers $1, 2, \ldots, n$. Thus $|X| = n \cdot 2^{n-1}$. On the other hand, there are $\binom{n}{k}$ k-element subsets $B \subseteq A$ and each such set contributes k elements (of the form (x, B)) to X. Therefore, there are $k \cdot \binom{n}{k}$ members of X of the form (x, B) where $|B| = k$ and $x \in B$. Thus $|X| = \sum_{k=1}^{n} k\binom{n}{k} = n \cdot 2^{n-1}$.

5.4.15. Hint: $x + y = x(1 + y/x)$.

Section 5.5

5.5.1. (a) The characteristic equation is $r^2 - r - 2 = 0$. The roots of this equation are $r = 2$ and $r = -1$, and thus by Theorem 5.5.1(i), the general solution is $a_n = \alpha_1 \cdot 2^n + \alpha_2(-1)^n$. From the initial conditions, we find that $1 = \alpha_1 + \alpha_2$ and $2 = 2\alpha_1 - \alpha_2$ from which we conclude that $\alpha_1 = 1$ and $\alpha_2 = 0$. Thus the solution to the recurrence is $a_n = 2^n$.

5.5.2. (a) Let $f(X) = \sum_{n=0}^{\infty} a_n X^n$. Since $a_n = 3a_{n-1}$ for $n \geq 1$, we have

$$\begin{aligned} f(X) &= a_0 + a_1 X + a_2 X^2 + \cdots + a_n X^n + \cdots \\ &= a_0 + 3 \cdot a_0 X + 3a_1 X^2 + \cdots + 3a_{n-1} X^n + \cdots \\ &= a_0 + 3 \cdot X\left(a_0 + a_1 X + \cdots + a_{n-1} X^{n-1} + \cdots\right) \\ &= a_0 + 3 \cdot X \cdot f(X). \end{aligned}$$

Therefore,

$$f(X)(1 - 3 \cdot X) = a_0$$

and

$$\begin{aligned} f(X) &= a_0/(1 - 3 \cdot X) = 1/(1 - 3X) \\ &= 1 + 3X + 3^2 X^2 + \cdots + 3^n X^n + \cdots. \end{aligned}$$

Therefore, $a_n = 3^n$.

5.5.3.

n	1	2	3	4	5
RB_n	1	2	3	5	8

In general, when the robot moves n meters, the last step is either of length 1 m or of length 2 m. Thus the robot moves n m by moving $(n - 1)$ m and then 1 m or by moving $(n - 2)$ m and then 2 m. Therefore, $\mathrm{RB}_n = \mathrm{RB}_{n-1} + \mathrm{RB}_{n-2}$.

5.5.6. (a) Let Q_n be the set of arrangements of nonattacking indistinguishable rooks on an $n \times n$ board which are symmetric about the diagonal, D, running from lower left to upper right. Then $Q_n = A_n \cup B_n$ where A_n is the set of arrangements in which the lower left corner is occupied and B_n is the set of arrangements in which the lower left corner is unoccupied. Clearly, $A_n \cap B_n = \emptyset$.

Let $\alpha \in A_n$. Thus α is an arrangement of n nonattacking indistinguishable rooks on an $n \times n$ board that is symmetric about D in which the lower left corner is occupied. If the left column and bottom row of the board are removed, then the resulting arrangement of rooks on the

$(n-1) \times (n-1)$ board is also symmetric about D. Conversely, each arrangement of nonattacking rooks on an $(n-1) \times (n-1)$ board determines an element $\alpha \in A_n$ by the addition of a column (on the left) and a row (on the bottom) to create an $n \times n$ board and by the placement of a rook in the lower left corner of the resulting $n \times n$ board. Thus $|A_n| = Q_{n-1}$.

Let $\beta \in B_n$. Then β contains a rook in row k of column 1 where $k < n$ and by symmetry in column k of row n. By removing rows k and n and columns 1 and k, we obtain a symmetric arrangement on an $(n-2) \times (n-2)$ board. Conversely, each arrangement on an $(n-2) \times (n-2)$ board determines an $n \times n$ symmetric arrangement by adjoining rows k and n and columns 1 and k. Since k varies from 1 to $n-1$, each $(n-2) \times (n-2)$ symmetric arrangement determines $(n-1)$ $n \times n$ symmetric arrangements. Therefore, $|B_n| = (n-1)Q_{n-2}$. Thus $|S_n| = Q_n = Q_{n-1} + (n-1)Q_{n-2}$.

Section 5.6

5.6.1. In a simple graph, each edge is determined by a set of two distinct vertices. Since G has ν vertices, G has at most $\binom{\nu}{2}$ edges.

5.6.3. (b) $|E(K_n)| = \binom{n}{2}$ since each pair of vertices does determine an edge of K_n.

5.6.4. (a) $K_{2,4}$:

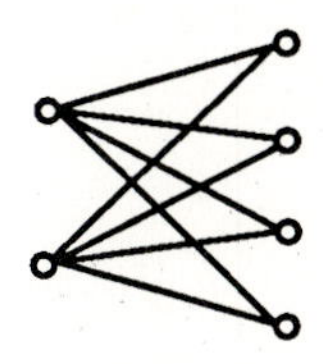

(b) Note that $|E(K_{2,4})| = 8$, $|E(K_{2,5})| = 10$, and $|E(K_{3,5})| = 15$.

5.6.5. (a) After experimenting with $n = 2, 3, 4, 5$, we might conjecture that K_n contains an Euler circuit if and only if n is odd. We can prove this conjecture using Theorem 5.6.1.

(b) K_n has a Hamilton circuit for all $n \geq 2$.

5.6.8. (a) With S as defined in the problem, we determine $|S|$ as follows: For each vertex v, there are $\deg(v)$ edges e such that e is incident to v: For each $v \in V$, let $S_v = \{(v, e) | \ e \text{ is incident to } v\}$. Then $|S_v| = \deg(v)$ and $|S| = \Sigma_{v \in V}|S_v| = \Sigma_{v \in V}\deg(v)$. Next, for each $e \in E$, let $V_e = \{(v, e) | e \text{ is incident to } v\}$. Then $|V_e| = 2$ and $|S| = \Sigma_{e \in E}|V_e| = \Sigma_{e \in E}2 = 2 \cdot \epsilon$.

5.6.12. (a) Proof by contradiction (outline). Suppose G is disconnected. Then there exist two subsets, V_1 and V_2, of V such that $V_1 \cup V_2 = V$, $V_1 \cap V_2 = \emptyset$, and no vertex of V_1 is connected by a path to any vertex of V_2. Now either $|V_1| \leq \nu/2$ or $|V_2| \leq \nu/2$. Without loss of generality, we may assume that $|V_1| = \nu/2$. Let $v \in V_1$. Then $\deg(v) \leq |V_1| - 1 < (\nu - 1)/2$, which contradicts the hypothesis.

Miscellaneous Exercises

3. Let M be the set of married women in the survey; let E be the set of employed women; let B be the set of blond women. We must find $|M \cap E \cap B|$. By Inclusion–Exclusion,

$$|M \cap E \cap B| = |M \cup E \cup B| - |M| - |E| - |B|$$

$$+ |M \cap E| + |M \cap B| + |E \cap B|.$$

Now $|M| = |E| = |B| = 200$, $|M \cap E| = 104$, $|M \cap B| = 116$, and $|E \cap B| = 93$. Also, $|M \cup E \cup B| + |(M \cup E \cup B)^c| = |M \cup E \cup B| + |M^c \cap E^c \cap B^c| = 400$ and hence $|M \cup E \cup B| = 347$. Thus $|M \cap E \cap B| = 60$.

5. Hint: Compare with Example 5.2.9.

7. Let $n = 2 \cdot k + 1$. Thus k of the integers $a_1, \ldots, a_n$ are even and $k + 1$ of these integers are odd. It follows that at least one of the integers $a_1, a_3, \ldots, a_{2k+1} = a_n$ is odd. Let a_i be this integer. Then $a_i - i$ is even and the given product is even.

Chapter 6

Section 6.1

6.1.1. (a) An associative and commutative binary operation.
(b) A binary operation.
(c) A binary operation.
(d) An associative, commutative binary operation that has an identity and for which inverses exist.
(e) Not a binary operation.
(f) A binary operation.

6.1.2. (a) Let $*$ be a binary operation of a set A. Suppose e_1 and e_2 are identities for $*$. We show $e_1 = e_2$. Since e_2 is an identity for $*$, $e_1 = e_1 * e_2$. But e_1 is also an identity for $*$ and hence $e_1 * e_2 = e_2$. Therefore, $e_1 = e_2$.

6.1.3. (a) Let us try to construct a simple example. Let $A = \{a, b\}$ and consider the operation $*_1$ defined by the table:

	a	b
a	b	a
b	a	a

Since $a * b = b * a$, $*$ is commutative. Since $a * a = b \neq a$, $*$ is not idempotent. To show $*$ is not associative, consider $a *(a * b)$ and $(a * a) * b$: $a *(a * b) = a * a = b$ while $(a * a) * b = b * b = a$. Thus $*$ is commutative but neither associative nor idempotent.

6.1.4. (a) There are four idempotent binary operations on $\{a, b\}$. To see this, consider how an operation table can be formed for an idempotent operation on $\{a, b\}$. The entries $a * a$ and $b * b$ must be a and b, respectively. Each of the entries $a * b$ and $b * a$ can be either a or b. Thus there are $2 \cdot 2 = 4$ ways to fill in both $a * b$ and $b * a$.
(b) There are 3^6 idempotent binary operations of $\{a, b, c\}$.

6.1.8. Since

$$\det\begin{bmatrix} 2 & & & 0 \\ & 1 & & \\ & & \ddots & \\ 0 & & & 1 \end{bmatrix} = \begin{bmatrix} 1 & & & 0 \\ & \ddots & & \\ & & 1 & \\ 0 & & & 2 \end{bmatrix} = 2,$$

det is not injective and therefore is not an isomorphism.

Section 6.2

6.2.4. Proof by mathematical induction. For $n \in \mathbf{Z}$, $n \geq 1$, let $S(n)$ be the statement: If p is prime and $a \in \mathbf{Z}$ and $p | a^n$, then $p | a$.

Basis step Clearly, $S(1)$ holds.

Inductive step Suppose $S(n)$ holds and show $S(n + 1)$ holds. Thus we suppose that p is prime, and that $p | a^{n+1}$ where $a \in \mathbf{Z}$. We must show that $p | a$. Now $a^{n+1} = a^n \cdot a$ and thus $p | a^n \cdot a$. By Lemma 6.2.2, $p | a^n$ or $p | a$. But if $p | a^n$, then by inductive hypothesis $p | a$. Therefore, if $p | a^{n+1}$, then $p | a$.

Thus by PMI, the given statement is proved.

6.2.8. (a) $\sigma(p) = 1 + p$, $\sigma(p^2) = 1 + p + p^2$, and $\sigma(p^a) = 1 + p + \cdots + p^a = (p^{a+1} - 1)/(p - 1)$.

6.2.10. If $a, b \in \mathbf{Z}$ and $a | b$, then $\gcd(a, b) = |a|$.

6.2.12. (a) $\operatorname{lcm}(6, 14) = 42$.
(b) $\operatorname{lcm}(91, 55) = 91 \cdot 55$.

6.2.13. Express a and b as a product of positive primes: $a = p_1^{c_1} \cdot \cdots \cdot p_r^{c}$ and $b = q_1^{d_1} \cdot \cdots \cdot q_s^{d_s}$ where the $p_i \neq p_j$ for $i \neq j$ and $q_i \neq q_j$ for $i \neq j$. Since $\gcd(a, b) = 1$, $p_i \neq q_j$ for $1 \leq i \leq r$ and $1 \leq j \leq s$. Thus $k^2 = a \cdot b = p_1^{c_1} \cdot \cdots \cdot p_r^{c_r} q_1^{d_1} \cdot \cdots \cdot q_s^{d_s}$ where the primes $p_1, \ldots, p_r, q_1, \ldots, q_s$ are all distinct. By Theorem 6.2.4, each c_i is even and each d_j is even. Therefore, for $1 \leq i \leq r$, $c_i = 2e_i$ where $e_i \in \mathbf{Z}^+$, and for $1 \leq j \leq s$, $d_j = 2f_j$ where $f_j \in \mathbf{Z}^+$. Thus $a = a_1^2$ and $b = b_1^2$ where $a_1 = p_1^{e_1} \cdot \cdots \cdot p_r^{e_r}$ and $b_1 = q_1^{f_1} \cdot \cdots \cdot q_s^{f_s}$.

6.2.14. Proof by contradiction (outline). Suppose $\sqrt{3} \in \mathbf{Q}$. Then $\sqrt{3} \in \mathbf{Q}$. Then $\sqrt{3} = a/b$ where $a, b \in \mathbf{Z}^+$. Hence $3 = a^2/b^2$ and $a^2 = 3 \cdot b^2$. Factor a and b into primes and count the number of times the prime 3 appears on each side. In a^2, 3 appears an even number of times. In $3 \cdot b^2$, 3 appears an odd number of times. Since $a^2 = 3 \cdot b^2$, a contradiction of the Unique Factorization Theorem is obtained. Thus $\sqrt{3} \notin \mathbf{Q}$.

6.2.19. Suppose $\sqrt[3]{3} \in \mathbf{Q}$. Then $\sqrt[3]{3} = a/b$ where $a, b \in \mathbf{Z}^+$. It follows that $a^3 = 3 \cdot b^3$. Now factor a and b into positive primes. In the resulting prime factorization of a^3, 3 appears $3 \cdot m$ times for some integer m; in the resulting prime factorization of $3 \cdot b^3$, 3 appears $3 \cdot n + 1$ times for some integer n. Since $3 \cdot m \neq 3 \cdot n + 1$ for all $m, n \in \mathbf{Z}$, 3 does not appear the same number of times in the prime factorizations of $a^3 = 3 \cdot b^3$. This contradiction of Theorem 6.2.1 means that $\sqrt[3]{3} \notin \mathbf{Q}$.

Section 6.3

6.3.1. (c) By distributivity and commutativity,

$$\begin{aligned}(a - b) \cdot (a + b) &= (a - b) \cdot a + (a - b) \cdot b \\ &= a \cdot a - b \cdot a + a \cdot b - b \cdot b \\ &= a^2 - ab + ab - b^2 = a^2 - b^2.\end{aligned}$$

(d) If y and z are multiplicative inverses of x, then $y = y \cdot 1 = y \cdot (x \cdot z) = (y \cdot x) \cdot z = 1 \cdot z = z$.

6.3.2. (a) $\{[1], [3]\}$ is the set of units in $\mathbf{Z}_4$.
(c) $\{[1], [5]\}$ is the set of units in $\mathbf{Z}_6$.
(d) $\{[1], [2], [3], [4], [5], [6]\}$ is the set of units in $\mathbf{Z}_7$.

6.3.3. (a) $\{[2]\}$ is the set of zero divisors in $\mathbf{Z}_4$.
(c) $\{[2], [3], [4]\}$ is the set of zero divisors in $\mathbf{Z}_6$.
(d) $\varnothing$ is the set of zero divisors in $\mathbf{Z}_7$.

6.3.5. (a) We prove that if $f(X) \in \mathbf{R}[X]$ is nonconstant, then $f(X)$ is not a unit in $\mathbf{R}[X]$. If $f(X) \in \mathbf{R}[X]$ is nonconstant, then $\deg(f(X)) > 0$. Thus for all $g(X) \in \mathbf{R}[X]$, $\deg(f(X) \cdot g(X)) = \deg(f(X)) + \deg(g(X)) \geq \deg(f(X)) > 0 = \deg(1)$. Thus for all $g(X) \in \mathbf{R}[X]$, $f(X) \cdot g(X) \neq 1$, and f is not a unit in $\mathbf{R}[X]$.

(c) Let $f(X) = X^3 - 3$. If $f(X)$ is reducible in $\mathbf{Q}[X]$, then $f(X) = g(X) \cdot h(X)$ where $\deg(g(X)) > 0$ and $\deg(h(X)) > 0$. Thus either $\deg(g(X)) = 1$ and $\deg(h(X)) = 2$ or $\deg(g(X)) = 2$ and $\deg(h(X)) = 1$. We have $X^3 - 3 = (aX + b)(cX^2 + dX + e)$ where $a, b, c, d, e \in \mathbf{Q}$ and $a \neq 0$. Then

$$(-b/a)^3 - 3 = \big(a(-b/a) + b\big)\big(c \cdot (-b/a)^2 + d(-b/a) + e\big) = 0;$$

i.e., $(-b/a)^3 = 3$. Therefore, $\sqrt[3]{3}$ is irrational, a contradiction of Exercise 6.2.19.

6.3.7. If $a \cdot b = a \cdot c$, then $a \cdot (b - c) = 0$. Since R is an integral domain and $a \neq 0$, $b - c = 0$ and hence $b = c$.

6.3.11. We prove that if $f(X), g(X) \in \mathbf{R}[X]$ are nonzero polynomials, then $f(X) \cdot g(X)$ is nonzero. Since $f(X) \neq 0$ and $g(X) \neq 0$, $\deg(f(X)) \geq 0$ and $\deg(g(X)) \geq 0$. Therefore, $\deg(f(X) \cdot g(X)) = \deg(f(X)) + \deg(g(X)) \geq 0$ and $f(X) \cdot g(X) \neq 0$.

Section 6.4

6.4.1. Hint: Remember the quadratic formula.

6.4.3. Let R be an integral domain having n elements where $n \in \mathbf{Z}^+$. To show that R is a field, we must prove that if $a \in R$ and $a \neq 0$, then there exists $b \in R$ such that $a \cdot b = 1$. Consider the set $A = \{a, a^2, \ldots, a^{n+1}\}$. Since $A \subseteq R$ and $|R| = n$, the elements of A are not all distinct. Thus $a^i = a^j$ where $i \neq j$ and $1 \leq i, j \leq n + 1$. Suppose $i < j$. Then $a^i = a^j = a^i \cdot a^{j-1}$ and by Exercise 6.3.7, it follows that $a^{j-i} = 1$. Thus if $b = a^{j-i-1}$, then $a \cdot b = 1$.

6.4.4. Hint: The proof of Theorem 6.3.3 carries over from $\mathbf{R}[X]$ to $F[X]$ almost word for word.

6.4.5. Hint: Mimic the proof of Theorem 6.3.2.

6.4.9. Hint: If $n \in \mathbf{Z}^+$ is not prime, then $n = a \cdot b$ where $1 < a, b < n$. Consider $[a] \cdot [b]$ in $\mathbf{Z}_n$.

6.4.10. (a) $X^2 + [1] = (X + [4])(X - [4])$ in $\mathbf{Z}_{17}[X]$.

6.4.13. Hint: Look back at the proof that there are infinitely many prime numbers in $\mathbf{Z}$ (Theorem 1.3.5).

Section 6.5

6.5.2. (a) Sym(R) has four elements: the identity, reflection about each of the two lines that bisect the rectangle, and rotation by π about the center of the rectangle.

(c) Sym(T_1) has two elements: the identity and reflection about the line L. See the figure.

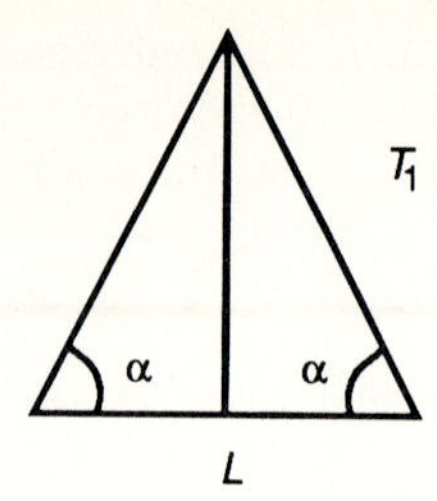

6.5.4. (f) Notice that $(g \cdot h) \cdot (h^{-1} \cdot g^{-1}) = g \cdot (h(h^{-1} \cdot g^{-1})) = g \cdot ((h \cdot h^{-1}) \cdot g^{-1}) = g \cdot e \cdot g^{-1} = g \cdot g^{-1} = e$ and that $(h^{-1} \cdot g^{-1}) \cdot (g \cdot h) = e$. Thus by part (b) of this exercise, $h^{-1} \cdot g^{-1} = (g \cdot h)^{-1}$.
(g) This result follows from (f) and (e).

6.5.6. First we show that r is one-to-one. Suppose that $r(g) = r(h)$. Then $g \cdot a = h \cdot a$. Hence $g = g \cdot e = g \cdot (a \cdot a^{-1}) = (g \cdot a) \cdot a^{-1} = (g \cdot a) \cdot a^{-1} = (h \cdot a) \cdot a^{-1} = h \cdot (a \cdot a^{-1}) = h \cdot e = h$. Thus $r(g) = r(h)$ implies that $g = h$, and r is one-to-one. Next we show that r is onto. Let $z \in S \cdot a$. Then $z = g \cdot a$ for some $g \in S$ and hence $z = g \cdot a = r(g)$. Thus r is onto. Therefore, r is a one-to-one correspondence or, if you will, a bijection.

6.5.13. (a) Using the notation of Example 6.6.5, $H_1 = \{e, f_2\}$, $H_2 = \{e, f_3\}$, and $H_3 = \{e, f_6\}$ are subgroups of S_3 having order 2.
(b) The partition determined by the equivalence relation $\equiv_{H_1}$ is $P = \{H_1, \{f_3, f_5\}, \{f_4, f_6\}\}$. For example, $f_3 \equiv_{H_1} f_5$ since $f_3 \cdot f_5^{-1} = f_3 \cdot f_4 = f_2$.

6.5.16. Let g be a group of prime order. Let H be a subgroup of G. Then by Lagrange's Theorem, $|H| \,|\, |G|$. But $|G|$ is prime, which means that $|H| = 1$ or $|H| = |G|$. In the first case, $H = \{e\}$, while in the second case, $H = G$. Thus G is a simple group.

Section 6.6

6.6.2. We must show that for all $x, y \in G$, $(g \circ f)(x \cdot y) = (g \circ f)(x) \cdot (g \circ f)(y)$. Now

$$\begin{aligned}(g \circ f)(x \cdot y) &= g(f(x \cdot y)) \\ &= g(f(x) \cdot f(y)) \text{ since } f \text{ is a homomorphism} \\ &= g(f(x)) \cdot g(f(y)) \text{ since } g \text{ is a homomorphism} \\ &= (g \circ f)(x) \cdot (g \circ f)(y).\end{aligned}$$

6.6.3. Hint: Consider $f(2) = f(1 + 1)$.

6.6.5. One example: $f([x]) = [2x]$.

6.6.6. From the table

$[x]$	[0]	[1]	[2]	[3]	[4]	[5]	[6]	[7]	[8]	[9]	[10]	[11]
$f([x])$	[0]	[4]	[8]	[0]	[4]	[8]	[0]	[4]	[8]	[0]	[4]	[8]

we see that $\text{Ker}(f) = \{[0],[3],[6],[9]\}$ and $\text{Ran}(f) = \{[0],[4],[8]\}$.

6.6.7. (b) For the group $\{\pm 1\}$ under multiplication and $n = 1$ or 3 (or any odd integer), f_n is an isomorphism. Also for the group $\mathbf{R}^+$ under multiplication and n odd, f_n is an isomorphism.

6.6.10. Let G be an infinite cyclic group. Then there exists $g \in G$ such that for each $x \in G$, there exists $n \in \mathbf{N}$ such that $x = g^n$. If $g^n = e$ for some $n > 0$, then $G \subseteq \{e, g, \ldots, g^{n-1}\}$ and G is finite. Thus $g^n \neq e$ for all $n > 0$. It follows that $g^n \neq e$ for all $n < 0$. From these observations, it follows that for $n, m \in \mathbf{Z}$ with $n \neq m$, $g^n \neq g^m$. Therefore, for each $x \in G$ there exists a unique $n \in \mathbf{Z}$ such that $x = g^n$. Thus the function $f\colon G \to \mathbf{Z}$ defined by $f(g^n) = n$ is well defined and is a bijection. Since $f(g^n \cdot g^m) = f(g^{n+m}) = n + m = f(g^n) + f(g^m)$, f is an isomorphism. Thus G is isomorphic to $\mathbf{Z}$.

Miscellaneous Exercises

2. (c) When $n\mathbf{Z}$ acts on $\mathbf{Z}$, the set of equivalence classes is $\{n\mathbf{Z}, n\mathbf{Z} + 1, \ldots, n\mathbf{Z} + (n - 1)\}$.

3. (a) Proof by contradiction. Suppose $\log_4(10) \in \mathbf{Q}$. Thus $\log_4(10) = a/b$ where $a, b \in \mathbf{Z}$. Since $\log_4(10) > 0$, we can assume that $a > 0$ and $b > 0$. Since $\log_4(10) = a/b$, $4^{a/b} = 10$, which implies that $4^a = 10^b$. But since $5 \mid 10^b$ and $5 \nmid 4^a$, we have a contradiction.

10. (a) Yes.
    (b) Consider $\mathbf{Z}_2 \times \mathbf{Z}_2$.

15. $\{[p],[2p],\ldots,[(p - 1)p]\}$.

16. $\mathbf{Z}_{p^2}$ and $\mathbf{Z}_p \times \mathbf{Z}_p$ are two nonisomorphic groups of order p^2.

Chapter 7

Section 7.1

7.1.1. Let $A = \{n \in \mathbf{N} \mid s(n) \neq n\}$. Then $0 \in A$ by Axiom P3. Suppose $n \in A$; we show $s(n) \in A$. Suppose instead that $s(n) \notin A$. Then $s(s(n)) = s(n)$. By Axiom P4, $s(n) = n$, which means that $n \notin A$. Therefore, if $n \in A$, then $s(n) \in A$. By Axiom P5, $A = \mathbf{N}$ and $s(n) \neq n$ for all $n \in A$.

7.1.2. (a) Define $0 \cdot a = 0$ and for $k \neq 0$, $k \cdot a = n \cdot a + a$ where $k = s(n)$.

7.1.4. Define s': $E \to E$ to be the function $s'(n) = s(s(n)) = n + 2$ where $s:\mathbf{N} \to \mathbf{N}$ is the successor function on $\mathbf{N}$. Axiom P1 holds for E with the element 0. Since $s(n) \neq 0$ for all $n \in \mathbf{N}$, $s'(n) \neq 0$ for all $n \in E$. Also, since s is one-to-one, s' is one-to-one. Thus Axioms P2–P4 hold for E. Finally, if A' is a subset of E such that (i) $0 \in A'$ and (ii) $s'(n) \in A'$ wherever $n \in A'$. Let $A = \{k \in \mathbf{N} | k \in A' \text{ or } s(k) \in A'\}$. Then (i) $0 \in$ A and (ii) if $k \in A$, then either $k \in A'$ and hence $s'(k) = s(s(k)) \in A'$, which implies that $s(k) \in A$; or $s(k) \in A'$, which implies that $s(k) \in A$. Thus since $\mathbf{N}$ satisfies P1–P5 with successor function s, $A = \mathbf{N}$. But $A = \{k \in \mathbf{N} | k \in A'\} \cup \{k \in \mathbf{N} | s(k) \in A'\} = A' \cup \{k | s(k) \in A'\}$ and $A' \cap \{k \in \mathbf{N} | s(k) \in A'\} = \varnothing$. Also, $A' = \{k \in \mathbf{N} | k \in A'\} \subseteq E$ and $E \cap \{k \in \mathbf{N} | s(k) \in A'\} = \varnothing$. Therefore, $A' = E$.

Section 7.2

7.2.2. (a) Let $A = \{m \in \mathbf{N} | \text{For all } n \in \mathbf{N},\ n + m^+ = n^+ + m\}$. (i) $0 \in A$ since $n + 0^+ = n + 1 = n^+ = n^+ = 0$. (ii) Suppose $m \in A$. We show $m^+ \in A$: $n + (m^+)^+ = (n + m^+)^+$ by Definition 7.2.1 (ii); in addition, $n + m^+ = n^+ + m$ by inductive hypothesis. Therefore, $(n + (m^+))^+ = (n^+ + m)^+ = n^+ + m^+$. Thus $m^+ \in A$. By Axiom P5, $A = \mathbf{N}$ and the statement is proved.

7.2.3. Proof by contradiction. Suppose $n, m \in \mathbf{N}$ with $n \neq 0$ and $n + m = 0$. Since $n \neq 0$, $n = k^+$ for some $k \in \mathbf{N}$. Therefore, $0 = n + m = m + n = m + k^+ = (m + k)^+ = s(m + k)$. But the conclusion contradicts Axiom P3 of Definition 7.1.1. Therefore, if $n, m \in \mathbf{N}$ and $n \neq 0$, then $n + m \neq 0$.

7.2.7. (a) Suppose $n < m$. We show $n^+ \leq m$. Since $n < m$, there exists $k \in \mathbf{N}$, $k \neq 0$ such that $m = n + k$. Since $k \neq 0$, $k = j^+$ for $j \in \mathbf{N}$. Thus $m = n + k = n + j^+ = (n + j)^+ = (j + n)^+ = j + n^+$. If $j = 0$, then $n^+ = m$. If $j \neq 0$, then $n^+ < m$. In either case, we conclude that $n^+ \leq m$.

7.2.8. Let A satisfy the hypothesis of the statement and suppose $A \neq \mathbf{N}$. Let $B = A^c = \mathbf{N} - A$. But $B \neq \varnothing$ and hence by WOP, B has a least element k. Since $0 \in A$, $k \neq 0$. Thus $k = n^+$ for some $n \in \mathbf{N}$. Since $n < k$, $n \notin B$ and therefore $n \in A$. In fact, if $j \leq n$, then $j \in A$ since $n^+ = k$ is the least element of $B = \mathbf{N} - A$. Since $0, 1, \ldots, n \in A$, $n^+ = k \in A$ by assumption. But then $k \in A \cap A^c = \varnothing$, which is impossible. Thus $B = \varnothing$ and $A = \mathbf{N}$.

7.2.9. (b) Proof by contradiction. Suppose that $n, m, k \in \mathbf{N}$ with $k \neq 0$ and $n \cdot k \leq m \cdot k$ and that $n \nleq m$. Then by Theorem 7.2.10, $m < n$, which means that there exists $j \in \mathbf{N}$, $j \neq 0$, such that $m + j = n$. Therefore, $(m + j) \cdot k = m \cdot k + j \cdot k = n \cdot k$. Since $k \neq 0$ and $j \neq 0$, $j \cdot k \neq 0$, which implies that $m \cdot k < n \cdot k$. The conclusion contradicts the assumption that $n \cdot k \leq m \cdot k$. Therefore, $n \leq m$.

7.2.10. (Outline.) Let $A = \{n | n \geq 2 \text{ and } n \text{ is not divisible by a prime}\}$. If $A \neq \varnothing$, then A has a least element n_0. First note that n_0 is not prime; thus $n_0 = a \cdot b$ where $1 < a < n_0$ and $1 < b < n_0$. Since $a < n_0$, $a \notin A$ and a

has a prime factor that is also a factor of n_0. Thus $A = \varnothing$ and the statement is proved.

7.2.13. Hint: Let $A = \{k \mid \text{If } n, m \in \mathbf{N} \text{ and } n + k = m + k, \text{ then } n = m\}$. Show $A = \mathbf{N}$ using Axiom P5.

Section 7.3

7.3.6. Let $x, y \in \mathbf{Z}$ and $z \in \mathbf{Z}^+$. First we prove that if $x < y$, then $x \cdot z < y \cdot z$. If $x < y$, then $y - x \in \mathbf{Z}^+$. Hence $(y - x) \cdot z \in \mathbf{Z}^+$ by Theorem 7.3.4. But $(y - x) \cdot z = (y + (-x)) \cdot z = y \cdot z + (-x) \cdot z = y \cdot z - x \cdot z$ by Exercise 7.3.4(d). Therefore, $x \cdot z < y \cdot z$. For the converse, we prove that if $x \not< y$, then $x \cdot z \not< y \cdot z$. If $x \not< y$, then by Theorem 7.3.4(iii) $y \leq x$ and hence $x - y \geq 0$. Therefore, $(x - y) \cdot z = x \cdot z - y \cdot z \geq 0$, which means that $y \cdot z \leq x \cdot z$. Therefore, if $x \not< y$, then $x \cdot z \not< y \cdot z$.

7.3.7. Let $x = [(n, 0)]$. Then $x + [(0, n)] = [(n, 0)] + [(0, n)] = [(n, n)] = [(0, 0)] = 0$. Therefore, $-x = [(0, n)]$.

7.3.8. (a) Let $x, y \in \mathbf{Z}$. Then

$$\begin{aligned} -(x + y) &= (-1)(x + y) \text{ by Exercise 7.3.4(c)} \\ &= (-1) \cdot x + (-1) \cdot y \\ &= (-x) + (-y) \in \mathbf{Z}^+ \text{ by Theorem 7.3.4(i)}. \end{aligned}$$

Therefore, $x + y \in \mathbf{Z}^-$.

Section 7.4

7.4.1. (b) Let $x = a/b$ where $a, b \in \mathbf{Z}$ and $b \neq 0$. If $b < 0$, then $-b > 0$ and $x = -a/-b$ since in $\mathbf{Z}$ $(a) \cdot (-b) = -(a \cdot b) = (-b \cdot a) = b \cdot (-a)$.

7.4.6. Suppose $x, y \in \mathbf{Q}$ and $x \cdot y = 0$. Then, writing $x = a/b$ and $y = c/d$ where $a, b, c, d \in \mathbf{Z}$, we have $0 = x \cdot y = a \cdot c/b \cdot d$. Thus in $\mathbf{Z}$, $a \cdot c = 0 \cdot b \cdot d = 0$. But by Theorem 7.3.3(9), if $a \cdot c = 0$ for $a, c \in \mathbf{Z}$, then either $a = 0$ or $c = 0$. Therefore, either $x = a/b = 0$ or $y = c/d = 0$.

Second proof. Suppose that $x \cdot y = 0$ and $x \neq 0$. Then $0 = x^{-1} \cdot 0 = x^{-1} \cdot (x \cdot y) = (x^{-1} \cdot x) \cdot y = 1 \cdot y = y$.

7.4.7. (c) Proof by contradiction. Suppose $x \in \mathbf{Q}^+$ and $x^{-1} \notin \mathbf{Q}^+$. Therefore, either $x^{-1} = 0$ or $x^{-1} \in \mathbf{Q}^-$. Since $x \cdot x^{-1} = 1$, $x^{-1} \neq 0$. If $x^{-1} \in \mathbf{Q}^-$, then $1 = x \cdot x^{-1} \in \mathbf{Q}^-$ by Exercise 7.4.7(b).

(d) Suppose $x, y \in \mathbf{Q}^+$. If $x < y$, then $y - x \in \mathbf{Q}^+$ and hence $x^{-1} \cdot (y - x) = x^{-1} \cdot y - 1 \in \mathbf{Q}^+$. Therefore, $y^{-1} \cdot (x^{-1} \cdot y - 1) = x^{-1} - y^{-1} \in \mathbf{Q}^+$ and $x^{-1} > y^{-1}$. We can prove the converse similarly or as follows: If $x^{-1} > y^{-1}$, then $(x^{-1})^{-1} < (y^{-1})^{-1}$ as we just proved. But $(x^{-1})^{-1} = x$ for all $x \in \mathbf{Q} - \{0\}$. Therefore, $x < y$.

7.4.9. (a) Let $x \in \mathbf{Q}$ and write $x = a/b$ where $b > 0$. If $a \leq 0$, then $x < 1$. If $a > 0$, then $x = a/b < a + 1$, since $(a + 1) - a/b = (a \cdot b + b - a)/b = (a \cdot (b - 1) + b)/b$ and $a \cdot (b - 1) + b \geq b > 0$.
(b) (Outline.) By the proof of part (a), for each $x \in \mathbf{Q}$ there exists $n_0 \in \mathbf{Z}^+$ such that $x < n_0$. By induction, we can prove that for all $n \in \mathbf{Z}^+$, $n < 2^n$. Thus $x < n_0 < 2^{n_0}$.

Section 7.5

7.5.3. Let $\epsilon \in \mathbf{Q}^+$ be given. We must find $N \in \mathbf{N}$ such that if $n, m > N$, then $|1/n - 1/m| < \epsilon$. Now $|1/n - 1/m| = |(m - n)/m \cdot n|$. Since $-n \leq m - n \leq m$, $-1/m = -n/mn \leq (m - n)/m \cdot n \leq m/m \cdot n = 1/n$. Thus $|(m - n)/m \cdot n| \leq \max(1/n, 1/m)$. Thus if we choose $N \in \mathbf{N}$ such that $N > 1/\epsilon$ (by Exercise 7.4.9 such an N exists), then for $n, m > N$, $|1/n - 1/m| < \max(1/n, 1/m) < 1/N < \epsilon$ by Exercise 7.4.7(d).

7.5.5. Hint: Notice that $(n - 1)/(n + 1) = [(n + 1) - 2]/(n + 1) = (n + 1)/(n + 1) - 2/(n + 1) = 1 - 2/(n + 1)$.

7.5.6. The sequences in Exercise 7.5.1–7.5.3 are null sequences.

7.5.7. (a) Let $\epsilon \in \mathbf{Q}^+$ be given. Let $N = |0|$. Then for $n > N$, $|a_n - b_n| = |1 - 1| = 0 < \epsilon$. Thus $\{a_n | n \in \mathbf{N}\} \simeq \{b_n \in \mathbf{N}\}$.

7.5.8. (a) Let $\{a_n | n \in \mathbf{N}\}$ be a null sequence and let $r \in \mathbf{Q}$. If $r = 0$, then $\{r \cdot a_n = 0 | n \in \mathbf{N}\}$ is clearly a null sequence. Suppose $r \neq 0$. Let ϵ be an arbitrary positive real rational number. We must find $N \in \mathbf{N}$ such that for all $n > N$, $|r \cdot a_n| < \epsilon$. Recall that $|r \cdot a_n| = |r| \cdot |a_n|$. Let $\epsilon' = \epsilon/|r|$. Since $\{a_n\}$ is a null sequence, there exists $N \in \mathbf{N}$ such that for all $n > N$, $|a_n| < \epsilon'$. Therefore, for $n > N$, $|r \cdot a_n| = |r| \cdot |a_n| < |r| \cdot \epsilon/|r| = \epsilon$, which proves that $\{a_n | n \in \mathbf{N}\}$ is a null sequence.

7.5.11. (a) Hint: Use the Triangle Inequality and the fact that $|a_n - a_m| = |a_n - r + r - a_m|$.

Section 7.6

7.6.2. Let $s_1 = \{a_n | n \in \mathbf{N}\}$ be Cauchy and let $s_2 = \{b_n | n \in \mathbf{N}\}$ be null. To show that $s_1 \cdot s_2 = \{a_n \cdot b_n | n \in \mathbf{N}\}$ is null, we must show that for every $\epsilon \in \mathbf{Q}^+$, there exists $N \in \mathbf{N}$ such that $|a_n \cdot b_n| < \epsilon$ for all $n > N$. Since s_1 is Cauchy by Lemma 7.6.2, there exists $M \in \mathbf{Q}$ such that $|a_n| < M$ for all $n \in \mathbf{N}$. Thus for all $n \in \mathbf{N}$, $|a_n \cdot b_n| = |a_n| \cdot |b_n| < M \cdot |b_n|$. Since s_2 is null, there exists $N \in \mathbf{N}$ such that for all $n > N$, $|b_n| < \epsilon/M$. Therefore, if $n > N$, then $|a_n \cdot b_n| < M \cdot |b_n| < M \cdot (\epsilon/M) = \epsilon$.

7.6.3. For property 3, let $r' = \{-a_n | n \in \mathbf{N}\}$ where $r = \{a_n | n \in \mathbf{N}\}$. Since r is Cauchy, r' is Cauchy and $r + r' = 0$. The other properties follow from the corresponding statements in $\mathbf{Q}$.

7.6.4. (a) Hint: Look at the solution for Exercise 7.6.2.
(b) Hint: Try some simple examples.

Section 7.7

7.7.1. (a) The sequence $\{a_n | a_n = r \text{ for all } n\}$ is positive if and only if there exist $\delta \in \mathbf{Q}^+$ and $N \in \mathbf{N}$ such that $a_n = r > \delta$ for all $n > N$. But this condition occurs if and only if $r \in \mathbf{Q}^+$.

7.7.2. If $x, y \in \mathbf{R}^+$, then $x = [\{a_n | n \in \mathbf{N}\}]$ and $y = [\{b_n | n \in \mathbf{N}\}]$ and there exist $\delta_1 \in \mathbf{Q}^+$ and $N_1 \in \mathbf{N}$ and $\delta_2 \in \mathbf{Q}^+$ and $N_2 \in \mathbf{N}$ such that $a_n > \delta_1$ for all $n > N_1$ and $b_n > \delta_2$ for all $n > N_2$. Let $\delta = \delta_1 \cdot \delta_2$ and $N = \max\{N_1, N_2\}$. Then for $n > N$, $a_n \cdot b_n > \delta_1\delta_2 = \delta$. Hence $x \cdot y \in \mathbf{R}^+$.

7.7.6. By definition, if $x = [\{a_n | n \in \mathbf{N}\}] \in \mathbf{R}^+$, then there exist $\delta \in \mathbf{Q}^+$ and $N \in \mathbf{N}$ such that $a_n > \delta$ for all $n > N$. Let $r = [\{b_n | b_n = \delta/2 \text{ for all } n \in \mathbf{N}\}]$. Then $r \in \mathbf{Q}$ and for all $n > N$, $a_n - b_n > \delta - \delta/2 = \delta/2$. Hence $\{a_n - b_m | n \in \mathbf{N}\}$ is a positive Cauchy sequence. Since $[\{a_n - b_n | n \in \mathbf{N}\}] = x - r$, $x - r \in \mathbf{R}^+$ and $x > r$.

7.7.7. (Outline.) Consider three cases: $x > 0$, $x = 0$, $x < 0$. By Theorem 7.7.2, $x^2 > 0$ if $x > 0$. If $x < 0$, then $x^2 = (-x) \cdot (-x) > 0$. If $x = 0$, then $x^2 = 0$. Therefore, in every case, $x^2 \geq 0$.

Section 7.8

7.8.2. (a) Suppose x and y are least upper bounds for a set $A \subseteq \mathbf{R}$. We show that $x = y$ by proving that $x \leq y$ and $y \leq x$. Since x is a least upper bound for A and y is an upper bound for A, $x \leq y$. The same argument shows that $y \leq x$. Thus $x = y$.
(b) Let x be the least upper bound for A and let $y < x$. If, for all $a \in A$, $a \leq y$, then y is an upper bound for A; thus there exists $a \in A$ such that $y < a$. Since x is an upper bound for A, $a \leq X$. Therefore, $y < a \leq x$.

7.8.5. Let $\epsilon \in \mathbf{Q}^+$ be given. Then by Exercise 7.4.9, there exists $N \in \mathbf{N}$ such that $2^N > 1/\epsilon$. Since $10^n > 2^n$ for all $n \in \mathbf{N}$, $10^N > 1/\epsilon$. Therefore, if $n, m > N$ with $n > m$, then $0 \leq s_n - s_m = (a_1/10 + \cdots + a_m/10^m + \cdots + a_n/10^n) - (a_1/10 + \cdots + a_m/10^m) = a_{m+1}/10^{m+1} + \cdots + a_n/10^n \leq 9(1/10^{m+1} + \cdots + 1/10^n) \leq 9/10^{m+1}(1 + \cdots + 1/10^{n-m}) \leq 18/10^{m+1} < 2/10^m \leq 1/10^{m-1}$. Since $m > N$, $m - 1 \geq N$ and hence $1/10^{m-1} \leq 1/10^n < \epsilon$. Therefore, for $n, m > N$, $|s_n - s_m| < \epsilon$.

7.8.7. (a) .4. (b) $.\overline{285714}$. (c) $.22\cdots = .\overline{2}$. (d) $.1212\cdots = .\overline{12}$.

7.8.9. Let $x = .237171 = .23\overline{71}$. Then $100 \cdot x = 23.\overline{71}$ and $99 \cdot x = 100 \cdot x - x = 23.\overline{71} - .23\overline{71} = 23.48$. Thus $9900 \cdot x = 2348$ and $x = 2348/9900 = 587/2475$.

Miscellaneous Exercises

1. Let $\{x_n | n \in \mathbf{N}\}$ be a bounded increasing sequence of real numbers. Since the set $\{x_n\}$ is bounded, this set has a least upper bound x. We show that $\lim_{n \to \infty} x_n = x$; in other words, we show that for every $\epsilon \in \mathbf{R}^+$ there exists $N \in \mathbf{N}$ such that for all $n > N$, $|x_n - x| < \epsilon$. Since $x - \epsilon < x$, by Exercise 7.8.1(b), there exists $N \in \mathbf{N}$ such that $x - \epsilon < x_N \leq x$. Notice that $0 \leq x - x_N < \epsilon$. Since $\{x_n | n \in \mathbf{N}\}$ is an increasing sequence, for $n > N$, $x_N \leq x_n \leq x$. Therefore, for $n > N$, $0 \leq x - x_n = |x - x_n| < \epsilon$. In other words, $\lim_{n \to \infty} x_n = x$.

2. (a) Hint: Suppose x and y are limits of $\{x_n | n \in \mathbf{N}\}$ such that $x \neq y$. Let $\epsilon = |x - y|/4$ and, using the assumption that x and y are limits for $\{x_n | n \in \mathbf{N}\}$, derive the conclusion that $\{x_n | x \in \mathbf{N}\}$ is not Cauchy.

4. (Outline.) Suppose that Axioms P1–P4 and WOP hold. To show that Axiom P5 holds, let A be a subset of $\mathbf{N}$ such that $0 \in A$ and $n + 1 \in A$ whenever $n \in A$ and show that $A = \mathbf{N}$. To do so, argue by contradiction. If $A \neq \mathbf{N}$, then $A^c = \mathbf{N} - A \neq \varnothing$ and by WOP, A^c has a least element m. Since $0 \in A$, $m \neq 0$ and $m = k + 1$ where $k \in \mathbf{N}$. Since $k < m$, $k \notin A^c$ and hence $k \in A$. But if $k \in A$, then by assumption $k + 1 = m \in A$. Therefore, $m \in A \cap A^c = \varnothing$, which is a contradiction.

Index

A

abelian group 368
addition principle 260
adjacent 304
algebraic number 240
algebraically closed field 443
algebraic structure 322
archimedean property
  in **Q** 416
  in **R** 439
associativity 323
axiom of choice 243, 249ff
axiom system 24, 25
axioms 24
  field 26–27

B

biconditional 4, 10
binary operation 105, 132, 197, 321
  associative 323
  commutative 323
  idempotent 323
binary string 101
binomial coefficient 275, 278ff
binomial theorem 283ff
boundedness 27, 441

C

Cantor–Bernstein Theorem 245, 247ff
cardinality (cardinal number)
  215, 244ff
Cartesian product 132
Cauchy sequence 421
characteristic equation 294
circuit 303ff
  Euler 305
  Hamilton 313
combination 273
commutative ring with unity 337ff
commutativity 323
comparable elements 158
complete set of representatives 175
completeness 28, 441ff
complex numbers 350ff
composite integer 330
composition of functions 189
composition of relations 151
conditional 4, 7
congruence modulo *n* 171
conjunction 4
connected graph 307
containment of sets 97
contradiction 12, 45
contrapositive 8, 44
converse 9
countable set 216, 230
counterexamples 62
cyclic group 376, 382

D

degree of a polynomial 343, 356
degree of a vertex 308
DeMorgan's laws 116
denseness 415, 438
denumerable set 216, 227ff
derangement 319
diagram of a poset 160ff
diagonal argument 137, 240

diagonal set 137
disconnected graph 307
discrete structures 255ff
disjoint sets 106
disjunction 4, 5
distributivity 323
division theorem (algorithm) 55, 345, 357
divisor 44, 330, 341
domain of a function 183

E

edge 300
element of a set 95
empty (null) set 96
end of an edge 304
equality of sets 97
equinumerous sets 209
equivalence class 172
equivalence relation 144, 167
equivalent Cauchy sequences 424
Euler circuit 303
existential quantifier 15

F

factor 44, 330
families of sets 120
Fibonacci sequence 184, 290
field 649ff
 of complex numbers 350ff
 of real numbers 20ff, 389ff
 ordered 33
finite set 97, 210, 216, 219ff
Fubini principle 266
function 133, 144, 183
 bijective or one-to-one correspondence 185
 injective or one-to-one 185
 onto or surjective 185
fundamental theorem of algebra 359

G

generating function 295ff
graph 300ff
 bipartite 316
 complete 316
 connected 307
 directed 302
 disconnected 307
 simple 300
greatest common divisor 334
greatest lower bound 159
group 361ff
 abelian (commutative) 368
 cyclic 376, 382
 finite 367
 infinite 367
 simple 369, 377
 symmetric 366
 symmetry 365

H

Hamilton circuits 313ff
Hasse diagrams 161
heuristics 70ff
homomorphism 325, 376

I

idempotent 323
identity element 323
implication 8
incident 304
inclusion–exclusion principle 220, 225, 261, 263
incomparable elements 158
indexing set 120
infinite set 210
integral domain 344
integers 21, 329ff, 403ff
inverse element 323
inverse function 192
inverse image 196
inverse relation 150
irrational number 22, 46
irreducible element 343
isomorphic 29, 391
isomorphism 325, 378

K

kernel 380
Koenigsberg bridge problem 257, 303

L

Lagrange's theorem 373
least common multiple 336
least upper bound 28, 159, 441
linear recurrence relation 292
logical equivalence 7
lower bound 159

M

mathematical induction 50ff
model 29
modulo 174
multiplication principle 263

N

natural number 21, 391ff
negation 4, 5
node 300
null sequence 423
null (empty) set 96

O

operations on sets 105ff
  complement 108
  difference 107
  intersection 106
  union 107
order of an element 375
order relations
  in **N** 399ff
  in **Q** 414ff
  in **R** 438ff
  in **Z** 408ff
ordered field 33
ordered pair 128
ordering 22
  partial 143, 156
  total 158

P

partial ordering 143, 156
partially ordered set (poset) 156
partition 176
Pascal's triangle 283ff
path 305
  Euler 305
  Hamilton 313
Peano axioms 391
permutation 270, 271, 272
pigeonhole principle 222, 267
polynomial ring 339, 353ff
positive Cauchy sequence 435
postulates 24
power set 117, 245ff
predicate 14
prime integer 47, 330
problem representation 73ff
proof methods 37ff
  case analysis 58
  contradiction 45
  contraposition 44
  direct 41
  mathematical induction 50

R

range of a function 183
rational number 21, 411ff
recurrence relation 289
recursion theorem 393
recursive sequence 191, 289
related problems 77
real number 20ff, 418ff
relation 140, 146
  antisymmetric 142
  asymmetric 142
  composition of 151
  equivalence 144, 167
  identity 141
  inverse 150
  irreflexive 142
  order 143
  reflexive 141
  symmetric 142
  transitive 142
ring 337ff
root of a polynomial 357
Russell paradox 95, 104

S

sequence 421
  Cauchy 421
  null 423

set 95
  countable 216, 227
  denumerable 216, 227ff
  finite 97, 210, 216
  infinite 210
  uncountable 216, 237ff
singleton 96
sketch of a relation 148
strict partial ordering 157
strong induction 51
subset 97
subgroup 372ff

T

tautology 12
transcendental number 241
traveling salesman problem 313
tree 319
trichotomy law
  in **N** 400
  in **R** 438
  in **Z** 408

U

unary operation 107, 321
uncountable set 216, 237ff
unique factorization domain 344
unique factorization theorem in **Z** 332
unit 342
universal quantifier 15
upper bound 27, 159

V

variable 14
Venn diagram 99
vertex 308

W

well defined 340, 407
well-formed formula 133
well-ordered set 164
well-ordering principle in **N** 399

Z

Zermelo–Fraenkel axioms of set theory 102ff, 123, 133ff
zero divisor 344
Zorn's Lemma 251